IET ENERGY ENGINEERING 121

Monitoring and Control Using Synchrophasors in Power Systems with Renewables

Other volumes in this series:

Volume 1	**Power Circuit Breaker Theory and Design** C.H. Flurscheim (Editor)
Volume 4	**Industrial Microwave Heating** A.C. Metaxas and R.J. Meredith
Volume 7	**Insulators for High Voltages** J.S.T. Looms
Volume 8	**Variable Frequency AC Motor Drive Systems** D. Finney
Volume 10	**SF_6 Switchgear** H.M. Ryan and G.R. Jones
Volume 11	**Conduction and Induction Heating** E.J. Davies
Volume 13	**Statistical Techniques for High Voltage Engineering** W. Hauschild and W. Mosch
Volume 14	**Uninterruptible Power Supplies** J. Platts and J.D. St Aubyn (Editors)
Volume 15	**Digital Protection for Power Systems** A.T. Johns and S.K. Salman
Volume 16	**Electricity Economics and Planning** T.W. Berrie
Volume 18	**Vacuum Switchgear** A. Greenwood
Volume 19	**Electrical Safety: A guide to causes and prevention of hazards** J. Maxwell Adams
Volume 21	**Electricity Distribution Network Design, 2nd Edition** E. Lakervi and E.J. Holmes
Volume 22	**Artificial Intelligence Techniques in Power Systems** K. Warwick, A.O. Ekwue and R. Aggarwal (Editors)
Volume 24	**Power System Commissioning and Maintenance Practice** K. Harker
Volume 25	**Engineers' Handbook of Industrial Microwave Heating** R.J. Meredith
Volume 26	**Small Electric Motors** H. Moczala *et al.*
Volume 27	**AC–DC Power System Analysis** J. Arrillaga and B.C. Smith
Volume 29	**High Voltage Direct Current Transmission, 2nd Edition** J. Arrillaga
Volume 30	**Flexible AC Transmission Systems (FACTS)** Y.-H. Song (Editor)
Volume 31	**Embedded Generation** N. Jenkins *et al.*
Volume 32	**High Voltage Engineering and Testing, 2nd Edition** H.M. Ryan (Editor)
Volume 33	**Overvoltage Protection of Low-Voltage Systems, Revised Edition** P. Hasse
Volume 36	**Voltage Quality in Electrical Power Systems** J. Schlabbach *et al.*
Volume 37	**Electrical Steels for Rotating Machines** P. Beckley
Volume 38	**The Electric Car: Development and future of battery, hybrid and fuel-cell cars** M. Westbrook
Volume 39	**Power Systems Electromagnetic Transients Simulation** J. Arrillaga and N. Watson
Volume 40	**Advances in High Voltage Engineering** M. Haddad and D. Warne
Volume 41	**Electrical Operation of Electrostatic Precipitators** K. Parker
Volume 43	**Thermal Power Plant Simulation and Control** D. Flynn
Volume 44	**Economic Evaluation of Projects in the Electricity Supply Industry** H. Khatib
Volume 45	**Propulsion Systems for Hybrid Vehicles** J. Miller
Volume 46	**Distribution Switchgear** S. Stewart
Volume 47	**Protection of Electricity Distribution Networks, 2nd Edition** J. Gers and E. Holmes
Volume 48	**Wood Pole Overhead Lines** B. Wareing
Volume 49	**Electric Fuses, 3rd Edition** A. Wright and G. Newbery
Volume 50	**Wind Power Integration: Connection and system operational aspects** B. Fox *et al.*
Volume 51	**Short Circuit Currents** J. Schlabbach
Volume 52	**Nuclear Power** J. Wood
Volume 53	**Condition Assessment of High Voltage Insulation in Power System Equipment** R.E. James and Q. Su
Volume 55	**Local Energy: Distributed generation of heat and power** J. Wood
Volume 56	**Condition Monitoring of Rotating Electrical Machines** P. Tavner, L. Ran, J. Penman and H. Sedding
Volume 57	**The Control Techniques Drives and Controls Handbook, 2nd Edition** B. Drury
Volume 58	**Lightning Protection** V. Cooray (Editor)
Volume 59	**Ultracapacitor Applications** J.M. Miller

Volume 62 **Lightning Electromagnetics** V. Cooray
Volume 63 **Energy Storage for Power Systems, 2nd Edition** A. Ter-Gazarian
Volume 65 **Protection of Electricity Distribution Networks, 3rd Edition** J. Gers
Volume 66 **High Voltage Engineering Testing, 3rd Edition** H. Ryan (Editor)
Volume 67 **Multicore Simulation of Power System Transients** F.M. Uriate
Volume 68 **Distribution System Analysis and Automation** J. Gers
Volume 69 **The Lightening Flash, 2nd Edition** V. Cooray (Editor)
Volume 70 **Economic Evaluation of Projects in the Electricity Supply Industry, 3rd Edition** H. Khatib
Volume 72 **Control Circuits in Power Electronics: Practical issues in design and implementation** M. Castilla (Editor)
Volume 73 **Wide Area Monitoring, Protection and Control Systems: The enabler for smarter grids** A. Vaccaro and A. Zobaa (Editors)
Volume 74 **Power Electronic Converters and Systems: Frontiers and applications** A.M. Trzynadlowski (Editor)
Volume 75 **Power Distribution Automation** B. Das (Editor)
Volume 76 **Power System Stability: Modelling, analysis and control** A.A. Sallam and B. Om P. Malik
Volume 78 **Numerical Analysis of Power System Transients and Dynamics** A. Ametani (Editor)
Volume 79 **Vehicle-to-Grid: Linking electric vehicles to the smart grid** J. Lu and J. Hossain (Editors)
Volume 81 **Cyber-Physical-Social Systems and Constructs in Electric Power Engineering** S. Suryanarayanan, R. Roche and T.M. Hansen (Editors)
Volume 82 **Periodic Control of Power Electronic Converters** F. Blaabjerg, K.Zhou, D. Wang and Y. Yang
Volume 86 **Advances in Power System Modelling, Control and Stability Analysis** F. Milano (Editor)
Volume 87 **Cogeneration: Technologies, optimisation and implementation** C.A. Frangopoulos (Editor)
Volume 88 **Smarter Energy: From smart metering to the smart grid** H. Sun, N. Hatziargyriou, H.V. Poor, L. Carpanini and M.A. Sánchez Fornié (Editors)
Volume 89 **Hydrogen Production, Separation and Purification for Energy** A. Basile, F. Dalena, J. Tong and T.N. Veziroğlu (Editors)
Volume 90 **Clean Energy Microgrids** S. Obara and J. Morel (Editors)
Volume 91 **Fuzzy Logic Control in Energy Systems with Design Applications in MATLAB®/Simulink®** İ.H. Altaş
Volume 92 **Power Quality in Future Electrical Power Systems** A.F. Zobaa and S.H.E.A. Aleem (Editors)
Volume 93 **Cogeneration and District Energy Systems: Modelling, analysis and optimization** M.A. Rosen and S. Koohi-Fayegh
Volume 94 **Introduction to the Smart Grid: Concepts, technologies and evolution** S.K. Salman
Volume 95 **Communication, Control and Security Challenges for the Smart Grid** S.M. Muyeen and S. Rahman (Editors)
Volume 96 **Industrial Power Systems with Distributed and Embedded Generation** R. Belu
Volume 97 **Synchronized Phasor Measurements for Smart Grids** M.J.B. Reddy and D.K. Mohanta (Editors)
Volume 98 **Large Scale Grid Integration of Renewable Energy Sources** A. Moreno-Munoz (Editor)
Volume 100 **Modeling and Dynamic Behaviour of Hydropower Plants** N. Kishor and J. Fraile-Ardanuy (Editors)
Volume 101 **Methane and Hydrogen for Energy Storage** R. Carriveau and D.S.-K. Ting
Volume 104 **Power Transformer Condition Monitoring and Diagnosis** A. Abu-Siada (Editor)
Volume 106 **Surface Passivation of Industrial Crystalline Silicon Solar Cells** J. John (Editor)

Volume 107 **Bifacial Photovoltaics: Technology, applications and economics** J. Libal and R. Kopecek (Editors)
Volume 108 **Fault Diagnosis of Induction Motors** J. Faiz, V. Ghorbanian and G. Joksimović
Volume 110 **High Voltage Power Network Construction** K. Harker
Volume 111 **Energy Storage at Different Voltage Levels: Technology, integration, and market aspects** A.F. Zobaa, P.F. Ribeiro, S.H.A. Aleem and S.N. Afifi (Editors)
Volume 112 **Wireless Power Transfer: Theory, technology and application** N. Shinohara
Volume 115 **DC Distribution Systems and Microgrids** T. Dragičević, F. Blaabjerg and P. Wheeler
Volume 117 **Structural Control and Fault Detection of Wind Turbine Systems** H.R. Karimi
Volume 119 **Thermal Power Plant Control and Instrumentation: The control of boilers and HRSGs, 2nd Edition** D. Lindsley, J. Grist and D. Parker
Volume 120 **Fault Diagnosis for Robust Inverter Power Drives** A. Ginart (Editor)
Volume 123 **Power Systems Electromagnetic Transients Simulation, 2nd Edition** N. Watson and J. Arrillaga
Volume 124 **Power Market Transformation** B. Murray
Volume 125 **Wind Energy Modeling and Simulation, Volume 1: Atmosphere and plant** P. Veers (Editor)
Volume 126 **Diagnosis and Fault Tolerance of Electrical Machines, Power Electronics and Drives** A.J.M. Cardoso
Volume 128 **Characterization of Wide Bandgap Power Semiconductor Devices** F. Wang, Z. Zhang and E.A. Jones
Volume 129 **Renewable Energy from the Oceans: From wave, tidal and gradient systems to offshore wind and solar** D. Coiro and T. Sant (Editors)
Volume 130 **Wind and Solar Based Energy Systems for Communities** R. Carriveau and D.S.-K. Ting (Editors)
Volume 131 **Metaheuristic Optimization in Power Engineering** J. Radosavljević
Volume 132 **Power Line Communication Systems for Smart Grids** I.R.S. Casella and A. Anpalagan
Volume 139 **Variability, Scalability and Stability of Microgrids** S.M. Muyeen, S.M. Islam and F. Blaabjerg (Editors)
Volume 145 **Condition Monitoring of Rotating Electrical Machines** P. Tavner, L. Ran and C. Crabtree
Volume 146 **Energy Storage for Power Systems, 3rd Edition** A.G. Ter-Gazarian
Volume 147 **Distribution Systems Analysis and Automation, 2nd Edition** J. Gers
Volume 155 **Energy Generation and Efficiency Technologies for Green Residential Buildings** D. Ting and R. Carriveau (Editors)
Volume 157 **Electrical Steels, 2 Volumes** A. Moses, K. Jenkins, Philip Anderson and H. Stanbury
Volume 160 **Microgrids for Rural Areas: Research and case studies** R.K. Chauhan, K. Chauhan and S.N. Singh (Editors)
Volume 172 **Lighting Interaction with Power Systems, 2 Volumes** A. Piantini (Editor)
Volume 905 **Power System Protection, 4 Volumes**

Monitoring and Control Using Synchrophasors in Power Systems with Renewables

Edited by
Innocent Kamwa, Chao Lu and Lipeng Zhu

The Institution of Engineering and Technology

Published by The Institution of Engineering and Technology, London, United Kingdom

The Institution of Engineering and Technology is registered as a Charity in England & Wales (no. 211014) and Scotland (no. SC038698).

First published 2020

The Institution of Engineering and Technology
Michael Faraday House
Six Hills Way, Stevenage
Herts, SG1 2AY, United Kingdom

www.theiet.org

British Library Cataloguing in Publication Data
A catalogue record for this product is available from the British Library

ISBN 978-1-78561-477-4 (Hardback)
ISBN 978-1-78561-478-1 (PDF)

Typeset in India by MPS Limited

Contents

About the editors

Innocent Kamwa is a Research Scientist at Hydro-Quebec Research Institute, specializing in system dynamics, power grid control, wind energy, and storage integration studies. He became Chief Scientist for smart grid in 2009, and Head of Power System and Mathematics in 2014, overseeing the Hydro-Quebec Network Simulation Centre, and Acting Scientific Director of IREQ in 2016. He holds Adjunct Professor positions at McGill University and Université Laval.

Chao Lu is an Associate Professor of Electrical Engineering Department at Tsinghua University. His research interests include complex system analysis and control in smart grid, wide-area measurement systems, and applications of energy storage systems. He has published more than 130 academic papers and been awarded ten academic and scientific awards. He is an IEEE senior member, and also serves on the boards of key journals, including IET Generation, Transmission & Distribution.

Lipeng Zhu received his Ph.D. degree in Electrical Engineering from Tsinghua University, Beijing, China, in 2018. He is currently working as a Senior Research Assistant in The Center for Electrical Energy Systems, The University of Hong Kong, Hong Kong, China. His research interests mainly include PMU data analysis, spatial-temporal data analytics, and data-driven power system stability and control.

Preface

In recent years, the ever-increasing penetration of distributed and/or aggregated renewables have brought a series of unprecedented challenging problems into modern power systems. For example, the significant volatility, intermittency, and uncertainty of renewables have made it extremely difficult to ensure secure yet reliable system operation and control. Fortunately, with the fast development of wide-area synchrophasor measurement technologies, the advanced phasor measurement unit (PMU)/wide-area measurement system (WAMS) infrastructures provide immense opportunities to address these challenges. In general, PMU/WAMS can capture full-scale snapshots of system dynamics at a high resolution, helping to identify system dynamic characteristics and prevent/correct risky conditions during online monitoring. With the main scope in system online monitoring and control, this book aims to address emerging ideas, new technologies, and advanced applications of WAMS-based monitoring and control in modern power grids with high renewables.

In particular, this book will present the state-of-the-art PMU/WAMS technologies and innovative applications from the following three aspects: (1) wide-area measurement, modeling, and identification; (2) wide-area online dynamics monitoring; and (3) wide-area control and decision making. Researchers from all over the world have contributed to this book to comprehensively present recent advances of PMU/WAMS in these aspects. However, we would like to emphasize that it is impossible to cover all the PMU/WAMS related applications for wide-area monitoring and control in this book. In fact, many other promising applications are still in their researching and developing stages.

Although significant progress has been made by far, the high potential beneath PMU/WAMS still remains to be explored. For instance, the valuable information underlying massive high-resolution PMU data could be fully leveraged to better monitor and control the future power networks via advanced techniques and tools such as data analytics and machine learning. In future power grids accommodating more and more renewables, the PMU/WAMS technologies are hopeful to play a more crucial role in helping to operate the grids in a safer, more reliable and cleaner way. Many unsolved problems in today's power system monitoring and control could be tackled by developing new PMU/WAMS technologies and designing new WAMS implementation paradigms. We hope this book can inspire the exploration of more innovative PMU/WAMS technologies and applications not only in the academic research community but also in the industrial context.

Acknowledgments

We are grateful to those who helped us to complete this book. We would like to express our special gratitude to all the contributors, who contributed their enlightening research efforts and findings to this book without any reservation. Our sincere thanks go to the IET editorial team assisting us to edit the book, including Olivia Wilkins, Christoph von Friedeburg, Nicki Dennis, etc. Their patience in coordinating the book project during the whole period is heartily appreciated. We are also thankful to our colleagues and graduate students for suggesting the improvements and corrections of this book.

Dr Innocent Kamwa acknowledges the support of National Science and Engineering Research Council of Canada under the Discovery Grant: "Harnessing Flexible Resources and Wide-Area Sensing to Enable Smarter Power Grids that Automatically Combat Transmission Bottlenecks and Prevent Blackouts." Dr Chao Lu acknowledges the support of National Key Research and Development Program of China (under Grant No. 2017YFB0902801) and National Natural Science Foundation of China (under Grant Nos. 51677097 and U1766214). Dr Lipeng Zhu acknowledges the support of the Research Grants Council of Hong Kong Special Administrative Region of China under the General Research Fund (GRF) through Project No. 17208817 and under the Theme-based Research Scheme (TRS) through Project No. T23-701/14-N.

Chapter 1

Introduction to synchrophasor measurements in modern power systems with renewables

Innocent Kamwa[1], Chao Lu[2] and Lipeng Zhu[3]

1.1 Emerging trends of WAMS in modern power systems

In recent decades, with the fast development of advanced information and communication technologies (ICTs), modern power systems have significantly enhanced their capability of online situational awareness, especially for system-wide dynamics identification, online monitoring and stability control. Taking the wide-area measurement systems (WAMS) [1] for instance, system-wide dynamic behaviors can be measured by synchronous phasor measurement units (PMUs) at a high resolution. With full-scale snapshots of system dynamics synchronously measured and captured in real time, the WAMS is capable of largely helping to identify system dynamic features and mitigate risky conditions during online monitoring. With the help of this powerful platform, it is prospective to operate modern power systems in a more secure, economic, and efficient way. Under such circumstances, a wide variety of promising WAMS applications have been developed for system online monitoring and control, for example, wide-area dynamics modeling [2–4], high-resolution state estimation [5–7], wide-area event detection [8–10], wide-area fault location [11–13], dynamic stability assessment (DSA) [14–17], online stability control [18–21], etc.

On the other hand, the ever-increasing integration of distributed and/or aggregated renewables has brought a series of new issues into modern power systems, posing unprecedented challenges to system operation and control. To list a few, the significant volatility, intermittency, and uncertainty of renewables [22] have made it extremely difficult to carry out safe yet economic system schedule and dispatch. In some countries such as China with large-scale wind farms connected to grids through long-distance transmission lines, the low-inertia

[1]Hydro-Québec/IREQ, Power System Simulation and Evolution, Varennes, Quebec, Canada
[2]Department of Electrical Engineering, Tsinghua University, Beijing, China
[3]Department of Electrical and Electronic Engineering, The University of Hong Kong, Hong Kong, China

characteristic of wind generators is likely to cause frequency oscillation problems [23]. Moreover, as renewables are generally integrated into the grids via power electronics devices, the interactions between fast-acting power electronics converters and conventional dynamic components (e.g., synchronous generators and dynamic loads) could induce new stability issues such as voltage stability and sub-/supersynchronous oscillations [24]. Fortunately, the aforementioned advanced tool and platform, that is, WAMS, provides immense opportunities to address these challenging issues associated with high renewables. With the main scope in system online monitoring and control, this book aims to comprehensively address emerging ideas, new technologies and advanced applications of WAMS-based monitoring and control in modern power grids in the presence of high renewables.

The rest of this chapter is organized as follows. First, we will provide a brief overview on the basic WAMS functionalities and the development of WAMS in practical power grids. Then, we will present various successful and promising WAMS applications in modern power grids, with emphasis on online dynamic monitoring and control. Finally, we will introduce the organization of the whole book and concisely describe the main contents in individual parts and chapters.

1.2 Overview of WAMS functionality and development

1.2.1 WAMS functionality and architecture

In general, the WAMS deployed in a specific power grid is composed of a large number of dispersed PMUs and phasor data concentrators (PDCs). These devices are hierarchically linked together via wide-area communication networks. From the perspective of substations, how PMU measurement and communication are realized is illustrated in Figure 1.1. As can be seen, PMUs generally take raw inputs from merging units and switchgear control units complying with IEC 61850-9-2 SV and GOOSE protocols. After phasor calculations, the PMUs transmit synchrophasors at a fast rate (up to 100/120 frames per second for 50/60 Hz power systems) toward a substation PDC based on the IEEE C37.118 standard. Individual PDCs further transmit time-stamped PMU data to the dispatching and control center, where a WAMS central station is configured.

The WAMS communication network generally has a high demand for communication reliability, efficiency, and safety. Therefore, key performance requirements such as network time delay, convergence time, and cybersecurity must be guaranteed. To this end, some countries with advanced smart grid infrastructures have constructed special communication networks for WAMS. Taking the State Grid Corporation of China (SGCC), for example, according to the principle of dual independent backbone networks and multiple layers integration, an exclusive dispatching data network based on dedicated communication channels (synchronous digital hierarchy + IP + optical fiber), that is, the State Grid Dispatching Digital network (SPDnet), has been applied [25]. The network utilizes highly reliable core equipment, together with double master controls. A three-level border gateway protocol route is used to realize a multiprotocol label switching virtual private

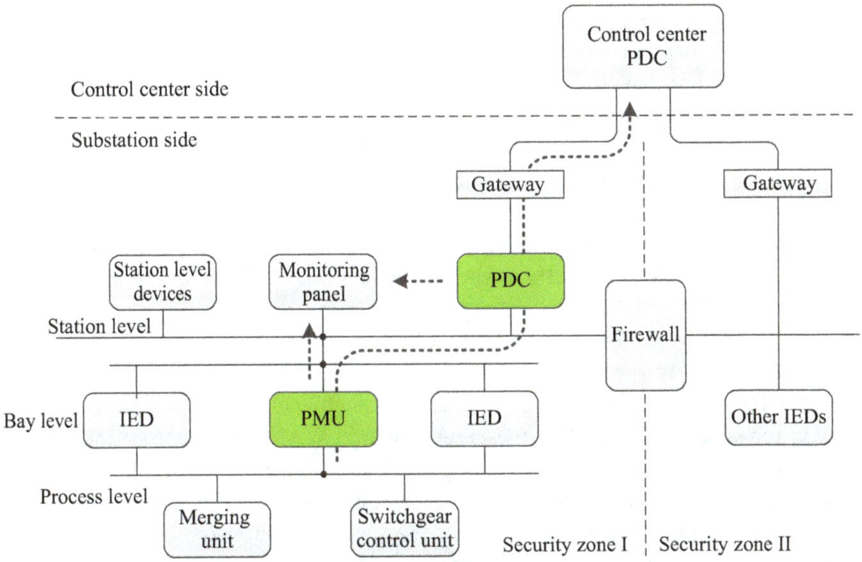

Figure 1.1 PMU functions realized in smart substation structure

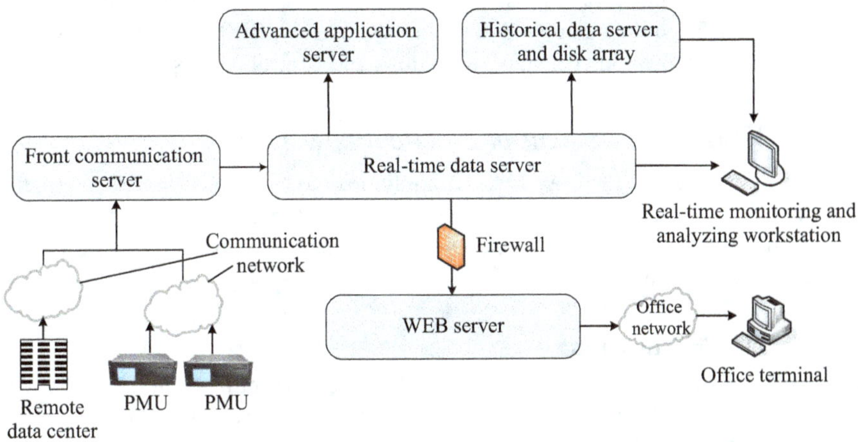

Figure 1.2 Framework of WAMS central structure

network, through which different types of network tasks can be isolated and the safety of dispatching tasks can be maintained.

The general framework of a WAMS central station is shown in Figure 1.2. In addition to the measurement terminals (PMUs) and the communication network, the WAMS central station involves several servers, including front communication, web, historical data, real-time data, and advanced application servers. Among these servers, the real-time database and advanced application servers are the key components. The real-time database techniques oriented to power system equipment are developed based on the common information model (CIM) criterion, wide-area distributed data storage and direct location methods. The scope of data access is extended from local to remote areas, which meets the need for highly efficient real-time data sharing over the entire network. Advanced techniques such as fixed time duration, equal data time interval, and Hash table index are used to satisfy the requirements of storage and searching of massive data. In China, the processing capability of the real-time database reaches the level of 5 million events per second.

The time synchronization management system is another important infrastructure for WAMS. In many countries, the global position system (GPS) is widely applied to provide the time reference in power grids, with a precision of 6–12 ns. In China, a fast-developing navigation satellite system called Beidou is also being adopted for synchronizing PMU measurements and WAMS communication, with a precision of 20–100 ns. With the help of the unified time reference system (e.g., GPS or Beidou), dispersed sites and devices, including power grid dispatch and control stations, monitoring systems in substation level, and synchronized IEDs inside substations, run coordinately and synchronously to realize efficient online monitoring. Different synchronization protocols are adopted within the WAMS. The network time protocol is applied for equipment at the station-control level; at the bay level and process level, optical fiber pulse per second (PPS) or inter-range instrumentation group B (IRIG-B) codes are applied to set time. Currently, IEEE Standard 1588-v2 is applied to communications between substations and control centers.

1.2.2 WAMS development in practical power grids

Over past decades, the PMU/WAMS technologies have attracted increasing attention and achieved fast developments in practical power grids around the world. The developments of WAMS infrastructures in some representative large-scale power grids are briefly introduced here.

1.2.2.1 WAMS in China

China has one of the largest scale PMU/WAMS deployment maps in the world. Up to now, more than 3,000 PMUs have been configured in two major power grids, that is, SGCC and China Southern Power Grid (CSG) [26]. Both of them have deployed large numbers of PMUs as part of a nationwide mandate requiring that all substations in 500 kV networks and above and generators larger than 100 MW be monitored by PMUs. In China, all of the national, inter-regional, and provincial dispatching centers have been deployed with WAMS central stations. Some of the control centers in metropolitan power grids, such as Guangzhou and Shenzhen, have also designed and are in the process of constructing their own WAMS stations. With the enhancement of synchronized phasor measurement technologies, current major practice for PMU development in China has two main goals: supporting smart substation techniques and improving measurement accuracy based on the latest standards for PMUs.

Approximately 100 PMUs supporting IEC 61850 have been commissioned in recent years. With the fast development of smart substation technologies, over 3,000 smart substations have been put into operation, covering voltage levels from 110 kV to 750 kV. Most of these smart substations are distributed in low-voltage networks. About 100 high-voltage substations are planned to be equipped with the synchronized phasor measurement function.

1.2.2.2 WAMS in North America

Since the 2003 large-scale blackout in North America, power grids in the United States and Canada have made remarkable progress in deploying PMU/WAMS infrastructures and developing various advanced WAMS applications, so as to enhance the grids' observability and visibility. At present, about 2,000 PMUs have been commissioned by some major transmission system operators or regional utilities [26], including North-East, Midwest, Texas, WECC, etc. The practical experiences in North America show that PMU/WAMS technologies can be very useful for both real-time situational awareness and offline power grid performance analysis. The practical power grids in North America have significantly benefited from advanced applications with synchrophasor measurements, for example, model and parameter identification, online state estimation, and wide-area protection and control.

In 2007, a collaborative organization called North American SynchroPhasor Initiative (NASPI) was launched. It aims to enhance power system reliability and visibility by fostering the use and application of WAMS. Funded by the U.S. Department of Energy (DOE), it has been working toward advancing the development and deployment of PMU devices, phasor data sharing, WAMS research and analysis, and practical WAMS applications.

1.2.2.3 WAMS in other countries/regions

In addition to China and North America, WAMS technologies are also progressing in many other countries and regions around the world. For example, in Europe, PMU/WAMS infrastructures have almost been deployed in every country [26]. In particular, some developed countries such as Italy, Denmark, Swiss, and Switzerland are planning to install more PMUs to modernize their power grids for intelligent online monitoring and control. In South America, Brazil has undertaken practical WAMS-related projects to realize WAMS-based offline analysis and online monitoring. In Mexico, PMU/WAMS technologies have been put into use for more than 20 years; WAMS has been successfully utilized to monitor low-frequency oscillations and disturbances in daily operation. In India, nearly 1,500 PMUs are planning to be configured for WAMS deployment, so as to implement wide-area state estimation, parameter calibration and estimation, remedial corrective control, etc.

The boom of PMU/WAMS technologies also witnessed in many other countries and regions, for example, South Africa, Japan, and Russia. However, here we do not intend to list all the countries involving fast WAMS deployments and applications. Interested readers may refer to [26,27] for more information.

1.3 WAMS applications in modern power grids

1.3.1 Wide-area modeling and identification

In practical power grids, many postdisturbance PMU measurements have shown significant differences between the simulation results and real system responses, mostly caused by errors in the system models and parameters. PMU/WAMS based modeling and validation approaches are capable of providing reliable and precise field measurement data that contain valuable dynamic information. Based on PMU measurement data collected from actual grids, model validation and parameter identification can be performed to improve the precision of system modeling. Taking generator parameter identification, for instance, system identification or parameter optimization approaches can be leveraged to fulfill the identification task. Taking the profiles of generator terminal and excitation voltages as the inputs, the objective can be set to minimizing the error between the simulated generator current and the current measured by the PMU. With well-identified parameters, system dynamic components such as generators can be modeled and calibrated with a higher precision, thereby contributing to more accurate system dynamics modeling, simulation, and analysis.

1.3.2 Power system dynamics monitoring

1.3.2.1 Wide-area state estimation

Traditional state estimation (SE) is implemented in the dispatching and control center using measurements remotely collected from the Supervisory Control and Data Acquisition (SCADA) system. Conventionally, identifying topology errors is a major challenge for SCADA-based state estimation. With hybrid measurements acquired from both the WAMS and SCADA systems, wide-area state estimation (WASE) provides an alternative solution by prefiltering bad data using a detailed model and redundant local measurements. In WASE, the zero impedance subnetwork, with circuit breakers (CBs) as zero-impedance branches, is handled at each voltage level separately. In traditional SE, the topology must be determined first to build nodal equations using Ohm's law. With WASE, only the analog measurements from the PMU and SCADA systems are used to first solve the local SE for each voltage level using Kirchhoff's current law. The estimated power flows across each CB are then used to check for potentially bad CB status data. In this process, the detected bad analog and topology data are decoupled, making them minimally susceptible to the measurement errors of CB status. Bad topology data can be further identified by using methods such as a hypothesis test.

With a small problem size, the complete utilization of fast-sampled PMU measurements becomes possible. This condition benefits WASE in terms of accuracy and reliability. In China, the WASE method has been implemented in East China Power Grid and other power utilities. A field test reveals that data quality can be greatly improved by WASE. The average computing time for WASE is 20 ms, which is feasible for real-time applications.

The above WASE takes an implicit assumption that system network parameters remain constant, and it is often utilized to estimate algebraic variables, for example, phasors of networked bus voltages and currents. In fact, if more complicated system dynamics such as transients are taken into account, dynamic state estimation (DSE) can be carried out to further estimate system state variables such as rotor angles and speeds of synchronous generators [28]. With dynamic states well estimated in real time, more reliable wide-area system monitoring and control can be achieved in various dynamic conditions and transient scenarios.

1.3.2.2 Low-frequency oscillation monitoring

Low-frequency oscillation (LFO) is always a threat for the interconnected power grids of China. A comprehensive WAMS-based solution can be developed for the monitoring and assessment of LFO stability. Such a solution is expected to include three functions: (1) real-time warning regarding harmful oscillation based on ringdown signals identification; (2) early warning about small oscillation based on statistics; and (3) identification of oscillation modes and mode shapes based on the power system's ambient data. With this solution, the weakly damped oscillation modes can be detected early, and system operators can make effective decisions based on the identified information. The Prony algorithm is the basic method for the detection and identification of LFO based on ringdown signals after disturbances. This algorithm has already been implemented in practical power grids. A number of advanced algorithms, such as the Hilbert–Huang transform and the estimation of signal parameters via rotational invariance techniques can be used for LFO monitoring as well. Given that disturbances in real power systems are relatively rare, the identification of oscillation based on ambient data seems more practical for early warning and decision making.

1.3.2.3 Online disturbance and fault detection

When a disturbance occurs in power systems, system operators in the dispatching and control centers receive numerous warnings and information on the sequence of events. However, this information usually only contains breaker trips and protection relay actions. System operators cannot directly determine the kind of fault that actually led these disturbances to occur. The causes can generally be identified by analyzing the fault recording wave files after the fault events. However, this analysis is time consuming. Based on the real-time PMU data, online detection methods for different kinds of faults or disturbances, for example, short circuit faults, generator breaker trips, phase shift failures of DC converters, and islanding, can be designed and implemented. In some cases, the message of protective relay actions cannot be transmitted to the dispatching and control center because of issues such as communication traffic under system faults. However, with the dynamic responses measured by PMUs, the fault can be efficiently detected and localized. System operators can quickly determine what kind of fault occurred and on which part of the power grid. This information is considerably important for the system operators to decide whether further countermeasures should be taken to maintain safe operations.

1.3.2.4 Assessing generator ancillary services

Ensuring safe and reliable generator operation requires ancillary services such as speed control and excitation control. When the system frequency deviates from the nominal frequency over a certain threshold, the unit power output and frequency curve can be analyzed to assess their contribution to the primary frequency regulation. Based on the active power and the frequency measurement provided by the generator's PMU during a frequency disturbance, the performance parameters of the generator's primary frequency regulation can be calculated. These parameters include delay time, response time, frequency dead zone, speed variation ratio, and contribution energy. Monitoring of a generator's excitation control can be carried out in a similar way. A generator's primary frequency regulation function is extremely important for maintaining the frequency of a power grid within a secure range when it suffers an active power deficit, such as the tripping of a large generator. In a number of tests on some generators' primary frequency regulation function, even though it seemed that this function had switched on, it was actually switched off. The system frequency response may be deteriorated in this condition. For this reason, every generator with the primary frequency regulation function switched on should be monitored and assessed.

1.3.2.5 Real-time wind farm monitoring

Wind turbines in China occasionally suffer from cascading tripping induced by internal or external wind farm faults. Such cascading tripping may occur when groups of wind farms are located close to each other, and events like these put considerable pressure on utility companies. However, because of commercial secrecy, utility companies have no access to wind turbine monitoring data. Based on PMUs installed at step-up substations, an intelligent alarm system can be designed to detect when cascading tripping events occur, trace the entire process, and subsequently assess the impact. The cascading failures of wind farms can last from several seconds to as much as a minute and are accompanied by sudden voltage dips because of contingencies, the continuous loss of active power, and the increase in voltage. Meanwhile, under the most violent fluctuation scenarios, the magnitude and speed of a wind farm's output power ramps can reach 25% of the installed capacity within 5 min; these values are far below those under cascading scenarios, in which the magnitude and speed can reach 100% of installed capacity within 5 s. The intelligent alarm system can detect large voltage dips at the common coupling points of wind farms as start-up conditions and then identify the power ramps within a minute-level time window.

1.3.3 Wide-area control and protection

1.3.3.1 Wide-area damping control

Efficiently suppressing interarea LFO requires that global dynamics for different areas be considered. PMUs are expected to play a crucial role in real-time and continuous wide-area damping control (WADC). In designing WADC systems, the following issues should be taken into account carefully: (1) selection of feedback signals and control sites; (2) controller structure, parameters, and adaptability; and (3) modeling and compensation of random time delay in the communication system. The world's first WADC system commissioned by the CSG in 2008 is a successful demonstration project that operates in a particularly complicated power grid. Based on calculations for observability and controllability, six PMUs across three provinces spanning over 1,000 km as well as three HVDC links with capacities of more than 11 GW were chosen for inclusion in this system. The controllers were coordinated to damp two dominant interarea oscillation modes that limited long-distance power delivery.

Some new oscillations with high frequencies (about 5 Hz) caused by time delays and the characteristics of the quick response of HVDC were investigated, and a solution using a low-pass filter was proposed. Based on the improved online Prony identification method, the controller parameters can be adaptively adjusted according to changes in the oscillation frequency. The WADC system was implemented in a real-time system and carefully tested using the CSG's real-time digital simulation (RTDS) platform, which comprises more than ten racks of RTDS and real HVDC control and protection cubicles. After field debugging and trial operations, the performance of the system was validated in 2008 and 2009 through the manual block and deblock of three different HVDC links and by tripping a 500-kV AC tie line. The field test results show that commissioning the WADC system has the potential to increase the damping ratio of the dominant modes from 5% to more than 15%. This increase indicates a power transfer limitation enhancement of 650 MW in the CSG. In the United States, by modulating the transmission power on the 846-mile Pacific DC Intertie, a real-time wide-area damping control system is designed to damp the North-South B mode in the North American Western Interconnection System. Actual test results showed that this project is able to achieve 4%–5% improvement in the damping ratio of the North-South B mode [29].

1.3.3.2 Wide-area protection scheme

Existing protection systems are decentralized control systems that mainly use local information to detect power system faults and abnormal states. To solve the problem of coordinating among different backup protection systems, wide-area current differential backup protection can be carried out. In this way, faults can be efficiently identified and cleared, thus the time for fault clearance being shortened. In the case of power loss after fault clearance at a bus or substation, fast recovery can be achieved under a wide-area protection system (WAPS). Following this idea, an actual WAPS project has been implemented at seven 110-kV substations in the Duyun district of Guizhou, China. The communication structure of the system is designed in a ring-like form. The system consists of two (one for redundancy) master stations and seven subsidiary control stations. A master station can analyze the mode and status of a power grid, determine fault locations, and send corresponding commands to control stations. A subsidiary control station is responsible for acquiring electrical and state quantities and issuing protection and control functions that only require local substation information. The coordination among backup protection systems can be realized by using the WAPS to isolate faulty components and accelerate fault clearance. Field disturbance recordings show that the fault clearing time for a circuit breaker failure can be decreased from 1.0–2.0 s to 0.2 s with the aid of the WAPS. Based on this system, a fast auto-switching of wide-area backup power supply can be fulfilled. In Canada, starting from decades ago, the Hydro-Québec Research Institute has been striving to develop a wide-area special protection scheme (SPS) for the Hydro-Québec grid [30]. This SPS is capable of not only increasing system power transfer limits with respect to transient stability but also protecting the system against voltage instability after severe contingencies with large voltage drop [31].

1.4 Structure of the book

The above introduction gives an overview of recent developments of WAMS and a number of typical WAMS applications, some of which have already been successfully applied to practical power grids. Focusing on wide-area monitoring and control in modern power systems with renewables, this book will present the state-of-the-art PMU/WAMS technologies and innovative WAMS applications from multiple perspectives. With this overall target, the whole book is structured as the following three parts.

Part I: Wide-area measurement, modeling, and identification

Chapter 2, *Renewable power system interharmonies monitoring based on PMUs*, by Hao Liu, Tianshu Bi, Sudi Xu, and Jue Li, explores the influence of signals that contain the fundamental component and interharmonics on synchrophasors acquired by PMUs in renewable power systems. An interharmonic monitoring method based on complex numbers of PMU measurements is proposed. Numerical tests with field PMU data collected from an actual event in the power grid of western China where large-scale wind farms are located show that the proposed interharmonics method is able to monitor the interharmonics precisely.

Chapter 3, *Causality analysis in phasor measurement unit algorithms*, by Wenchao Meng, Xiaoyu Wang, and Innocent Kamwa, introduces a novel group delay compensation method for PMU algorithms without shifting time stamps. For PMUs using such group delay compensations, they are defined as causal PMUs in this chapter. The impact of causality on the performance of PMU algorithms is first investigated, and then how causal PMUs compensate for the group delay caused by digital filters is analyzed. It is found that causal PMUs normally reduce latency by group delay compensation. This could be very helpful for wide-area control and protection applications where latency could be more important than accuracy.

Chapter 4, *Comprehensive modeling and parameter identification of wind farms based on wide-area measurement systems*, by Yinfeng Wang, Chao Lu, Zhanhe Li, and Lipeng Zhu, develops a novel methodology for PMU-based modeling and parameter identification of wind farms. To improve the efficiency of the identification procedure, trajectory sensitivity analysis is combined with improved genetic algorithm (IGA) to fulfill the identification task. It is found that system responses with identified parameters based on WAMS are more accurate and reasonable than those with the empirical parameters.

Chapter 5, *Load modeling based on synchrophasor measurements*, by Xinran Zhang and David J. Hill, investigates approaches to load modeling based on synchrophasor measurement in distribution networks with PV generation and energy storage. It extensively discusses the contributors' new approaches to ambient signal-based load modeling, parameter identification of composite demand-side models with renewables, and the clustering of uncertain load model parameters. With more accurate parameters identified, these approaches would further benefit power system stability analysis and control which are largely impacted by dynamic load behaviors.

Part II: Wide-area online dynamics monitoring

Chapter 6, *Coherency identification of generators in interconnected power systems with renewables*, by Zhenzhi Lin, Fushuan Wen, Yi Ding, Yusheng Xue, and Shengyuan Liu, presents an agglomerative hierarchical clustering based coherency identification method for interconnected power systems with aggregated renewable sources. Various similarity indices are presented for wide-area coherency identification of generators, and the CRiteria Importance Through Intercriteria Correlation method is utilized to integrate these indices. Realistic case studies on the Guangdong power system in South China with a recorded oscillation event show that the proposed method is able to well identify coherent generators.

Chapter 7, *A fast state estimator for systems including limited number of PMUs*, by Ting Wu, Chi Yung Chung, and Innocent Kamwa, proposes a new SE method to improve computational efficiency and maintain high estimation accuracy simultaneously, especially for large-scale networks with multiple bad data. Three-stage SE composed of two WLS linear problems and a nonlinear transformation is proposed to process the SCADA data, while the PMU observable states are calculated by a linear estimator. Both bad SCADA and PMU measurements are identified and processed efficiently, which can dramatically reduce the overall SE implementation time.

Chapter 8, *Response-based event detection for one shot wide-area stability controls*, by Steven M. Rovnyak, Daniel W. Longbottom, Diana C. Vasquez, and Maryam N. Nilchi, performs dynamic event detection with two decision trees (DT) to trigger one-shot control in response to loss of synchronism. One DT utilizes an index based on PMU measurements of bus magnitudes to detect the clearance of severe short-circuit events. Another DT is employed to predict the loss of synchronism using PMU measurements of bus angles immediately after such dynamic events. This approach enables control to be taken earlier and the resulting scheme has a strong capability to stabilize the system against transient events.

Chapter 9, *Online short-term voltage stability monitoring based on wide-area trajectory data analytics*, by Lipeng Zhu and Chao Lu, develops a new framework for online short-term voltage stability monitoring from the perspective of wide-area

PMU data analytics. Faced with the practical class imbalance and cost-sensitive issues, a time series shapelet classification based imbalance learning machine is systematically designed to derive a reliable and explainable data-driven stability assessment model. An incremental learning strategy is embedded into the learning machine, which results in persistent enhancement of the model's adaptability and robustness during online monitoring.

Chapter 10, *A hierarchical data-driven method for short-term voltage stability assessment of power systems*, by Yan Xu, Zhao Yang Dong, and Rui Zhang, designs an intelligent system to tackle the problems of classifying transient voltage collapse and predicting unacceptable dynamic voltage deviation sequentially. The intelligent system is based on ensemble learning of random-weights neural networks and is implemented in an off-line training, real-time application, and on-line updating pattern. Numerical test results show its superiority in both learning speed and accuracy over some state-of-the-art learning algorithms.

Part III: Wide-area control and decision making

Chapter 11, *Impact of data quality on synchrophasor-based control to minimize wind curtailment*, by Anurag Srivastava, Vignesh Venkata Gopala Krishnan, Shyam Gopal, Shikhar Pandey, and David Bakken, discusses the design, modeling, and implementation of a PMU-based remedial action scheme (RAS) for minimizing wind power curtailment. A mathematical model for RAS is formulated and its real-time validation using a testbed comprising of RTDS, PMUs, and controllers is described. Sources of anomalies in PMU data, their detection, and the effect of data anomaly on RAS operation are also discussed.

Chapter 12, *Power system controlled islanding schemes*, by Lei Ding and Yichen Guo, provides a detailed demonstration of PMU-based controlled islanding schemes by systematically solving three problems: (1) how to predict the generator groups, that is, how to group generators; (2) where to split the system; and (3) when to split the system. A generator grouping algorithm based on system linearization and stable/unstable equilibrium point (SEP/UEP) analysis is proposed. The islanding boundary is figured out with the help of spectral clustering. Feasible islanding time intervals are identified using the controlling UEP method.

Chapter 13, *Intelligent voltage regulation scheme of active distribution networks with renewable energy integration*, by Yongxi Zhang, Fengji Luo, and Zhao Yang Dong, proposes an intelligent distributed voltage control scheme using multiple battery energy storage systems (BESS) for distribution networks with high renewable energy penetration. It takes into consideration multiple possible weather scenarios to deal with the renewable uncertainty. A fully distributed control scheme based on the distributed-alternating direction method of multipliers algorithm (D-ADMM) is designed to optimally regulate multiple BESS for voltage control.

Chapter 14, *Analysis and compensation of multiple PMU data delays for frequency control in islanded microgrids*, by Shichao Liu, Xiaoyu Wang, and Peter Xiaoping Liu, investigates the effect of multiple PMU communication delays on the secondary frequency control of an islanded microgrid with distributed generators. To improve the performance of the microgrid when multiple PMU delays exist, a gain scheduling approach is proposed. Theoretical analysis and numerical tests on the Canadian distribution system show that PMU delays can adversely impact the microgrid secondary frequency control and the designed gain schedulers can improve the control scheme's robustness to communication delays.

Chapter 15, *Adaptive wide-area power oscillation damper design for photovoltaic plant considering delay compensation*, by Wei Yao, Yu Shen, Jinyu Wen, and Haibo He, proposes an adaptive wide-area power oscillation damper (WPOD) based on goal representation heuristic dynamic programming (GrHDP) algorithm for PV plants to enhance damping of the concerned interarea mode. By using GrHDP, the adaptive WPOD is implemented in a model-free manner and has adaptability to various operating conditions. An adaptive delay compensator (ADC) is employed to compensate the wide-area communication delay.

Chapter 16, *Phasor measurement unit based wide-area control: integration, control, and challenges*, by Ahmed S. Musleh, S.M. Muyeen, Ahmed Al-Durra, and Innocent Kamwa, provides a reliable and secure wide-area control approach for controlling the voltage of modern power systems. PMU measurements are used to design a wide-area controller for the FACTS devices which can enhance the voltage stability of power grids. A complete hardware-in-the-loop (HIL) system is developed using RTDS and a real-world voltage source converter (VSC) based FACTS device; it shows the effectiveness of PMU-based wide-area control.

References

[1] De La Ree J., Centeno V., Thorp J. S., and Phadke A. G. "Synchronized phasor measurement applications in power systems." *IEEE Transactions on Smart Grid*. 2010;1(1):20–27.

[2] Chakrabortty A., Chow J. H., and Salazar A. "A measurement-based framework for dynamic equivalencing of large power systems using wide-area phasor measurements." *IEEE Transactions on Smart Grid*. 2010;2(1):68–81.

[3] Ge Y., Flueck A. J., Kim D. K., Ahn J. B., Lee J. D., and Kwon D. Y. "An event-oriented method for online load modeling based on synchrophasor data." *IEEE Transactions on Smart Grid*. 2015;6(4):2060–2068.

[4] Fan L., and Wehbe Y. "Extended Kalman filtering based real-time dynamic state and parameter estimation using PMU data." *Electric Power Systems Research*. 2013;103:168–177.

[5] Chakrabarti S., and Kyriakides E. "PMU measurement uncertainty considerations in WLS state estimation." *IEEE Transactions on Power Systems*. 2009;24(2):1062–1071.

[6] Korres G. N., and Manousakis N. M. "State estimation and bad data processing for systems including PMU and SCADA measurements." *Electric Power Systems Research*. 2011;81(7):1514–1524.

[7] Wu T., Chung C. Y., and Kamwa I. "A fast state estimator for systems including limited number of PMUs." *IEEE Transactions on Power Systems*. 2017;32(6):4329–4339.

[8] Xie L., Chen Y., and Kumar P. R. "Dimensionality reduction of synchrophasor data for early event detection: linearized analysis." *IEEE Transactions on Power Systems*. 2014;29(6):2784–2794.

[9] Ge Y., Flueck A. J., Kim D.-K., *et al.* "Power system real-time event detection and associated data archival reduction based on synchrophasors." *IEEE Transactions Smart Grid*. 2015;6(4):2088–2097.

[10] Kim D. I., Chun T. Y., Yoon S. H., Lee G., and Shin Y. J. "Wavelet-based event detection method using PMU data." *IEEE Transactions on Smart Grid*. 2015;8(3):1154–1162.

[11] Jiang J. A., Yang J. Z., Lin Y. H., Liu C. W., and Ma J. C. "An adaptive PMU based fault detection/location technique for transmission lines. I. Theory and algorithms." *IEEE Transactions on Power Delivery*. 2000;15 (2):486–493.

[12] Jiang Q., Li X., Wang B., and Wang H. "PMU-based fault location using voltage measurements in large transmission networks." *IEEE Transactions on Power Delivery*. 2012;27(3):1644–1652.

[13] Azizi S., and Sanaye-Pasand M. "A straightforward method for wide-area fault location on transmission networks." *IEEE Transactions on Power Delivery*. 2015;30(1):264–272.

[14] He M., Vittal V., and Zhang J. "Online dynamic security assessment with missing PMU measurements: a data mining approach." *IEEE Transactions on Power Systems*. 2013;28(2):1969–1977.

[15] Li M., Pal A., Phadke A. G., and Thorp J. S. "Transient stability prediction based on apparent impedance trajectory recorded by PMUs." *International Journal of Electrical Power & Energy Systems*. 2014;54:498–504.

[16] Wang B., Fang B., Wang Y., Liu H., and Liu Y. "Power system transient stability assessment based on big data and the core vector machine." *IEEE Transactions on Smart Grid*. 2016;7(5):2561–2570.

[17] Dasgupta S., Paramasivam M., Vaidya U., and Ajjarapu V. "Real-time monitoring of short-term voltage stability using PMU data." *IEEE Transactions on Power Systems*. 2013;28(4):3702–3711.

[18] Xie X., Xiao J., Lu C., and Han Y. "Wide-area stability control for damping interarea oscillations of interconnected power systems." *IEE Proceedings on Generation, Transmission and Distribution*. 2006;153(5):507–514.

[19] Terzija V., Valverde G., Cai D., *et al.* "Wide-area monitoring, protection, and control of future electric power networks." *Proceedings of the IEEE*. 2010;99(1):80–93.

[20] Mokhtari M., Aminifar F., Nazarpour D., and Golshannavaz S. "Wide-area power oscillation damping with a fuzzy controller compensating the continuous communication delays." *IEEE Transactions on Power Systems*. 2012;28(2):1997–2005.

[21] Vahidnia A., Ledwich G., and Palmer E. W. "Transient stability improvement through wide-area controlled SVCs." *IEEE Transactions on Power Systems*. 2015;31(4):3082–3089.

[22] Eriksen P. B., Ackermann T., Abildgaard H., Smith P., Winter W., and Garcia J. R. "System operation with high wind penetration." *IEEE Power and Energy Magazine*. 2005;3(6):65–74.

[23] Ghosh S., Kamalasadan S., Senroy N., and Enslin J. "Doubly fed induction generator (DFIG)-based wind farm control framework for primary frequency and inertial response application." *IEEE Transactions on Power Systems*. 2015;31(3):1861–1871.

[24] Virulkar V. B., and Gotmare G. V. "Sub-synchronous resonance in series compensated wind farm: a review." *Renewable and Sustainable Energy Reviews*. 2016;55:1010–1029.

[25] Lu C., Shi B., Wu X., and Sun H. "Advancing China's smart grid: phasor measurement units in a wide-area management system." *IEEE Power and Energy Magazine*. 2015;13(5):60–71.

[26] Phadke A. G., and Bi T. "Phasor measurement units, WAMS, and their applications in protection and control of power systems." *Journal of Modern Power Systems and Clean Energy*. 2018;6(4):619–629.

[27] Phadke A. G., Volskis H., de Moraes R. M., *et al.* "The wide world of wide-area measurement." *IEEE Power and Energy Magazine*. 2008;6(5):52–65.

[28] Zhao J., Gomez-Exposito A., Netto M., *et al.* "Power system dynamic state estimation: motivations, definitions, methodologies and future work." *IEEE Transactions on Power Systems*. 2019;34(4):3188–3198.

[29] Pierre B. J., Wilches-Bernal F., Schoenwald D. A., *et al.* "Design of the Pacific DC intertie wide area damping controller." *IEEE Transactions on Power Systems*. 2019;34(5):3594–3604.

[30] Kamwa I., Beland J., Trudel G., *et al.* "Wide-area monitoring and control at Hydro-Québec: past, present and future." *Proceedings of the IEEE Power Engineering Society General Meeting*, 2006, 1–12.

[31] Ghahremani E., Heniche A., Perron M., *et al.* "A detailed presentation of an innovative local and wide-area special protection scheme to avoid voltage collapse: from proof of concept to grid implementation." *IEEE Transactions on Smart Grid*. 2019;10(5):5196–5211.

Part I

Wide-area measurement, modeling, and identification

Chapter 2

Renewable power system interharmonics monitoring based on PMUs

Hao Liu[1], *Tianshu Bi*[1], *Sudi Xu*[1] *and Jue Li*[1]

With the integration of more distributed or aggregated renewables and high-voltage direct current (HVDC) transmission, power electronic devices are widely deployed in power system [1–3]. This significantly changes the dynamic and operating characteristics of the power system, with the result that it is facing more and more stability and reliability challenges [4–6]. For example, the interharmonics introduced by the power electronic devices could cause the subsynchronous oscillation (SSO) and other further reliability problems [7,8].

SSO events in renewable integration have occurred all over the world [9], including the Electric Reliability Council of Texas in 2009 [10], Oklahoma Gas and Electric Company in 2011 [11], and China Jibei Power Grid in 2013 and 2014 [12]. From July 2015, interharmonics have been detected about 100 times in western China which has a high concentration of renewables and HVDC transmission lines. During the most severe event, the interharmonics that started around the wind farms propagated to the thermal generators 300 km away, and caused SSO. The torsional stress relays tripped three thermal generators. More than 1 MW of power was lost. Then, the frequency of the power grid of western China dropped to 49.91 Hz. The affected area was more than 130,000 km^2. From July 2015 to July 2016, there were 100 SSO events that collectively resulted in the tripping of over 6,000,000 kW of wind capacity.

The synchronization and high-reporting rates of phasor measurement units (PMUs) make them an effective tool to monitor the appearance, propagation, and distribution of interharmonics. Further actions can be taken to control and eliminate the interharmonics based on PMUs measurements. However, PMUs are designed to measure the fundamental component [13]. The frequency of the interharmonics was from 10 to 110 Hz, which can be divided into subsynchronous interharmonics and supersynchronous interharmonics [14–16]. The published PMU algorithms are unsuitable for making such measurements [17–31]. These interharmonics cause error in the measurement of the fundamental component in addition to threatening the operating security of the power system.

[1]State Key Lab of Alternate Electric Power System With Renewable Energy Sources (North China Electric Power University), Changping District, Beijing, China

In this charter, the influence of signals that contain the fundamental component and interharmonics on phasors are explored. An interharmonic monitoring method based on complex numbers of phasors measured by PMUs is proposed. Then, the sample data and phasor measurements measured by PMUs during an SSO event in western China are studied.

2.1 Influence of interharmonics on phasor data

In this section, the phasor expressions of the power signals, which contain the fundamental component and the interharmonics, are derived. It is found that both subsynchronous and supersynchronous interharmonics will result in the subsynchronous oscillation in phasor measurements. The mechanism frequency transmutation between the interharmonic and the phasor oscillation is revealed. Then, the recording data and the phasor measurements obtained by PMUs during the subsynchronous oscillation in the area of China with a high concentration of wind farms and HVDC transmission lines are investigated.

2.1.1 Phasor expression of signal contains single interharmonic

Under dynamic conditions, the pure stationary sinusoidal formula is not suitable even for an envelope of the signal within one calculation data window. The correct dynamic fundamental signal is expressed as [32]

$$x(t) = \sqrt{2}X_{\mathrm{m}}(t)\cos\left(2\pi f_0 t + \left(2\pi\int_0^t \Delta f(t)\mathrm{d}t + \varphi_0\right)\right) \tag{2.1}$$

where $x(t)$ is the instantaneous value of the power signal, $X_{\mathrm{m}}(t)$ is the magnitude, φ_0 is the initial phase, f_0 is the nominal frequency, and $\Delta f(t)$ is the frequency deviation. The magnitude and frequency of the signal are the functions with respect to time. Then (2.1) is represented by a complex number $\dot{X}$ known as its dynamic phasor representation [13]:

$$\begin{aligned} x(t) \leftrightarrow \dot{X} &= X_{\mathrm{m}}(t)\mathrm{e}^{\mathrm{j}\left(2\pi\int_0^t \Delta f(t)\mathrm{d}t + \varphi_0\right)} \\ &= X_{\mathrm{m}}(t)\angle\left(2\pi\int_0^t \Delta f(t)\mathrm{d}t + \varphi_0\right) \\ &= X_{\mathrm{m}}(t)\left[\cos\left(2\pi\int_0^t \Delta f(t)\mathrm{d}t + \varphi_0\right) + \mathrm{j}\sin\left(2\pi\int_0^t \Delta f(t)\mathrm{d}t + \varphi_0\right)\right] \end{aligned} \tag{2.2}$$

It is customary to suppress the term $\mathrm{e}^{\mathrm{j}(2\pi f_0 t)}$ in the expression above, with the understanding that the reference coordinate rotates with the angular frequency of $2\pi f_0$. Therefore, the measured frequency by using the rotating coordinate is the difference between real frequency of the power system and the rated one. The further frequency measurement transformation is needed to obtain the real frequency of the power system. In the following part, the power signal is regarded as steady-state signal to simplify the derivation, indicating that

$$x(t) = \sqrt{2}X_{m0}\cos(2\pi f_0 t + \varphi_0) \tag{2.3}$$

where X_{m0} denotes the static magnitude. Thus, its phasor can be expressed as

$$\dot{X}_0 = X_{m0}\angle\varphi_0 \tag{2.4}$$

Then, suppose the interharmonic signal is

$$x_{ih}(t) = \sqrt{2}X_{m1}\cos(2\pi f_1 t + \varphi_{10}) \tag{2.5}$$

where X_{m1}, f_1, and φ_1 are the amplitude, frequency, and initial phase of the interharmonic signal. Then, its phasor expression $\dot{X}_1$ is [13]

$$\dot{X}_1 = X_{m1}\angle\varphi_1 = X_{m1}\angle(2\pi(f_1 - f_0)t + \varphi_{10}) \tag{2.6}$$

When the signal is the subsynchronous interharmonic or supersynchronous one, f_1 need not be an integer multiple or divisor of f_0. As shown in (2.6), the phasor of the interharmonic signal rotates with the speed of $2\pi\Delta f_1 = 2\pi(f_1 - f_0)$.

If the input signal contains the rated component and an interharmonic component, the expression is shown as follows:

$$x(t) = \sqrt{2}X_{m0}\cos(2\pi f_0 t + \varphi_0) + \sqrt{2}X_{m1}\cos(2\pi f_1 t + \varphi_{10}) \tag{2.7}$$

Then, its phasor expression $\dot{X}_{c1}$ is the superposition of the two components. In the real power system, all the phasors in this chapter is expressed in a synchronously rotating coordinate. The relationship among three phasors is shown in the

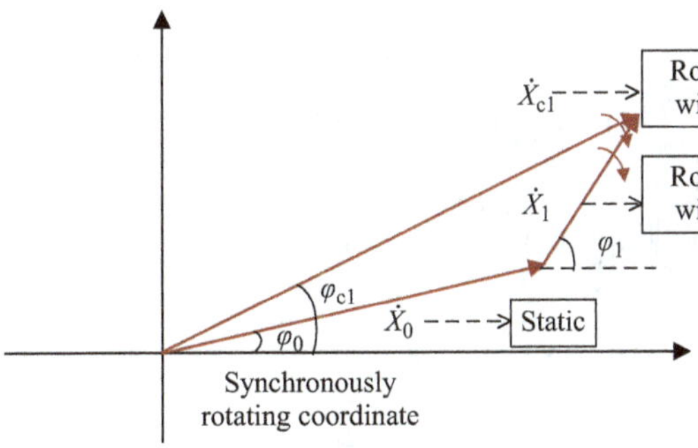

Figure 2.1 Relationship of three phasors

synchronously rotating coordinate (Figure 2.1). It can be seen that $\dot{X}_0$ is static since it rotates with the coordinate at the same speed, while $\dot{X}_1$ rotates with the speed of $2\pi\Delta f_1$. Therefore, $\dot{X}_{c1}$ rotates with the same period. According to the cosine theorem, the formula of $\dot{X}_{c1}$ is given as

$$\begin{cases} \dot{X}_{c1} = X_{mc1}\angle\varphi_{c1} = \dot{X}_0 + \dot{X}_1 \\ \quad = \sqrt{X_{m0}^2 + X_{m1}^2 + 2X_{m0}X_{m1}\cos(2\pi\Delta f_1 t)}\angle\beta \\ \beta = \left(\text{actan}\left(\dfrac{X_{m0}\sin(2\pi\Delta f_1 t)}{X_{m1} + X_{m1}\cos(2\pi\Delta f_1 t)}\right)\right) \end{cases} \tag{2.8}$$

It can be observed that both the magnitude and phase angle are the functions with respect to time.

Now suppose the input contains the interharmonic whose X_{m1} is 5.773 V and f_1 is 20 Hz, and the rate signal whose X_{m0} is 57.73 V and f_0 is 50 Hz, then the magnitude and phase angle of the superimposed phasor are outlined in Figure 2.2. It can be observed that the superimposed phasor oscillates with the subsynchronous frequency of $\Delta f_1 = 30$ Hz. However, both of them are not pure but approximate sinusoidal curves, which can also be concluded from (2.8). That is also the reason why there are other frequency components in the magnitude and the phase angle spectrum, as shown in Figure 2.2(b) and (d). Therefore, the subsynchronous or

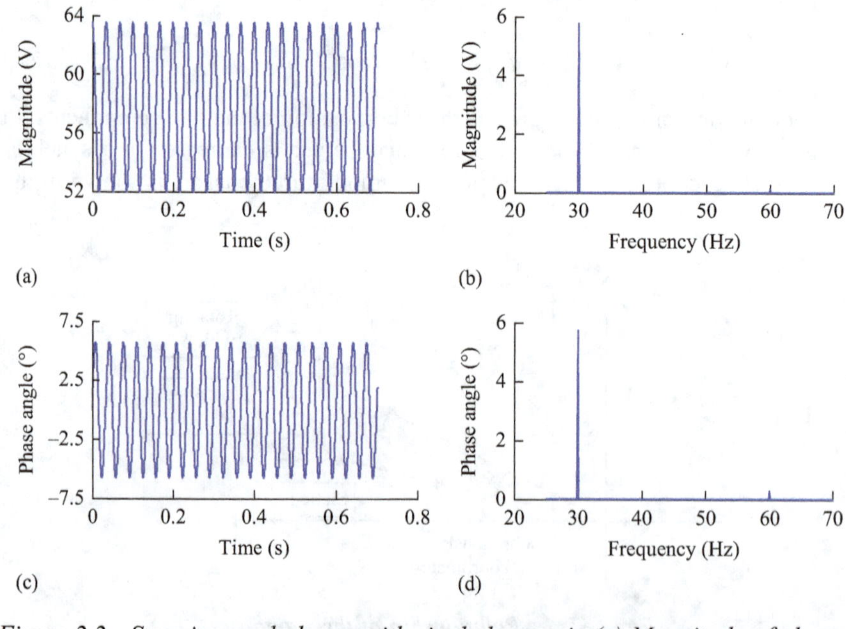

Figure 2.2 Superimposed phasor with single harmonic (a) Magnitude of phasor (b) Spectrum of magnitude (c) Phase angle of phasor (d) Spectrum of phase angle

supersynchronous harmonic will result in the subsynchronous oscillation of the superimposed phasor.

2.1.2 Phasor expression of signal contains multiple interharmonics

In practice, several interharmonics exist simultaneously in the transmission system. As an example, a signal with two interharmonics is given as

$$x(t) = \sqrt{2}X_{m0}\cos(2\pi f_0 t + \varphi_0) + \sqrt{2}X_{m1}\cos(2\pi f_1 t + \varphi_{10}) + \sqrt{2}X_{m2}\cos(2\pi f_2 t + \varphi_{20}) \tag{2.9}$$

where X_{m2}, f_2, and φ_{20} are the magnitude, frequency, and initial phase of the second interharmonic. To simplify the derivation process, φ_0 is set as 0°.

The superimposed phasor $\dot{X}_{c2}$ can be obtained by

$$\dot{X}_{c2} = X_{mc2}\angle\varphi_{c2} = \dot{X}_0 + \dot{X}_1 + \dot{X}_2 = \dot{X}_{c1} + \dot{X}_2 \tag{2.10}$$

where X_{mc2} and φ_{c2} are the magnitude and phase angle of $\dot{X}_{c2}$.

Since $\dot{X}_1$ and $\dot{X}_2$ keep rotating in the synchronously rotating coordinate, while $\dot{X}_0$ is static, the position of $\dot{X}_{c2}$ keeps changing. Figure 2.3 shows six possible superimposition scenarios. Since $\dot{X}_{c1}$ and φ_{c1} can be calculated by (2.8), the key

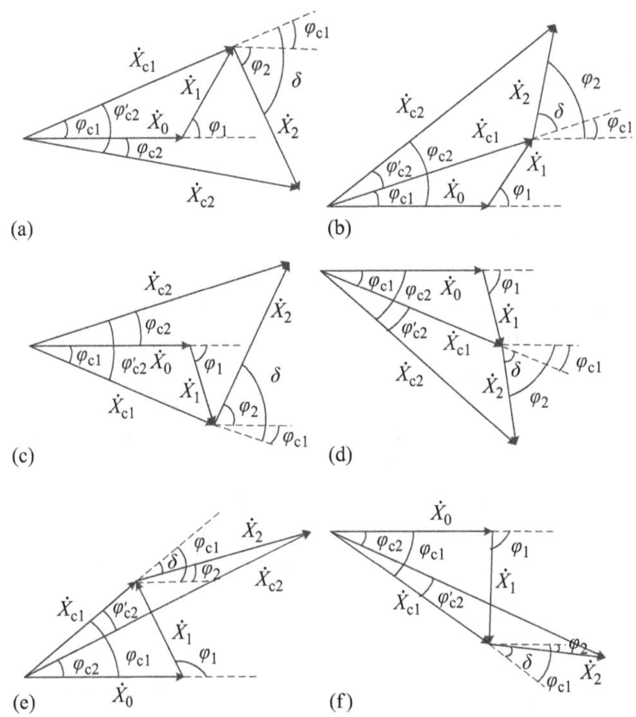

Figure 2.3 Different scenarios for the superimposing of $\dot{X}_0$, $\dot{X}_1$, and $\dot{X}_2$

point to obtain X_{mc2} and φ_{c2} is to calculate φ'_{c2} and δ by the triangle that is formed by $\dot{X}_{c1}$, $\dot{X}_2$, and $\dot{X}_{c2}$, as shown in (2.11) and (2.12):

$$X_{mc2} = \sqrt{X_{mc1}^2 + X_{m2}^2 - 2X_{mc1}X_{m2}\cos(\pi - \delta)} \tag{2.11}$$

$$\varphi'_{c2} = \angle\left(\text{actan}\left(\frac{X_{m2}\sin\delta}{X_{mc1} + X_{m2}\cos\delta}\right)\right) \tag{2.12}$$

However, the formulas of δ and φ_{c2} are different for different scenarios in Figure 2.3, which are given respectively form (2.13) to (2.18):

$$\begin{cases} \delta = \varphi_2 + \varphi_{c1} \\ \varphi_{c2} = \varphi_{c1} - \varphi'_{c2} \end{cases} \tag{2.13}$$

$$\begin{cases} \delta = \varphi_2 - \varphi_{c1} \\ \varphi_{c2} = \varphi'_{c2} + \varphi_{c1} \end{cases} \tag{2.14}$$

$$\begin{cases} \delta = \varphi_2 + \varphi_{c1} \\ \varphi_{c2} = \varphi'_{c2} - \varphi_{c1} \end{cases} \tag{2.15}$$

$$\begin{cases} \delta = \varphi_2 - \varphi_{c1} \\ \varphi_{c2} = \varphi_{c1} - \varphi'_{c2} \end{cases} \tag{2.16}$$

$$\begin{cases} \delta = \varphi_{c1} - \varphi_2 \\ \varphi_{c2} = \varphi'_{c2} + \varphi_{c1} \end{cases} \tag{2.17}$$

$$\begin{cases} \delta = \varphi_{c1} - \varphi_2 \\ \varphi_{c2} = \varphi'_{c2} - \varphi_{c1} \end{cases} \tag{2.18}$$

Now, the phase angle of the phasor is considered to be positive, when the phasor advances $\dot{X}_0$; the phase angle of the phasor is considered to be negative when the phasor lags $\dot{X}_0$. Therefore, for (2.13), $\dot{X}_{c1}$ advances $\dot{X}_0$ and $\dot{X}_2$ lags $\dot{X}_0$. Therefore, φ_2 and δ are negative, φ_{c1} is positive. δ can be obtained by (2.19). Then, φ'_{c2} computed by (2.12) is negative, because $X_{m2}\sin\delta/(X_{mc1} + X_{m2}\cos\delta)$ is negative. φ_{c2} can be obtained by (2.20). For (2.15), $\dot{X}_{c1}$ lags $\dot{X}_0$ and $\dot{X}_2$ advances $\dot{X}_0$. Therefore, φ_2 and δ are positive, and φ_{c1} is negative. δ can be obtained by (2.19) as well. Then, φ'_{c2} computed by (2.12) is positive, because $X_{m2}\sin\delta/(X_{mc1} + X_{m2}\cos\delta)$ is positive. φ_{c2} can be obtained by (2.20). It can be verified that formulas of δ and φ_{c2} are the same as (2.19) and (2.20) for other scenarios. The detail process is not outlined here:

$$\delta = \varphi_2 - \varphi_{c1} \tag{2.19}$$

$$\varphi_{c2} = \varphi'_{c2} + \varphi_{c1} \tag{2.20}$$

Consequently, $\dot{X}_{c2}$ is expressed as

$$\begin{aligned}\dot{X}_{c2} &= \sqrt{X_{mc1}^2 + X_{m2}^2 - 2X_{mc1}X_{m2}\cos(\pi - \varphi_2 + \varphi_{c1})} \\ &\quad \angle\left(\varphi_{c1} + \operatorname{actan}\left(\frac{X_{m2}\sin(\varphi_2 - \varphi_{c1})}{X_{mc1} + X_{m2}\cos(\varphi_2 - \varphi_{c1})}\right)\right) \\ &= \sqrt{X_{mc1}^2 + X_{m2}^2 - 2X_{mc1}X_{m2}\cos(\pi - 2\pi\Delta f_2 t + \varphi_{c1})} \\ &\quad \angle\left(2\pi\Delta f_1 t + \operatorname{actan}\left(\frac{X_{m2}\sin(2\pi\Delta f_2 t - \varphi_{c1})}{X_{mc1} + X_{m2}\cos(2\pi\Delta f_2 t - \varphi_{c1})}\right)\right)\end{aligned} \tag{2.21}$$

Figure 2.4 depicts the magnitude and phase angle of the superimposition of the rated component, 10% interharmonic of 20 Hz and 20% interharmonic of 70 Hz. It can be stated that the major frequency components are 20 Hz and 30 Hz. The 20 Hz component is introduced by the interharmonic of 70 Hz, while the 30 Hz component is introduced by the interharmonic of 20 Hz. This can be distinguished by their respective magnitude. In particular, when the f_1 and f_2 are symmetric with f_0, φ_{10}, and φ_{20} are symmetric with φ_0, and $X_{m1} = X_{m2}$, $\dot{X}_{c2}$ will be modulated magnitude only.

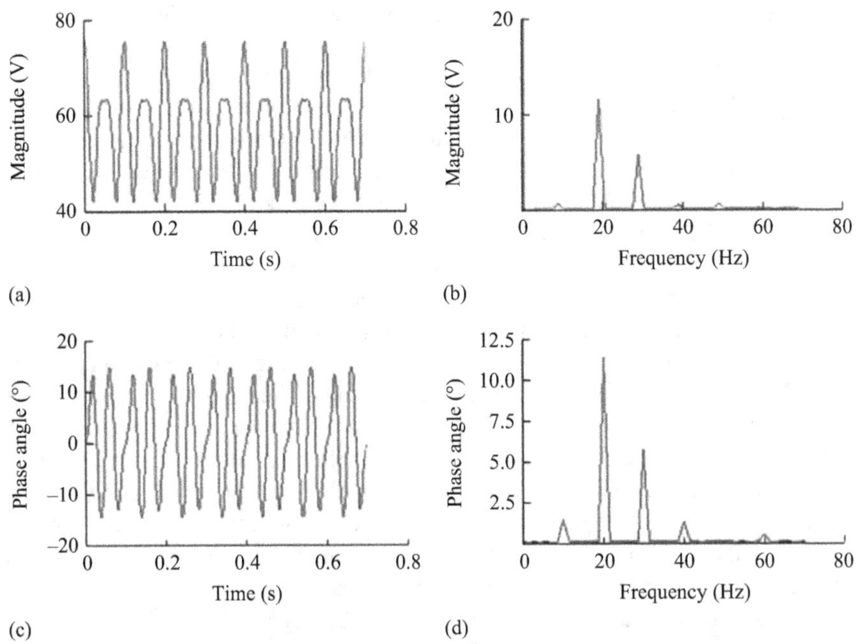

Figure 2.4 Superimposed phasor with two interharmonics (a) Magnitude of phasor (b) Spectrum of magnitude (c) Phase angle of phasor (d) Spectrum of phase angle

More generally, if more than two interharmonics exit in the input signal, the mechanism of frequency transmutation between the interharmonic and the phasor oscillation frequencies is the same as that illustrated by Figure 2.2.

The PMU is designed to measure the fundamental component rather than interharmonics. But if the interharmonics are not eliminated, their existence results in the fundamental component modulating at the subsynchronous frequency. There are two strategies to dispose of the interharmonics. The first one is to remove all of them by filters, and calculate the fundamental phasor only, and the second one is to calculate the superimposed phasor that contains the fundamental component and interharmonics. By the second way, we can monitor and detect the subsynchronous oscillation caused by the interharmonics with the phasor measurements of PMUs. Then corresponding controls can be carried out to prevent further faults, just like the low-frequency oscillation detection based on PMUs.

2.1.3 Subsynchronous oscillation analysis based on PMU measurements

In this section, a subsynchronous oscillation that occurred in western China where there is a large concentration of wind farms and HVDC transmission lines is studied using the recorded PMU data. Results of the analysis show the severe consequences of the interharmonics and verify the correctness of the theoretical derivation.

2.1.3.1 Subsynchronous oscillations distribution

From July 2015, several subsynchronous oscillations caused by the interharmonics occurred in western China, which resulted in several thermal power generations and the tripping-off of wind farms. PMUs installed in the area tracked the oscillation processes by both recording the signal instant value and measuring the phasor. The total PMU data are over 40 GB and cover over 140 transmission lines. The subsynchronous oscillations that occurred in September are exemplary, so the phasor amplitude of PMU data on that day is analyzed.

Figure 2.5 shows the geographical wiring diagram of the area with the main oscillation components by histograms. In the figure, the different colors of the histograms represent different oscillation frequencies. The height of the histograms bars indicates the oscillation magnitudes. The larger the oscillation magnitude is, the higher the histogram bars are. The exact oscillation frequencies are outlined under the histograms, and the oscillation magnitudes and their respective proportions of the fundamental one are outlined above the histograms. It can be seen that all the oscillation frequencies are around 10.2, 20.6, and 30.7 Hz. The highest oscillation magnitude reached 32.88% of the fundamental component.

The substations A and B are connected with the large scale of wind farms. The substation H is the convertor station. It can be seen that the oscillation components around wind farms are bigger than those at other places. It means that there are more subsynchronous and supersynchronous harmonics around wind farms. The subsynchronous oscillations are propagated along the transmission lines that cover the voltage levels from 35 kV to 750 kV. The total affected area is over

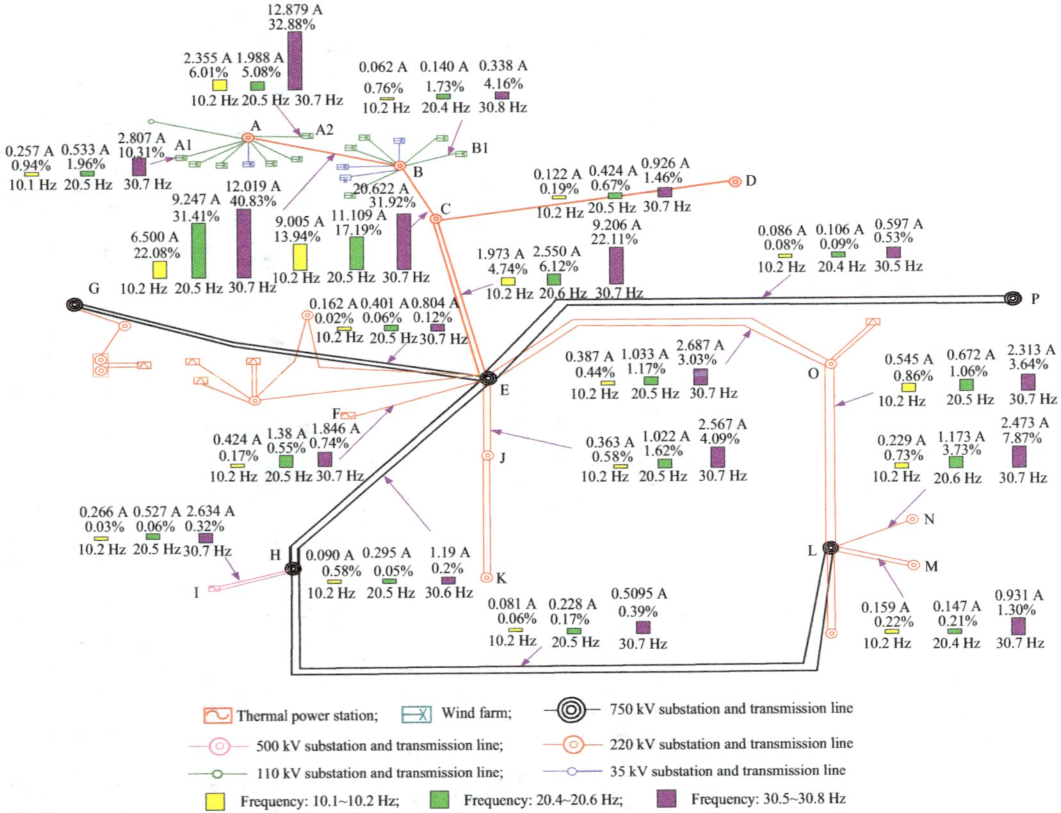

Figure 2.5   Oscillation analysis based on PMU measurements

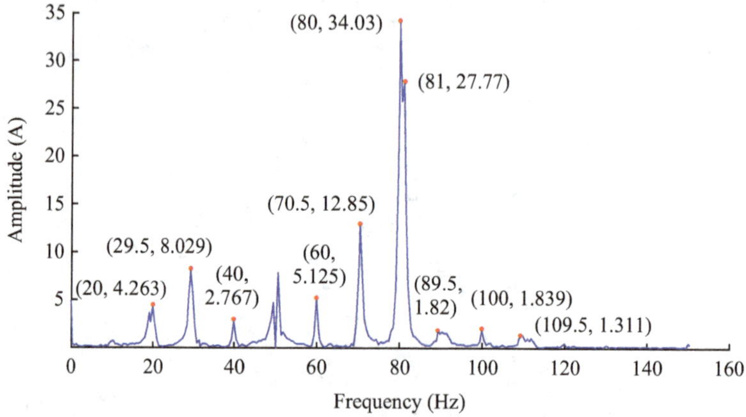

Figure 2.6 Spectrum analysis of the current instant values of the transmission line C–E

130,000 km^2. Farther away and at higher voltage levels, the subsynchronous oscillation magnitudes become smaller. However, the especially high magnitude of the subsynchronous oscillations around the wind farms does not mean that it is the wind farms that cause the interharmonics and the subsynchronous oscillation. The exact reason is complicated, and further studies are necessary.

Consequently, the large scale of renewable energy and HVDC introduce a large number of interharmonics. They result in the subsynchronous oscillation which can be propagated widely and threaten the security of the power system.

2.1.3.2 Interharmonics analysis

In addition of phasor measurements, PMUs also recorded the instant values of the signals for some transmission lines. Figure 2.6 shows the spectrum analysis of the current instantaneous values of the transmission line C–E, while Figure 2.7 shows the spectrum analysis of current phasor amplitudes from the PMU of the same transmission during the same period. To show the interharmonics and their impacts on the phasor measurements clearly, the fundamental component is not plotted in Figures 2.6 and 2.7.

Then, all the frequencies that cannot be ignored appeared in Figures 2.6 and 2.7 are shown in Figure 2.8 in two rows. According to the mechanism of frequency transmutation between the harmonic and the phasor oscillation illustrated by Figure 2.2, all the oscillation components can be explained by the corresponding harmonics. For example, the phasor oscillation component at 10 Hz is caused by the interharmonics of 40 and 60 Hz, because their differences between the rated frequency are both 10 Hz. One oscillation component can be caused by multiple subsynchronous or supersynchronous harmonics. Therefore, the derivations in Sections 2.1.1 and 2.1.2 are verified by the real PMU measurements and recorded data. Since it is feasible to identify the oscillation from the active power, Figure 2.9 depicts the spectrum analysis of the active power of the same line during the time

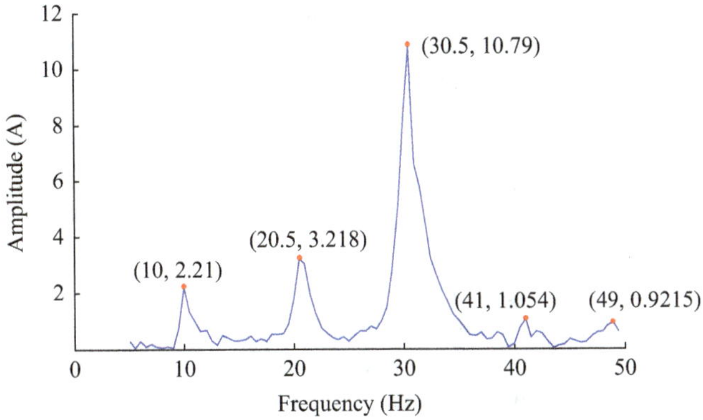

Figure 2.7 Spectrum analysis of the current phasor of the transmission line C–E

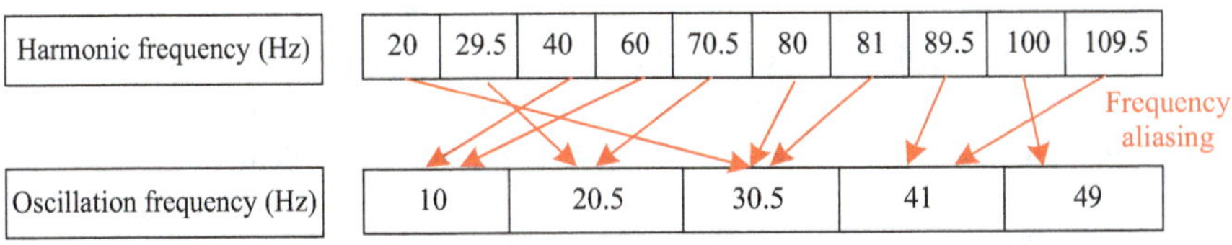

Figure 2.8 Corresponding relation between the interharmonic and the oscillation frequencies

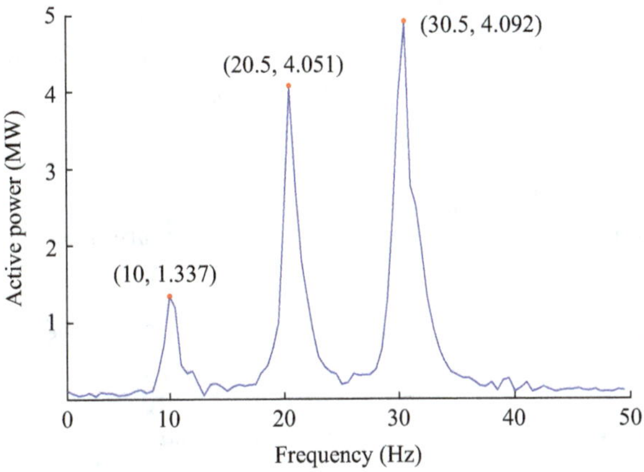

Figure 2.9 Spectrum analysis of the active power of the transmission line C–E

window measured by PMUs. It can be seen that the oscillation frequencies are the same as Figure 2.7.

It may seem that the 109.5 Hz interharmonic appears to be an exception. According to the mechanism, such an interharmonic should result in phasor oscillation at 59.5 Hz. But, the highest reporting rate of PMUs in China is 100 Hz, which cannot meet the Shannon sampling theorem. Therefore, the frequency aliasing occurs, the phasor oscillation frequency becomes 40.5 Hz. Since the frequency aliasing is irreversible, it will be considered that there is more interharmonic of 109.5 Hz than there should be.

Consequently, when the frequency aliasing occurs, phasor measurements cannot restore the original interharmonics accurately. Measures should be taken to prevent the frequency aliasing from occurring, especially in the case of the subsynchronous oscillation. For example, a digital filter in the PMU algorithm can eliminate the oscillation components which may lead to the frequency aliasing.

As shown in Figure 2.8, one subsynchronous oscillation frequency component of the phasor amplitudes may be caused by two interharmonics that are symmetrical about the rated frequency. Based on the phasor amplitude, we can only tell if there are subsynchronous oscillation that may be caused by the subsynchronous or supersynchronous interharmonics. But the exact magnitude of the subsynchronous and supersynchronous interharmonics cannot be obtained. Further algorithms for the interharmonic monitoring based on PMUs are needed to achieve that. In addition, the fast-changing phasor makes precise measurement more difficult. Because DFT assume that the phasor is static in the data window. The faster the phasor changes, the bigger the measurement error is. Therefore, relevant phasor measurement algorithms which can track the phasor modulated in subsynchronous frequency precisely are also needed.

2.2 Interharmonics monitoring method

As shown above, the interharmonics would cause both the instantaneous power and the phasor to oscillate with the frequency of Δf. Therefore, it cannot be distinguished whether the harmonic is subsynchronous or supersynchronous from the phasor amplitude. Therefore, another method is needed to monitor the interharmonics based on PMUs.

Now, rewrite the phasor containing two interharmonics using complex numbers:

$$\begin{aligned}\dot{X}_{c2} &= X_{m0}[\cos\varphi_0 + \mathrm{j}\sin\varphi_0] + X_{m1}[\cos(2\pi(f_0 - f_1)t) + \mathrm{j}\sin(2\pi(f_0 - f_1)t)] \\ &\quad + X_{m2}[\cos(2\pi(f_0 - f_2)t) + \mathrm{j}\sin(2\pi(f_0 - f_2)t)] \\ &= X_{m0}\cos\varphi_0 + X_{m1}\cos(2\pi(f_0 - f_1)t) + X_{m2}\cos(2\pi(f_0 - f_2)t) \\ &\quad + \mathrm{j}[X_{m0}\sin\varphi_0 + X_{m1}\sin(2\pi(f_0 - f_1)t) + X_{m2}\sin(2\pi(f_0 - f_2)t)] \\ &= r(t) + \mathrm{j}i(t)\end{aligned} \tag{2.22}$$

where f_1 is the frequency of the subsynchronous harmonic, f_2 is the frequency of the supersynchronous harmonic and they are symmetrical about f_0, $r(t)$ is the real part of $\dot{X}_{c2}$, and $i(t)$ is the imaginary part of $\dot{X}_{c2}$.

The Fourier transforms of $r(t)$ and $i(t)$ are

$$R(f) = \frac{1}{2}X_{m0}[\delta(f-0)+\delta(f+0)] + \frac{1}{2}X_{m1}[\delta(f-(f_0-f_1)) \\ +\delta(f+(f_0-f_1))] + \frac{1}{2}X_{m2}[\delta(f-(f_0-f_2)) + \delta(f+(f_0-f_2))] \tag{2.23}$$

$$\begin{aligned} I(f) &= \frac{1}{j2}X_{m0}[\delta(f-0)-\delta(f+0)] \\ &\quad + \frac{1}{j2}X_{m1}[\delta(f-(f_0-f_1)) - \delta(f+(f_0-f_1))] \\ &\quad + \frac{1}{j2}X_{m2}[\delta(f-(f_0-f_2)) - \delta(f+(f_0-f_2))] \end{aligned} \tag{2.24}$$

Since $(f_0-f_1)+(f_0-f_2)=0$, so (2.23) and (2.24) are derived as

$$\begin{aligned} R(f) &= \frac{1}{2}X_{m0}[\delta(f-0)+\delta(f+0)] \\ &\quad + \frac{1}{2}X_{m1}[\delta(f-(f_0-f_1)) + \delta(f-(f_0-f_2))] \\ &\quad + \frac{1}{2}X_{m2}[\delta(f-(f_0-f_2)) + \delta(f-(f_0-f_1))] \\ &= \frac{1}{2}X_{m0}[\delta(f-0)+\delta(f+0)] \\ &\quad + \frac{1}{2}(X_{m1}+X_{m2})[\delta(f-(f_0-f_1)) + \delta(f-(f_0-f_2))] \end{aligned} \tag{2.25}$$

$$\begin{aligned} I(f) &= \frac{1}{j2}X_{m0}[\delta(f-0)-\delta(f+0)] \\ &\quad + \frac{1}{j2}X_{m1}[\delta(f-(f_0-f_1)) - \delta(f-(f_0-f_2))] \\ &\quad + \frac{1}{j2}X_{m2}[\delta(f-(f_0-f_2)) - \delta(f-(f_0-f_1))] \\ &= \frac{1}{j2}X_{m0}[\delta(f-0)+\delta(f+0)] \\ &\quad + \frac{1}{j2}(X_{m1}-X_{m2})[\delta(f-(f_0-f_1)) - \delta(f-(f_0-f_2))] \end{aligned} \tag{2.26}$$

The Fourier transform of $\dot{X}_{c2}$ is

$$\begin{aligned} X(f) &= R(f) + jI(f) \\ &= X_{m0}\delta(f-0) + X_{m1}\delta(f-(f_0-f_1)) \\ &\quad + X_{m2}\delta(f-(f_0-f_2)) \end{aligned} \tag{2.27}$$

As we can see from (2.27), we can obtain 0 Hz, $f_0 - f_1$, and $f_0 - f_2$. The magnitude of $f_0 - f_1$ is equal to that of the f_1 interharmonic and the magnitude of $f_0 - f_2$ is the same as that of the f_2 interharmonic.

The fast Fourier transform (FFT) is used to do the spectrum analysis. The frequency resolution is directly related to the length of the data window. Their relationship is shown as

$$f_r = \frac{F_s}{N} \tag{2.28}$$

where f_r is the frequency resolution, F_s is the reporting rate of the PMU, and N is the number of the phasor in one data window. It can be concluded that the longer the data window is, the higher the resolution is. Suppose $F_s = 100$ Hz and $N = 100$ (the data window of 1 s), then $f_r = 1$ Hz. To achieve the resolution of 0.1 Hz, a data window of 10 s is required. However, according to the field data analysis, the frequency and the magnitude of the interharmonic may change with the time. A long data window such as 10 s may result in calculation errors. Therefore, the zero-padding method is applied to eliminate fence effect with a short data window.

Now, suppose the signal has two components, whose frequencies are 20.3 Hz and 30.6 Hz, and magnitudes are 20 V and 30 V, respectively. Again, $F_s = 100$ Hz, and $N = 100$, the spectrum analysis without and with zero-padding method is shown in Figure 2.10, respectively. With the zero-padding, for the component of 20.3 Hz, the frequency analyzed error is 0 Hz (0.3 Hz without zero-padding), the magnitude analyzed error is 0.82 V (1.88 V without zero-padding); for the component of 30.6 Hz, the frequency analyzed error is 0 Hz (0.4 Hz without zero-padding), the magnitude analyzed error is 0.49 V (6.87 V without zero-padding). We can see that both the frequency and the magnitude analyzed error are decreased dramatically. To reduce the effect of the sidelobe, some window functions such as Hann and Hamming window can be used into FFT method to suppress the sidelobe.

Therefore, if there are multiple frequency components in a signal, the lack of the frequency resolution results in magnitude estimation errors as well. By the

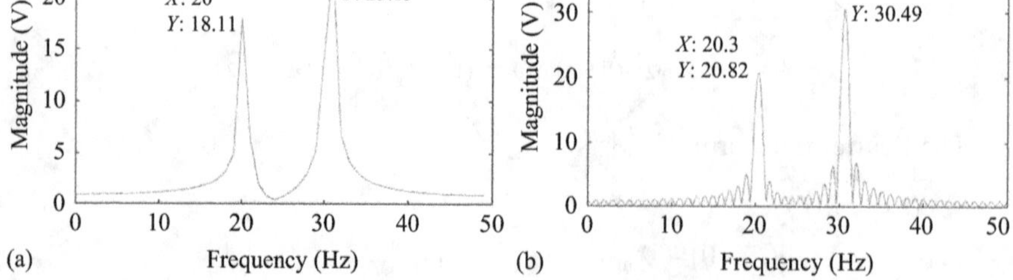

Figure 2.10 Frequency spectrum analysis: (a) without zero-padding and (b) with zero-padding

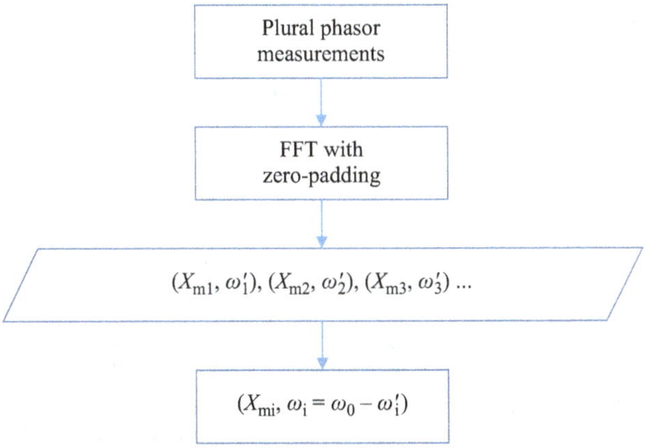

Figure 2.11 Flowchart of interharmonics monitoring method

Table 2.1 Interharmonics frequency calculation based on phasor

Real frequency (Hz)	Data window of 10 s		Data window of 1 s		Data window of 1 s with zero-padding	
	Frequency measurement (Hz)	Frequency error (Hz)	Frequency measurement (Hz)	Frequency error (Hz)	Frequency measurement (Hz)	Frequency error (Hz)
12.1	12.2	0.1	12	0.1	12.1	0
20.3	20.3	0	20	0.3	20.3	0
31.8	31.8	0	32	0.2	31.8	0
50	50	0	50	0	50	0
56.4	56.4	0	56	0.4	56.4	0
79.7	79.7	0	80	0.3	79.7	0
87.9	87.9	0	88	0.1	87.9	0

zero-padding, both the frequency and the magnitude extraction accuracy can be improved.

Consequently, instead of doing the Fourier transform by using the magnitude or phase angle of phasor measurements, we can analyze the frequency spectrum of the complex phasor directly. After getting the frequency and magnitude of different components, frequency shifting can be used to obtain the real frequency of the interharmonics. The flowchart of the algorithm is outlined in Figure 2.11.

Now, suppose there are interharmonics of 12.1 Hz (5 V), 20.3 Hz (5 V), 31.8 Hz (10 V), 56.4 Hz (12.5 V), 79.7 Hz (7.5 V), and 87.9 Hz (10 V) in the signal of 50 Hz (50 V), in which 12.1 Hz and 87.9 Hz, 20.3 Hz, and 79.7 Hz are symmetrical with 50 Hz. The reporting rate of PMU is 100 Hz. Three data windows are used: 10 s, 1 s without zero-padding, and 1 s with zero-padding to 10 s. Then, the interharmonics that is calculated by the proposed method are shown in Tables 2.1 and 2.2. It can be seen that the accuracy of 10 s is very high. However, the long data

Table 2.2 Interharmonics magnitude calculation based on phasor

Real magnitude (V)	Data window of 10 s		Data window of 1 s		Data window of 1 s with zero-padding	
	Magnitude measurement (V)	Magnitude error (%)	Magnitude measurement (V)	Magnitude error (%)	Magnitude measurement (V)	Magnitude error (%)
5	5	0	5.29	5.80	5.443	8.86
5	5	0	4.459	10.82	5.724	14.48
10	10	0	9.467	5.33	10.56	5.60
50	50	0	50.03	0.06	50.03	0.06
12.5	12.5	0	9.567	23.46	12.72	1.76
7.5	7.5	0	6.525	13.00	7.893	5.24
10	10	0	10.26	2.60	10.35	3.50

window would result in calculation errors when the frequency and magnitude of interharmonics change with time. When the data window is decreased to 1 s, the frequency resolution drops to 1 Hz. Therefore, the frequency calculation and magnitude errors become larger. After the zero padding, the frequency calculation accuracy is improved dramatically, ever higher in one case than using a data window of 10 s, but the magnitude errors still exist.

2.3 Case study

In this section, the samples and phasor measurements recorded by PMUs during two SSO events caused by interharmonics that occurred in a power grid of western China with a large concentration of renewables and HVDC transmission lines are analyzed and studied.

The geographical wiring diagram of the power grid impacted by interharmonics is outlined in Figure 2.12. Substations A, B, D, E, and F are connected to large scale wind farms. Substation H is a converter station. The tables in Figure 2.12 are the interharmonics calculation results based on current samples, phasor magnitudes, instantaneous power, and phasors of complex number.

The first rows of the tables in Figure 2.12 are the spectrum analysis results of current samples. Since the current samples are not processed by other algorithms, they contain most of the interharmonics, except for the ones whose frequencies are higher than the Nyquist frequency of the sample rate. The analysis results based on the current samples are taken as the references. Shown as in Figure 2.10, the closer to the wind farms the line is, the larger the proportion of the interharmonic is. Taking the results of line B–C as an example, there are interharmonics of 25 and 75.1 Hz in the current. These interharmonics are propagated to a very long distance throughout different voltage levels of the power system.

The third and the fourth rows are the spectrum analysis results of current phasor amplitudes and instantaneous power, respectively. In line B–C, there is a constant component (0 Hz) and a component of 25.1 Hz in the two rows,

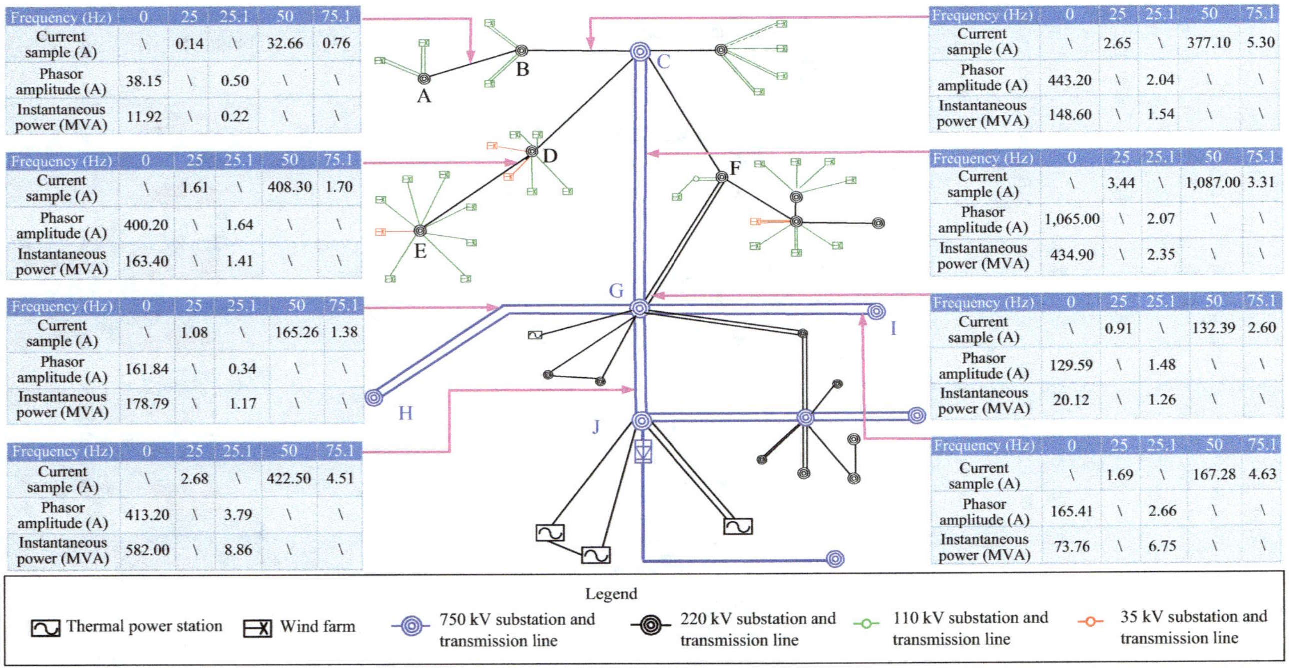

Frequency (Hz)	0	25	25.1	50	75.1
Current sample (A)	\	0.14	\	32.66	0.76
Phasor amplitude (A)	38.15	\	0.50	\	\
Instantaneous power (MVA)	11.92	\	0.22	\	\

Frequency (Hz)	0	25	25.1	50	75.1
Current sample (A)	\	1.61	\	408.30	1.70
Phasor amplitude (A)	400.20	\	1.64	\	\
Instantaneous power (MVA)	163.40	\	1.41	\	\

Frequency (Hz)	0	25	25.1	50	75.1
Current sample (A)	\	1.08	\	165.26	1.38
Phasor amplitude (A)	161.84	\	0.34	\	\
Instantaneous power (MVA)	178.79	\	1.17	\	\

Frequency (Hz)	0	25	25.1	50	75.1
Current sample (A)	\	2.68	\	422.50	4.51
Phasor amplitude (A)	413.20	\	3.79	\	\
Instantaneous power (MVA)	582.00	\	8.86	\	\

Frequency (Hz)	0	25	25.1	50	75.1
Current sample (A)	\	2.65	\	377.10	5.30
Phasor amplitude (A)	443.20	\	2.04	\	\
Instantaneous power (MVA)	148.60	\	1.54	\	\

Frequency (Hz)	0	25	25.1	50	75.1
Current sample (A)	\	3.44	\	1,087.00	3.31
Phasor amplitude (A)	1,065.00	\	2.07	\	\
Instantaneous power (MVA)	434.90	\	2.35	\	\

Frequency (Hz)	0	25	25.1	50	75.1
Current sample (A)	\	0.91	\	132.39	2.60
Phasor amplitude (A)	129.59	\	1.48	\	\
Instantaneous power (MVA)	20.12	\	1.26	\	\

Frequency (Hz)	0	25	25.1	50	75.1
Current sample (A)	\	1.69	\	167.28	4.63
Phasor amplitude (A)	165.41	\	2.66	\	\
Instantaneous power (MVA)	73.76	\	6.75	\	\

Figure 2.12 Interharmonics distribution

respectively. The constant components are the measurements of fundamental inputs. The components of 25.1 Hz are caused by the interharmonics of both 25 and 75.1 Hz, which verify the correctness of the theoretical deviation. Again, it shows the phasor amplitudes and instantaneous power cannot distinguish whether the interharmonic is a subsynchronous or supersynchronous one.

The proposed interharmonics monitoring method has been implemented in the main station of wide-area measurement system (WAMS) in this area. However, all the phase angle measurements of line A–B are 0°. Therefore, there are no interharmonics calculation results based on the phasors of complex numbers in the table of line A–B. It can be seen that the proposed method can distinguish the subsynchronous and supersynchronous interharmonics correctly. The detailed error analysis is shown in Figure 2.13. The analysis results of the current samples are taken as references. The x-axes are the names of lines. The histograms are the calculated frequencies. It can be observed that the largest error is within 0.1 Hz. The magnitude errors are plotted by the curves in different colors. The grey one is

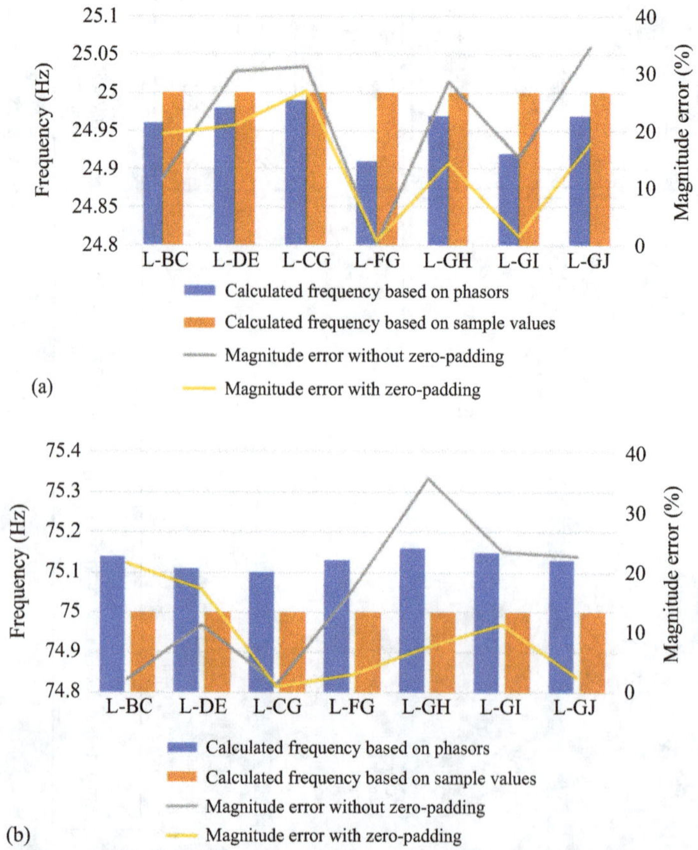

Figure 2.13 Calculation errors analysis of interharmonics: (a) subsynchronous interharmonics and (b) supersynchronous interharmonics

the error of the method without zero padding, while the orange one is the error with zero padding. It shows that the zero padding decreases the error dramatically. The largest magnitude error is 25%, while most of them are under 15%. The errors are larger than they are in the simulation because the field data has more noise, and the phasor measurements have errors. As the frequency of the interharmonics deviates further from 50 Hz, the phasor oscillates faster and the errors become larger. In addition, some of the interharmonics are very small and little absolute errors can result in big relative errors. However, the accuracy of the proposed interharmonics monitoring method is good enough for the online monitoring and alert.

Table 2.3 and Figure 2.14 outline the calculation results of another SSO event. It can be observed that the maximum frequency difference is 0.4 Hz, whereas the maximum magnitude difference is 8.1%.

Table 2.3 Interharmonics monitoring based on samples and phasor

Frequency measurement based on samples (Hz)	Frequency measurement based on phasor (Hz)	Frequency difference (Hz)	Magnitude measurement based on samples (A)	Magnitude measurement based on phasor (A)	Magnitude difference (%)
20.5	20.6058	0.1058	4.26	4.193	1.577
25	25.0412	0.0412	1.473	1.545	4.094
30.5	30.5496	0.0496	8.03	7.776	3.171
40	39.6861	0.3139	2.77	2.784	0.498
60	60.3035	0.3035	5.12	5.180	1.175
70	69.7079	0.3921	12.85	12.011	6.527
75	74.9874	0.0126	2.846	3.077	8.114
80	79.5997	0.4003	34.03	33.326	2.068

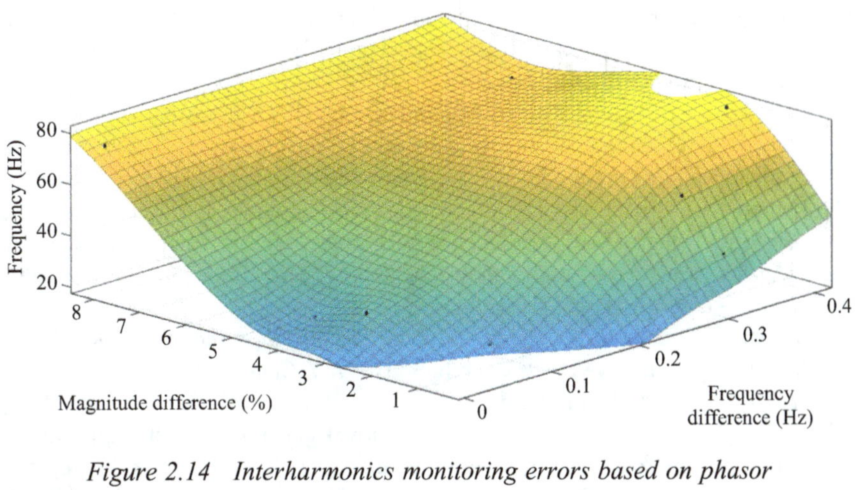

Figure 2.14 Interharmonics monitoring errors based on phasor

2.4 Conclusion

In this charter, the impacts of interharmonics on synchrophasor measurements are revealed. The phasor expression of a signal that contains the fundamental component and single or multiple interharmonics is proposed. It is found that the interharmonics can cause the subsynchronous oscillation of the phasor measurements. The oscillation frequency of phasor measurements is the difference between the interharmonic and the fundamental component. The phasor measurements obtained by PMUs during a subsynchronous oscillation from a real power grid that has a large concentration of wind farms and HVDC transmission lines are analyzed. The geographical wiring diagram with the supersynchronous oscillation distribution shows that more interharmonics are located near the wind farms. In addition, the interharmonics can transmit many voltage levels, leading to affecting a wider area. Furthermore, the spectrum analysis on the phasor demonstrates the correctness of the theoretical derivation. The monitoring of interharmonics using PMUs is possible and necessary.

Then, an interharmonic monitoring method based on complex phasor is proposed. This method has been implemented in the main station of WAMS to monitoring the interharmonics in a very wide area. Instantaneous current measurements and phasor measurements from installed PMUs during an SSO event from a real power system that has a large concentration of wind farms and HVDC transmission lines are analyzed. The analysis results verify that the proposed interharmonics method is able to measure the interharmonics precisely.

References

[1] Sun, J., Li, M., Zhang, Z., *et al.* "Renewable energy transmission by HVDC across the continent: system challenges and opportunities." *CSEE Journal of Power and Energy Systems*. 2017;3(4):353–364.

[2] Hafeez, K., and Khan, S. A. "High voltage direct current (HVDC) transmission: future expectation for Pakistan." *CSEE Journal of Power and Energy Systems*. 2019;5(1):82–86.

[3] Mitra, B., Chowdhury, B., and Manjrekar, M. "HVDC transmission for access to off-shore renewable energy: a review of technology and fault detection techniques." *IET Renewable Power Generation*. 2018;12(13):1563–1571.

[4] Magdy, G., Shabib, G., Elbaset, A. A., *et al.* "A novel coordination scheme of virtual inertia control and digital protection for microgrid dynamic security considering high renewable energy penetration." *IET Renewable Power Generation*. 2019;13(3):462–474.

[5] Skea, J., Anderson, D., Green, T., *et al.* "Intermittent renewable generation and maintaining power system reliability." *IET Generation, Transmission & Distribution*. 2008;2(1):82–89.

[6] Fan, M., Sun, K., Lane, D., *et al.* "A novel generation rescheduling algorithm to improve power system reliability with high renewable energy penetration." *IEEE Transactions on Power Systems*. 2018;33(3):3349–3357.

[7] Tao, W., Yao, H., Dawei, Z., *et al.* "Mechanism of sub-synchronous interaction caused by inter-harmonics in DFIG." *Journal of Engineering*. 2017;2017(13): 892–896.

[8] Alizade, A., and Noshahr, J. B. "Evaluating noise and DC offset due to inter-harmonics and supra-harmonics caused by back-to-back converter of (DFIG) in AC distribution network." *CIRED - Open Access Proceedings Journal*. 2017;2017(1):629–632.

[9] Wu, M., Xie, L., Cheng, L., *et al.* "A study on the impact of wind farm spatial distribution on power system sub-synchronous oscillations." *IEEE Transactions on Power Systems*. 2016;31(3):2154–2162.

[10] Belkin, P. "Event of 10-22-09." *Proceedings of 2010 CREZ Technical Conference*, Council of Texas, 2010.

[11] Wan, Y. H. "Synchronized phasor data for analyzing wind power plant dynamic behavior and model validation." *Proceedings of 2013 National Renewable Energy Laboratory*, Golden, CO, USA, 2013.

[12] Wu, M., Sun, R., Cheng, L., *et al.* "Parameter sensitivity analysis for sub-synchronous control interactions in wind-integrated power systems." *Proceedings of 2014 CIGRE Grid of the Future Symposium*, Houston, TX, USA, 2014.

[13] *IEEE/IEC International Standard - Measuring relays and protection equipment - Part 118-1: Synchrophasor for power systems - Measurements*, IEC/IEEE 60255-118-1:2018, December 2018.

[14] Wang, J., Gao, S., Ji, Z., *et al.* "Time/frequency domain modelling for grid-connected MMC sub-synchronous/super-synchronous oscillation in PV MVDC power collection and integration system." *IET Renewable Power Generation*. 2019;13(1):57–66.

[15] Bi, T., Li, J., Zhang, P., *et al.* "Study on response characteristics of grid-side converter controller of PMSG to sub-synchronous frequency component." *IET Renewable Power Generation*. 2017;11(7):966–972.

[16] Tan, Y., Chen, R., Wang, M., *et al.* "High-accurate estimation method of sub-synchronous and super-synchronous harmonic parameters on PMU." *Journal of Engineering*. 2017;2017(13):1275–1279.

[17] Macii, D., Petri, D., and Zorat, A. "Accuracy analysis and enhancement of DFT-based synchrophasor estimators in off-nominal conditions." *IEEE Transactions on Instrumentation and Measurement*. 2012;61(10):2653–2664.

[18] Belega, D., and Petri, D. "Accuracy analysis of the multicycle synchrophasor estimator provided by the interpolated DFT algorithm." *IEEE Transactions on Instrumentation and Measurement*. 2013;62(5):942–953.

[19] Romano, P., and Paolone, M. "Enhanced interpolated-DFT for synchrophasor estimation in FPGAs: theory, implementation, and validation of a PMU prototype." *IEEE Transactions on Instrumentation and Measurement*. 2014;63(12):2824–2836.

[20] Premerlani, W., Kasztenny, B., and Adamiak, M. "Development and implementation of a synchrophasor estimator capable of measurements under dynamic conditions." *IEEE Transactions on Power Delivery*. 2008;23(1):109–123.

[21] Mai, R. K., Fu, L., Dong, Z., *et al*. "An adaptive dynamic phasor estimator considering DC offset for PMU applications." *IEEE Transactions on Power Delivery*. 2011;26(3):1744–1754.

[22] Petri, D., Fontanelli, D., and Macii, D. "A frequency-domain algorithm for dynamic synchrophasor and frequency estimation." *IEEE Transactions on Instrumentation and Measurement*. 2014;63(10):2330–2340.

[23] Zhan, L., Liu, Y., Culliss, J., *et al*. "Dynamic single-phase synchronized phase and frequency estimation at the distribution level." *IEEE Transactions on Smart Grid*. 2015;6(4):2013–2022.

[24] Zhan, L., Liu, Y., and Liu, Y. "A Clarke transformation-based DFT phasor and frequency algorithm for wide frequency range." *IEEE Transactions on Smart Grid*. 2018;9(1):67–77.

[25] de la O Serna, J. A. "Dynamic phasor estimates for power system oscillations." *IEEE Transactions on Instrumentation and Measurement*. 2007;56(5): 1648–1657.

[26] Platas-Garza, M. A. P. a. J., and de la O Serna, J. A. "Dynamic phasor and frequency estimates through maximally flat differentiators." *IEEE Transactions on Instrumentation and Measurement*. 2010;59(7):1803–1811.

[27] Castello, P., Liu, J., Muscas, C., *et al*. "A fast and accurate PMU Algorithm for P+M class measurement of synchrophasor and frequency." *IEEE Transactions on Instrumentation and Measurement*. 2014;63(12):2837–2845.

[28] Huang, C., Xie, X., and Jiang, H. "Dynamic phasor estimation through DSTKF under transient conditions." *IEEE Transactions on Instrumentation and Measurement*. 2017;66(11):2929–2936.

[29] Sadinezhad, I., and Agelidis, V. G. "Real-time power system phasors and harmonics estimation using a new decoupled recursive-least-squares technique for DSP implementation." *IEEE Transactions on Industrial Electronics*. 2013;60(6):2295–2308.

[30] Vejdan, S., Sanaye-Pasand, M., and Malik, O. P. "Accurate dynamic phasor estimation based on the signal model under off-nominal frequency and oscillations." *IEEE Transactions on Smart Grid*. 2017;8(2):708–719.

[31] Chauhan, K., Reddy, M. V., and Sodhi, R. "A novel distribution-level phasor estimation algorithm using empirical wavelet transform." *IEEE Transactions on Industrial Electronics*. 2018;65(10):7984–7995.

[32] Bi, T., Liu, H., Feng, Q., *et al*. "Dynamic phasor model-based synchrophasor estimation algorithm for M-class PMU." *IEEE Transactions on Power Delivery*. 2015;30(3):1162–1171.

Chapter 3

Causality analysis in phasor measurement unit algorithms

Wenchao Meng[1], Xiaoyu Wang[2] and Innocent Kamwa[3]

In order to compensate for the group delay, existing phasor measurement unit (PMU) algorithms usually shift time stamps backward. Such schemes make PMU output dependent on past samples, and thus these PMUs are defined as noncausal PMUs in this chapter. The main drawback of noncausal PMUs is that they may result in high latency that will impair performance of PMU applications. For this, in this chapter, we introduce a novel group delay compensation method without shifting time stamps, and the PMUs using such group delay compensations are defined as causal PMUs. We also investigate the impact of causality on performance of PMU algorithms. We firstly analyze how causal and noncausal PMUs compensate for the group delay caused by digital filters and it is found that causal PMUs normally reduce latency by the group delay while noncausal PMUs will not. Then, the impact of causality is investigated by using the Hydro Quebec Research Institute (IREQ) PMU model and the IEEE PMU model.

3.1 Background and motivation

The smart grids are regarded as the next generation of power grids aimed at improving efficiency, security, reliability, and environmental friendliness. These new capabilities are made possible by integrating various advanced technologies such as information and communications technologies (ICT) and computer processing into the power grid [1–4]. As a phasor data source, the PMU technology is commonly used in dynamic monitoring, protection, and control of multiarea networks [5–8].

The first standard regarding PMU is IEEE Std 134-1995 [9]. It is later replaced by IEEE Std C37.118-2005 [10], in which the magnitude and phase angle limits of PMU measurement error are set using a combined metric total vector error (TVE).

[1]Department of Systems and Computer Engineering, Carleton University Ottawa, Ontario, Canada
[2]Department of Electronics, Carleton University, Ottawa, Ontario, Canada
[3]Hydro-Quebec/IREQ, Power Systems and Mathematics, Varennes, Quebec, Canada

However, PMU performance requirements under dynamic conditions are not taken into account in this standard. For this, IEEE Std. C37.118.1-2011 [11] has been released.

Many PMU algorithms have been proposed based on various signal processing technologies such as discrete Fourier transform (DFT) [12–14], Taylor series [15–17], combined least-squares methods with Kalman filtering [18–21]. However, most existing PMUs are focused on improving estimation accuracy while the causality is rarely considered. The causality is well-known concept in signal processing field. Most of existing PMUs compensate for the group delay by shifting time stamps backward, which may result in a high latency that will impair PMU performance [22,23]. For example, a high latency may make a wide-area closed-loop feedback control system unstable [22,23]. Therefore, causal PMUs without introducing high latency should be investigated.

3.2 Group delay caused by PMUs

Phasor representation of sinusoidal signals is commonly used in AC power system analysis. The sinusoidal waveform is defined as [11]

$$\begin{aligned} x(t) &= \sqrt{2}X_m(t)\cos\left(2\pi\int_0^t f_c(t)dt + \phi_0\right) \\ &= \sqrt{2}X_m(t)\cos\left(2\pi\int_0^t (f_0 + \Delta f(t))dt + \phi_0\right) \\ &= \sqrt{2}X_m(t)\cos\left(2\pi f_0 t + 2\pi\int_0^t \Delta f(t)dt + \phi_0\right) \end{aligned} \tag{3.1}$$

where $x(t)$ is the instant value, $X_m(t)$ is the magnitude, $f_c(t)$ is the fundamental frequency in Hz, f_0 is the nominal frequency in Hz (50 or 60), $\Delta f(t)$ is the frequency deviation in Hz, and ϕ_0 is the initial phase angle in radians.

The sinusoid can be expressed as below:

$$\begin{aligned} x(t) &= \mathrm{Re}\left\{\sqrt{2}X_m(t)e^{j\left(2\pi f_0 t + \left(2\pi\int_0^t \Delta f(t)dt + \phi_0\right)\right)}\right\} \\ &= \mathrm{Re}\left\{e^{j(2\pi f_0 t)}\sqrt{2}X_m(t)e^{j\left(2\pi\int_0^t \Delta f(t)dt + \phi_0\right)}\right\} \end{aligned} \tag{3.2}$$

where Re stands for the real part of a complex value. Note that the synchrophasor representation for a given signal as shown in (3.1) is defined as a complex number with the magnitude being magnitude and the phase angle being relative to a cosine function at the nominal system frequency $2\pi f_0$. Because of this, in the synchrophasor representation, the term $e^{j2\pi f_0 t}$ in (3.2) associated with the nominal system frequency is usually omitted. The synchrophasor representation is thus given by

$$\begin{aligned} x(t) \leftrightarrow X(t) &= X_m(t)e^{j\left(2\pi\int_0^t \Delta f(t)dt + \phi_0\right)} \\ &= X_m(t)\cos\left(2\pi\int_0^t \Delta f(t)dt + \phi_0\right) \\ &\quad + jX_m(t)\sin\left(2\pi\int_0^t \Delta f(t)dt + \phi_0\right) \end{aligned} \tag{3.3}$$

For a given set of samples of a power signal $x(i)$, the PMU is used to measure its corresponding synchrophasor estimate $\widehat{X}(i)$ with i being the sample index. Due to inevitable noises, different digital filters will be applied to $\widehat{X}(i)$ as shown in Figure 3.1.

A digital filter is a system that performs mathematical operations on a sampled, discrete-time signal to reduce or enhance certain aspects of that signal. Suppose the length of a digital filter is N, the digital filter will store N samples into a buffer and update it when a new sample arrives. The output of phasor estimation block can be written as

$$\widetilde{X}(i) = \widetilde{X}_m(i)e^{j\left(2\pi\widetilde{\Delta f}(i)i+\widetilde{\phi}_0\right)} \tag{3.4}$$

where $\widetilde{X}_m$ stands for the magnitude estimate at time instant i, $\widetilde{\Delta f}$ denotes the frequency deviation estimate, and $\widetilde{\phi}_0$ denotes the initial phase angle estimate. The phasor estimate (3.4) in discrete-time is obtained from the theoretical phasor in continuous-time (3.3) by replacing each true value with its estimate and replacing time t with the sample index i. In practice, in order to improve the accuracy of estimating phasor, various digital filters capable of reducing noise are commonly used to process the output of phasor estimate block. Suppose that the phase response of digital filter is $-\theta(\omega)$, the output of digital filter can be expressed as

$$\begin{aligned} \widehat{X}(i) &= \widehat{X}_m(i)e^{j\left[2\pi\widetilde{\Delta f}(i)i+\widetilde{\phi}_0-\theta\left(2\pi\widetilde{\Delta f}(i)\right)\right]} \\ &= \widehat{X}_m(i)e^{j\left[2\pi\widetilde{\Delta f}(i)\left(i-\frac{\theta\left(2\pi\widetilde{\Delta f}(i)\right)}{2\pi\widetilde{\Delta f}(i)}\right)+\widetilde{\phi}_0\right]} \\ &= \widehat{X}_m(i)e^{j\left[2\pi\widetilde{\Delta f}(i)(i-\tau)+\widetilde{\phi}_0\right]} \end{aligned} \tag{3.5}$$

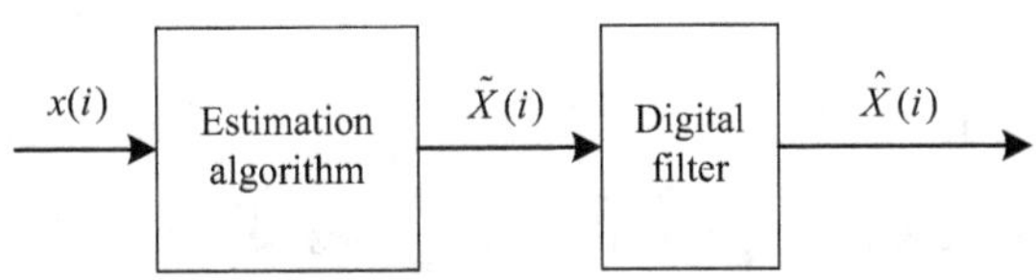

Figure 3.1 Synchrophasor estimation in PMUs

where $\widehat{X}(i)$ denotes the magnitude estimate after the filter and τ denotes the group delay.

We can observe one fact from (3.5) that the output signal after the digital filter is shifted in time with respect to the input by the group delay τ. Obviously, the time stamps of PMU should reflect the time instant when the corresponding event happens. Therefore, the PMU usually shifts the time instant to align the computed phasor with the true phasor.

3.3 Typical PMUs and group delay compensation

3.3.1 IREQ FIR PMU

The main idea of IREQ FIR approach [24] is passing the power signal through a K-taps FIR bandpass filter, whose coefficients can be written analytically with respect to the filter center frequency:

$$h_{BP}(K,f_c) = h_{\cos}(k,f_c) + jh_{\sin}(k,f_c), \ k = 0, 1, 2, \dots, K-1 \tag{3.6}$$

The K-taps FIR bandpass filter is able to retune its center frequency according to the incoming frequency f_c provided that f_c is known. It is worth noting that the filtering is adaptive but not the sampling.

The frequency shifting based on a local reference phasor can make the time-varying phasor stationary. Another interesting property of this filter is that the group delay can be written in number of samples as

$$D = \frac{K-1}{2} \tag{3.7}$$

Inspired by the exponentially modulated (EM) filter bank theory [25,26], the following filter bank definition is utilized to design the bandpass filter:

$$H_k^e(n) = \frac{2}{\sqrt{K}} h_p(n) \exp\left[j\frac{2\pi\gamma_i}{K}\left(k+\frac{1}{2}\right)\left(n+\frac{K+2}{4}\right)\right] \tag{3.8}$$

where $n = 1, \dots, 2K-1$, $k = 0, 1, \dots, 2K-1$, h_p stands for the impulse response coefficients of a linear-phase low-pass FIR prototype filter and γ_i the scaling factor. The number of filter cells K is chosen such that the length of the prototype filter is equal to $2K$. Meanwhile, the center frequency of ith filter will be located at the fundamental frequency if the scaling factor γ_i is properly designed. More specifically, the scaling factor γ_i is chosen as

$$\gamma_i = \frac{K}{N_p\left(i+\frac{1}{2}\right)} \tag{3.9}$$

where N_p denotes the sampling rate, that is, samples per cycle of the fundamental.

The IREQ FIR M-class PMU is implemented using an FIR prototype based on a classical four-cycle Taylor window of the form

$$h_M(n) = \frac{1}{2K}\sum_{h=0}^{8}(-1)^h a_h \cos\left(2\pi h \frac{n}{2K}\right)$$
$$n = 0, 1, \ldots, 2K-1 \tag{3.10}$$

with

$$a_0 = 1.0, a_1 = -1.03538, a_3 = 0.0824936,$$
$$a_4 = -0.00116197, a_5 = -0.00188862,$$
$$a_6 = -0.00123387, a_7 = -0.000671595,$$
$$a_8 = -0.000275885 \tag{3.11}$$

The IREQ FIR P-class PMU is implemented using an FIR prototype derived from

$$h_p(n) = K + (1-K)W_h(n), n = 0, 1, 2\ldots, K-1 \tag{3.12}$$

where $K = 0.5189$ and W_h is the Hanning window [25].

3.3.2 IEEE C37 PMU

The IEEE C37 PMU is based on DFT which is widely used for phasor estimation in PMUs due to its low computational requirements [27,28]. In the IEEE C37 PMU, the sampled signal is correlated with quadrature waveforms at the nominal frequency, which is followed by a low pass filter of at least one period of the fundamental frequency of the form [11]

$$X(i) = \frac{\sqrt{2}}{\text{Gain}}\sum_{k=-N/2}^{N/2} x(i+k)W_{(k)}\exp(-j(i+k)T\omega_0) \tag{3.13}$$

where $\omega_0 = 2\pi f_0$ is the nominal power system frequency in radians, N denotes the filter order, and $W(k)$ denotes the low-pass filter coefficients.

The IEEE C37 P class PMU is built using the low-pass filter with coefficients given by

$$W(k) = \left(1 - \frac{2}{N+2}|k|\right) \tag{3.14}$$

For example, let $N = 44$, the coefficients of filter used in P class are demonstrated in Figure 3.2.

From (3.14), we have that the filter used in IEEE C37-P is a fixed-length triangular-weighted symmetric filter, designed to provide good performance at the nominal system frequency. The P class provides excellent harmonic rejection performance at the nominal frequency. However, for signals with off-nominal frequency, the harmonics are not suppressed very well.

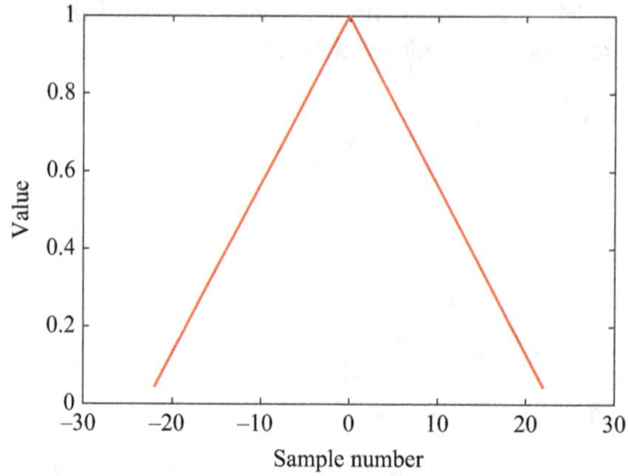

Figure 3.2 P class filter coefficient example $N = 28$

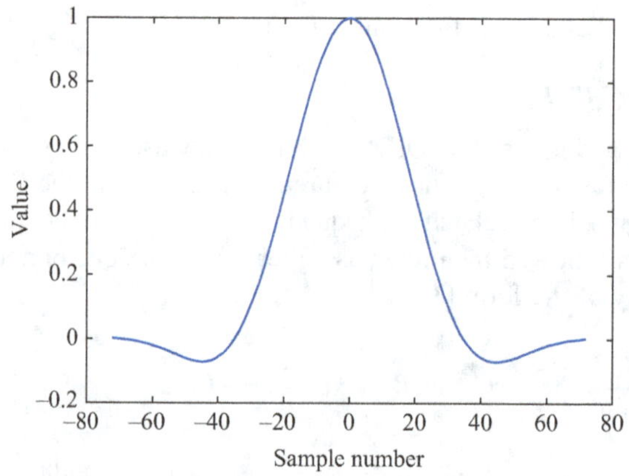

Figure 3.3 M class filter coefficient example $(F_s = 60, F_{\text{sampling}} = 1{,}440,$ and $N = 144)$

The IEEE C37 M class PMU is built using the low-pass filter with coefficients given by Figure 3.3.

$$W(k) = \frac{\sin\left(2\pi \times \frac{2F_{fr}}{F_{sampling}} \times k\right)}{2\pi \times \frac{2F_{fr}}{F_{sampling}} \times k} h(k) \tag{3.15}$$

where $k = -N/2{:}N/2$ with N being the filter order, $F_{sampling}$ denotes the sampling frequency, $h(k)$ denotes the hamming function, and F_{fr} is the low-pass filter reference frequency. The filter orders and reference frequency F_{fr} are given by Section C.6 in [11]. The filter used in M class PMU is a FIR filter having substantially greater length as opposed to the P class filter. This is because that the M class PMU has requirements for the passband and stopband. More details can be referred to [11]. As a consequence, in noncausal versions, the M class PMU has longer reporting delays than the P class PMU due to the time shift group delay compensation scheme. However, the M class PMU can produce somewhat higher accuracy.

3.3.3 Noncausal delay compensation

In order to compensate for the group delay, existing PMU algorithms shift time stamp backward to align the computed and true phasor. For example, for a PMU algorithm with group delay being τ, the time tag of all estimates at time instant $i+\tau$ should be set equal to i as shown below:

$$\widehat{X}_{nc}(i) = \widehat{X}(i+\tau) \tag{3.16}$$

where nc denotes noncausal. The compensation method is also depicted in Figure 3.4. It has to be noticed that shifting time stamps backward will cause latency that may affect PMU performance. For a PMU with the group delay being τ, a total of $T*\tau$ latency will be induced by the group delay compensation. In this chapter, we will introduce the well-known causality concept in the field of signal processing into the PMU field. In signal processing, a causal system utilizes only previous and present samples while a noncausal system also uses future samples. Specifically, the definition of noncausal PMU is given as

Definition 1: A noncausal PMU is an algorithm which needs to shift time stamps to align the computed and true phasor.

3.3.4 Causal group delay compensation

The causal PMU is an algorithm which embeds a group delay correction and outputs a computed phasor already aligned with the true phasor. The causal PMU assumes a stationary phasor over the group delay duration, and advances the phasor by the group delay. Specifically, the trick is to find an estimate of frequency deviation, and then use this value to compensate for the group delay in the frequency domain as follows:

$$\widehat{X}_c(i) = \widehat{X}(i)e^{j\left(2\pi\widehat{\Delta f}T\tau\right)} \tag{3.17}$$

where c means "causal," T is the sampling interval, and $\widehat{X}_c(i)$ is the compensated phasor estimate. Notice that the time instant i in (3.5) implies iT, that is to say, the

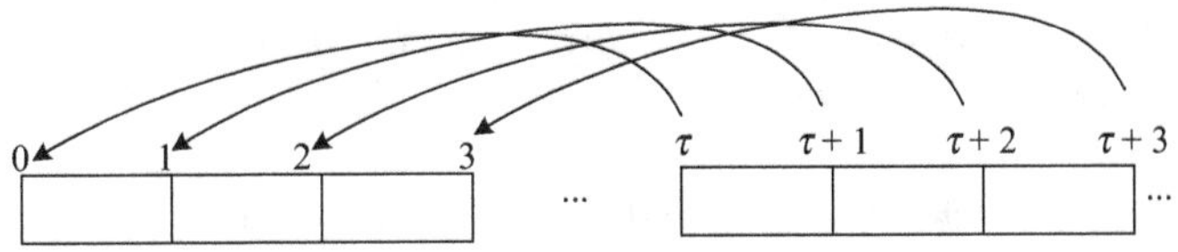

Figure 3.4 Group delay compensation in the noncausal PMU

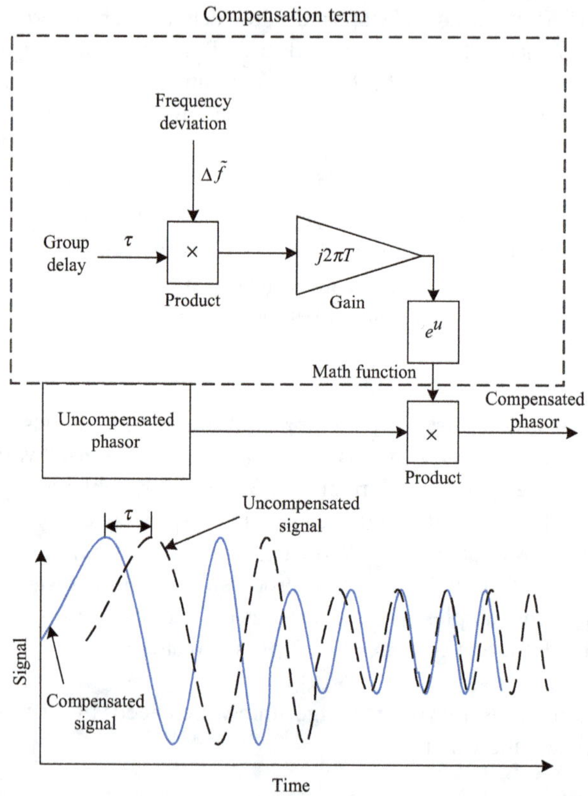

Figure 3.5 Group delay compensation in the causal PMU

sampling interval T has been omitted for simplicity. Hence, implementing the group delay compensation (3.17) should take the sampling interval into account. The group delay compensation of causal PMU is summarized in Figure 3.5.

The properties of causal PMUs can be summarized as

Property 1: *In contrast to noncausal PMUs, the causal PMUs are able to reduce the latency by the group delay.*

Property 2: *The group delay can be compensated by the causal PMUs for signals at steady state. It implies that the causal PMUs can provide almost similar estimation accuracy to their noncausal counterpart.*

Property 3: *For signals with variable magnitudes, the causal PMUs are failed to compensate the group delay.*

Theorem 1: *For a signal with constant frequency deviation over the group delay period, the proposed group delay compensation method (3.17) can compensate for the group delay completely.*

Proof. Substituting the compensation method (3.17) into (3.5), and adding the omitted time interval T, we have

$$\begin{aligned}\widehat{X}_c(i) &= \widehat{X}(i)e^{j(2\pi\widehat{\Delta f}T\tau)} \\ &= \widehat{X}_m(i)e^{j\left[2\pi\widehat{\Delta f}(i-\tau)+\widehat{\phi}_0\right]}e^{j\left(2\pi\widehat{\Delta f}T\tau\right)} \\ &= \widehat{X}_m(i)e^{j\left[2\pi\widehat{\Delta f}(iT-\tau T)+\widehat{\phi}_0\right]}e^{j\left(2\pi\widehat{\Delta f}\tau T\right)} \\ &= \widehat{X}_m(i)e^{j\left[2\pi\widehat{\Delta f}iT+\widehat{\phi}_0\right]}\end{aligned} \tag{3.18}$$

Aimed at studying the impact of frequency estimate error on the causal phasor, the true frequency and the frequency estimation error is firstly separated as follows:

$$\begin{aligned}\widehat{\omega}_T &= 2\pi[\Delta f(i) + \mathrm{FE}(i)] \\ &= 2\pi\Delta f(i) + 2\pi\mathrm{FE}(i)\end{aligned} \tag{3.19}$$

where $\Delta f(i)$ is the true frequency deviation at time instant i and FE is the frequency estimate error in Hz. The phase error can then be expressed as

$$\begin{aligned}\phi_{error} &= 2\pi\mathrm{FE}(i)\tau T \\ &= \kappa[x][y]\Delta = f(x,y)\end{aligned} \tag{3.20}$$

The phase error is a two-variable function which can be plotted in two dimensions. Suppose that the P class PMU has a group delay of 4, and the M class PMU has a group delay of 10. By recalling the formula (3.20), we can predict the impact of FE on the group delay cancellation.

3.4 PMU performance assessment

3.4.1 Overview of the developed IEEE PMU MATLAB®/Simulink® model

The main structure of the developed PMU model is shown in Figure 3.6. It mainly has three blocks including phasor signal, PMU, and plot blocks. The main task of phasor signal is to generate three-phase test signals. These test signals are then delivered to the PMU block which is responsible for estimating corresponding phasors, and providing other important information such as FE and ROCOF. Finally, all this information are exported to the plot block for plotting results.

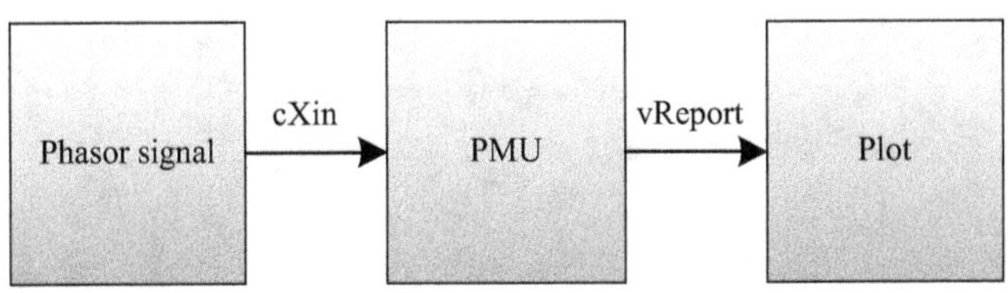

Figure 3.6 Main structure of the developed PMU MATLAB/Simulink model

The phasor signal block can also simulate signals with various features such as harmonic, amplitude step, phase step, amplitude modulation, and phase modulation. Features of the test signals can be set in the block parameters as follows.

Block output:

- CXin: three-phase AC signal.
- VTheory: the theoretical features of the test signal whose format is given in Table 3.1.

The PMU block simulates a PMU device and reports phasor estimations. The PMU parameters are set in the block parameters as shown in Figure 3.7. Our developed PMU block can plot the results to verify the performance of PMU in terms of TVE, FE, ROCOF, and so on as shown in Figure 3.8. Finally, the Simulink model is shown in Figure 3.9.

Table 3.1 Format of VTheory

Timestamp	Phasor A	Phasor B	Phasor C	Phasor PosSeq	Frequency	ROCOF

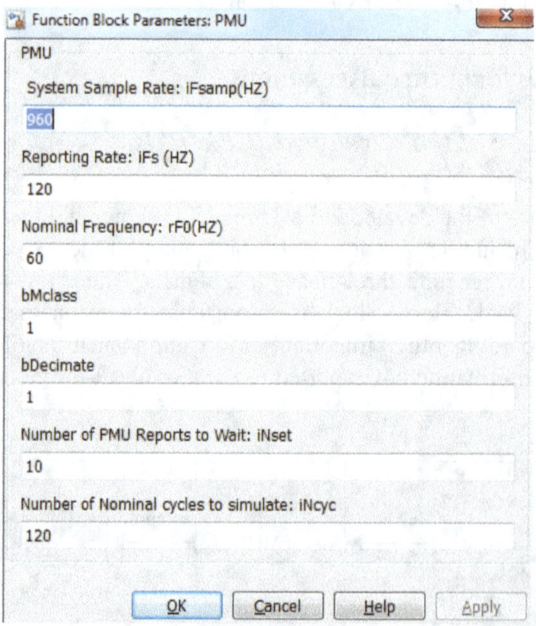

Figure 3.7 Parameter settings of PMU block

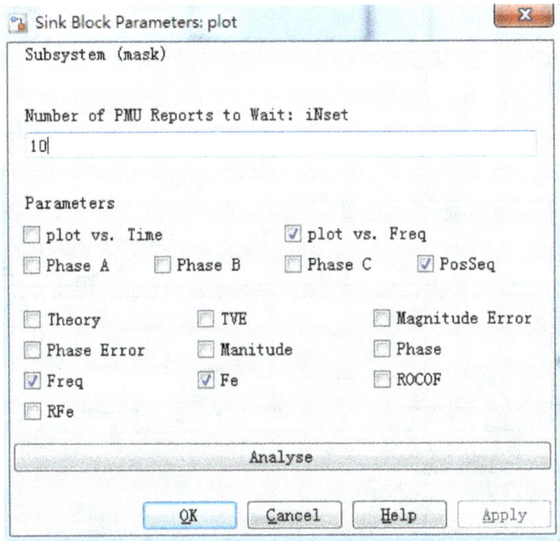

Figure 3.8 Parameter settings of plot block

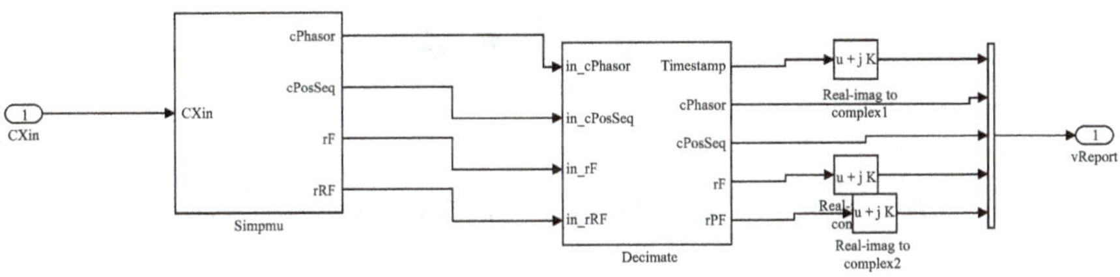

Figure 3.9 Developed PMU Simulink model

3.4.2 Time lead of causal PMUs

The simulation results of causal and noncausal PMUs in the presence of a close three-phase-to-ground fault will be presented for the well-known New-England test system IEEE 39-bus test system (see Figure 3.10) carried out in SimPowerSystem. It is assumed that a three-phase-to-ground short-circuit is applied to bus 2 just at the secondary of the G30 step-up transformer at $t = 4$ s and is cleared after 10 cycles $(10/60 = 0.1667$ s) at $t = 4.1667$ s.

Figure 3.11 depicts the voltage estimates of G30 using the IREQ PMUs. It can be observed that, for angle estimation, the causal PMUs provide advanced estimates in comparison with that of noncausal PMUs, and show larger overshoots as well. However, for magnitude estimation, the causal and noncausal PMUs using the same PMU estimation algorithm provide almost the same magnitude estimates. For example, the causal IREQ FIR-M provides almost the same magnitude estimates to the noncausal IREQ FIR-M. Similar results can also be observed in the IEEE PMUs as shown in Figure 3.12.

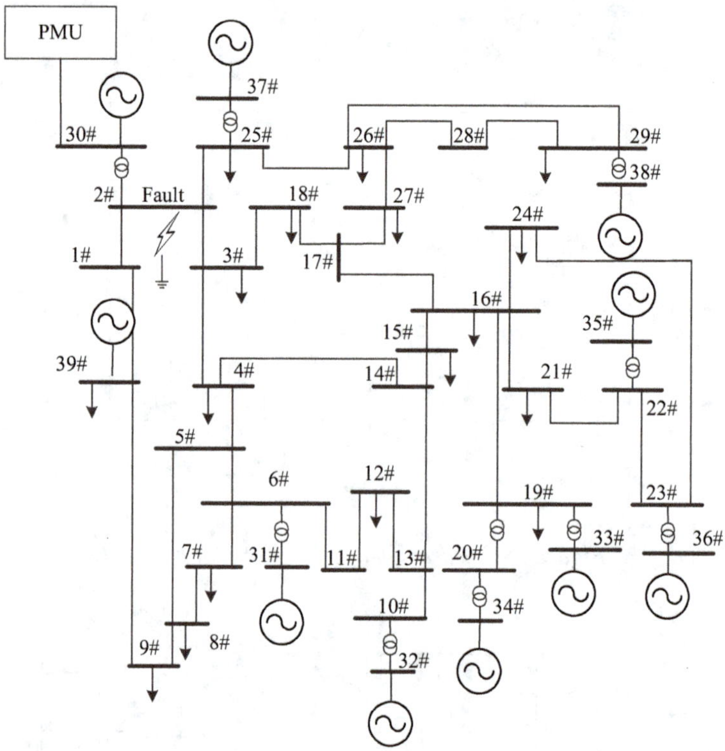

Figure 3.10 New-England 39-bus system with PMU installed for G30

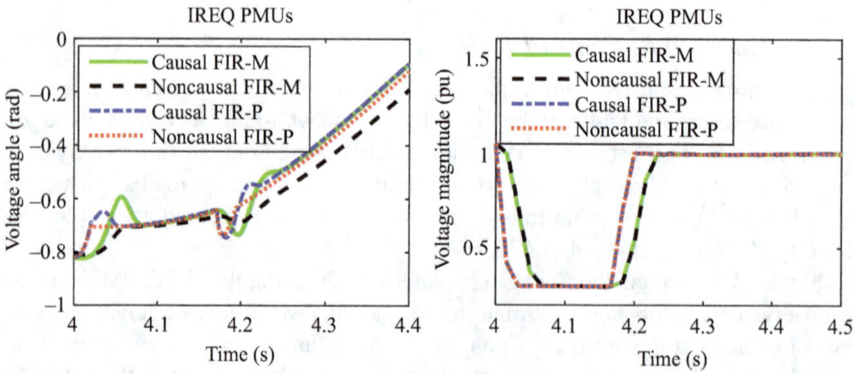

Figure 3.11 Voltage estimations of G30 using IREQ PMUs

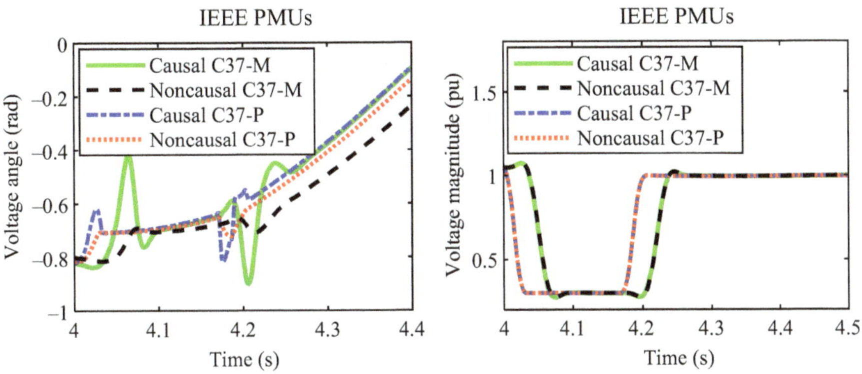

Figure 3.12 Voltage estimations of G30 using IEEE PMUs

3.4.3 Steady state test with different fundamental frequencies

Aimed at verifying the effectiveness of causal PMUs in compensating for the group delay at steady state, we have conducted the steady state test under different fundamental frequencies. Based on the IEEE standard [11], for M class PMUs, the frequency range is $f_0 \pm 5$ Hz, and for P class PMUs, the frequency range is $f_0 \pm 2$ Hz. The results are summarized in Figure 3.13, which can be observed that the causal PMUs can provide a similar TVE to the noncausal version by reducing the latency.

3.4.4 Modulation test

The modulation test is usually used to reflect variations in magnitude and phase angle of the voltage waveforms that occur during a stable power swing. Such a test is conducted by individually applying sinusoidal amplitude and phase-modulated signals to PMUs. The magnitude modulation depth is chosen as 10% and the phase-angle modulation depth is chosen as 0.1 rad specified in [11]. According to the IEEE standard [11], the modulation frequency is chosen up to 2 Hz for P class PMUs and 5 Hz for M class PMUs.

The comparison of different PMUs under magnitude modulation inputs is shown in Figure 3.14. It can be observed that the causal PMUs have larger TVE than their noncausal counterparts. Furthermore, the comparison of different PMUs under phase modulation inputs is also given in Figure 3.15. Although the causal PMUs still have larger TVE than the noncausal ones, the performance in causal PMUs has been largely improved compared with those in the magnitude modulation test.

3.4.5 Frequency ramp

Because of changes in load and generation imbalances and faults, the power frequency might operate in constantly changing frequency. Therefore, we also investigate the impact of causality on synchrophasor estimation in case of

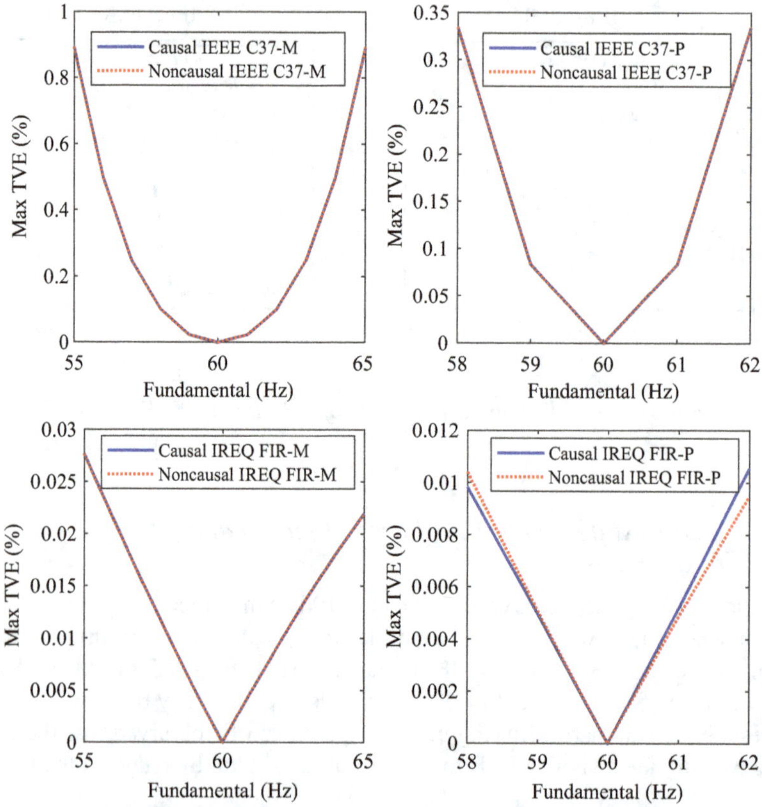

Figure 3.13 Steady state performance of PMUs at different fundamental frequencies

frequency excursion. In the frequency test, for M class PMUs, the frequency ramp range is $f_0 \pm 5$ Hz, and for P class PMUs, the frequency range is $f_0 \pm 2$ Hz. The change profile is shown in Figure 3.16.

Note that the phasor magnitude is constant in the frequency ramp test. By referring to properties of causal PMUs, the group delay in this situation can be compensated. From Figure 3.17, we can see that all the IREQ PMUs can satisfy the requirement of IEEE standard. The performance of IEEE PMUs is depicted in Figure 3.18, and we can observe that causal IEEE C37-M PMU exceeds the requirement a little while other PMUs can also satisfy the requirement.

3.4.6 Step test

In practice, the power system may suffer a sudden input change because of load switching and faults. For this, we have conducted the step test including magnitude step test and phase step test. Specifically, the magnitude is stepped by 10% of the nominal magnitude and the phase is stepped by 10° phase angle at $t = 5$ s.

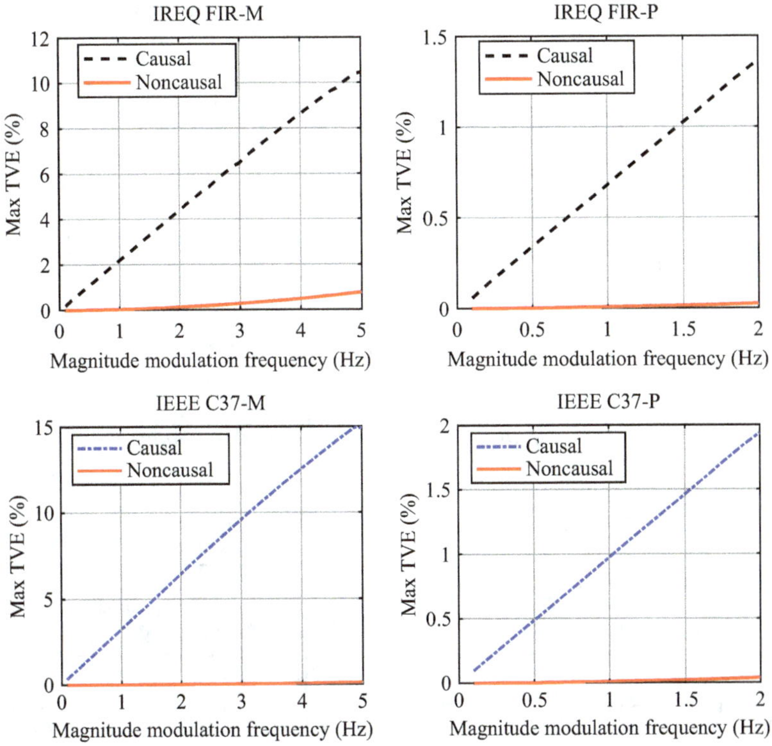

Figure 3.14 Comparison of different PMUs under magnitude modulation inputs

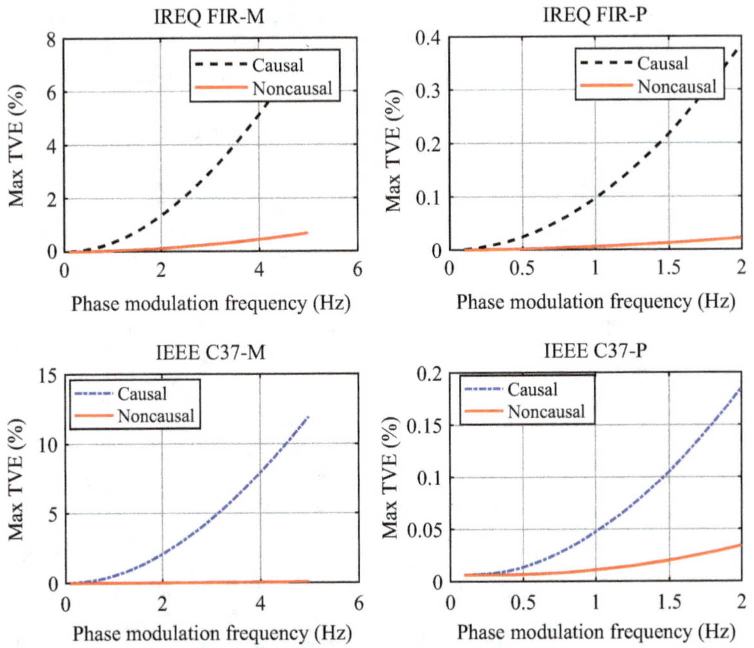

Figure 3.15 Comparison of different PMUs under phase modulation inputs

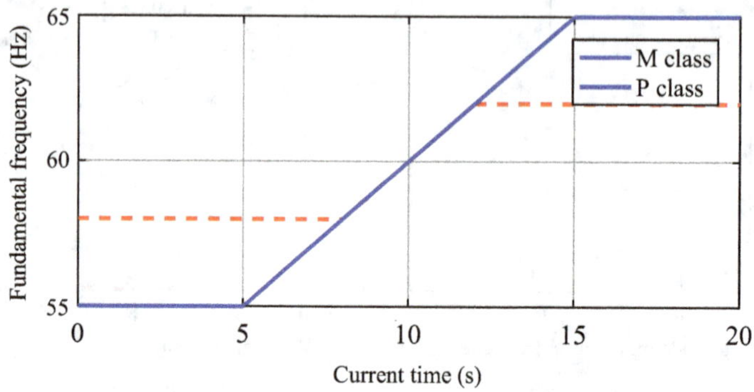

Figure 3.16 Fundamental frequency variation under frequency ramp test

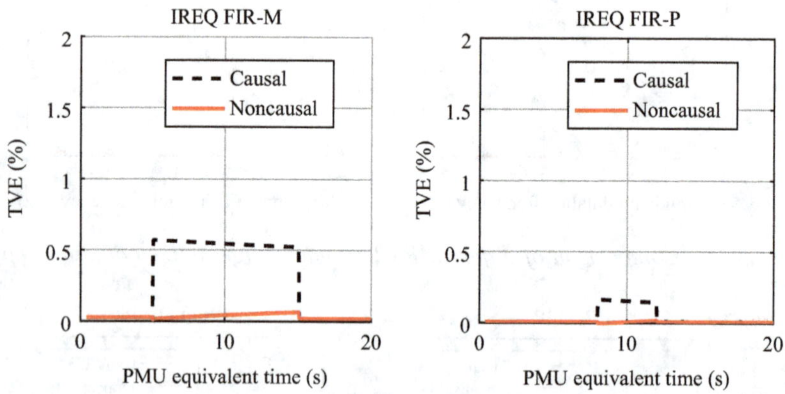

Figure 3.17 Performance of IREQ PMUs under frequency ramp test

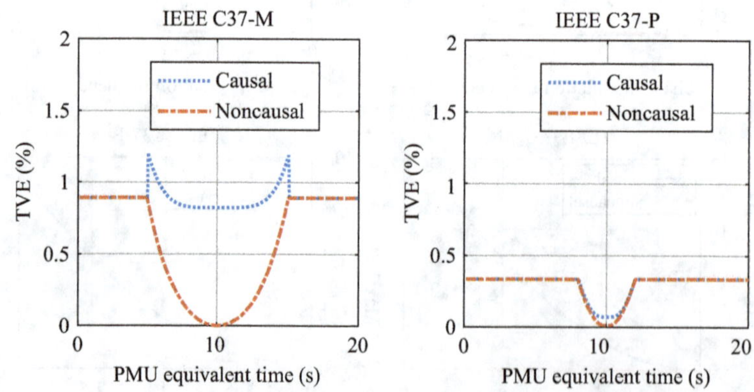

Figure 3.18 Performance of IEEE PMUs under frequency ramp test

Note that when the phase step occurs, there is a rapid change in frequency deviation estimate. Because of this, a large overshoot will be caused in the causal group delay compensation as shown in Figure 3.19. in which the causal PMUs show larger overshoots as opposed to their noncausal counterparts. The performance of PMUs in the magnitude step test is demonstrated in Figure 3.20. In this situation, the frequency deviation is zero, and thus its estimate is almost zero too. This makes the group delay in causal PMUs impossible. The noncausal PMUs can still compensate the group delay because of its independence of frequency deviation. This result can be verified in Figure 3.20. It can be observed that the magnitude estimates in causal and noncausal PMUs have almost the same profile but with different time shifts. In other words, the noncausal PMUs have more accurate time tags than the causal PMUs.

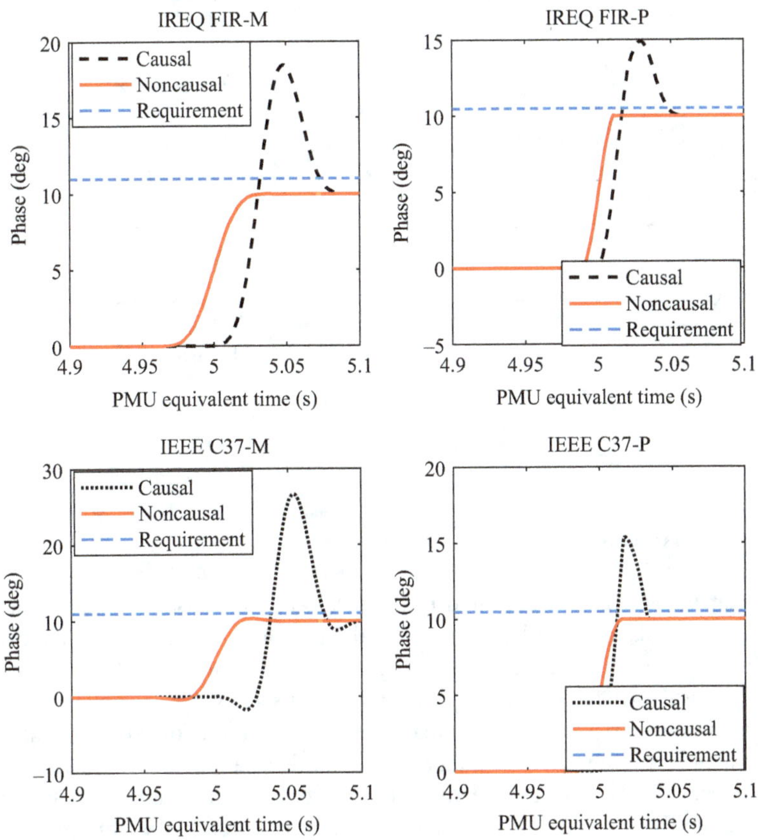

Figure 3.19 PMU response under phase step test

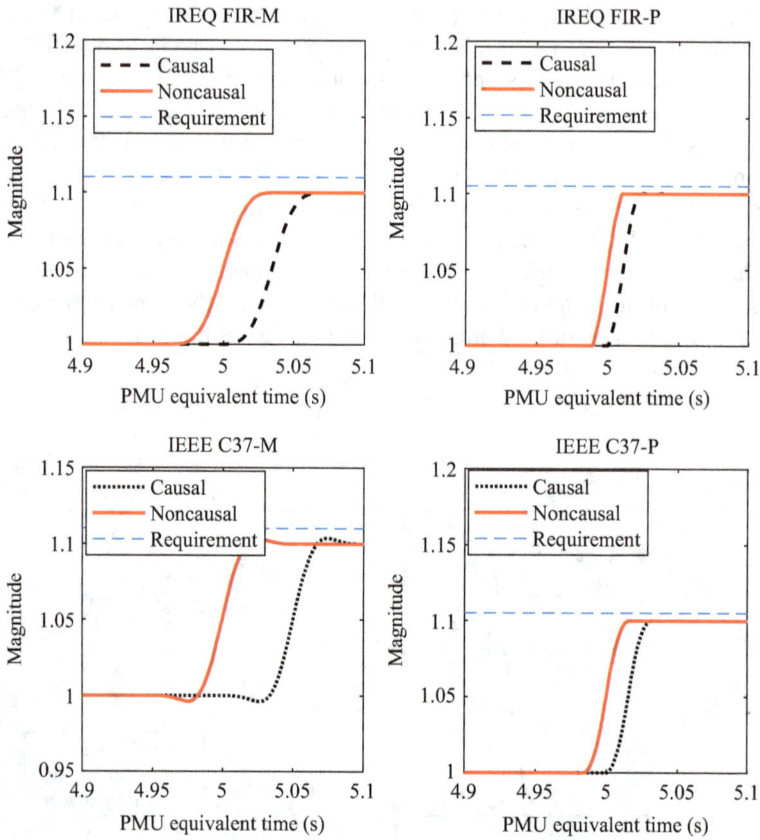

Figure 3.20 PMU response under magnitude step test

3.5 Conclusions

In this paper, we have investigated how causal and noncausal PMUs compensate for the group delay caused by PMUs digital filters and the impact of causality on PMU algorithms has been studied using IREQ FIR and IEEE C37 models. In general, causal PMUs can reduce the latency caused by the group delay, which plays an important role in wide-area control and protection applications since latency may be more important than accuracy in these applications. Obviously, the performance of causal PMUs needs to be improved in the future, but the causal PMUs cannot be expected to show the same performance to the noncausal PMUs expect for the reducing latency. In fact, the performance of causal PMUs can be improved with the help of ROCOF [29]. In the current IEEE Std, only metrics for noncausal PMUs are considered. In the future, the metrics for causal PMUs should be included.

References

[1] Moslehi K, and Kumar R. A reliability perspective of the smart grid. *IEEE Trans Smart Grid*. 2010;1(1):57–64.
[2] Meng W, and Wang X. Distributed energy management in smart grid with wind power and temporally coupled constraints. *IEEE Trans Ind Electron*. 2017;64(8):6052–6062.
[3] Kansal P, and Bose A. Bandwidth and latency requirements for smart transmission grid applications. *IEEE Trans Smart Grid*. 2012;3(3):1344–1352.
[4] Meng W, Wang X, and Liu S. Distributed load sharing of an inverter-based microgrid with reduced communication. *IEEE Trans Smart Grid*. 2018;9(2): 1354–1364.
[5] Ghahremani E, and Kamwa I. Local and wide-area PMU-based decentralized dynamic state estimation in multi-machine power systems. *IEEE Trans Power Syst*. 2016;31(1):547–562.
[6] Rouhani A, and Abur A. Linear phasor estimator assisted dynamic state estimation. *IEEE Trans Smart Grid*. 2018;9(1):211–219.
[7] Kamwa I, Grondin R, and Hebert Y. Wide-area measurement based stabilizing control of large power systems: a decentralized/hierarchical approach. *IEEE Trans Power Syst*. 2001;16(1):136–153.
[8] Meng W, Wang X, Fan B, *et al*. Adaptive non-linear neural control of wide-area power systems. *IET Gener Transm Distrib*. 2017;11(18):4531–4536.
[9] IEEE Standard for Synchrophasor Measurements for Power Systems; 1995.
[10] IEEE Standard for Synchrophasors for Power Systems; 2006.
[11] IEEE Standard for Synchrophasor Measurements for Power Systems; 2011.
[12] Kamwa I, and Grondin R. PMU configuration for system dynamic performance measurement in large, multiarea power systems. *IEEE Trans Power Syst*. 2002;17(2):385–394.
[13] Romano P, and Paolone M. Enhanced interpolated-DFT for synchrophasor estimation in FPGAs: theory, implementation, and validation of a PMU prototype. *IEEE Trans Instrum Meas*. 2014;63(12):2824–2836.
[14] Routtenberg T, Xie Y, Willett RM, *et al*. PMU-based detection of imbalance in three-phase power systems. *IEEE Trans Power Syst*. 2015;30(4):1966–1976.
[15] Ren J, and Kezunovic M. A hybrid method for power system frequency estimation. *IEEE Trans Power Del*. 2012;27(3):1252–1259.
[16] Ren J, and Kezunovic M. An adaptive phasor estimator for power system waveforms containing transients. *IEEE Trans Instrum Meas*. 2012;27(2): 735–745.
[17] Castello P, Lixia M, Muscas C, *et al*. Impact of the model on the accuracy of synchrophasor measurement. *IEEE Trans Instrum Meas*. 2012;61(8):2179–2188.
[18] Sadinezhad I, and Agelidis VG. Real-time power system phasors and harmonics estimation using a new decoupled recursive-least-squares technique for DSP implementation. *IEEE Trans Ind Electron*. 2013;60(6):2295–2308.

[19] Kamwa I, and Grondin R. Fast adaptive schemes for tracking voltage phasor and local frequency in power transmission and distribution systems. *IEEE Trans Power Del.* 1991;7(2):789–795.
[20] Chen C, Chang G, Hong R, *et al.* Extended real model of Kalman filter for time-varying harmonics estimation. *IEEE Trans Power Del.* 2010;25(1):17–26.
[21] Alves D, and Coelho R. An adaptive algorithm for real-time multi-tone estimation and frequency tracking of non-stationary signals. *IEEE Trans Nucl Sci.* 2011;58(4):1582–1587.
[22] Wang S, Meng X, and Chen T. Wide-area control of power systems through delayed network communication. *IEEE Trans Contr Syst Technol.* 2012; 20(2):495–503.
[23] Wu H, Tsakalis KS, and Heydt G T. Evaluation of time delay effects to wide-area power system stabilizer design. *IEEE Trans Power Syst.* 2004; 19(4):1935–1941.
[24] Kamwa I, Pradhan AK, and Joos G. Adaptive phasor and frequency-tracking schemes for wide-area protection and control. *IEEE Trans Power Del.* 2011;26(2):744–753.
[25] Lyons RG. *Understanding Digital Signal Processing.* Upper Saddle River, NJ: Prentice-Hall, 2011.
[26] PMU System Testing and Calibration Guide. NASPI report of the Performance and Standards Task TeamPSTT. Available from: https://www.naspi.org/File.aspx?fileID=555
[27] Roscoe AJ, Abdulhadi IF, and Burt GM. P and M class phasor measurement unit algorithms using adaptive cascaded filters. *IEEE Trans Power Del.* 2013;28(3):1447–1459.
[28] Bi T, Liu H, Feng Q, *et al.* Dynamic phasor model-based synchrophasor estimation algorithm for M-class PMU. *IEEE Trans Power Del.* 2015;30(3): 1162–1171.
[29] Meng W, Wang X, Wang Z, *et al.* Impact of causality on performance of phasor measurement unit algorithms. *IEEE Trans Power Syst.* 2018;33(2): 1555–1565.

Chapter 4

Comprehensive modeling and parameter identification of wind farms based on wide-area measurement systems

Yinfeng Wang[1]*, Chao Lu*[1]*, Zhanhe Li*[1] *and Lipeng Zhu*[2]

With intermittence, unpredictability, and stochastics of wind power largely introduced into power systems, power system stability analysis and control is in urgent need of reliable wind farm models. Considering the superiority of wide-area measurement systems (WAMS), this chapter develops a novel methodology for practical synchrophasor measurement-based modeling and parameter identification of wind farms. For the sake of preserving basic structural characteristics and control patterns simultaneously, a comprehensive wind farm model is constructed elaborately. To improve the efficiency of the identification procedure, dominant parameters are classified and selected by trajectory sensitivity analysis. Furthermore, an improved genetic algorithm (IGA) is proposed to strengthen the capability of global optimization. The test results on the WECC benchmark system and the standard 36-bus system of Chinese Electric Power Research Institute (EPRI) demonstrate the effectiveness and reliability of the proposed modeling and identification methodology.

4.1 Introduction

It is indubitable that wind power generation develops rapidly, as pursuing more usage of clean energy is a common goal around the world. However, large-scale wind farms are usually located in remote areas, which makes them difficult to adjust reactive power and to withstand voltage surge. The increasing penetration of wind power into power systems has brought tremendous challenges to system operation and maintenance especially in the sense that power system stability analysis and control seem to be largely dependent on the reliability of wind farm models.

However, the work of building wind farm models is complicated. To begin with, there are various kinds and classification methods of wind power generation. With respect to the general types of motor, cage induction generator, doubly fed

[1]Department of Electrical Engineering, Tsinghua University, Beijing, China
[2]Department of Electrical and Electronic Engineering, The University of Hong Kong, Hong Kong, China

induction generator (DFIG), and permanent magnet synchronous generator are commonly applied. With regard to control modes, wind turbines can function with variable or fixed pitch, and the speed of wind turbines can also be variable or fixed. Concerning driving modes, there are direct drive wind turbine power generators and wind power generators driven by gearbox. Besides, there are many wind generator manufacturers throughout the world. Due to the lack of standardized and unified standards, the internal structures, parameters, and characteristics of different types of wind generators vary greatly.

As a result, we need to build different models to do simulation analysis. Although some current simulation software has included different wind farm models, many problems are still difficult to tackle. For example, in the early stage of planning a wind farm, the specific type of wind power generators usually has not been determined yet. The establishment of different wind farm models requires researchers to have a deep understanding of the internal structures and parameters of various types of wind power generators. The secrecy of manufacturers about their key technologies also makes it difficult to design a satisfactory model. Nevertheless, if the modeling issue is not treated felicitously, relevant misleading analysis and control measures may improperly hamper the absorption of large-scale wind power [1–3]. From the above analysis, we can learn that it is imperative to study the general modeling method of wind farms and select different parameters to approximately simulate the dynamic characteristics of wind farms according to the characteristics of different manufacturers' products.

For the past few decades, several kinds of wind farm models have been established to study those problems such as transient stability risk assessment [4], subsynchronous resonance [5], operational outage [6], and so on. The research communities have made various efforts, mainly involving two categories of approaches, to cope with the modeling issue. Some researchers attempt to classify and cluster the wind turbine generators (WTGs) based on some dynamic parameters or response characteristics in a large wind farm [7–10]. However, the effectiveness of those clustering criteria including input wind speed, rotor slip, and rotor current is difficult to be theoretically verified. The others make the whole wind farm equivalent to a single wind generator, which focuses primarily on the whole wind farm's response behaviors, yet omitting the complex location information and electrical connections [11–13]. However, both kinds of approaches attempting to meet the parameter identification requirements establish reduced-order and simplified wind farm models at the cost of deteriorating some control effects and dynamic responses [14,15].

In addition to the unreliable models, inaccurate model parameters further exacerbate the modeling problem. Generally, the parameters for manufacturer-specific models are regarded as classified commercial data, and not suited for each kind of wind farm [16]. In the actual operation process, the internal parameters of the wind power generators will change with time and operational conditions. For example, frequency variation, excitation saturation, stray loss, and other variable factors will affect wind farm parameters. However, nowadays, the empirical parameters are widely used in practice by dispatchers for convenience. What is worse, almost all of the wind farms within a region are usually modeled with the unified parameters, which is clearly not in accordance with the practical scenarios. A survey of a regional power grid in China, as shown in Table 4.1, illustrates this ubiquitous problem. Empirical parameters-based models, ignoring the practical dynamics of wind farms, may lower the reliability of simulation results, or even cause potential hazards in power system security and stability analysis. Simulations with different empirical parameters result in quite different conclusions, as depicted in Figure 4.1. Therefore, instead of adopting empirical parameters, it is imperative to identify the parameters based on field or practical measurements.

The parameters of the wind farm model can be selected directly from the precise model provided by the manufacturer and adjusted by modeling experts. However, for some electromechanical transient parameters in the model, the manufacturer may not be able to provide or even directly measure them. On the other hand, because some modules are functionalized in the electrical model, or the connection relationship within the wind field is ignored, the model parameters may not be completely corresponding to the precise model. Therefore, parameter identification is irreplaceable to provide reference and guidance for the determination of the model parameters.

With the rapid development of wide-area measurement systems in recent years, phase measurement units (PMUs) and wide-area measurement systems (WAMS)-based modeling approaches have been widely used on various occasions to effectively reflect the devices' real dynamic characteristics, such as the modeling of generator, excitation system, and dynamic load [17–19]. Compared with SCADA, which mainly records steady-state data, PMUs are capable of providing reliable field measurement data that contains synchronous and dynamic information, which can meet the requirement of respectively modeling and identifying individual wind farms within a region.

Considering that the variable speed and pitch of DFIG can reduce the mechanical stress of wind turbine, maximize the capture of wind energy under different wind speed which renders wind turbine a wider speed range, reduce the impact of wind turbulence and tower shadow effect, and reduce the rapidity requirement for variable pitch system, DFIG wind farm is the major type of wind farms at present especially in large-scale ones. So DFIG wind farm is selected for modeling and parameter identification.

In this chapter, a wind farm modeling approach based on WAMS is introduced to deal with the unreliability and inaccuracy of conventional wind farm modeling methods. First, a dynamic equivalent model of DFIG wind farm is elaborately

Table 4.1 Model parameters adopted by wind farms in a regional power network simulation data in Northwest China Grid

Area	Area 1	Area 2	Area 3	Area 4
Total number of models	34	72	94	151
Number of models with unified parameters	33	72	89	125

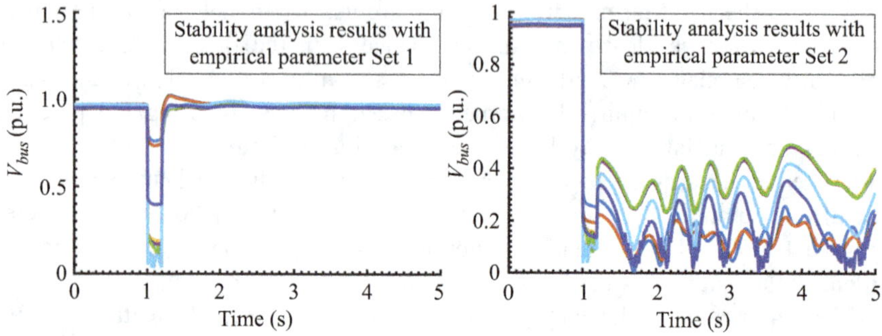

Figure 4.1 Comparison of dynamic responses with different empirical parameters

constructed to simulate the overall characteristics of the wind farm for system analysis. Then, dominant parameters are selected by trajectory sensitivity analysis to tackle the intricate and laborious identification issue of multiple parameters. Additionally, a method incorporating model simulation and the improved genetic algorithm (IGA) is employed to search the wind farm's optimal parameter combination based on the PMU data.

The rest of this chapter is organized as follows: Section 4.2 establishes the equivalent model. Section 4.3 presents the selection and classification method of dominant parameters. The IGA is proposed and used to identify the parameters in Section 4.4. Section 4.5 conducts tests on the WECC 9-bus system and the standard 36-bus system of Chinese Electric Power Research Institute (EPRI) to demonstrate the effectiveness and reliability of the modeling and identification methodology. Section 4.6 draws a conclusion.

4.2 Generic model of wind farm

The equivalent model based on the G.E. standard DFIG model, recommended by Western Electricity Coordinating Council (WECC) [20], is an s domain model and is established in the MATLAB®/Simulink. As shown in Figure 4.2, its main components are divided into four modules: two-mass shafting module, pitch angle control module, reactive control module, and wind farm interface module. The frequency control and reactive power droop control are omitted to simplify the model. The wind farm model is the s domain model. PI control unit and limiting link are used to describe the control law of wind farm in the following modules.

4.2.1 Two-mass shafting module

The mechanical transmission chain of the wind power generators consists of five parts: impeller, low-speed transmission shaft, gearbox, high-speed transmission shaft, and generator. The wind turbine shaft and generator shaft are not connected directly in DFIG but are connected by gearbox. The existence of gearbox makes the mechanical connection less stiff. As a result, the torsion angles of wind turbine

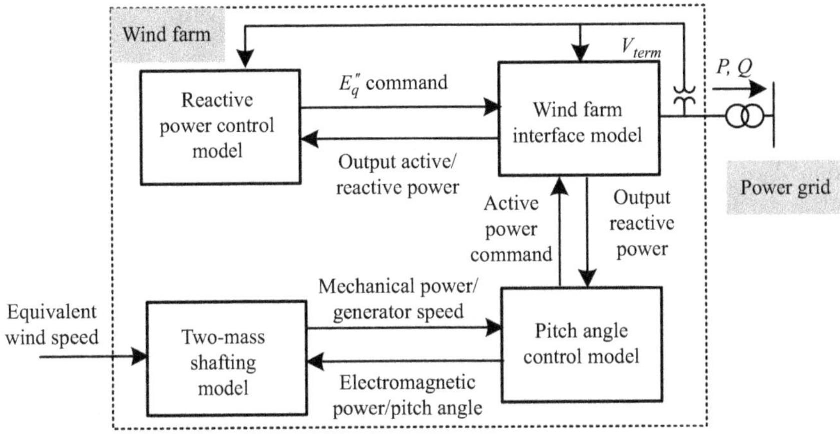

Figure 4.2 Schematic diagram of the wind farm module

shaft and generator shaft can be different to some extent, and there are some different dynamic response characteristics due to the distinction between structure and mass of wind turbine shaft and wind generator shaft. Therefore, two-mass module is closer to the actual conditions than one mass module. Ignoring the gearbox, two blocks with different inertia coefficients are used to represent the mechanical driving system, which is described as

$$\begin{cases} T_{wt}\dfrac{d\omega_{wt}}{dt} = \dfrac{P_{mech}}{\omega_{wt}} - K_{tg}\left(\theta_{wt} - \theta_g\right) - D_{tg}\left(\omega_{wt} - \omega_g\right) \\ T_g\dfrac{d\omega_g}{dt} = -\dfrac{P_{elec}}{\omega_g} + K_{tg}\left(\theta_{wt} - \theta_g\right) + D_{tg}\left(\omega_{wt} - \omega_g\right) \\ \dfrac{d\theta_{wt}}{dt} = \omega_{wt}, \dfrac{d\theta_g}{dt} = \omega_g \end{cases} \tag{4.1}$$

where T_{wt} and T_g represent the inertia coefficients of the equivalent wind turbine and generator, respectively. θ_{wt} and θ_g are the torsion angles. ω_{wt} and ω_g stand for the rotational speeds. P_{mech} and P_{elec} are mechanical and electromagnetic powers, respectively. K_{tg} is stiffness coefficient and D_{tg} is damping coefficient.

4.2.2 Pitch angle control module

The DFIG wind farms always adopt variable-speed constant-frequency (VSCF) control strategy. The pitch control plays an important role in DFIG wind farms. From aerodynamics analysis, we can learn that the wind turbines' input power will increase with the increment of the attack angle. Pitch angle is designed to change the attack angle as the sum of pitch angle and attack angle is constant in a given situation. When a wind turbine starts to function or when the speed of wind is lower than the rated speed but higher than the minimum speed, the pitch angle stays at 0 to maximize the attach angle, which leads to more input power. On the other hand, when a wind turbine is braking, the pitch angle stays at 90° in order to make wind offer braking torque. From the above illustration, it is not hard to conclude that pitch angle control is equivalent to active power control.

When the wind speed exceeds the rated speed or when the wind farm suffers from serious transient faults, the module shown in Figure 4.3 comes into play. The pitch angle control module consists of two parts: pitch control and pitch compensation. This control module ensures that the wind turbine's speed and power are both acceptable.

4.2.3 Reactive control module

The voltage of point of interconnection (POI) and reactive power are governed by closed-loop control strategy in this module. Constant power factor, constant reactive power, and constant voltage control modules are optional in reactive control. This module has good expansibility and can be changed according to different reactive power references. The diagram is shown in Figure 4.4.

4.2.4 Wind farm interface module

The whole wind farm is considered as a controllable current source in this module. However, the equivalent generator is not particularly modeled, because the rotor part and excitation system are separately included in the two-mass shafting module and the reactive control module. The impedances of the wind generators and the collection system are all considered in a wind farm's equivalent impedance. The diagram is shown in Figure 4.5.

Compared with the conventional reduced-order model in [14,15], the equivalent wind farm model composed of almost all the basic control units is able to better simulate the actual dynamic response, which will be verified by a comparison given in Section 4.5.

4.3 Selection of dominant parameters

The wind farm model established in Section 4.2 preserves the response features to a great extent. However, there are too many parameters needed to be identified. All

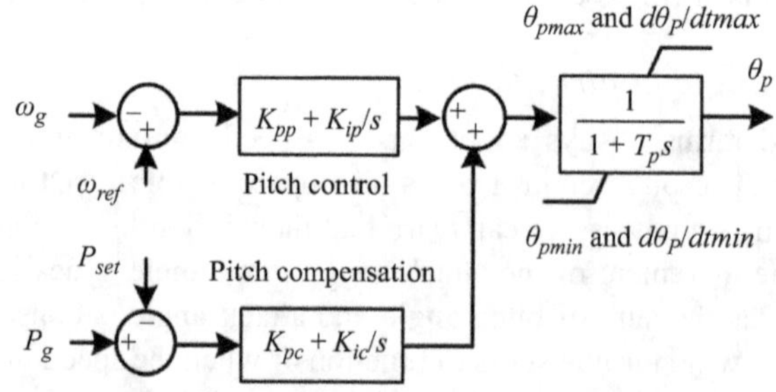

Figure 4.3 The pitch angle control module

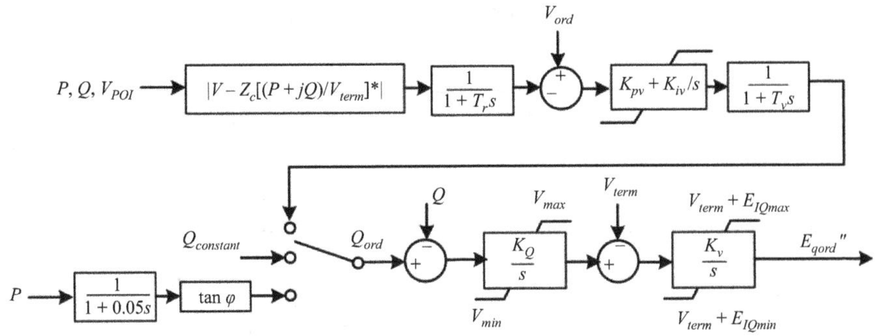

Figure 4.4 The reactive control model

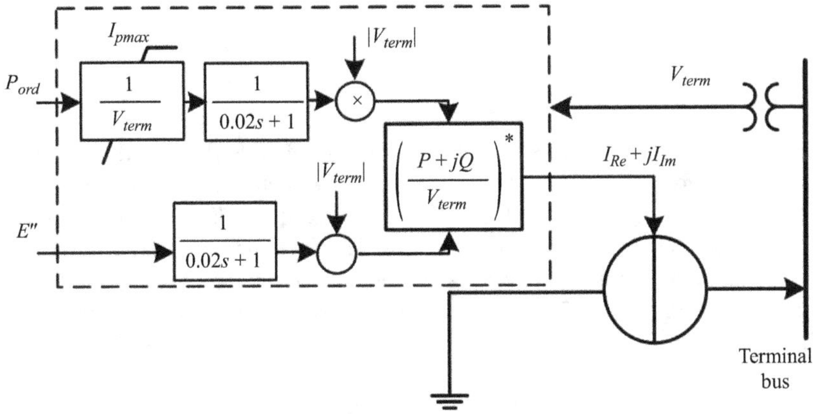

Figure 4.5 The wind farm interface model

the parameters and their empirical values [20] are listed in Table 4.2. It is obvious that not all of them have significant effects on the model response. Those parameters that dominate more in the model behaviors, namely dominant parameters, should be set accurately.

Identifiability generally means that the model parameters' property of being recognized uniquely based on the model's input and output variables. Most conventional parameter identifiability analysis approaches mainly rely on analytics. They can provide accurate analytical solutions based on linear models, but they are impractical for high-complexity models. Researches in [21,22] have found that there is inner relationship between the identifiability and sensitivity. Those parameters with high sensitivities can be identified easily. Hence the sensitivity analysis method is employed in this section to select the dominant parameters that are easier to be identified as well.

Table 4.2 The parameters of the wind farm model and their common empirical values in p.u.

Equivalent reactance (X_{eq})	0.55	Pitch angle PI control (K_{pp})	150
Integral coefficient of Q (K_{Qi})	0.05	Pitch angle PI control (K_{pi})	25.0
Integral coefficient of V (K_{Vi})	30.0	Inertia coefficient of generator (T_g)	1.14
Upper bound of E_q'' change (E_{IQmax})	0.30	Inertia coefficient of turbine (T_{wt})	7.64
Lower bound of E_q'' change (E_{IQmin})	–0.35	Stiffness coefficient of shaft (K_{tg})	1.25
Control mode of Q (Q_{flag})	Constant-voltage	Damping coefficient of shaft (D_{tg})	1.50

The nonlinear model of wind farm can be expressed as differential-algebraic equations

$$\begin{cases} \dfrac{\mathrm{d}}{\mathrm{d}t}\mathbf{x} = f(\mathbf{x}, \mathbf{y}, \mathbf{u}, \boldsymbol{\theta}) \\ \mathbf{y} = g(\mathbf{x}, \mathbf{u}, \boldsymbol{\theta}) \end{cases} \tag{4.2}$$

where the vector $\mathbf{x}$ represents the state variables (such as the q-axis subtransient electromotive force E_q'', rotor speed ω_g, etc.). The vector $\mathbf{u}$ is the input variables (such as the wind speed V_w and the POI bus voltage V_{term}). The vector $\boldsymbol{\theta}$ stands for the parameters. The vector $\mathbf{y}$ is the output variables of the wind farm model.

The parameter sensitivity reflects the output variables' level of sensitivity to the parameter variation. Thus, the parameter sensitivity in a certain state is defined as

$$S_{y_i/\theta_i} = \frac{\partial y_i / y_i}{\partial \theta_i / \theta_i} = \lim_{\Delta\theta_i \to 0} \frac{\frac{y_i(\theta_r, \theta_{i0} + \Delta\theta_i) - y_{i0}(\theta_r, \theta_{i0})}{y_{i0}(\theta_r, \theta_{i0})}}{\Delta\theta_i / \theta_{i0}} \tag{4.3}$$

where $S_{yi/\theta i}$ is the sensitivity of y_i related to θ_i. y_{i0} is the model output when the θ_i is θ_{i0}. $\Delta\theta_i$ is the change of θ_i. θ_r represents the other parameters except θ_i.

The parameter sensitivity can be expressed as an analytical expression, only if the model on an operation point is easy to be linearized. In this chapter, the parameter sensitivity is approximatively calculated by (4.3). Note that the variable $\mathbf{y}$ is a function of time, and the parameter sensitivities from different instances form a trajectory sensitivity curve.

Compared with the sensitivity, the trajectory sensitivity with dynamic features added can be devoted to analyzing the parameters' relation. The parameters whose trajectory sensitivity curves have the same phases or contrary phases couple with each other, so they cannot be independently identified.

The wind farm model established in Section 4.2 has many adjustable parameters, among which PI parameters can be determined by offline method. The active and reactive power outputs of the wind farm can be selected as the observation variables, and the trajectory sensitivity of each parameter can be calculated under gust disturbance and fault disturbance. In the following, only a few commonly used parameters are analyzed, and only the parameter trajectory sensitivity curve under fault disturbance is given. If necessary, the remaining parameters can also be analyzed according to this method.

To make comprehensive comparisons of those parameters' magnitudes of trajectory sensitivity more intuitively, the average of absolute values of each parameter's trajectory sensitivity in the observed time is computed by

$$\bar{S}_{y_i/\theta_i[t_1,t_2]} = \frac{1}{N}\sum_{k=0}^{N}\left[S_{y_i/\theta_i\left(t_1+k\cdot\frac{t_2-t_1}{N}\right)}\right] \tag{4.4}$$

where $\bar{S}_{y_i/\theta_i[t_1,t_2]}$ is the average sensitivity in the time domain from t_1 to t_2. And N is the number of samples. $S_{y_i/\theta_i\left(t_1+k\cdot\frac{t_2-t_1}{N}\right)}$ stands for the sensitivity at time instance $t_1 + k \times (t_2 - t_1)/N$.

Table 4.3 shows six parameters' mean value of trajectory sensitivity under different disturbances and Figure 4.6 shows parameter trajectory sensitivity under fault disturbance. From the result of calculation, we can learn that parameters' mean value of trajectory sensitivity has some gaps under different disturbances. X_{eq}, XIQ_{max} and other electromagnetic parameters' trajectory sensitivity is relatively higher under fault disturbance. T_g, K_{tg}, and other mechanical parameters' trajectory sensitivity is relatively higher under gust disturbance. In both cases, T_{wt} has good identifiability, while D_{tg} is hard to identify.

The procedure of selecting dominant parameters is divided into two steps. First, the independent parameters are selected by the phase of the trajectory sensitivity. Then, the parameters with larger amplitudes of trajectory sensitivity are treated as the dominant parameters.

4.4 Parameter identification based on IGA

4.4.1 Identification model

Essentially, model parameter identification is a procedure of optimization. The goal is to minimize the total square error of the actual output and the predicted output by reasonable adjustment of equivalent parameters under the constraints of state variables and structure parameters.

Table 4.3 Mean value of trajectory sensitivity under different disturbances

Disturbance	Observation variable	X_{eq}	XIQ_{max}	T_g	T_{wt}	K_{tg}	D_{tg}
Fault	P_g	0.107	0.021	0.012	0.129	0.012	0.001
	Q_g	0.315	0.047	0.007	0.299	0.030	0.002
Gust	P_g	0.000	0.000	0.019	0.135	0.039	0.000
	Q_g	0.016	0.000	0.069	0.486	0.240	0.000

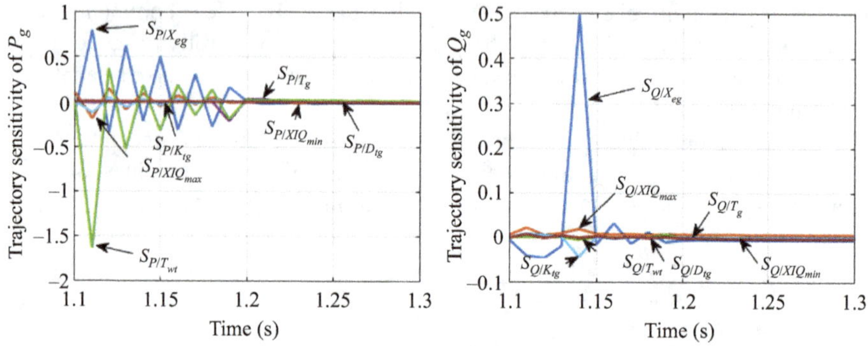

Figure 4.6 Parameter trajectory sensitivity under fault disturbance

Because of the extensive application of wide-area measurements, the active and reactive power injection, as well as the voltage at a terminal bus can be obtained credibly. In terms of the identification model here, the active and reactive power injection serve as the output variables, while the terminal voltage and the equivalent wind speed are chosen as the input variables. Note that the equivalent wind speed is calculated with the consideration of wake and tower-shadow effects, which is not a main focus in this chapter. There are more state variables in this model than the reduced-order models, owing to the complicated model construction, such as the pitch angle θ_p, the generator rotor speed ω_g, the wind turbine shaft speed ω_{wt}, and q-axis subtransient electromotive force E_q''. The dominant parameters selected for identification will be discussed in Section 4.5. Now they are represented as the vector $\boldsymbol{\theta}$. The model of parameter identification is given as

$$\min \int_{t_1}^{t_2} \left[|\widehat{P}(t) - P(t)|^2 + |\widehat{Q}(t) - Q(t)|^2 \right] dt$$
$$\text{s.t.} \begin{cases} \theta_{p\,min} \le \widehat{\theta}_p \le \theta_{p\,max}, E''_{q\,min} \le \widehat{E}''_q \le E''_{q\,max} \\ \omega_{g\,min} \le \widehat{\omega}_g \le \omega_{g\,max}, \omega_{wt\,min} \le \widehat{\omega}_{wt} \le \omega_{wt\,max} \\ \theta_{min} \le \widehat{\theta} \le \theta_{max} \end{cases} \tag{4.5}$$

where the variables with hats mean estimation values. The dominant parameters' boundaries θ_{min} and θ_{max} are determined by their empirical values, so as to guarantee the reasonability of the results.

4.4.2 Identification algorithm

In order to avoid complex calculation and the loss of dynamic characteristics due to the simplification and order reduction process in the modeling, genetic algorithm (GA) which less relies on the internal details of the wind farm model is taken into account. The off-line simulation process is applied to the parameter optimization process, which avoids the limitation that the traditional optimization method requires auxiliary information such as model derivative value to determine the search direction, and solves the identification and optimization problem of complex model parameters in Simulink.

The whole wind farm model is treated as a black box. Only the input information and output information are used to search for the optimal parameter combination that matches the dynamic response characteristic. Every parameter combination in GA corresponds to a chromosome, and each parameter in the combination is represented as a gene, as described in Figure 4.7. Each parameter to be identified is expressed with a 15-bit binary number, indicating a resolution of 0.001.

In order to enhance efficiency of identification, an improved genetic algorithm (IGA) is proposed based on the original version. Additionally, all the modifications based on the original GA aim to achieve a better balance between efficiency and reliability. The procedure of IGA, mainly involving four aspects, is shown in Figure 4.8.

4.4.2.1 Parameters initialization

Although the optimization results are less affected by the initial values in the GA, the closer to the optimal values the initial values are, the fewer iterations the optimization procedure costs. The population size is set to 100. In contrast to the original GA initializing parameters completely at random, here the IGA performs initialization under the guidance of feasible region, making it more probable to approach optimal states. In particular, the feasible region of each parameter is divided into 10 parts. Then 10 samples are randomly selected in each part. This method can improve the probability of finding the range including the optimal solution at first generation.

4.4.2.2 Good individuals selection

Considering that the objective function is non-negative and takes the minimum as the optimization objective, the fitness function is inversion of the objective function in (4.5). So, the parameter combinations that diminish the objective function have larger fitness.

To avoid the loss of the optimal genes in the traditional roulette selection strategy, the 10% parameters with largest fitness are directly inherited to the next generation. The remaining 90% individuals are chosen by the roulette selection strategy.

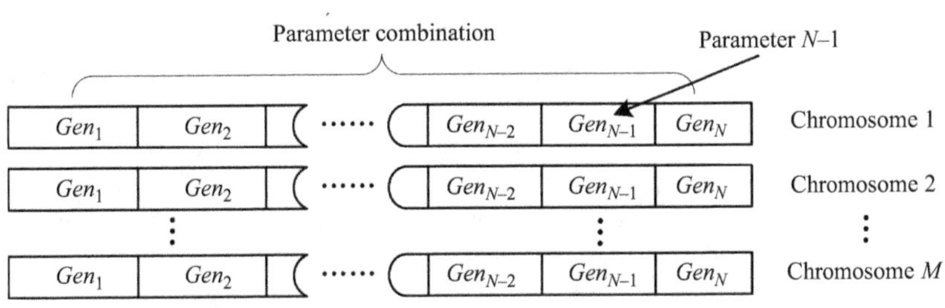

Figure 4.7 The representation of parameter combination in GA

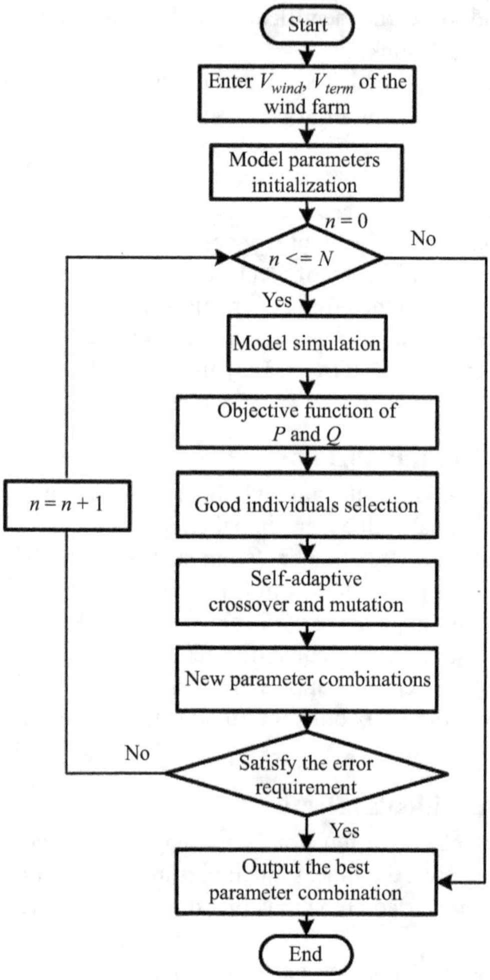

Figure 4.8 The procedure of parameter identification based on IGA

4.4.2.3 Self-adaptive crossover and mutation

The individuals in a generation are randomly paired. And the crossover operation is carried out on the individuals in each pair to generate offspring. The crossover probability P_c is used to control the frequency of crossover operation. A bigger P_c can improve the search capability in the space, while a smaller P_c can ensure the continuity of the search around the optimal value. So, a self-adaptive technique is utilized, adjusting P_c during the identification process by

$$P_c^n = P_c^{n-1} + \left(0.5 - P_c^0\right)/N \tag{4.6}$$

where n means the current number of iterations. P_c^n is the crossover probability at nth iteration. N is the maximum number of generations, which is 100 here. P_c^0 is the original value, which is set to 0.9 in this paper.

Multipoint mutation strategy is adopted in the mutation procedure. The purpose of mutation operation is to produce new individuals and expand the searching scope. Increasing the value of mutation probability P_m can avoid falling into local optimum, while may destroy the optimum solution space. The recommended value range of P_m is 0.001 to 0.1. The P_m is also adjusted in (4.7), according to the search results:

$$\begin{cases} P_m^n = P_m^{n-1} + (0.001 - P_m^0)/N, \text{if } \tau = 0 \\ P_m^n = P_m^0, \text{if } \tau = 1 \end{cases} \tag{4.7}$$

where P_m^n is the mutation probability at nth iteration. P_m^0 is set to 0.1. τ is the flag variable that decides whether to reset P_m^n as the original value. When the situation that the increment of the best fitness value of the current generation is less than 1% of the last generation's occurs ten times continually, τ is equal to 1. This means that the search range will change if better values cannot be found after repeatedly search.

4.4.2.4 Iterative identification

The model simulation is incorporated in the IGA. Every new parameter combination created by the genetic manipulation is passed to the model simulation to calculate its objective function value and fitness. Then good individuals are selected to be renewed in a new round of genetic manipulation. The optimal parameter combination is obtained in the process of iterative identification.

The algorithm process will stop when the output error is in the range of the allowable error or the number of iterations reaches its maximum limit. In the latter situation, the optimum results, the optimal parameter combination in the last interaction probably is not the global optimum, considering the influence of crossover and mutation. So the best parameter combination must be selected from the optima in all generations.

4.5 Case study

4.5.1 Model test based on the benchmark system

A model simulation test is conducted on a simple system to verify the dynamic response of the model, as shown in Figure 4.9. The test system is a benchmark system proposed by the WECC's Wind Generator Modeling Group [23]. There are 20 DFIGs that are rated at 1.5 MW in the wind farm. The standard capacity of this system is 33 MVA. The power grid is represented by an infinite voltage source. The parameters of lines and transformers are given on per unit value.

Different forms of disturbances are imposed to test the response characteristics of the electrical control parts and the mechanical transmission parts. Several

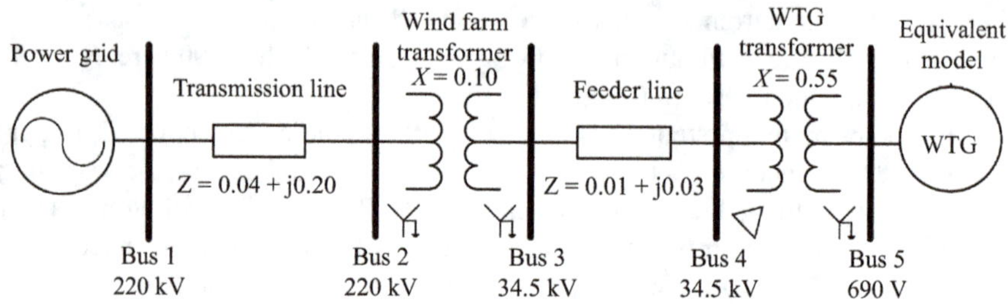

Figure 4.9 The benchmark test system

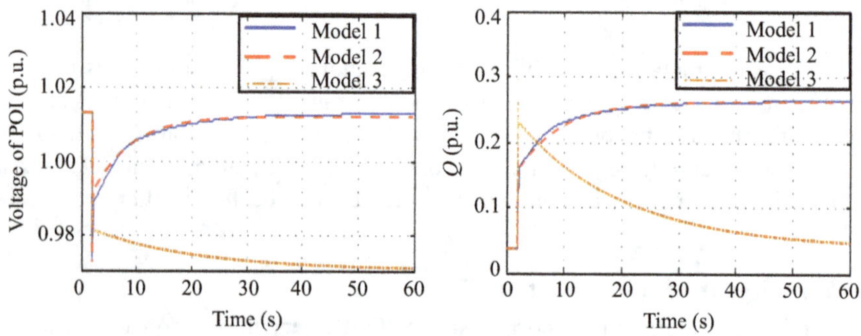

Figure 4.10 The voltage and the reactive power at Bus 2

comparisons are made among the G.E. standard model (model 1), the equivalent model (model 2), and the traditional reduced-order model (model 3).

4.5.1.1 Test with grid voltage disturbance

A small step change of the grid voltage is carried out to test the control effects of reactive power and voltage control. The setting is that the voltage of grid drops from 1.05 p.u. to 0.97 p.u. at 1.0 s. The voltage and the reactive power at Bus 2 (POI) are shown in Figure 4.10.

The test results show that the equivalent model can timely increase reactive power to provide assistance for the voltage recovery, which is as same as the standard model when the model is under constant-voltage control mode. Whereas the traditional model, which always uses constant reactive power control mode by setting the reactive power reference, is helpless for the voltage drop. It is obvious that the property of DFIGs is lost, although the model is simplified.

4.5.1.2 Test on the wind speed disturbance

The pitch angle control and regulation of active power are tested in this part by a wind ramp. The wind farm works under the rated condition. A wind ramp starts from 11.8 m/s at 10.0 s and reaches 17.8 m/s. Then, the wind speed remains steady for 30.0 s. Its rise time and fall time are both 30.0 s. The pitch angle and active power are shown in Figure 4.11.

It is evident that the traditional reduced-order model that ignores the pitch angle control exhibits some undesirable characteristics in the period of wind disturbance. However, the equivalent model has almost the same features as the standard model, which can approximately maintain the output power as a constant. The results demonstrate that the active control and pitch angle control are effective and better tally with the practical conditions.

4.5.2 *Validation of the identification algorithm*

The identification algorithm based on the IGA is verified on the CEPRI 36 bus system, as shown in Figure 4.12. The wind farm containing 240 × 1.5 MW DFIGs is connected to the system at Bus 8. A PMU is installed at Bus 31 to measure the dynamic information. The simulation data including amplitudes and phase angles of bus voltages, active power, and reactive power outputs are acquired to simulate PMU measurements. The standard capacity of this system is 100 MVA. Bus 1 is treated as a slack point.

4.5.2.1 Parameters selection and classification

It is known that different dynamic modes may be stimulated by different disturbances [24]. In the model verification, it can be seen that the voltage and reactive power control may significantly affect the dynamic process when a transient disturbance occurs. Analogously, the same relationship exists between the pitch angle control and the wind disturbance. So, the parameters may have different observability under different disturbances.

There will be two kinds of disturbances in this section to classify and select the dominant parameters. One kind disturbance is a three-phase short-circuit fault at Bus 31 that lasts for 0.1 s (Case 1). The other kind disturbance is a gust that starts at 2.0 s with an initial speed of 14 m/s and lasts for 20 s. The maximum of the speed is 20 m/s (Case 2). The parameters' trajectory sensitivity curves under both kinds of disturbances are given in Figures 4.13 and 4.14. The parameters of PI controllers are assumed to be similar to individual wind turbines. And they are also less

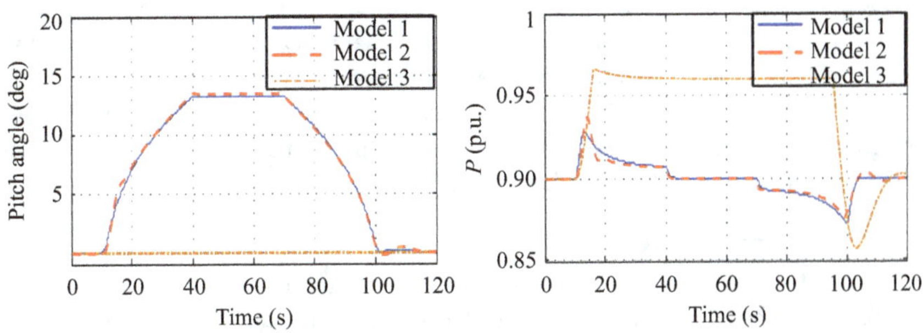

Figure 4.11 The pitch angle and active power of the wind farm

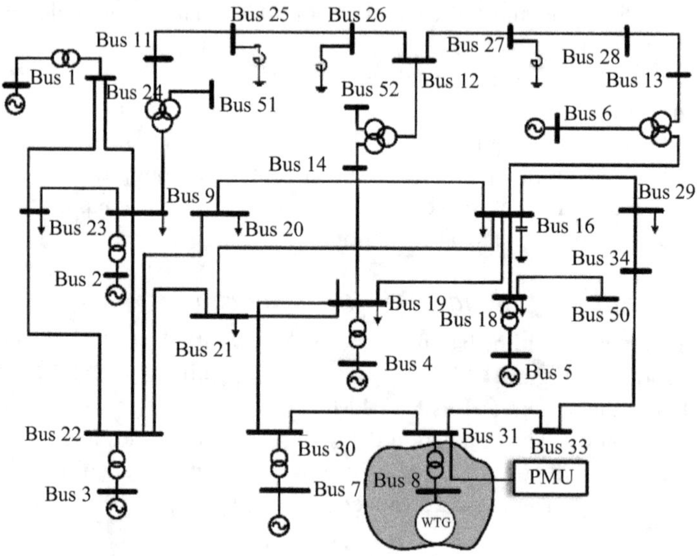

Figure 4.12 The CEPRI 36 bus system

affected by the wind farms' operating conditions, service life, etc. So, they are preset and regarded as unchanged. Only the parameters that are sensitive and variable, such as X_{eq}, E_{IQmax}, E_{IQmin}, T_g, T_{wt}, K_{tg}, and D_{tg} are analyzed in this section. The parameters X_{eq}, E_{IQmax}, and E_{IQmin} that are more associated with electrical response characteristics are referred to as electrical parameters. Meanwhile, the parameters T_g, T_{wt}, K_{tg}, and D_{tg} are referred to as mechanical parameters.

From the results, it is found that those parameters are uncoupled to each other except that some parameters' sensitivities are too small to identify. In order to compare the identifiability of those parameters, the average sensitivities are listed in Table 4.4. Every number in the table represents the corresponding sensitivity with respect to P or Q.

It indicates that the parameters playing an important role at the voltage and reactive power control such as X_{eq} and E_{IQmax} are easier to be identified under the transient fault. The parameters such as T_g, T_{wt}, and K_{tg} have some identifiable degrees in both cases, but they are relatively more sensitive under the gust disturbance. However, it is difficult to identify E_{IQmin} and D_{tg} in either case. Therefore, X_{eq} and E_{IQmax} are identified under the transient fault, while T_g, T_{wt}, and K_{tg} are identified under the gust disturbance.

4.5.2.2 Parameter identification

The parameter identification method can play an effective role when there is a disturbance in the wind farm operation. Different parameters are identified under different situations. The simulation experiment is also in the CEPRI 36 bus system with the same disturbances in Case 1 and Case 2.

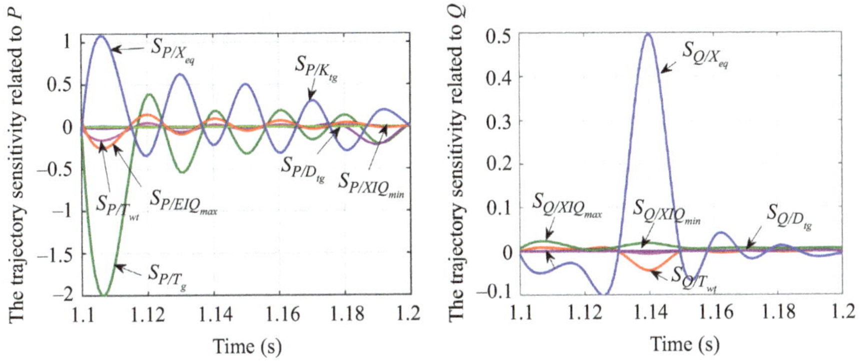

Figure 4.13 The trajectory sensitivity curves of the parameters under the transient fault

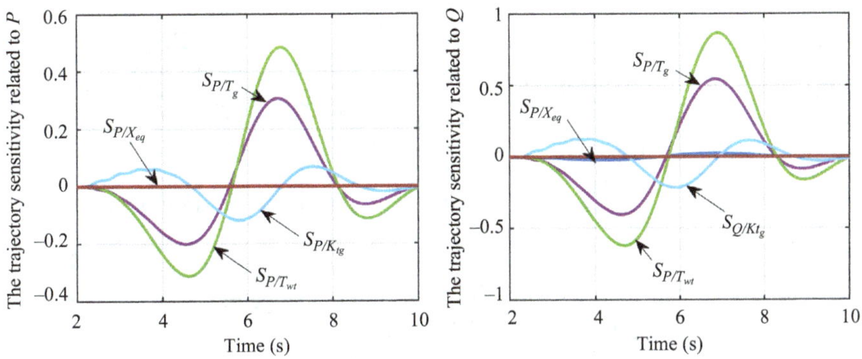

Figure 4.14 The trajectory sensitivity curves of the parameters under the gust disturbance

Table 4.4 Average sensitivities of the parameters

Case		X_{eq}	E_{IQmax}	E_{IQmin}	T_g	T_{wt}	K_{tg}	D_{tg}
Case 1	P	0.363	0.441	0.006	0.077	0.049	0.016	0.001
	Q	0.101	0.011	0.003	0.019	0.006	0.003	0.000
Case 2	P	0.001	0.000	0.000	0.088	0.140	0.034	0.000
	Q	0.009	0.000	0.000	0.165	0.261	0.064	0.000

The POI bus is probably not equipped with PMU, just as this case in Figure 4.12. If the resistance and ratio of the transformer between Bus 8 and Bus 31 are known, the information of the wind farm can be calculated from the PMU measurements by some state estimation methods.

The identified values of those parameters based on IGA are compared with the actual values in Table 4.5. It is obvious that the identified values are rather close to

Table 4.5 Comparison of parameter identification results

Parameter	X_{eq}	E_{IQmax}	T_g	T_{wt}	K_{tg}
Actual value	0.500	0.400	1.500	8.600	1.420
Identified value	0.489	0.387	1.519	8.608	1.437
Relative error	2.20%	3.25%	1.27%	0.12%	1.20%

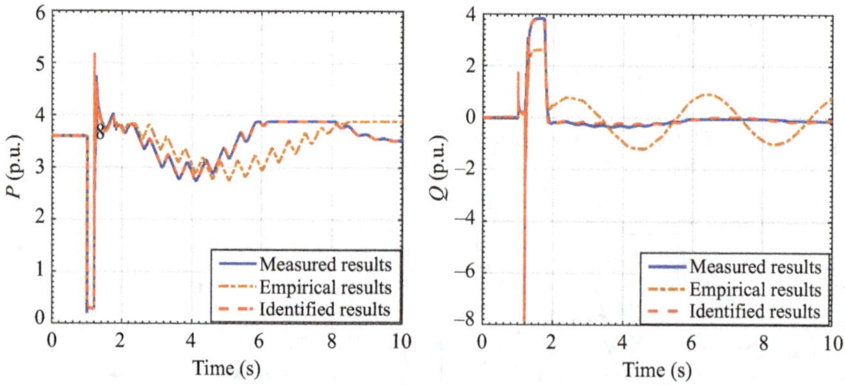

Figure 4.15 Comparison of responses under the transient fault

the actual values. The relative errors are all within the range of the errors permitted. The results verify the effectiveness of the method.

Furthermore, the identification results are shown in Figures 4.15 and 4.16, where the response results of measured data, identified data, and the outputs based on empirical parameters are comprehensively compared.

The comparisons show that the response based on identification is in general agreement with the practical measurement results, whereas the response based on empirical parameters is greatly different. It further suggests that power system security analysis using empirical parameters of wind farm models is likely to deviate from the actual situation. Moreover, it reveals the parameters identified from measurement data of WAMS are of desirable reliability.

To validate the advantages of the IGA related to the traditional GA, the iterative processes of the two kinds algorithms in both identification cases are shown in Figure 4.17, where the objective function values are given at same iterations.

It is shown that the convergence rate of the errors is greatly improved in the IGA with the adaptive crossover operator and mutation operator. Therefore, the IGA proposed is effective to enhance the global search capability of GA during the procedures of both identification cases.

To prove that the sensitivity is a good index for the identifiability, the parameters with small sensitivities, such as E_{IQmin} and D_{tg}, were added into the

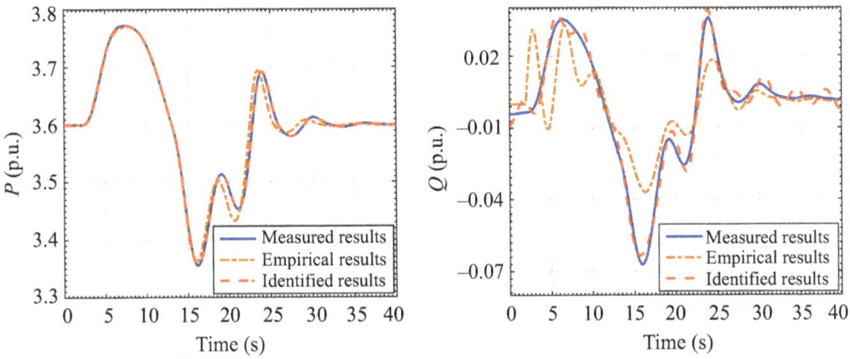

Figure 4.16 Comparison of responses under the gust disturbance

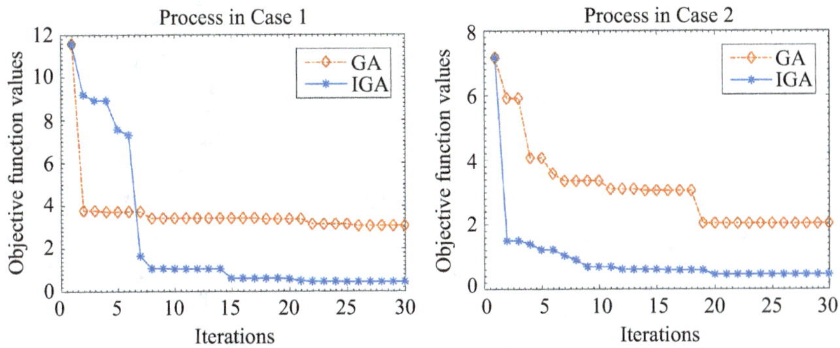

Figure 4.17 Comparisons between the GA and IGA

parameter set for identification. The fitting degrees of active power and identified results of X_{eq} and E_{IQmin} in five experiments are listed in Table 4.6.

In some high fitting degree situations, the values of X_{eq} are relative steady, while the values of E_{IQmin} are randomly distributed in its feasible region, which shows that those parameters with small sensitivities are difficult to be accurately identified. So, they are usually specified as empirical values. It also implies that the parameters selection based on sensitivity can effectively remove those intractable parameters and improve the identification efficiency.

4.5.2.3 Parameter validation

However, there may exist several sets of parameters that produce similar responses from the model. So, the parameters identified by the IGA need to be further selected by the responses under stochastic wind that frequently appears in practice (Case 3).

To validate the effectiveness of the optimized model parameters, a new set of parameters (Set B) producing comparable responses as the previously identified

Table 4.6 The identified values of parameters and fitting degrees

Number	1	2	3	4	5
X_{eq}	0.489	0.491	0.486	0.498	0.501
E_{IQmin}	–0.481	–0.205	–0.303	–0.428	–0.187
Fitting degree of P	98.7%	98.9%	98.6%	99.4%	99.2%

Table 4.7 The values of parameter Set B

Parameter	X_{eq}	E_{IQmax}	T_g	T_{wt}	K_{tg}
Set B	0.779	0.317	1.676	7.602	1.358

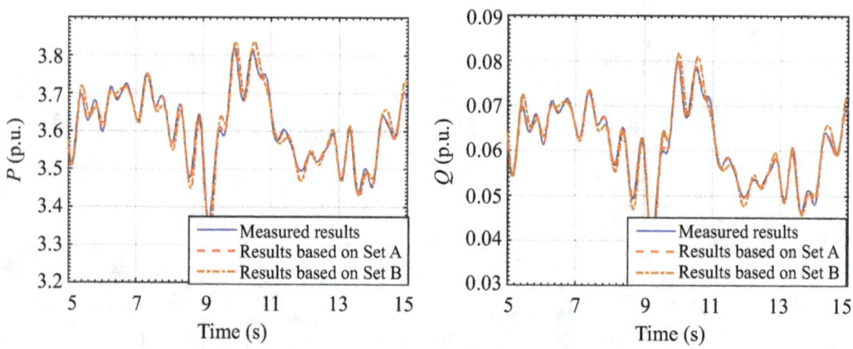

Figure 4.18 Comparison of responses based on Set A and Set B in Case 3

parameters (Set A, see Table 4.5) is obtained from identification processes by expanding the feasible regions of parameters, as listed in Table 4.7.

The measured results and the responses based on the parameter Set A and Set B are compared in Figure 4.18.

The fitting degree of active power based on the parameters of Set A is 98.55%, while the counterpart based on Set B is 92.15%. Meanwhile, the fitting degrees of reactive power corresponding to Set A and B are 97.32% and 91.08%, respectively. It is obvious that the results based on parameters of Set A stay much closer to the practical measurements than the results based on parameters of Set B in Case 3. So, the parameters of Set A are selected as the optimal parameters. Such a verification also reveals that the optimal parameters should be well adaptive under other disturbance occasions.

4.6 Conclusion

In this chapter, a wind farm modeling and parameter identification approach based on measured data of PMUs is proposed to solve the problem of unreliable models and the inaccurate parameters. An equivalent wind farm model involving basic structural characters and control principles is built to effectively simulate the actual response of the wind farm. Dominant parameters are selected and classified by trajectory sensitivity analysis under two kinds of disturbances, which helps to tackle the intricate and laborious identification issue of multiple parameters. Besides, an improved generic algorithm, enhancing the capability of globally and efficiently searching optimal parameters, is proposed. In particular, the genetic manipulation is optimized with the self-adaptive cross and selection operators, attempting to find a balance between efficiency and reliability.

The test on the WECC benchmark system under voltage steps and wind ramps demonstrates that the response of this model is coincided with the standard model, which can reflect the real dynamic characteristics better than the traditional model. In addition, the validation of the identification algorithm is carried out on the CEPRI 36-bus system, which suggests the model responses with identified parameters based on WAMS are more accurate and reasonable than responses with the empirical parameters. Then the IGA is compared with the traditional GA, which validates the better global search capability of this method. Finally, the optimal parameters are selected and validated under the stochastic wind condition in case of multisolution situations.

References

[1] Usaola J., Ledesma P., Rodriguez J. M., *et al.* "Transient stability studies in grids with great wind power penetration: modelling issues and operation requirements." *2003 IEEE Power Engineering Society General Meeting (IEEE Cat. No. 03CH37491)*, Toronto, Ontario, Canada, IEEE, 2003, pp. 1534–41

[2] Shi L., Sun S., Yao L., Ni Y., and Bazargan M. "Effects of wind generation intermittency and volatility on power system transient stability." *IET Renew Power Gen.* 2014;8(5):509–21

[3] Chowdhury M. A., Shen W., Hosseinzadeh N., and Pota H. R. "Transient stability of power system integrated with doubly fed induction generator wind farms." *IET Renew Power Gen.* 2015;9(2):184–94

[4] Lu M., Jiakun F., Jinyu W., and Weihua L. "Transient stability risk assessment of power systems incorporating wind farms." *J Mod Power Syst Cle.* 2013;1(2):134–41

[5] Jing L., and Zhang X.-P. "Impact of increased wind power generation on subsynchronous resonance of turbine-generator units." *J Mod Power Syst Cle.* 2016;4(2):219–28

[6] Lin C., Manjun L., Yuanzhang S., and Yi D. "A multi-state model for wind farms considering operational outage probability." *J Mod Power Syst Cle.* 2013;1(2):177–85

[7] Ali M., Ilie I., Milanovic J. V., and Chicco G. "Wind farm model aggregation using probabilistic clustering." *IEEE T Power Syst.* 2013;28(1):309–16

[8] Pan X., Zhang C., Ju P., Wu F., Jin Y., and Zhang Y. "Coherency-based dynamic equivalence of wind farm composed of doubly fed induction generators." *Power Syst Technol.* 2015;39(3):621–7

[9] Zou J., Peng C., Xu H., and Yan Y. "A Fuzzy clustering algorithm-based dynamic equivalent modeling method for wind farm with DFIG." *IEEE T Energy Conver*. 2015;30(4):1329–37

[10] Haiqiang Z., Ping J., Yusheng X., and Jie Z. "Probabilistic equivalent model of DFIG-based wind farms and its application in stability analysis." *J Mod Power Syst Cle*. 2016;4(2):248–55

[11] Brochu J., Larose C., and Gagnon R. "Validation of single- and multiple-machine equivalents for modeling wind power plants." *IEEE T Energy Conver*. 2011;26(2):532–41

[12] Xue F., Song X.-F., Chang K., Xu T.-C., Wu F., and Jin Y.-Q. "Equivalent modeling of DFIG based wind farm using equivalent maximum power curve." *2013 IEEE Power & Energy Society General Meeting*, Vancouver, BC, Canada, IEEE, 2013, pp. 1–5

[13] Zhang Y., Muljadi E., Kosterev D., and Singh M. "Wind power plant model validation using synchrophasor measurements at the point of interconnection." *IEEE T Sustain Energ*. 2015;6(3):984–92

[14] Andrés F., José C., and Camilo C. "A third order model for the doubly-fed induction machine." *Electr Pow Syst Res*. 2000;56(2):121

[15] Elkington K., and Ghandhari M. "Comparison of reduced order doubly fed induction generator models for nonlinear analysis." *2009 IEEE Electrical Power & Energy Conference (EPEC)*, Montreal, Quebec, Canada, IEEE, 2009, pp. 1–6

[16] Muljadi E., and Ellis A. "Validation of wind power plant models." *2008 IEEE Power and Energy Society General Meeting - Conversion and Delivery of Electrical Energy in the 21st Century*, Pennsylvania, USA, IEEE, 2008, pp. 1–7

[17] Tsai C., Lee W., Nashawati E., Wu C., and Lan H. "PMU based generator parameter identification to improve the system planning and operation." *2012 IEEE Power and Energy Society General Meeting*, San Diego, California, USA, IEEE, 2012. pp. 1–08

[18] Wang B., and Lu J. "A method for online analyzing excitation systems performance based on PMU measurements." *2011 International Conference on Advanced Power System Automation and Protection*, Beijing, China, IEEE, 2011, pp. 1538–42

[19] Zhang X., Lu C., Han Y., *et al.* "Ambient signal based load model parameter identification using optimization method." *2015 IEEE Power & Energy Society General Meeting*, Colorado, USA, IEEE, 2015, pp. 1–5

[20] Miller N. W., Price W. W., and Sanchez-Gasca J. J. *Modeling of GE Wind Turbine-Generators for Grid Studies*. New York, NY: GE International, Inc., Schenectady, 2004

[21] Pan X., Gao Y., Jin Y., and Ju P. "Parameter identification of drive system for fixed-speed wind power generation units." *Power Syst Technology.* 2013;37:1990–4

[22] Ju P., Guo L., Gao C., *et al.* "Frequency-domain sensitivities with application to power system modeling." *Proc CSEE*. 2010;30:19–24

[23] Price W. W., and Sanchez-Gasca J. J. "Simplified wind turbine generator aerodynamic models for transient stability studies." *2006 IEEE PES Power Systems Conference and Exposition*, Georgia, USA, IEEE, 2006, pp. 986–92

[24] Xueping P., Ping J., Qian X., Yongkang L., Feng W., and Yuging J. "A two-step method for estimating dfig parameters in a wind turbine and the measurement selection." *Proc CSEE*. 2013;33(13):116–26

Chapter 5

Load modeling based on synchrophasor measurements

Xinran Zhang[1] *and David J. Hill*[1]

Simulation-based approaches are an important branch of the research about power systems. To ensure that the simulation results are accurate enough to get meaningful conclusions, the models used in power system simulation should be as accurate as possible. For most power system components, including generators, transmission and distribution lines, transformers, flexible alternating current transmission system (FACTS), and so on, the models can be accurately built based on the details of the components.

However, modeling power load is different and more difficult. Unlike other power devices, power load contains different kinds of electrical devices. As a result, a load model is an aggregation of the models of these different devices. In addition, power load is changing all the time according to the demand of users, which makes load models to be uncertain and time varying [1]. With the uncertain, time-varying, and complicated properties, power load cannot be absolutely accurately represented with a detailed and unchanged model like other devices. Therefore, load modeling has been an important and difficult issue in the simulation of power systems [2].

In recent years, with the development of power system synchrophasor measurement techniques, the approaches of load modeling have been further developed. First, ambient signal-based load modeling methods have been proposed, through which load modeling can be conducted more frequently regardless of the existence of large disturbances. Second, with the increasing penetration of renewable energy in the demand side of the power systems, the models of renewables should also be considered in modeling power system demand side apart from power load. Third, the development of artificial intelligence technologies has brought new methods and tools for load modeling, which includes the application of clustering, deep learning, and so on.

In this chapter, the recent developed approaches about ambient signal-based load model parameter identification, identification of composite demand side

[1]Department of Electrical and Electronic Engineering, The University of Hong Kong, Pokfulam, Hong Kong

model with renewables, and the clustering of uncertain load model parameters have been introduced. To begin with, the approaches and structures of load models are introduced. Then, the method to identify load model parameters from ambient signals is proposed. Following this, the method to build the composite demand side model with PV generation and ES system is proposed. Afterward, the load model parameter clustering method to pick up representative load models from the identified load models' set is proposed. Finally, conclusions and future work directions are given.

5.1 Overview of load modeling

5.1.1 Approaches of building load models

For a satisfactory load model, the output results under the same input should follow the output measured from the real power system under all situations. There are mainly two categories of methodologies in load modeling, that is, component-based statistical synthesis method and measurement-based identification method.

The component-based approach was developed first in 1980s [3], which is to aggregate the models of the individual electrical devices connected to the same bus from the bottom to the top, forming an aggregated load model. In this approach, the knowledge of load composition is required, which can be obtained through survey or load monitoring. In contrast, the measurement-based approach is to identify the load model parameters from measurement data through a given load model structure by regarding the load model of one bus as a composite load model. In recent years, since power load is becoming more complicated and time varying, it is becoming more difficult to conduct component-based load modeling approach, which makes measurement-based load modeling approach more widely used.

For measurement-based identification method, the first step is to choose a load model structure, under which the load model parameters can be identified from measurement data. After designating the load model structure, load model parameters can be obtained from measurement data through curve fitting. An effective optimization scheme is presented for the measurement-based load modeling based on the sensitivity analysis of composite load model parameters [4]. An improved clone selection algorithm is proposed to improve the global optimization capability and identification efficiency in multidimensional function [5]. Apart from the composite load model, other load models also find their application, including local model network [6], second-order transfer function [7], and so on.

5.1.2 Load model structure

As introduced in Section 5.1.1, selecting a load model structure is the first step in measurement-based load modeling approach. In earlier measurement-based load modeling approaches, for the reason that only the magnitudes of the voltage and the current can be measured by traditional measurement devices, the static load models in which the phase angle measurement data are not necessary are widely developed, including the ZIP model, the exponential model, and so on [2]. However, the dynamic load models, of which the induction motor is a typical example, have been playing a more and more important role in modern power systems [8]. The phase angle is one of the input variables of the induction motor model, therefore its measurement is necessary for building the induction motor model [9]. Recently, with the accurate phase angle measurement data, phasor measurement unit (PMU) data have become a good data source for measurement-based load modeling [3]. As a result, more complicated load models can be built based on PMU measurement data. The composite model with static load and induction motor is most widely used as the load model [10], among which ZIP+IM [11] is a typical example.

In this chapter, the widely used composite load model with a static load part and an induction motor as the dynamic load part is chosen as the load model structure, the details of which will be introduced in Sections 5.1.2.1, 5.1.2.2, and 5.1.2.3.

5.1.2.1 Static load model

Various kinds of models can be chosen as the static load model, including the ZIP model, the exponential model, and the constant Z model. In this chapter, to simplify the model structure and allow focus on the newly proposed ambient signal-based method, the constant Z model is chosen as the static part model, forming the Z+IM composite load model. The selection of the constant Z model as the static part can also be supported by the fact that it is among the most commonly used static load models in power system simulation models in China. The proposed method has the potential ability to be applied for more complicated static load models.

5.1.2.2 Dynamic load model

The T-shape equivalent circuit of an induction motor is given in Figure 5.1, where X_s is the stator reactance, X_r is the rotor reactance, X_m is the excitation reactance, R_r is the rotor resistance, and s is the slip ratio. According to [8], the number of the parameters can be reduced through:

$$\begin{cases} X = X_s + X_m \\ X' = X_s + X_m // X_r \\ T_{d0} = (X_m + X_r)/(\omega_0 R_r) \end{cases} \tag{5.1}$$

where X is the rotor open circuit reactance, X' is the rotor transient reactance, T_{d0} is the rotor open circuit time constant, and ω_0 is the synchronous rotation angular

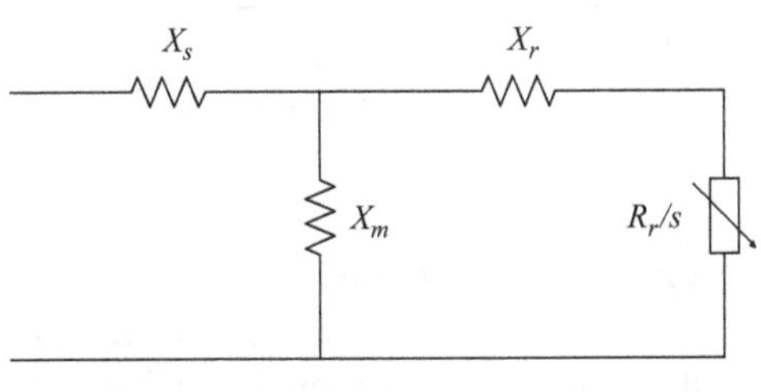

Figure 5.1 T-shape equivalent circuit of an induction motor

speed. Then, the dynamics of the induction motor can be described by the following third-order state space formula:

$$\begin{cases} \dfrac{dE_d}{dt} = \dfrac{1}{T_{d0}}\left(-E_d - (X - X')I_q\right) + s\omega_0 E_q \\ \dfrac{dE_q}{dt} = \dfrac{1}{T_{d0}}\left(-E_q + (X - X')I_d\right) - s\omega_0 E_d \\ \dfrac{ds}{dt} = \dfrac{1}{H_2}(T_m - T_e) \end{cases} \tag{5.2}$$

where $\dot{E} = E_d + jE_q$ is the phasor of the electromotive force, E_d and E_q are the d-axis and q-axis components. Similarly, $\dot{I} = I_d + jI_q$ is the phasor of the current. T_m is the mechanical torque, which is assumed to be constant during the identification period in this chapter. T_e is the electromagnetic torque, which can be calculated as follows:

$$T_e = E_d I_d + E_q I_q \tag{5.3}$$

The first and the second equations of (5.2) are about the electromagnetic dynamics of the induction motor, while the third equation is about the electromechanical dynamics. Therefore, X, X', and T_{d0} are called electromagnetic parameters, while H_2 and T_m are called mechanical parameters. The variables linking these two kinds of induction motor dynamics are the slip ratio s and the electromagnetic torque T_e.

The relationship among the voltage phasor, the current phasor, and the electromotive force phasor is given as follows:

$$\dot{U} = \dot{E} + jX\dot{I} \tag{5.4}$$

where $\dot{U} = U_d + jU_q$ is the voltage phasor. $\dot{U}$ can also be expressed in another form as $Ue^{i\theta}$, where U is the voltage magnitude and θ is the voltage phase angle. The power consumption P and Q can be calculated from E_d, E_q, U_d, and U_q as follows, which is the output formula of the induction motor:

$$\begin{aligned} P &= \frac{E_d U_q - E_q U_d}{X'} + \frac{U_d^2 + U_q^2}{R} \\ Q &= \frac{U_d^2 + U_q^2 - U_d E_d - U_q E_q}{X'} + \frac{U_d^2 + U_q^2}{X_l} \end{aligned} \tag{5.5}$$

where R and X_l are the resistance and the reactance of the constant Z static load part.

5.1.2.3 Simplification of Z+IM model

Z+IM model is the parallel of an induction motor and a constant impedance. The parallel of a motor and a reactance in the left part of Figure 5.2 can be further simplified according to Y-Δ transformation, after which the parallel can be replaced

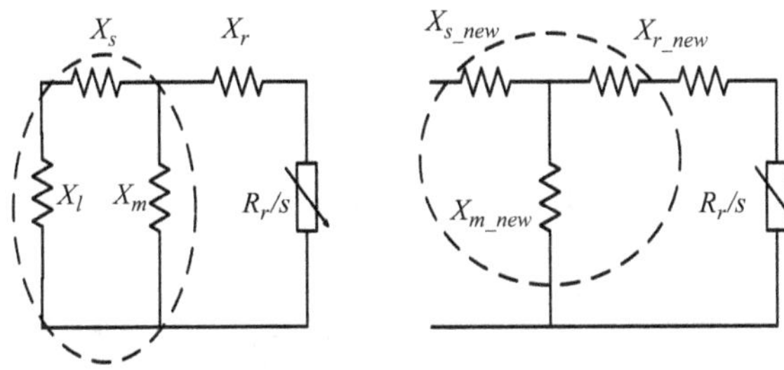

Figure 5.2 Simplification of Z+IM model through Y-Δ transformation

by an equivalent model. The relationship between the original parameters and the equivalent parameters is given as

$$\begin{cases} X_{new} = \dfrac{X_l X_s}{X_l + X_s + X_m} + \dfrac{X_l X_m}{X_l + X_s + X_m} = X//X_l \\ X'_{new} = \dfrac{X_l X_s}{X_l + X_s + X_m} + \left(\dfrac{X_m X_s}{X_l + X_s + X_m} + X_r\right) // \dfrac{X_l X_m}{X_l + X_s + X_m} = X'//X_l \\ T_{d0\_new} = \left(\dfrac{X_m X_s}{X_l + X_s + X_m} + X_r + \dfrac{X X_m}{X_l + X_s + X_m}\right) \Big/ (\omega_0 R_r) = T_{d0} \dfrac{X_l + X'}{X_l + X} \end{cases} \tag{5.6}$$

After this, the reactance part in Z is included in the equivalent motor, whose identification is no longer needed. Then, the parameters to be identified are X, X', T_{d0} and H_2 of the induction motor and the resistance R of the static load part.

5.2 Ambient signal-based load model parameters identification

In the area of measurement-based load modeling, there are mainly two approaches, that is, postfault response-based identification [4,7] and ambient signal-based identification. Postfault response is more widely studied, through which the dynamics of the load can be better excited. However, the weakness of this method is that whether identification can be conducted or not depends on the occurrence of large disturbances in power systems. However, the time-varying characteristics of load models require more frequent even periodical identification. Therefore, large disturbance independent analysis method is necessary to better meet the increasing practical demand.

In 1997, J. W. Pierre and D. J. Trudnowski found that ambient signal caused by change in load, transformer tap, and other small perturbation can be used in system dynamic analysis [9]. Since ambient signals always exist in system, daily output signals can be captured in real time through PMU measurement data; they are very suitable to be applied in power system dynamic stability monitoring and analysis

[10,11]. Since the dynamic property of load can also be probed by ambient signal, load model can be identified based on ambient signal, which is a large disturbance independent method. Ambient signal-based identification method is proposed in [12]. The advantage of the ambient signal-based load model identification method is that load model identification can be conducted more frequently, and therefore it is possible to better track the time-varying property of the power load. The weakness of the ambient signal-based method is that the magnitude of the disturbance is relatively smaller so that the identification results may not be as accurate for large disturbances as the postfault response-based identification results.

In this section, an ambient signal-based load model parameter identification method is proposed. With this method, load model parameter can be identified at any time regardless of the existence of fault, which can more frequently track the time-varying property of the electric power load compared with the traditional post large disturbance response-based approach.

5.2.1 Measurement data for identification

The measurement data used for load model parameter identification is shown in Figure 5.3. An input-output load model is identified from voltage amplitude (U), voltage phase angle (θ), active power (P), and reactive power (Q) measured from WAMS. U and θ are the input variables of the model, while P and Q are the output variables.

The task in this chapter is to identify the load model parameters from ambient signals measured by PMU during power system daily operation. Even under the same group of input voltage signals, different load model parameters will lead to different output power signals. Therefore, load model parameters can be identified from the PMU measurement signals by optimization method. For a given group of load model parameters, the output power signals can be predicted from the measured voltage signals. Then, the difference between the predicted output power signals and the measured power signals can be calculated. By regarding the load model parameters as the decision variables and the difference between the predicted output and the measured output as the objective function, the load model parameters can be identified through optimization.

5.2.2 Objective function: predicted error

In this chapter, an optimization method is proposed for parameter identification. The difference between the predicted output variables y_p and the measured output variables y_m is chosen as the objective function. While y_m can be directly measured by PMU, y_p should be calculated from the measured input signals and the parameter

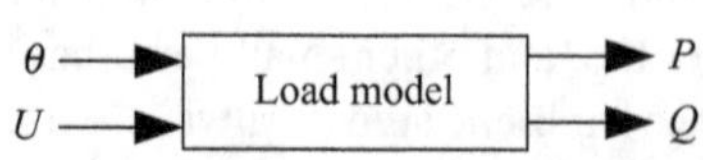

Figure 5.3 Measurement data used for load model identification

set (**S**). Therefore, the objective function is the function of the parameters to be identified, which meets the demand of an optimization problem:

$$z = \sum (y_p - y_m)^2 = f(S), S = \{X, X', T_{d0}, H_2, R\} \tag{5.7}$$

In the optimization process, all the members in S are given an initial value and then a value of function z can be calculated. Then, the optimization process is started to get smaller values of z. If z reaches its minimal value, the corresponding parameter set **S** is regarded as the identification results.

The most difficult part of this optimization problem is how to calculate y_p from input U, θ, and parameters in **S**, which is discussed in detail in this section.

From the output formula of motor in (5.5), it can be observed that if X' is regarded as a decision variable, which has a determined value, $E_d(k)$ and $E_q(k)$ can be calculated after solving a two-element linear equation group, as follows. The reason why $P(k)$ is replaced by $P(k)–U^2(k)/R$ is to remove static active power part caused by resistance R:

$$\begin{cases} E_d(k) = \dfrac{\left(P(k) - \dfrac{U^2(k)}{R}\right)X'U_d(k) + Q(k)X'U_q(k)}{U^2(k)} + U_q(k) \\ E_q(k) = \dfrac{Q(k)X'U_d(k) - \left(P(k) - \dfrac{U^2(k)}{R}\right)X'U_q(k)}{U^2(k)} + U_d(k) \end{cases} \tag{5.8}$$

$$\frac{\dot{E}(k) - \dot{E}(k-1)}{T} = \frac{1}{T_{d0}}(-\dot{E}(k) + j(X - X')\dot{I}(k)) - js(k)\omega_0\dot{E}(k) \tag{5.9}$$

$$s(k) = \frac{\dot{E}(k) - \dot{E}(k-1)}{T} + \frac{(-\dot{E}(k-1) + j(X - X')\dot{I}(k-1))T_{d0}}{j\dot{E}(k-1)\omega_0} \tag{5.10}$$

After the values of $E_d(k)$ and $E_q(k)$ are calculated, $s(k)$ can also be calculated from another form (5.9) of the first and second formulae in (5.2), as shown in (5.10). Then, $E_d(k)$, $E_q(k)$, and $s(k)$ are calculated as the function of X', which are functions of **S**.

Second, the values of the state variables at time $k + 1$ are predicted from the values of the state variables at time k. The predicted value of $E_d(k+1)$ and $E_q(k+1)$ can be calculated as follows:

$$\dot{E}(k+1) = \left(\frac{1}{T_{d0}}(-\dot{E}(k) + j(X - X')\dot{I}(k)) - js(k)\omega_0\dot{E}(k)\right)T + \dot{E}(k) \tag{5.11}$$

through which the impact of the decision variables X, X', and T_{d0} are included in the calculation of the objective function.

Third, the predicted value of $P(k + 1)$ and $Q(k + 1)$ can be calculated from the predicted value of $E_d(k + 1)$, $E_q(k + 1)$, the measured value of $U_d(k)$, $U_q(k)$, and X' according to (5.5). In this way, $P(k + 1)$ and $Q(k + 1)$ are predicted from the measured value of P, Q, U, and θ at time k, which are functions of the decision variables in **S**.

Finally, the objective function is the sum of the difference between measured value and predicted value of P and Q as

$$f(\mathbf{S}) = \sum (P_p(k) - P(k))^2 + (Q_p(k) - Q(k))^2 \tag{5.12}$$

The theoretical minimal value of $f(\mathbf{S})$ should be zero. The parameter set making $f(\mathbf{S})$ reaches its minimal value is regarded as the identification results.

5.2.3 Optimization algorithm: differential evolution

In this chapter, the differential evolution (DE) optimization method [13,14] is selected for optimizing the objective function to get the identification results. In DE algorithm, an individual is defined to be a vector that contains D elements. Each element in an individual is called a "gene." A set of individuals is defined to be a population. The generation is used to count how many times the evolution process has been conducted.

As an initial value independent optimization algorithm, DE does not require the initial value for parameters. Only the upper and lower bounds, that is, x_j^U and x_j^L are needed to start the optimization process. Therefore, for an optimization problem with D dimensions, the optimization problem can be described as follows:

$$\begin{aligned} &\min f(x_1, x_2, \ldots, x_D) \\ &\text{s.t. } x_j^L \le x_j \le x_j^U, j = 1, 2, \ldots, D \end{aligned} \tag{5.13}$$

Some definitions should be given to better understand DE algorithm. An individual is defined to be a vector that contains D elements, which may become the possible solution of the optimization problem after evolution process. Each element in an individual is called a "gene," which will evolve in the optimization process. A set of individuals is defined to be a population, the individuals in which will mutate with each other to form new individuals. The "generation" is used to count how many times the mutation, crossover, and selection process have been conducted.

There are four basic steps in differential evolution algorithm, that is, population initialization, mutation, crossover, and selection, which will be introduced in detail. In this chapter, *rand*(a, b) means a uniformly distributed stochastic number between a and b.

5.2.3.1 Population initialization

Population initialization is to randomly generate the initial population within the ranges of the parameters for further evolution, the process of which is given as follows:

$$\begin{aligned} &X_{i,j}(0) = rand(0,1)(X_j^U - X_j^L) + X_j^L \\ &i = 1, 2, \ldots, N; j = 1, 2, \ldots, D \end{aligned} \tag{5.14}$$

5.2.3.2 Mutation

A mutated individual is formed from three individuals of the last generation, where the differential vector of two individuals is weighted by the mutation factor and added to the third individual. All these three individuals are randomly selected from the population of the last generation. The mutation process is given as follows:

$$\boldsymbol{v}_i(G+1) = \boldsymbol{x}_{r_1}(G) + F(\boldsymbol{x}_{r_2}(G) - \boldsymbol{x}_{r_3}(G)) \tag{5.15}$$

where G means the generation of evolution, $r_1, r_2, r_3 \in [1, N]$ and $i \neq r_1 \neq r_2 \neq r_3$.

In the mutation process, one possible problem may occur that some of the genes in the mutated individual may be out of the range of the parameters. There are two possible solutions to this problem. One is to select three individuals and to generate the mutated individual again, while another is to select the range value of the gene as the corresponding gene value in the mutated individual. The second solution is used in the research of this chapter.

5.2.3.3 Crossover

If only mutation is conducted in the evolution process, all the individuals will be the linear combination of the first generation individuals no matter how many generations have been evolved. Therefore, crossover is necessary to avoid this situation. Crossover is to generate a random number between 0 and 1 and to compare this number with *CR*. If this number is larger, crossover will be carried out and the gene value of the last generation will be used. Otherwise, crossover will not be carried out and the mutated gene value will be used. To ensure that at least one gene value is different from the last generation, for the *i*th individual in the population, no crossover will be carried out on the *i*th gene of this individual. The crossover process is given as follows:

$$\boldsymbol{u}_{i,j}(G+1) = \begin{cases} \boldsymbol{v}_{i,j}(G+1) & \text{if } rand(0,1) \leq CR \mid j = i \\ \boldsymbol{x}_{i,j}(G) & \text{if } rand(0,1) > CR \ \& \ j \neq i \end{cases} \tag{5.16}$$

5.2.3.4 Selection

After mutation and crossover, the objective function values of the new individual and the last generation individual should be compared in order to select the better individual as the one in the new generation. Then, the selection process is conducted as given as follows:

$$\boldsymbol{x}_i(G+1) = \begin{cases} \boldsymbol{u}_i(G+1) & \text{if } f(\boldsymbol{u}_i(G+1)) \leq f(\boldsymbol{x}_i(G)) \\ \boldsymbol{x}_i(G) & \text{otherwise} \end{cases} \tag{5.17}$$

In each generation of the evolution process, each individual in the population will experience such mutation, crossover, and selection process so as to keep the better individuals in the next generation population and to eliminate the "bad" individuals. In this way, the whole optimization process can be led to the global optimal solution.

There are altogether four algorithm parameters in the DE algorithm, that is, the population scale N, the mutation factor F, the crossover rate CR, and the termination generations M. The explanations and typical ranges for these four algorithm parameters are given in Table 5.1.

5.2.4 Case study

5.2.4.1 System description

The simplified 500 kV network of the Guangdong Power Grid, China is used for the load model identification and verification. The structure of this simulation system is given in Figure 5.4. This system is simplified from the practical Guangdong Power Grid and the whole 500 kV network is kept. This system includes 83 buses and 97 lines. The load models of 16 load buses are identified in this section. PMUs are installed on all the load buses, the measurement data of which is used for load model identification. The other load buses are also set to be the composite load model containing a static part and an induction motor. The static

Table 5.1 Algorithm parameters in differential evolution optimization

	Name	Explanation	Typical range or value
N	Population scale	Number of individuals in a population	5–10 times of optimization dimensions
F	Mutation factor	The factor of the differential item in each mutation process	0.1–2
CR	Crossover rate	The probability of crossover for each individual	0.1–0.9
M	Termination generation	The number of generations for evolution termination	10,000, 20,000

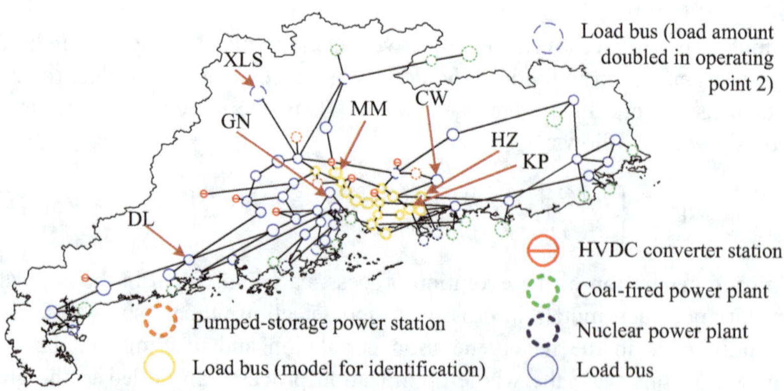

Figure 5.4 Structure of Guangdong Power Grid

load model is chosen to be constant Z to reduce complexity and focus on new method. The simulation time is 10 s with the time step being 0.01 s.

The following contributions of the proposed ambient signal based load model parameter identification method are to be validated by the case study results. First, the load model parameters can be identified from ambient signals. Second, the identification results of the load model parameters are accurate in repeated load model parameter identification cases. Third, the load model parameters identified from ambient signals have a good extrapolation ability, which can be used in the simulation cases of other kinds of disturbance.

5.2.4.2 Parameter identification

Sixteen load models are identified using ambient signals in this section. To generate the ambient signal disturbance on the test bus, the low pass filtered white noise signal is added on the active power of 30 other load buses. As a result, ambient signal occurs on the voltage of the load buses, which further leads the active and reactive power to oscillate.

Firstly, an example is given to validate that load model parameters can be identified from ambient signals. The identification process of Bus HZ is taken as an example. The load model parameters of this bus are [0.2162 0.0739 0.0959 0.1311] for $\{X\ X'\ T_{d0}\ R\}$, while the real values are [0.2159 0.0745 0.0961 0.1307]. In the optimization process, the data between 4 s and 6 s are used, which contain 200 measured data points. The values of the parameters converge to the real values after many generations of evolution, during which the value of $f(\mathbf{S}_{\text{opt}})$ also decreases toward 0.

The comparison between the measured and the predicted active power is shown in Figure 5.5. The predicted P and Q are calculated based on the identified load model. The fitting degrees of both the active power and the reactive power are better than 99.99%.

5.2.4.3 Identification accuracy

Second, the ambient signal-based load model parameter identification is repeated to validate the identification accuracy of load model parameters. The ambient

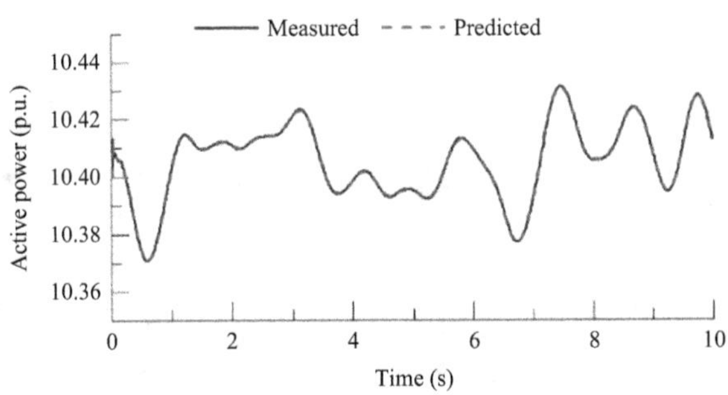

Figure 5.5 Active power response under ambient signal disturbance of Bus HZ

signal-based load model identification is repeated for ten times with ten different groups of ambient signals. Then, the load model parameters for the 16 load buses are also identified for 10 times accordingly. After achieving the ten groups of identification results, the per unit identification error for each result is calculated as

$$\text{Error} = \frac{S_{\text{identified}} - S_{\text{real}}}{S_{\text{real}}} \tag{5.18}$$

where $S_{\text{identified}}$ is the identified parameter value and S_{real} is the real parameter value.

If the identification results of all the parameters are around $\pm 10\%$ range of the real value, it is regarded as a successful identification case. After the first round of identification, there are 26 failed cases among all the 160 cases. In the 26 failed cases, the failure of 25 cases are caused by the inaccurate identification of T_{d0}, which matches the identifiability analysis results. Then, the identification process is repeated on the 26 failed cases. After the second round of identification, there are six failed cases left, which are all cleared after the third round of identification.

Then, the boxplots of the identification errors are given in Figure 5.6, while the per unit average and the maximal abs errors of the identification results are given in Table 5.2. The average identification time is 325.33 s, with the longest identification time being 1258.70 s. The fitting degrees of the active reactive power for all the 16 load buses in 10 times repeated identification are better than 99.99%. All these results show that the proposed identification algorithm has the ability to accurately identify the load model parameters from the ambient signal data.

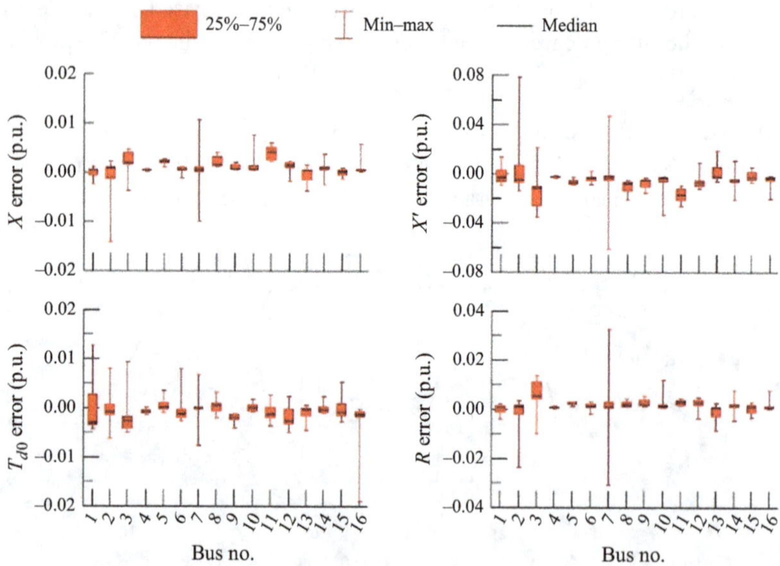

Figure 5.6 Identification error boxplot of load model parameters

Table 5.2 Identification error of load model parameters

	X	X'	T_{d0}	R
Max abs error	0.0140	0.0787	0.0188	0.0326
Mean abs error	0.0017	0.0088	0.0018	0.0032

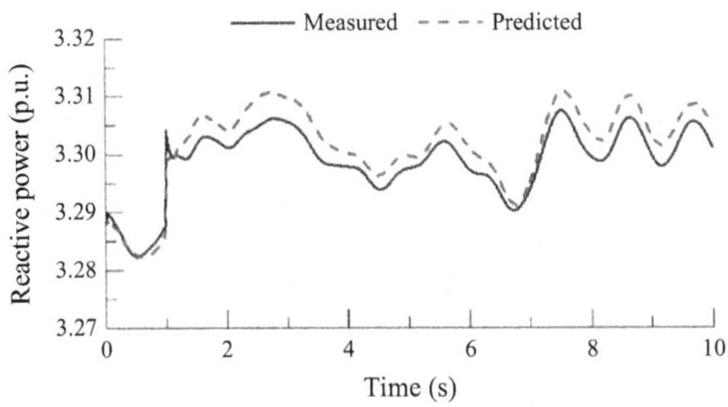

Figure 5.7 Reactive power response after load cut disturbance of Bus MM

5.2.4.4 Model extrapolation ability validation

Third, the extrapolation ability of the identified load model from ambient signals is validated. To validate the extrapolation ability of the identified load model, it should be further validated under different operation situations. Two different situations are considered in this part, that is, the load cut disturbance and the short circuit fault. To better validate the model, the simulation in this part is conducted in another operating point. Compared with the operating point in Section 5.2.4.2, the active power load of Bus GN, Bus DL, and Bus XLS is doubled. In addition, for each load bus, the worst results among the ten times identification are applied in the validation process.

First, the identified model is validated under load cut disturbance. At 1.0 s, 30% of the load on Bus FY is cut from the power system to generate large disturbance. The simulation time length is 10 s to observe postdisturbance response. The predicted P and Q are calculated based on the identified load model. The average fitting degrees of the active and reactive power for all the 16 load buses are 99.79% and 98.88%, respectively. Among all the load buses, the worst fitting degree is 92.47%, which appears on the reactive power of Bus MM, as shown in Figure 5.7. It is validated that the load model can predict P and Q accurately under load cut disturbance.

Second, the identified model is validated under short circuit fault. A three-phase to ground short circuit fault happens on Bus CW at 1.0 s, which is cleared at 1.1 s. The simulation time length is 10 s to observe postdisturbance response. The predicted P and Q are calculated based on the identified load model. The average fitting degrees of the active and reactive power for all the 16 load buses are 99.56%

and 99.54%, respectively. Among all the load buses, the worst fitting degree is 95.94%, which appears on the active power of Bus KP, as shown in Figure 5.8. It is validated that the load model can predict P and Q accurately under short circuit fault.

5.3 Identification of composite demand side model with renewables

In recent years, with the increasing penetration of distributed photovoltaic (PV) generation and energy storage (ES) system in the demand side of the power system [15], traditional load model identification technology can no longer provide an adequate demand side model due to the limitations in model structure. Meanwhile, the dynamic properties of distributed PV and ES are different from that of typical load models, which may lead to the deviation in dynamic performance simulation of power system demand side [16,17]. Therefore, demand side structures which accommodate new technologies including distributed PV generation and ES system are necessary to provide a better demand side model for power system dynamic simulation [18].

In this chapter, a model structure with both traditional load and new elements, that is, PV and ES, is proposed and identified based on ambient signal and disturbance data in power systems, as shown in Figure 5.9. The basic procedure of load model parameter identification is based on the method in Section 5.2, which has been further developed into the composite demand side model with load, PV and ES. The optimization algorithm has been improved to the differential evolution algorithm to provide faster and more reliable optimization results. In addition, the combined identification of ambient signal-based and fault response-based is proposed in this chapter.

Because the composite demand side model built in this chapter is mainly applied in power system electromechanical transient dynamic simulation, there is no need to include the detailed dynamic processes in the composite demand side model. The ignored dynamic properties mainly include the electromagnetic

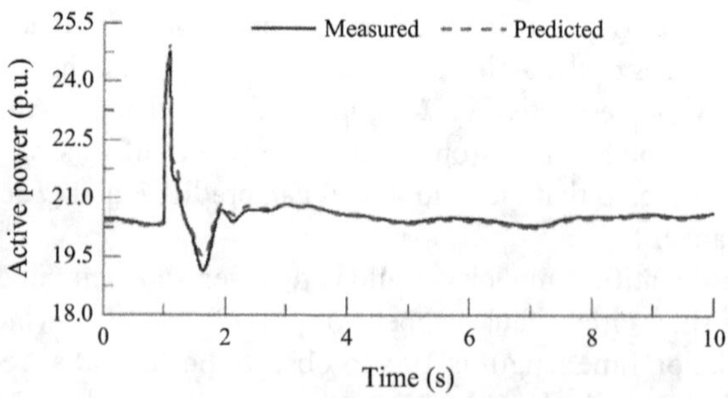

Figure 5.8 Active power response after short circuit fault of Bus KP

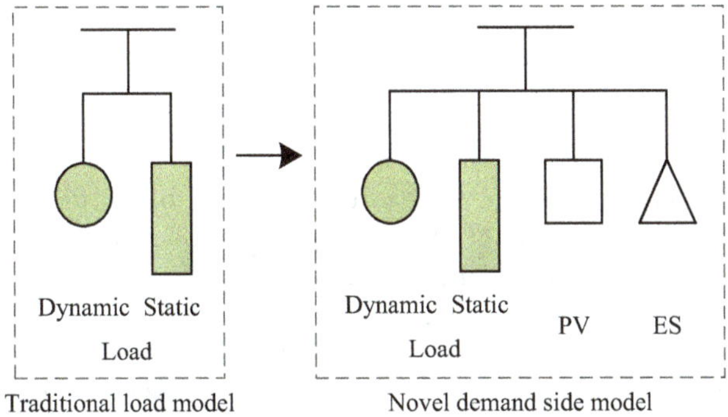

Figure 5.9 Structure of composite demand side model

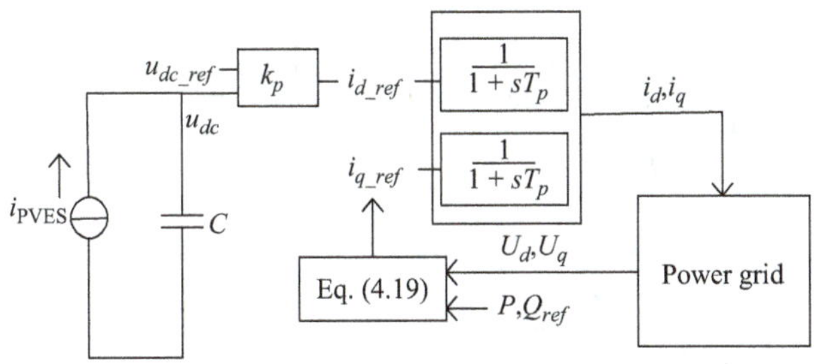

Figure 5.10 Uniform simplified model for PV and ES

transient dynamics of power electronic components, the dynamics of the PV cells, and the battery cells. Therefore, an assumption is made that the output power of the PV cells and the battery cells remain unchanged during the period of model identification. Under this assumption, the composite demand side model only captures how the active power and reactive power will response with the change of the bus voltage, which is mainly caused by small disturbances or faults in the rest of the power system. Other cases, such as the harmonics, the changes in frequency, and three-phase unbalanced conditions are the unsuitable applications.

5.3.1 Model structure of renewables

5.3.1.1 Uniform simplified model for PV and ES

Since the DC/AC inverter is the key part in modeling PV and ES and the control strategies for PV and ES are similar, a uniform simplified model can be used for describing the dynamics of PV and ES.

Then we note: u_{dc} here is the voltage of the capacitor in Figure 5.10, which is an important state variable in the control process. i_d and i_q are the d-axis and q-axis current magnitudes, and i_{d_ref} and i_{q_ref} are their reference values given by the controller. The relationship among i_d, i_q, U_d, U_q, P, and Q are given as follows:

$$\begin{bmatrix} i_d \\ i_q \end{bmatrix} = \begin{bmatrix} U_d & U_q \\ U_q & -U_d \end{bmatrix}^{-1} \begin{bmatrix} P \\ Q \end{bmatrix} \tag{5.19}$$

In this model, u_{dc} is mainly decided by the output of the PV cell and is also affected by the dynamics of the power grid. The power regulator decides the reference value for i_d, while the reference value for i_q is calculated from the reference value of Q and the voltage of the PV generation bus, U_d and U_q. Then, the current regulator controls the value of i_d and i_q to their reference value. The PI control strategy is selected for both the power regulator and the current regulator.

Then, the current regulator is also simplified as a delay link. This simplified model can describe most of the dynamics of the detailed PV generation model in [19] mainly by simplifying the dynamics between i_{d_ref}, i_{q_ref} and i_d, i_q. Therefore, it has been used in many power system simulation tools such as MATLAB® PSAT [20].

Based on the detailed PV and ES models in [19], a uniform simplified model for PV and ES is given and the state space formula is given by

$$\begin{cases} \dfrac{du_{dc}}{dt} = \dfrac{i_{\text{PVES}} - P/u_{dc}}{C} \\ i_{d\_ref} = k_p\left(u_{dc\_ref} - u_{dc}\right) \\ \dfrac{di_d}{dt} = \dfrac{i_{d\_ref} - i_d}{T_d} \\ \dfrac{di_q}{dt} = \dfrac{i_{q\_ref} - i_q}{T_q} \end{cases} \tag{5.20}$$

In the power control loop, the active power P is decided by the voltage of the capacitor u_{dc} and the output current of PV cell or the battery i_{PVES}, while the reference value of reactive power Q_{ref} is based on the control strategy. In this chapter, the Q control mode is analyzed as an example. Other control modes can be realized just by adding a controller between the feedback signals and $Q_{\_ref}$. After simplification, there are four parameters to be identified in this model structure, that is, C, k_p, T_p, and T_q. C is the capacitor parallel to the current source, k_p is the gain of the power regulator, T_p is the time constant of i_d, and T_q is the time constant of i_q.

5.3.1.2 Parallel of PV and ES model

In traditional load model identification, all the loads that are connected to the same bus are represented with one equivalent load model. Whether such equivalent method can be applied to PV/ES model depends on whether the parallel of PV/ES models can be represented by one single PV/ES model. Since the PV/ES model is a nonlinear model like the induction motor, the parallel model cannot be strictly derived from the formula. Therefore, an approximate parallel model is proposed here to represent the parallel of PV/ES models. To match the same parameter definition with the dynamic model for one PV/ES, the same formula is chosen as the model structure of the approximate parallel model. Then, the key problem is how to get the equivalent parameters for the approximate parallel model. Afterward, the equivalent model parameters are identified from the aggregated dynamic response of two PV or ES models. From many groups of identification results, an approximation of the equivalent parallel model parameters of two paralleled PV/ES models with the same active power output is given by

$$\begin{aligned} &C_{eq} = C_1 + C_2, K_{peq} = K_{p1} + K_{p2} \\ &T_{peq} = \frac{T_{p1} + T_{p2}}{2}, T_{qeq} = \frac{T_{q1} + T_{q2}}{2} \end{aligned} \tag{5.21}$$

where subscripts 1 and 2 represents the parameters for each single model, while *eq* represents the parameters for the equivalent model. This approximation can be applied when the parameters of two PV/ES models are at the same order of magnitude. For more paralleled PV/ES models with different active power output, (5.21) can be adjusted by considering the active power proportions.

The equivalent parameters for representing the parallel of multiple PV/ES models have been given in (5.21). The validation for this equivalent method should be validated through simulation results. In this part, Fault 1 and Fault 2 are used for test on two different groups of paralleled PV/ES model parameters, respectively. The parameters of the two paralleled PV/ES models and the equivalent parameters for the two groups are given in Table 5.3. The aggregated dynamics of two models and the dynamics of the equivalent model are compared in Figure 5.11. Conclusion can be drawn from the results that the equivalent method proposed in this chapter can give a good approximation.

5.3.2 Identification framework

5.3.2.1 Introduction of identification procedure

Based on the composite demand side model structure in Section 5.3.1. As in Figure 5.12, the amplitude and phase angle of voltage (U, θ) at the high voltage level bus are the input variables of the model, and the active power and the reactive power (P, Q) at the high voltage level bus are the output variables. Then, an input-

Table 5.3 PV/ES model parameters

Group	No.	C	K_p	T_p	T_q
1	1	0.01	3	0.002	0.002
	2	0.02	5	0.006	0.006
	Equivalent	0.03	8	0.004	0.004
2	1	0.01	10	0.007	0.007
	2	0.01	12	0.003	0.003
	Equivalent	0.02	22	0.005	0.005

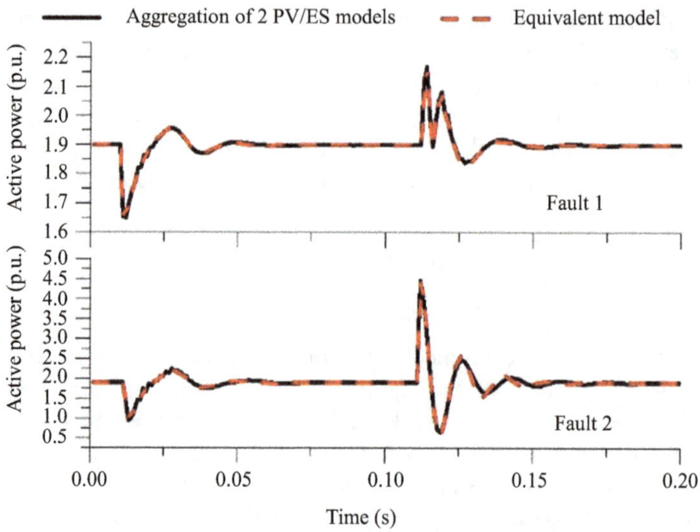

Figure 5.11 Dynamic performance of parallel of PV/ES model under Fault 2

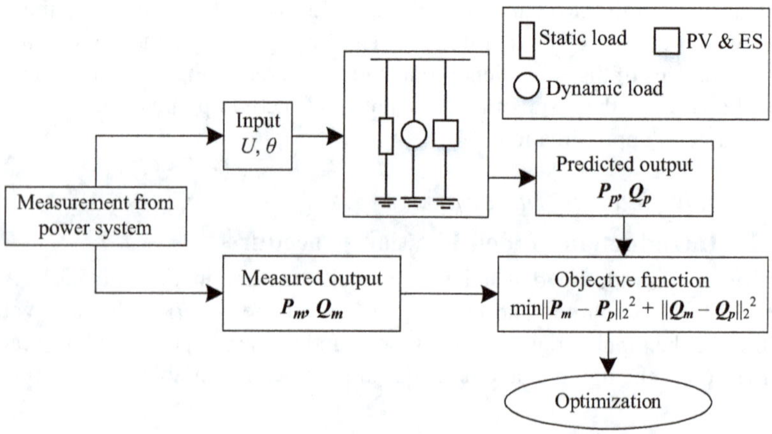

Figure 5.12 Calculation of objective function for optimization

output model is formed. For a satisfactory set of composite demand side model parameters, the predicted output results by the model should follow the measured output data from the real power system under the same input data in all conditions.

5.3.2.2 Measurement used for identification

Apart from the measurements of P, Q, U, and θ from the high voltage bus, PV and ES systems also have their own measurement systems to report the output of P *and* Q, which can also be applied in the composite demand side model parameter identification. On the contrary, since load is more widely and diversely distributed in the distribution network, the P and Q measurement for load can hardly be accurately measured from the low voltage buses discretely.

Therefore, the measurements of P, Q, U, and θ from the high voltage bus together with the measurements of P and Q for PV and ES systems are used in identification.

Since PV and ES systems have been modeled as current sources, the current injection to the high voltage bus can be controlled at the low voltage PV/ES bus, which will not be impacted by the existence of the distribution lines. Therefore, the power injection of the PV/ES to the high voltage bus can be measured at the low voltage PV/ES bus. Then, the power share of PV and ES can be calculated by summing the power measurements of PV and ES and then the equivalent model can be identified. The power share of load can be calculated by subtracting the share of PV and ES from the total power measurements from the high voltage bus and then the load parameters can be identified. The whole process is given in Figure 5.13. It can be concluded that the model for load and the model for PV/ES are identified separately in this process.

5.3.2.3 Calculation of objective function

To identify the model parameters, the objective function for parameter optimization should be calculated. The objective function in this chapter is the difference between the measured output and the predicted output by the identified model, as shown in Figure 5.12. The optimized parameters should make the objective function close to 0 so that the predicted output (P, Q) can follow the measured output under the same input (U, θ). Since the output is already known from measurement, the aim here is to calculate the predicted output from the measurement.

Based on the model structure in Figure 5.11, the objective function for PV/ES model identification can be obtained. The state variables of the PV/ES model

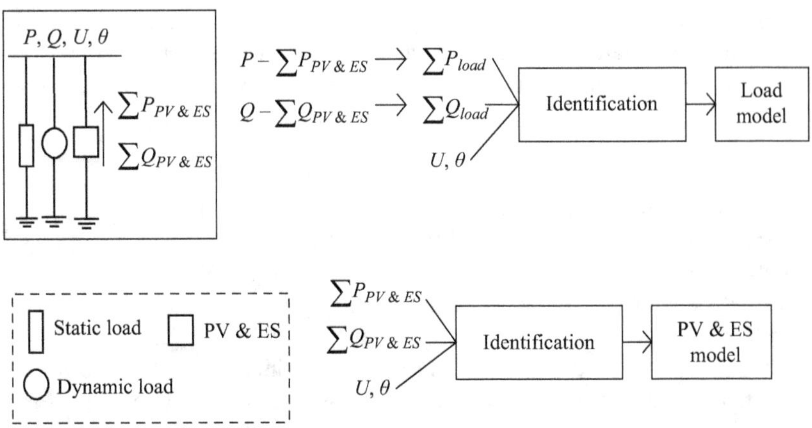

Figure 5.13 Measurements used for identification and the identification process

include i_{d_ref}, i_{q_ref}, i_d, i_q, and u_{dc}, while the output variables of the PV/ES model include P and Q.

The first step is to get the initial value of the state variables. At the initial state, i_{d_ref}, i_{q_ref} should be the same as i_d, i_q, which could be calculated from the initial values of P, Q, U, and θ through (5.19). For u_{dc} and u_{dc_ref}, as the values of DC side are p.u. value, the initial value of u_{dc} and u_{dc_ref} can be set as 1 by adapting the base value of the DC voltage, and the base value of i_{PVES} can be adapted to ensure that the base value of power in the DC side is the same as that in the AC side.

The second step is to predict the values of i_d, i_q from the measurement, and then to predict P and Q according to the model structure in Figure 5.10. There are three first-order dynamic components in this model, that is, the capacitor C and two time delay links, forming a third-order dynamic system, the formulae of which have been given in (5.20). The values of i_d and i_q can be predicted based on (5.20). Afterward, the values of P and Q can also be predicted from (5.19).

5.3.2.4 Parameter optimization

After the calculation of objective functions for load model and PV/ES model, the optimal model parameters can be obtained through the optimization process. Many different optimization methods have been applied in load model parameter identification. In this section, a differential evolution (DE) method is applied for parameter optimization. The details of DE can be found in Section 5.2.3.

5.3.3 Case study

5.3.3.1 System introduction

The simplified 500 kV network of the Guangdong Power Grid, China is used for the composite demand side model identification and verification. The structure of this simulation system is given in Figure 5.14. This system includes 83 buses, 97 lines, and 51 load buses. PMUs are installed on all the load buses, the measurement data of which is used for identification. PV and ES penetration is assumed in ten of the load buses. The other load buses are also set to be the composite load model containing a static part and an induction motor. We have chosen the constant Z as the static load model in order to reduce complexity and allow focus on PV and ES.

5.3.3.2 Model identification

As mentioned in Section 5.3.2, both ambient signal data and disturbance data are used for model identification. Two of the models containing both load and PV/ES are used for identification, which are the models on Bus FY and Bus DP. In this part, 15 independent groups of white noises are injected on the active power and reactive power of 15 load buses in order to generate ambient signal on the voltage of the buses with the models to be identified, from which X, X', T_{d0}, and R can be identified. The ambient signal always exists during the simulation period. A three-phase to ground short circuit fault happens on Bus CW at 6.0 s, from which H_2, K_p, C, T_p, and T_q can be identified.

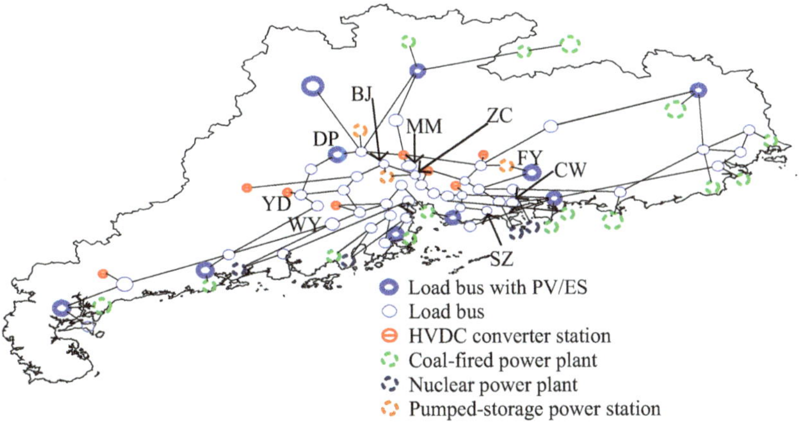

Figure 5.14 Structure of Guangdong Power Grid

Bus DP is taken as an example to illustrate the identification process. First, the ambient signal data from 0 s to 5 s is used for the identification of the load model parameters (except H_2), which is captured and recorded from PMU for offline identification. The measurement data of U, θ, P, and Q of the load part is used for identification. With U, θ being the input data, the parameters of X, X', T_{d0}, and R are optimized through DE algorithm to make the difference between the predicted P, Q and the measured P, Q to the least. Second, the fault data from 5.9 s to 6.4 s is used for the identification of the PV/ES parameters and H_2. The measurement data of U, θ, P, and Q of the PV/ES part is used for PV/ES parameter identification, while the measurement data of the load part is used for H_2 identification. In the same way as load, these parameters can also be obtained through the DE optimization process. After the optimization process, the P and Q of the load part and the PV/ES part are predicted with the identification results under the measured U and θ. Then, the predicted P and Q for each part (the predicted values in the figures) are compared with the measured P and Q (the measured values in the figures) to verify whether the model identified from the measurement data is accurate enough to follow the measured dynamics.

The identification results of DP and FY are given in Table 5.4. The comparison between the measured and the predicted power by the model of DP is given in Figure 5.15.

5.3.3.3 Model validation

To verify the generalization ability of the identified model, further simulation results are analyzed from other large disturbances in the system. In addition, to further convince the generalization ability of the identified model, the simulation in this part is conducted under another operation point, where the load of four buses, namely BJ, SZ, WY, and YD, are doubled.

Table 5.4 Identification results

Parameters	X	X'	T_{d0}	R	H_2
DP	0.6418	0.3278	0.1525	0.7144	3.0733
FY	0.7315	0.4259	0.1292	0.4576	2.0121
Parameters	K_p	C	T_p	T_q	
DP	10.1195	0.0508	0.0050	0.0050	
FY	2.0652	0.0105	0.0050	0.0050	

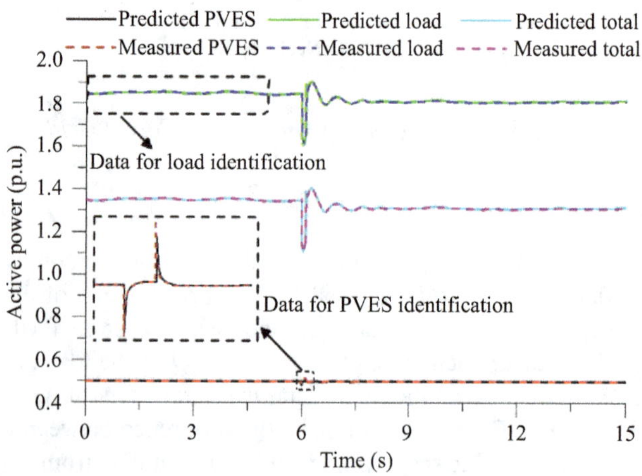

Figure 5.15 Active power comparison of the measured and predicted values (Bus DP)

The short circuit fault on Bus ZC is analyzed here. The ambient signal always exists during the simulation period. The comparison of dynamics between the measured and predicted active power on Bus FY is given in Figure 5.16. The identified model can well predict the dynamics under another short circuit fault.

The sudden load increase on Bus MM is analyzed. The load on this bus is suddenly increased by 100% at 6 s. The ambient signal always exists during the simulation. The comparison of dynamics between the measured and predicted active power on Bus FY is given in Figure 5.17. The identified model can well predict the dynamics under the sudden load increase.

5.4 Clustering of load model identification results

Through ambient signal-based load model parameter identification, load model parameters can be more frequently identified, which is an important improvement. However, it also brings new problems to be settled. From the results of ambient

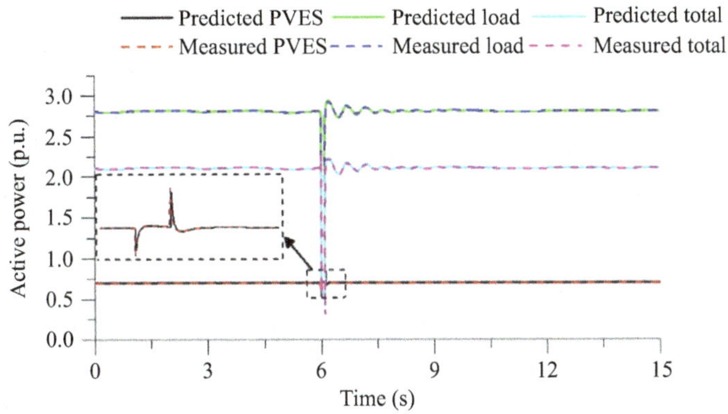

Figure 5.16 Model validation under another short circuit fault (Bus FY)

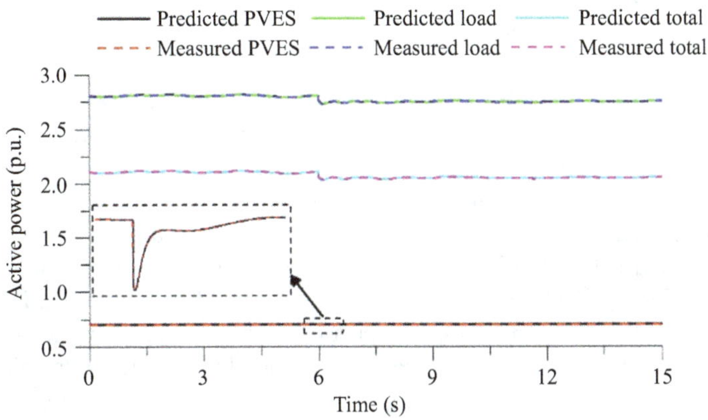

Figure 5.17 Model validation under a sudden load increase (Bus FY)

signal-based load modeling, many different load models can be built at different time points within a period. Such a large number may lead to potential problems. For one thing, although the total number of load models identified is very large, many elements in the load model set may be similar to each other and it is not necessary to keep all of them in the load model set. For another thing, in the simulation cases where load models are applied, that is, power system stability analysis, optimal power flow, and unit commitment, it is neither necessary nor possible to conduct the simulation on all the different load models. Several representative load models are enough to represent the diversity of all the load models. In such situation, similar load models should be represented as one.

Therefore, the method to pick up representative load models from the identified load models' set is needed to settle this problem. In this chapter, clustering method is applied in picking up the representative load models from the large load models' set. Clustering analysis is the task of grouping a set of objects in such a way that the objects in the same cluster are more similar to each other than to those in other clusters, which belongs to the scope of data mining [21]. A postfault response-based clustering method of uncertain load model parameters is proposed in this section. The postfault response-based load model distance definition is given. The K-medoids clustering algorithm is applied for the clustering process and the average cluster radius-based k selection method is proposed. The effectiveness of the proposed clustering method is validated in the simulation case in WSCC 9 bus system. Based on this clustering method, the impact of uncertain load models on power system stability and control can be considered in future research.

5.4.1 Definition of distance

In the clustering process, the distances between different elements play an important role, because the aim of clustering is to pick up the elements with small distances to each other into one cluster. In this section, the definition of load model parameter distance is given.

5.4.1.1 Data for distance calculation: postfault response

Before the calculation of load model parameter distance, what kind of data should be used for distance calculation needs to be pointed out first.

The intuitive choice should be calculating the Euclidean distance between the parameters directly. However, such a choice may lead to potential problems. On one hand, when calculating the Euclidean distance between the parameters, how the weights for X, X', T_{d0}, and R should be given is difficult to decide as these parameters have different degrees of impact on load dynamics. On the other hand, two groups of different load model parameters may have similar dynamic properties and should be clustered into one group, even if the Euclidean distance of the parameters may be very big. Therefore, the data used for the distance calculation of the load model parameters should not be the Euclidean distance between the parameters.

Therefore, the data for distance calculation should be reconsidered. As the aim for clustering load models is to find some representative load models in dynamic performance simulation, the dynamic performance curves can be selected as the data for distance calculation. The load models with similar dynamic performance after the same fault or disturbance can be clustered into one group in this way.

Afterward, one further selection should be made between ambient signal dynamic response and postfault response. Although both can be applied in load model parameter identification, postfault response is more important in power system stability analysis simulation as most instability situations are caused by faults instead of ambient signal. Therefore, postfault response is selected as the data for distance calculation.

In this way, a one to one mapping between the load models' set and the postfault response's set can be built. The distance between two postfault response curves can be regarded as the distance between two load models. In the clustering process, the representative postfault response (RPFR) curves are selected from the postfault response curves' set, and the corresponding load models for the representative postfault response curves are then selected as the representative load models (RLM). This process is given in Figure 5.18.

5.4.1.2 Postfault response generation and distance calculation

To cluster the load models, the postfault response curves under the same fault for all the elements in the uncertain load model set should be generated. The generation process in this chapter is to insert each of the load models in the set into a small simulation system, at the same load bus and the same static operation point, and then to perform the same fault for each element. Then, the postfault response curves generated in this way can be used for the distance calculation.

As stated above, the static operation points for each postfault response simulation case are the same. Further explanation needs to be given about this statement. Since the different load models are inserted into the simulation system one by one at one load bus in the simulation system, the active and reactive power at this load bus cannot both be kept the same because of the difference in load models. Then, when talking about "keep the static operation point the same", it means the generation and load at other buses remain the same.

Then, the next step is to calculate the distance between two time series curves. The Euclidean distance D between the curves are used here, the calculation of which is given as follows:

$$D(y_1, y_2) = \sum_t (y_1(t) - y_2(t))^2 \tag{5.22}$$

where $y_1(t)$ and $y_2(t)$ are the two time series data for distance calculation.

5.4.2 Clustering algorithm

5.4.2.1 K-medoids clustering algorithm

K-medoids clustering algorithm is a typical clustering algorithm, which is based on the partition aiming at selecting k representative elements in a set [22]. The distance between the elements and the representative elements is minimized to ensure

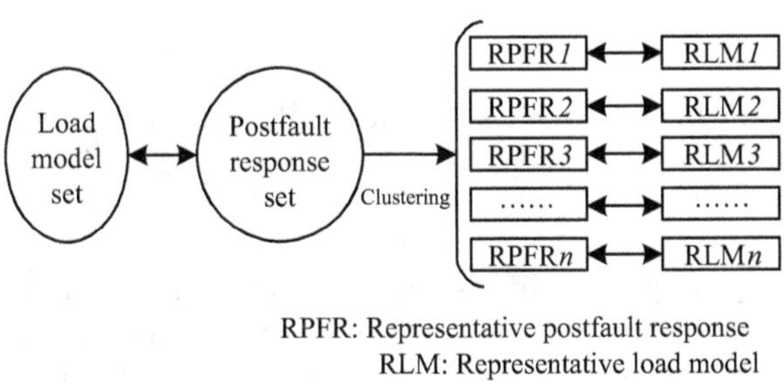

Figure 5.18 Clustering of load models through postfault response

that the objects in the set are divided into the nearest cluster represented by the representative element.

K-medoids clustering algorithm is similar to K-means algorithm. Nevertheless, K-medoids can be regarded as an improvement of K-means for the reason that K-means uses the average of the elements in the cluster as the cluster center while the cluster center in K-medoids is one of the elements in the cluster. Therefore, K-medoids algorithm has a better robustness toward outliers. In this chapter, since a one to one mapping should be built from the RPFR to the RLM, the representative elements should be the original elements in the set. Therefore, K-medoids clustering algorithm is more suitable for solving this problem.

For a set S, with n elements $e_1, e_2, \ldots, e_n$, the basic process for K-medoids is explained as follows, where $D(e_i, e_j)$ represents the distance between e_i and e_j. $C_1, C_2, \ldots, C_k$ represents the k clusters and e_{r_i} represents the ith representative element, also known as the cluster center of C_i.

1. Initialization

   Select k elements from the whole set as the initial cluster centers, namely, $e_{r\_1}, e_{r\_2}, \ldots, e_{r\_k}$.
2. Distance calculation

   Calculate the distance between each element in S and each cluster center, and divide all the elements in S into the nearest cluster based on the distance calculated. In this way, k clusters, namely $C_1, C_2, \ldots, C_k$, is formed. This step can be explained as follows:

$$N_i = \arg\min_{j=1}^{k} D(e_i, e_{r\_j}) \tag{5.23}$$

   where N_i refers to the cluster number to which e_i belongs to.
3. Cluster center update

   For each cluster, calculate the distance between each two elements in the cluster and then calculate the total distance between one element in the cluster and all the other elements in the cluster. Then, choose the element with the minimal total distance to be the updated cluster center. This step can be explained as follows:

$$e_{ri} = \arg\min_{e_j \in C_i} \sum_{e_m \in C_i, m \neq n} D(e_j, e_m) \tag{5.24}$$

4. Finish judgement

   If the updated cluster centers are exactly the same as the ones before updated, then finish the clustering process. Otherwise, repeat (2)–(3) again.

5.4.2.2 Selection of *k*

After introducing the process of K-medoids clustering algorithm, one problem remained to be solved is how many clusters should the elements be divided into. This is a difficult problem in almost all clustering algorithms. This number is often selected manually after the clustering results reviewed by human.

One possible way may help in the selection of k is to judge by the radius of the clusters. The radius of C_i is defined to be R_i as follows:

$$R_i = \max_{e_j \in C_i} D(e_{r\_i}, e_j) \tag{5.25}$$

Then, the average R among all clusters is chosen to be the index to select k. Define AR_m to be the average radius of all clusters when k is chosen to be m and assume k_0 to be the most suitable k. When $m < k_0$, AR_m will decrease very apparently with the increase in m. When m reaches and becomes larger than k_0, which means one perfect cluster will be divided, the decrease speed will be much slower. Then, m can be selected in this way.

5.4.3 Case study

In this chapter, to generate different kinds of postfault responses for load model clustering, the WSCC 9 bus system is used for postfault response generation, the structure of which is given in Figure 5.19. This system contains three generators and three load buses. In the simulation process, two of the load buses, Bus 7 and Bus 8, use constant PQ models, while Bus 9 uses Z+IM model for postfault response record and load model clustering. A PMU is installed on Bus 9 to measure the data which are used for load model parameter identification. The load model parameters on Bus 9 are changed randomly and the postfault responses under different kinds of faults are generated and recorded. The data length used for postfault response analysis is 5.5 s. A three-phase short circuit fault on Bus 4 happens at 1 s and is cut on 1.25 s to generate the postfault response.

In the simulation process, 300 different load models are used for clustering. The method to generate these different load models is to generate 6 groups of 50 load models around 6 basic load models, with some random change for each load model in the group around the basic load models. The 50 random load models are generated by normal distribution, with the basic load models to be the average value and 3% of the parameter values to be the standard deviation. The details for load model generation is given in Table 5.5.

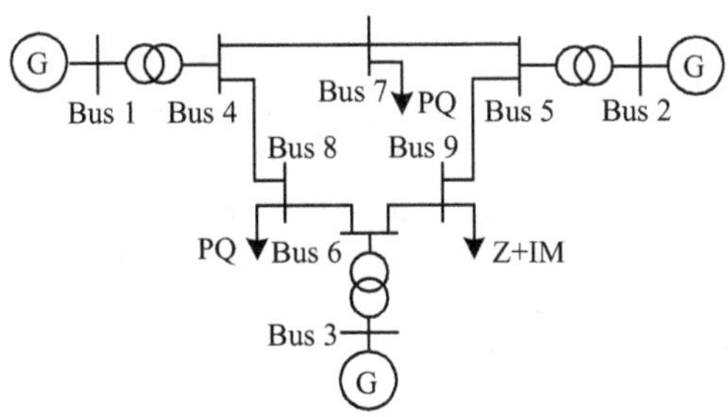

Figure 5.19 Structure of WSCC 9 bus system

Table 5.5 Basic load model parameters

Group	X	X'	T_{d0}	R	Load model number
1	0.73	0.42	0.13	0.46	1–50
2	0.85	0.12	0.22	0.35	51–100
3	0.65	0.38	0.17	0.41	101–150
4	0.80	0.15	0.14	0.36	151–200
5	0.70	0.25	0.22	0.50	201–250
6	0.90	0.35	0.24	0.60	251–300

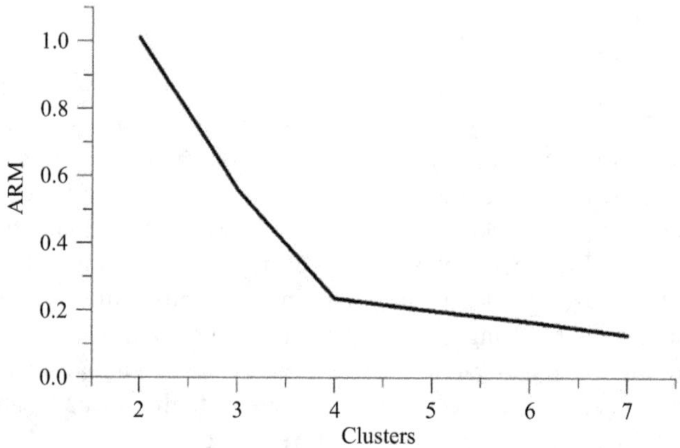

Figure 5.20 Average radius with the change in number of clusters

5.4.3.1 Load model clustering results

First, the number of k should be selected. As described in Section 5.4.2.2, the clustering with k changing from 2 to 7 are conducted and the average radius are recorded as the results in Figure 5.20. It can be observed that when k is smaller than 4, the average radius decreases very fast with the increase in k. On contrast, the decrease speed is much slower when k reaches 5. Therefore, k should be selected to be 4.

The clustering results are given in Figure 5.21. It can be observed that the elements in the same groups in Table 5.5 are all clustered into the same cluster except for only one exception. All the 300 postfault response curves together with the postfault response of the representative load models are given in Figure 5.22. Judging from the curves, the load models being divided into the same clusters do have similar postfault responses during and after the short circuit fault.

In addition, another fact worth attention is that as k is smaller than the number of groups, Group 1 and Group 6, Group 3 and Group 5, are clustered into two same clusters. This means that although the load model parameters are different, the postfault responses are similar between these groups. The postfault response of

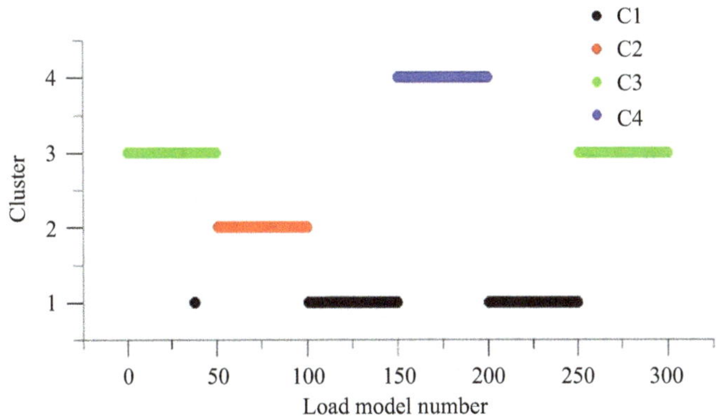

Figure 5.21 Clustering results (k = 4)

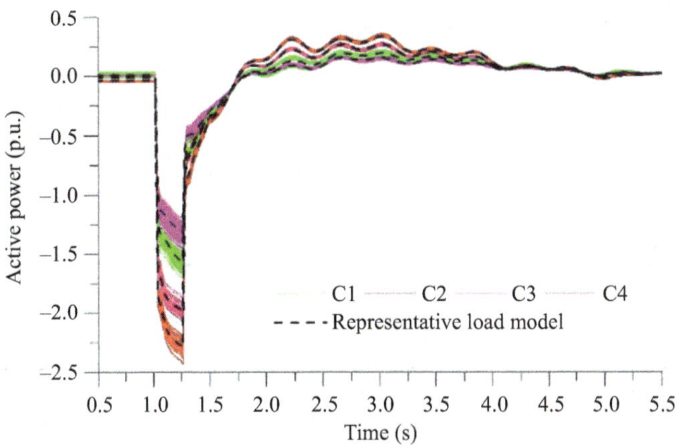

Figure 5.22 Postfault response curves of all 300 load models

these groups is further compared in Figure 5.23. It can be observed that the load models divided into the same cluster do have similar postfault response, although the load model parameters may be different. Therefore, it is reasonable to divide all the load models into four clusters instead of six.

Finally, the postfault response-based load model clustering is compared with the parameter value-based clustering, the results of which is given in Table 5.6. It can be concluded from the data that clustering based on parameter values cannot pick up the load models with similar dynamic properties into the same cluster since the clustering results are different from postfault response-based clustering. As the aim of load modeling is to describe the dynamic performance of load in power

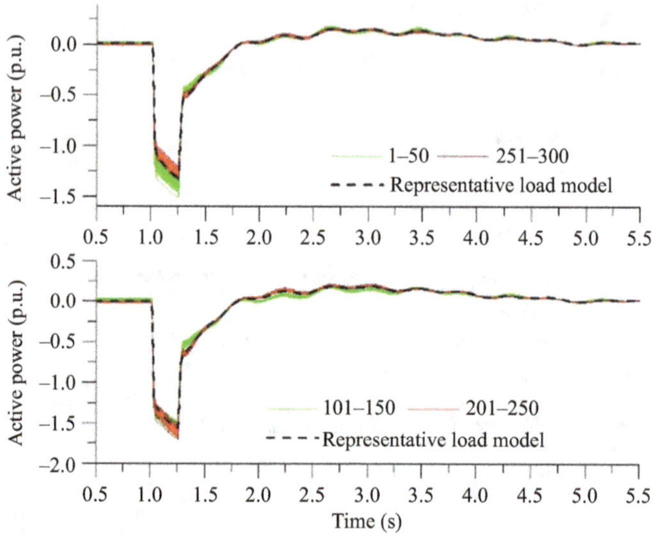

Figure 5.23 Postfault response curves of the models in one cluster

Table 5.6 Load model clustering results comparison

Load model number	**1–50**	**51–100**	**101–150**	**151–200**	**201–250**	**251–300**
By PFR, 4 clusters	3	2	1	4	1	3
By PFR, 6 clusters	3	2	1	4	5	6
By parameters, 4 clusters	1	2	3	4	4	1

PFR, postfault response.

system simulation, postfault response-based clustering can better describe the dynamic properties.

5.4.3.2 Clustering results validation under another fault

Another problem remains to be solved for postfault response-based load model clustering is whether the clustering results under one fault can be applied into another fault. If not, the postfault-based clustering will not have extrapolation and will have a limited application.

Therefore, a three-phase short circuit fault on Bus 6 happens at 3 s and is cut on 3.1 s to generate another group of postfault response, the results of which is given in Figure 5.24. The clustering results used in Figure 5.24 are the results got in Section 5.4.3.1 based on the fault happened on Bus 4. It can be concluded that the

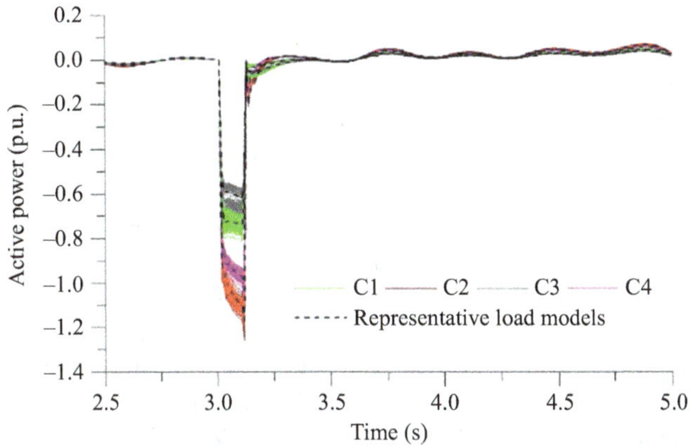

Figure 5.24 Postfault response curves under another fault

clustering results can still be applied and the load models in the same cluster still have similar postfault response even after another fault.

5.5 Conclusion and future work

In this chapter, approaches of load modeling based on synchrophasor measurement in distributed network are discussed. The approaches of building load models and load model structures are introduced first. Then, the method to identify load model parameters from ambient signals is proposed. In addition, the approach to identify the composite demand side model with renewables is discussed, in which PV and ES are included. Finally, the method to pick up representative load models from ambient signal-based load model identification results is proposed.

Following the approaches in this chapter, further work can be considered from the following aspects. First, since the disturbance magnitudes of ambient signals are relatively smaller, the impact of PMU measurement error on load model identification results is not neglectable, which should be analyzed in detail. Second, with the development of machine learning techniques, they can be further applied in load model identification and results processing, for the reason that large amounts of PMU measurement data can be obtained and applied in load modeling. Third, if the load models can be well built from PMU measurement data, the load models can be further applied in power system stability analysis and control, because load dynamics have a noticeable impact on power system stability.

Acknowledgments

This chapter was partially supported by a grant from the Research Grants Council of the Hong Kong Special Administrative Region under Theme-based Research Scheme through Project No. T23-701/14-N.

References

[1] Cui, M., Wang, J., Wang, Y., Diao, R., and Shi, D. "Robust time-varying synthesis load modeling in distribution networks considering voltage disturbances." *IEEE Transactions on Power Systems*. 2019;34(6):4438–4450.

[2] Arif, A., Wang, Z., Wang, J., Mather, B., Bashualdo, H., and Zhao, D. "Load modeling – a review." *IEEE Transactions on Smart Grid*. 2017;**9**(6):5986–99.

[3] Price, W.W., Wirgau, K.A., Murdoch, A., Mitache, J.V., Vaahedi, E., and El-Kady, M. "Load modeling for power flow and transient stability computer studies." *IEEE Transactions on Power Systems*. 1988;**3**(1):180–7.

[4] Son, S., Lee, S.H., Choi, D.-H., *et al.* "Improvement of composite load modeling based on parameter sensitivity and dependency analyses." *IEEE Transactions on Power Systems*. 2014;**29**(1):242–50.

[5] Yuan, R.-F., Ai, Q., and He, X. "Research on dynamic load modelling based on power quality monitoring system." *IET Generation, Transmission & Distribution*. 2013;**7**(1):46–51.

[6] Miranian, A., and Rouzbehi, K. "Nonlinear power system load identification using local model networks." *IEEE Transactions on Power Systems*. 2013; **28**(3):2872–81.

[7] Visconti, I.F., Lima, D.A., Costa, J.M.C.d.S., and Sobrinho N.R.d.B.C. "Measurement-based load modeling using transfer functions for dynamic simulations." *IEEE Transactions on Power Systems*. 2014; **29**(1):111–20.

[8] Li, L., Xie, X., Yan, J., and Han, Y. "Fast online identification of the dominant parameters of composite load model using Volterra model and pattern classification." *Proceedings of IEEE PES General Meeting*, New York, NY: IEEE, 2007, pp. 1–8.

[9] Pierre, J.W., Trudnowski, D.J., and Donnelly, M.K. "Initial results in electromechanical mode identification from ambient data." *IEEE Transactions on Power Systems*. 1997;**12**(3):1245–1.

[10] Zhang, J., Chung, C.Y., Zhang, S., and Han, Y. "Practical wide area damping controller design based on ambient signal analysis." *IEEE Transactions on Power Systems*. 2013;**28**(2):1687–96.

[11] Wu, C., and Zhang, J. "Applicability comparison of three algorithms for electromechanical mode identification." *Proceedings of IEEE PES General Meeting*, New York, NY: IEEE, 2012.

[12] Zhang, X., Lu, C., Han, Y., *et al.* "Ambient signal based load model parameter identification using optimization method." *Proceedings of IEEE PES General Meeting*; New York, NY: IEEE, 2015, pp. 1–5.

[13] Cheng, M.-Y., and Tran, D.-H. "Two-phase differential evolution for the multiobjective optimization of time–cost tradeoffs in resource-constrained construction projects." *IEEE Transactions on Engineering Management*. 2014;**61**(3):450–61.

[14] Rocca, P., Oliverfi, G., and Massa, A. "Differential evolution as applied to electromagnetics." *IEEE Antennas and Propagation Magazine*. 2011; **53**(1):38–49.

[15] Zhang, X., Lu, C., Liu, S., and Wang, X. "A review on wide-area damping control to restrain inter-area low frequency oscillation for large-scale power systems with increasing renewable generation." *Renewable & Sustainable Energy Reviews*. 2016;**57**:45–58.

[16] You, S., Kou, G., Liu, Y., *et al.* "Impact of high PV penetration on the inter-area oscillations in the U.S. Eastern interconnection." *IEEE Access*. 2017;**5**:4361–9.

[17] Pourbeik, P., Williams, S.E., Weber, J., *et al.* "Modeling and dynamic behavior of battery energy storage: a simple model for large-scale time-domain stability studies." *IEEE Electrification Magazine*. 2015;**3**(3):47–51.

[18] Zhang, X., Hill, D.J., and Lu, C. "Identification of composite demand side model with distributed photovoltaic generation and energy storage." *IEEE Transactions on Sustainable Energy*. 2020;11(1):326–336.

[19] Kawabe, K., and Tanaka, K. "Impact of dynamic behavior of photovoltaic power generation systems on short-term voltage stability." *IEEE Transactions on Power Systems*. 2015;**30**(6):3416–24.

[20] Tamimi, B., Canizares, C., and Bhattacharya, K. "System stability impact of large-scale and distributed solar photovoltaic generation: the case of Ontario, Canada." *IEEE Transactions on Sustainable Energy*. 2013;**4**(3):680–8.

[21] Rodriguez, A., and Laio, A. "Clustering by fast search and find of density peaks." *Science*. 2014;**344**(6191):1492–6.

[22] Xu, Y., Yue, G., and Mao, S. "User grouping for massive MIMO in FDD systems: new design methods and analysis." *IEEE Access*. 2014;**2**:947–59.

Part II

Wide-area online dynamics monitoring

Chapter 6

Coherency identification of generators in interconnected power systems with renewables

Zhenzhi Lin[1], Fushuan Wen[1], Yi Ding[1], Yusheng Xue[2] and Shengyuan Liu[1]

Identification of coherent generators (CGs) in power systems is one of the key steps in determining controlled islanding strategies. In this chapter, a wide-area measurement system (WAMS) and agglomerative hierarchical clustering (AHC) algorithm-based coherency identification method is presented for interconnected power systems with aggregated renewable sources. First, the trajectories measured by WAMS are transformed to the center of inertia (COI) based ones for better representing the dynamic behavior of a given power system, and ten dissimilarity indexes are presented for determining the similarity between the trajectories of any two generators. Second, a CRITIC (CRiteria Importance Through Intercriteria Correlation)-based method, in which entropy and the Spearman's rank correlation coefficient are integrated for reflecting the differences and correlations among multiple indexes, respectively, is presented to integrate the trajectory dissimilarity indexes. Next, the AHC algorithm is utilized to identify CGs. Finally, a modified New England–New York interconnected power system (NENYPS) with a large number of renewables, a simplified actual Western Interconnection power system (WECCPS) in North America, and the eastern part of Guangdong power system in China with a recorded oscillation event happened are utilized to demonstrate the proposed wide-area coherency identification methodology.

6.1 Introduction

With the ever-increasing high penetration of renewable generation sources, modern power systems are facing inevitable uncertainties which could lead to security and even stability problems, especially weakly damped oscillations [1]. Meanwhile,

[1]School of Electrical Engineering, Zhejiang University, Hangzhou, P.R. China
[2]State Grid Electric Power Research Institute, Nanjing, P.R. China

with more and more synchronized phasor measurement units (PMUs) deployed in practical power systems, some WAMS-based control strategies are presented and implemented in power system operation. The identification of coherent groups of generators is one of the key steps in determining controlled islanding strategies for preventing power systems from blackout. Thus, correct identification of coherent generators plays a key role in emergency control.

Several methods are available for identifying coherent generators and can be classified into two types, that is, the model-based and measurement-based methods [2]. In the model-based methods, a linearized model (i.e., a differential-algebraic equation set) is utilized to study electromechanical oscillation characteristics and then identify the coherency among generators. This type of method is with high computing efficiency, but the accuracy depends on the models and parameters of the power system concerned [2]. In the measurement-based methods, measured data of a power system are utilized for exhibiting the dynamic characteristics of the power system and following the changes of dynamics in real time, and detailed models and parameters of the power system are not required. In [3], the theory of the Koopman modes is utilized to identify CGs based on spectral analysis of the Koopman operator. A method for identifying CGs is proposed in [4] by using spectrum analysis of the deviations of rotor speeds. A method based on independent component analysis and WAMS is employed to identify CGs in [5]. A wavelet phase difference (WPD) approach is proposed in [6] to identify CGs by observing the movement of low-frequency electromechanical oscillations at different frequency bands and is applied to the European interconnected power system with actual wide-area measurement data. A dynamic mode decomposition (DMD) approach is presented in [7] to identify CGs by interpreting global dynamic behavior in terms of the spatial and temporal patterns associated with dynamic modes. In [8], a graph modeling-based real-time coherency identification methodology is presented by using the instantaneous values measured by WAMS. In [9], a unified wide-area controlled islanding framework is proposed to study "where" and "when" to carry out controlled islanding for power systems. In [2], a projection pursuit (PP)-based wide-area coherency detection method is proposed and employed in the actual China Southern power Grid (CSG). In [10], a WAMS-based coherency identification algorithm based on trajectory similarity and the entropy weight theory is developed, and then an optimization strategy for determining controlled islands presented by using the minimal cutset method. In [11], an intelligent out-of-step predictor for interarea oscillation is presented based on the slope of extended speed-acceleration locus curve. It is observed that most of these methods are used to identify CGs by considering one or two selected indexes directly or indirectly. However, there are many indexes which could reflect the dynamic behavior of power systems and the similarity of the trajectories measured by WAMS. Given this background, multiple indexes for determining the trajectory dissimilarity of generators are presented in this work first, and then CRITIC-based decision-making method and AHC algorithm proposed to identify CGs.

6.2 CRITIC and AHC algorithms for wide-area coherency identification

In a power system with renewable energy sources (such as wind farms), the renewable energy sources could be clustered into one or more equivalent renewable energy sources by a dynamic equivalent algorithm [12]. Hence, PMUs can be installed at the point with renewable energy sources connected to the power system concerned (i.e., the equivalent point of renewable energy sources), and then the equivalent renewable energy source could be regarded as a conventional generator in the WAMS-based coherency identification procedure. For a power system with multiple generators including equivalent renewable energy sources, the COI is defined as an assumed dynamic position measured by the weighted average of the instantaneous rotor angle or rotor speed of all synchronous generators in the power system, and can well represent the dynamic behaviors of the whole power system and coherent generator clusters [1]. In this work, the COI-based generators' trajectories are utilized to identify CGs in a given power system.

For a series of generators' trajectories $\mathbf{T_r} = \{T_{r1}, T_{r2}, T_{r3}, \ldots, T_{rm}, \ldots, T_{rM}\}$ $(1 \leq m \leq M)$ measured by WAMS, where M denotes the number of all generators and T_{rm} denotes the trajectory of the mth generator. Thus the original trajectories $\mathbf{T_r}$ could be transformed to the COI-based ones $\mathbf{T_r^{COI}}$, and T_{rm} in the COI coordinate could be described as $T_{rm}^{\text{COI}} = \{P_{m,1}^{\text{COI}}, P_{m,2}^{\text{COI}}, \ldots, P_{m,n}^{\text{COI}}, \ldots, P_{m,N}^{\text{COI}}\}$, where N denotes the number of the points sampled in each trajectory and $P_{m,n}^{\text{COI}}$ denotes the value of the nth point in the mth rotor angle or speed trajectory and expressed by the COI coordinate point $(T_n, \delta_{m,n}^{\text{COI}}$ or $\omega_{m,n}^{\text{COI}})$. T_n is the nth sampling time. $\delta_{m,n}^{\text{COI}}$ and $\omega_{m,n}^{\text{COI}}$ are respectively the nth values sampled in the COI-based rotor angle and rotor speed for the mth generator, and are represented by

$$\delta_{m,n}^{\text{COI}} = \delta_{m,n} - \frac{1}{\sum_{m=1}^{M} H_m} \sum_{m=1}^{M} H_m \delta_{m,n} \tag{6.1}$$

$$\omega_{m,n}^{\text{COI}} = \omega_{m,n} - \frac{1}{\sum_{m=1}^{M} H_m} \sum_{m=1}^{M} H_m \omega_{m,n} \tag{6.2}$$

where $\delta_{m,n}$, $\omega_{m,n}$, and H_m are respectively the rotor angle, rotor speed, and inertia constant of the mth generator.

6.2.1 COI-based trajectory dissimilarity for identifying CGs

By employing the COI-based rotor angle and speed trajectories, five types of indexes (with total ten indexes) for coherency identification are defined, including the location deviation, amplitude deviation, corner deviation, swing direction, and correlation dissimilarities of the trajectories [10,13,14].

1. Location deviation dissimilarity of trajectories

   In a power system, the ith and jth generators are regarded as CGs in the time interval $[0, T_N]$ if the following equation holds:

$$\max_{t\in[0,T_N]} \left|\Delta\delta_i - \Delta\delta_j\right| \le \varepsilon \tag{6.3}$$

   where ε is a given small positive constant; $\Delta\delta_i$ and $\Delta\delta_j$ are respectively rotor angle increments of the ith and jth generators. It can be seen from (6.3) that the relative distance of two trajectories is one of the most important criteria for identifying the coherency of generators, but the coherent results depend on ε to a great extent. In mathematics, the Hausdorff distance is usually utilized to measure how far two trajectories are from each other and is equal to the maximum one among all the distances from a point in one trajectory to the closest point in the other. Thus, the Hausdorff distance is utilized to measure the closeness of two angle trajectories and defined as

$$H\left(\delta_i^{\mathrm{COI}}, \delta_j^{\mathrm{COI}}\right) = \max\left\{h\left(\delta_i^{\mathrm{COI}}, \delta_j^{\mathrm{COI}}\right), h\left(\delta_j^{\mathrm{COI}}, \delta_i^{\mathrm{COI}}\right)\right\} \tag{6.4}$$

$$h\left(\delta_i^{\mathrm{COI}}, \delta_j^{\mathrm{COI}}\right) = \max_{1\le n_i\le N}\left\{\min_{1\le n_j\le N}\left[d\left(\delta_{i,n_i}^{\mathrm{COI}}, \delta_{j,n_j}^{\mathrm{COI}}\right)\right]\right\} \tag{6.5}$$

   where $h\left(\delta_i^{\mathrm{COI}}, \delta_j^{\mathrm{COI}}\right)$ is the Hausdorff distance of the trajectories of the ith generator to the jth one; $d\left(\delta_{i,n_i}^{\mathrm{COI}}, \delta_{j,n_j}^{\mathrm{COI}}\right)$ is the Euclidean metric between $\delta_{i,n_i}^{\mathrm{COI}}$ and $\delta_{j,n_j}^{\mathrm{COI}}$, where $\delta_{i,n_i}^{\mathrm{COI}}$ and $\delta_{j,n_j}^{\mathrm{COI}}$ are respectively the n_ith and n_jth values sampled in the ith and jth COI-based trajectories of rotor angles. Thus, the location deviation dissimilarity of any two trajectories can be defined as

$$S_1(i,j) = H\left(\delta_i^{\mathrm{COI}}, \delta_j^{\mathrm{COI}}\right) - d\left(\delta_{i,1}^{\mathrm{COI}}, \delta_{j,1}^{\mathrm{COI}}\right) \tag{6.6}$$

   where $d\left(\delta_{i,1}^{\mathrm{COI}}, \delta_{j,1}^{\mathrm{COI}}\right)$ is the Euclidean metric between $\delta_{i,1}^{\mathrm{COI}}$ and $\delta_{j,1}^{\mathrm{COI}}$, where $\delta_{i,1}^{\mathrm{COI}}$ and $\delta_{j,1}^{\mathrm{COI}}$ are respectively the first values sampled in the ith and jth COI-based trajectories of rotor angles.

   Similarly, the location deviation dissimilarity of the COI-based speed trajectory can be defined as

$$S_2(i,j) = H\left(\omega_i^{\mathrm{COI}}, \omega_j^{\mathrm{COI}}\right) - d\left(\omega_{i,1}^{\mathrm{COI}}, \omega_{j,1}^{\mathrm{COI}}\right) \tag{6.7}$$

$$H\left(\omega_i^{\mathrm{COI}}, \omega_j^{\mathrm{COI}}\right) = \max\left\{h\left(\omega_i^{\mathrm{COI}}, \omega_j^{\mathrm{COI}}\right), h\left(\omega_j^{\mathrm{COI}}, \omega_i^{\mathrm{COI}}\right)\right\} \tag{6.8}$$

$$h\left(\omega_i^{\mathrm{COI}}, \omega_j^{\mathrm{COI}}\right) = \max_{1\le n_i\le N}\left\{\min_{1\le n_j\le N}\left[d\left(\omega_{i,n_i}^{\mathrm{COI}}, \omega_{j,n_j}^{\mathrm{COI}}\right)\right]\right\} \tag{6.9}$$

   where $d\left(\omega_{i,1}^{\mathrm{COI}}, \omega_{j,1}^{\mathrm{COI}}\right)$ is the Euclidean metric between $\omega_{i,1}^{\mathrm{COI}}$ and $\omega_{j,1}^{\mathrm{COI}}$, where $\omega_{i,1}^{\mathrm{COI}}$ and $\omega_{j,1}^{\mathrm{COI}}$ are respectively the first values sampled in the ith and jth

COI-based trajectories of rotor speeds; $H\left(\omega_i^{\mathrm{COI}}, \omega_j^{\mathrm{COI}}\right)$ is the Hausdorff distance of the ith and jth trajectories of rotor speeds; $h\left(\omega_i^{\mathrm{COI}}, \omega_j^{\mathrm{COI}}\right)$ is the Hausdorff distance of the speed trajectory of the ith generator to the jth one; $d\left(\omega_{i,n_i}^{\mathrm{COI}}, \omega_{j,n_j}^{\mathrm{COI}}\right)$ is the Euclidean metric between $\omega_{i,n_i}^{\mathrm{COI}}$ and $\omega_{j,n_j}^{\mathrm{COI}}$, where $\omega_{i,n_i}^{\mathrm{COI}}$ and $\omega_{j,n_j}^{\mathrm{COI}}$ are respectively the n_ith and n_jth values sampled in the ith and jth COI-based trajectories of rotor speeds.

2. Amplitude deviation dissimilarity of trajectories

In actual power systems, the amplitude deviations of rotor angles or speeds among generators are also usually utilized to identify CGs. Thus, the angle amplitude deviation dissimilarity between any two generators' trajectories of rotor angles is defined for reflecting their coherency. For the ith and jth generators' trajectories of rotor angles, the angle amplitude deviation dissimilarity between them can be represented by

$$S_3(i,j) = \sqrt{\sum_{n=1}^{N}\left[\left(\delta_{i,n}^{\mathrm{COI}} - \delta_{j,n}^{\mathrm{COI}}\right) - \left(\delta_{i,1}^{\mathrm{COI}} - \delta_{j,1}^{\mathrm{COI}}\right)\right]^2} \tag{6.10}$$

where $\delta_{i,1}^{\mathrm{COI}}$ and $\delta_{j,1}^{\mathrm{COI}}$ are respectively the first values sampled in the ith and jth COI-based trajectories of rotor angles.

Similarly, the speed amplitude deviation dissimilarity is defined as the difference between any two generators' trajectories of rotor speeds and is represented by

$$S_4(i,j) = \sqrt{\sum_{n=1}^{N}\left(\omega_{i,n}^{\mathrm{COI}} - \omega_{j,n}^{\mathrm{COI}}\right)^2} \tag{6.11}$$

3. Corner deviation dissimilarity of trajectories

For a trajectory, its inner direction changes at any point sampled are described as the deviation between two neighboring intervals and is represented by

$$\beta_{i,n} = \begin{cases} \arccos\left(\dfrac{\overrightarrow{P_{i,n-1}^{\mathrm{COI}}P_{i,n}^{\mathrm{COI}}}\cdot\overrightarrow{P_{i,n}^{\mathrm{COI}}P_{i,n+1}^{\mathrm{COI}}}}{\|\overrightarrow{P_{i,n-1}^{\mathrm{COI}}P_{i,n}^{\mathrm{COI}}}\|\|\overrightarrow{P_{i,n}^{\mathrm{COI}}P_{i,n+1}^{\mathrm{COI}}}\|}\right), \\ if\, \delta_{i,n-1}^{\mathrm{COI}} + \delta_{i,n+1}^{\mathrm{COI}} \geq 2\delta_{i,n}^{\mathrm{COI}} \text{ or } \omega_{i,n-1}^{\mathrm{COI}} + \omega_{i,n+1}^{\mathrm{COI}} \geq 2\omega_{i,n}^{\mathrm{COI}} \\ -\arccos\left(\dfrac{\overrightarrow{P_{i,n-1}^{\mathrm{COI}}P_{i,n}^{\mathrm{COI}}}\cdot\overrightarrow{P_{i,n}^{\mathrm{COI}}P_{i,n+1}^{\mathrm{COI}}}}{\|\overrightarrow{P_{i,n-1}^{\mathrm{COI}}P_{i,n}^{\mathrm{COI}}}\|\|\overrightarrow{P_{i,n}^{\mathrm{COI}}P_{i,n+1}^{\mathrm{COI}}}\|}\right), \\ if\, \delta_{i,n-1}^{\mathrm{COI}} + \delta_{i,n+1}^{\mathrm{COI}} < 2\delta_{i,n}^{\mathrm{COI}} \text{ or } \omega_{i,n-1}^{\mathrm{COI}} + \omega_{i,n+1}^{\mathrm{COI}} < 2\omega_{i,n}^{\mathrm{COI}} \end{cases} \tag{6.12}$$

where $\overrightarrow{P_{i,n-1}^{\mathrm{COI}}P_{i,n}^{\mathrm{COI}}}$ and $\overrightarrow{P_{i,n}^{\mathrm{COI}}P_{i,n+1}^{\mathrm{COI}}}$ are the vectors from (T_{n-1}, $\delta_{i,n-1}^{\mathrm{COI}}$ or $\omega_{i,n-1}^{\mathrm{COI}}$) and ($T_n$, $\delta_{i,n}^{\mathrm{COI}}$ or $\omega_{i,n}^{\mathrm{COI}}$) to ($T_n$, $\delta_{i,n}^{\mathrm{COI}}$ or $\omega_{i,n}^{\mathrm{COI}}$) and ($T_{n+1}$, $\delta_{i,n+1}^{\mathrm{COI}}$ or $\omega_{i,n+1}^{\mathrm{COI}}$), respectively. It can be seen from (6.12) that $\beta_{i,n}$ is relative to intersection angles between $\overrightarrow{P_{i,n-1}^{COI}P_{i,n}^{COI}}$ and $\overrightarrow{P_{i,n}^{COI}P_{i,n+1}^{COI}}$. The value of $\beta_{i,n}$ is positive if $\delta_{i,n+1}^{COI}-\delta_{i,n}^{COI}\geq\delta_{i,n}^{COI}-\delta_{i,n-1}^{COI}$ or $\omega_{i,n+1}^{COI}-\omega_{i,n}^{COI}\geq\omega_{i,n}^{COI}-\omega_{i,n-1}^{COI}$ (i.e., $\delta_{i,n-1}^{COI}+\delta_{i,n+1}^{COI}\geq 2\delta_{i,n}^{COI}$ or $\omega_{i,n-1}^{COI}+\omega_{i,n+1}^{COI}\geq 2\omega_{i,n}^{COI}$); otherwise, the value of $\beta_{i,n}$ is negative.

Thus, two dissimilarity indexes for describing the corner deviation dissimilarity between the ith and jth generators' trajectories of rotor angles or rotor speeds are defined for measuring the differences of inner direction changes at all sampled points, respectively. The indexes of corner deviation dissimilarities between the ith and jth generators' trajectories of rotor angles and rotor speeds are respectively represented by

$$S_5(i,j)=\sum_{n=2}^{N-1}\frac{1}{N-2}\left|\beta_{i,n}^{\delta^{COI}}-\beta_{j,n}^{\delta^{COI}}\right| \tag{6.13}$$

$$S_6(i,j)=\sum_{n=2}^{N-1}\frac{1}{N-2}\left|\beta_{i,n}^{\omega^{COI}}-\beta_{j,n}^{\omega^{COI}}\right| \tag{6.14}$$

where $\beta_{i,n}^{\delta^{COI}}$, $\beta_{j,n}^{\delta^{COI}}$, $\beta_{i,n}^{\omega^{COI}}$, and $\beta_{j,n}^{\omega^{COI}}$ are the nth values of inner direction changes for the ith and jth trajectories of rotor angles and speeds, respectively.

4. Swing direction dissimilarity of trajectories

The swing direction dissimilarity is generally utilized to measure the oscillation direction deviations between any two trajectories. Thus, the swing direction dissimilarity between any two trajectories T_{ri}^{COI} and T_{rj}^{COI} at sampling point n can be expressed as

$$\vartheta_{ij,n}=\arccos\left(\frac{\overrightarrow{P_{i,n}P_{i,n+1}}\cdot\overrightarrow{P_{j,n}P_{j,n+1}}}{\|\overrightarrow{P_{i,n}P_{i,n+1}}\|\|\overrightarrow{P_{j,n}P_{i,n+1}}\|}\right),1\leq n\leq N-1 \tag{6.15}$$

where $\overrightarrow{P_{i,n}P_{i,n+1}}$ and $\overrightarrow{P_{j,n}P_{i,n+1}}$ are respectively the vectors of two trajectories T_{ri}^{COI} and T_{rj}^{COI} from sampling points n to $n+1$.

For the rotor angle trajectories, the swing direction dissimilarity of any two angle trajectories is formulated as

$$S_7(i,j)=\frac{1}{N-1}\sum_{n=2}^{N}\vartheta_{ij,n}^{\delta^{COI}} \tag{6.16}$$

where $\vartheta_{ij,n}^{\delta^{COI}}$ is the swing direction dissimilarity at the sampling point n between the trajectories T_{ri}^{COI} and T_{rj}^{COI} of rotor angles.

For the rotor speed trajectories, the swing direction dissimilarity of any two-speed trajectories is formulated as

$$S_8(i,j)=\frac{1}{N-1}\sum_{n=2}^{N}\vartheta_{ij,n}^{\omega^{COI}} \tag{6.17}$$

where $\vartheta_{i,j,n}^{\omega^{\mathrm{COI}}}$ is the swing direction dissimilarity at the sampling point n between the trajectories T_{ri}^{COI} and T_{rj}^{COI} of rotor speeds.

5. Correlation dissimilarity of trajectories

   The coherency of generators can be evaluated based on the correlation coefficient [11] because it shows the strength of linear association between the trajectories. In statistics, the Pearson's correlation coefficient ρ is utilized to measure the linear dependence between two variables and is equal to the covariance of the two variables divided by the product of their standard deviations. Thus, the correlation dissimilarity of the rotor angle trajectory can be defined as

$$S_9(i,j) = \rho_{\delta_i^{\mathrm{COI}},\delta_j^{\mathrm{COI}}} = \frac{\mathrm{cov}\left(\delta_i^{\mathrm{COI}},\delta_j^{\mathrm{COI}}\right)}{\sigma_{\delta_i^{\mathrm{COI}}}\sigma_{\delta_j^{\mathrm{COI}}}} = \frac{N\sum_{n=1}^{N}\left(\delta_{i,n}^{\mathrm{COI}}\delta_{j,n}^{\mathrm{COI}}\right) - \left(\sum_{n=1}^{N}\delta_{i,n}^{\mathrm{COI}}\right)\left(\sum_{n=1}^{N}\delta_{j,n}^{\mathrm{COI}}\right)}{\sqrt{N\sum_{n=1}^{N}\left(\delta_{i,n}^{\mathrm{COI}}\right)^2 - \left(\sum_{n=1}^{N}\delta_{i,n}^{\mathrm{COI}}\right)^2}\sqrt{N\sum_{n=1}^{N}\left(\delta_{j,n}^{\mathrm{COI}}\right)^2 - \left(\sum_{n=1}^{N}\delta_{j,n}^{\mathrm{COI}}\right)^2}} \tag{6.18}$$

where $\mathrm{cov}(\delta_i^{\mathrm{COI}},\delta_j^{\mathrm{COI}})$ is the covariance of the vectors δ_i^{COI} and δ_j^{COI}; δ_i^{COI} and δ_j^{COI} are the vectors of the trajectories T_{ri}^{COI} and T_{rj}^{COI} of rotor angles; $\sigma_{\delta_i^{\mathrm{COI}}}$ and $\sigma_{\delta_j^{\mathrm{COI}}}$ are the standard deviations of the vectors δ_i^{COI} and δ_j^{COI}, respectively.

   Similarly, the correlation dissimilarity of the rotor speed trajectory can be defined as

$$S_{10}(i,j) = \rho_{\omega_i^{\mathrm{COI}},\omega_j^{\mathrm{COI}}} = \frac{\mathrm{cov}\left(\omega_i^{\mathrm{COI}},\omega_j^{\mathrm{COI}}\right)}{\sigma_{\omega_i^{\mathrm{COI}}}\sigma_{\omega_j^{\mathrm{COI}}}} = \frac{N\sum_{n=1}^{N}\left(\omega_{i,n}^{\mathrm{COI}}\omega_{j,n}^{\mathrm{COI}}\right) - \left(\sum_{n=1}^{N}\omega_{i,n}^{\mathrm{COI}}\right)\left(\sum_{n=1}^{N}\omega_{j,n}^{\mathrm{COI}}\right)}{\sqrt{N\sum_{n=1}^{N}\left(\omega_{i,n}^{\mathrm{COI}}\right)^2 - \left(\sum_{n=1}^{N}\omega_{i,n}^{\mathrm{COI}}\right)^2}\sqrt{N\sum_{n=1}^{N}\left(\omega_{j,n}^{\mathrm{COI}}\right)^2 - \left(\sum_{n=1}^{N}\omega_{j,n}^{\mathrm{COI}}\right)^2}} \tag{6.19}$$

where $\mathrm{cov}(\omega_i^{\mathrm{COI}},\omega_j^{\mathrm{COI}})$ is the covariance of the vectors ω_i^{COI} and ω_j^{COI}; ω_i^{COI} and ω_j^{COI} are the vectors of the trajectories T_{ri}^{COI} and T_{rj}^{COI} of rotor speeds; $\sigma_{\omega_i^{\mathrm{COI}}}$ and $\sigma_{\omega_j^{\mathrm{COI}}}$ are the standard deviations of the vectors ω_i^{COI} and ω_j^{COI}, respectively.

   It can be seen from (6.6), (6.7), (6.10), (6.11), (6.13), (6.14), and (6.16)–(6.19) that ten dissimilarity indexes include the location deviation dissimilarity S_1 and S_2, amplitude deviation dissimilarity S_3 and S_4, corner deviation dissimilarity S_5 and S_6, swing direction dissimilarity S_7 and S_8, correlation dissimilarity S_9 and S_{10} of the COI-based rotor angle and speed trajectories, respectively. After these ten dissimilarity indexes for evaluating trajectory dissimilarity are attained, the trajectory dissimilarity identification is converted to a multicriteria decision-making (MCDM) problem. In this chapter, the CRITIC-based decision-making method is proposed for trajectory dissimilarity identification.

6.2.2 Trajectory dissimilarity decision-making based on the CRITIC method

For an M-generator power system, ten $M \times M$ matrices $\mathbf{S_k}$ of all trajectories can be attained, and $\mathbf{S_k} = \ldots$, where $k \in \{1, 2, \ldots, K\}$ and K is equal to 10 in this work. These ten dissimilarity indexes' matrices are symmetrical. However, they cannot be compared with each other directly due to dimensional problems. Thus, the normalization of these ten indexes should be done first. For indexes S_1, S_2, S_3, S_4, S_5, S_6, S_7, and S_8, the smaller the index value is, the less dissimilarity it will be (i.e., the more similarity it will be). For indexes S_9 and S_{10}, the larger the index value is, the less dissimilarity it will be. Based on the ideal point concept, the normalization of ten dissimilarity indexes can be formulated as

$$S_k^*(i,j) = \frac{S_k(i,j) - \min\limits_{i,j\in\{1,2,\ldots,M\}}\{S_k(i,j)\}}{\max\limits_{i,j\in\{1,2,\ldots,M\}}\{S_k(i,j)\} - \min\limits_{i,j\in\{1,2,\ldots,M\}}\{S_k(i,j)\}}, \quad k \in \{1,2,\ldots,8\} \tag{6.20}$$

$$S_k^*(i,j) = \frac{\max\limits_{i,j\in\{1,2,\ldots,M\}}\{S_k(i,j)\} - S_k(i,j)}{\max\limits_{i,j\in\{1,2,\ldots,M\}}\{S_k(i,j)\} - \min\limits_{i,j\in\{1,2,\ldots,M\}}\{S_k(i,j)\}}, \quad k \in \{9,10\} \tag{6.21}$$

where $S_k^*(i,j)$ denotes the normalized value of the kth index between the trajectories T_{ri}^{COI} and T_{rj}^{COI} of rotor angles or speeds.

Then, ten normalized indexes of a pair of trajectories (i, j) could be integrated as a synthesized dissimilarity index as

$$S_{SD}^*(i,j) = \sum_{k=1}^{K} w_k S_k^*(i,j) \tag{6.22}$$

where w_k is the weight of the kth index for identifying the similarity between two trajectories.

If two generators' trajectories are similar, the synthesized dissimilarity $S_{SD}^*(i,j)$ will have a small index value, and this means that these two generators can be classified into one coherent group; otherwise, two generators should be classified into different coherent groups. Thus, the synthesized dissimilarity index matrix $\mathbf{S_{SD}^*}$ can be utilized to identify CGs. Because $S_{SD}^*(i,j) = S_{SD}^*(j,i)$, the number of all generator pairs in $\mathbf{S_{SD}^*}$ is equal to $M(M-1)/2$. It can be seen from (6.22) that $\mathbf{S_{SD}^*}$ is impacted by ten dissimilarity matrices $\mathbf{S_k^*}$ and the index weights w_k to be determined, where $k \in \{1, 2, \ldots, K\}$. Different decision-making results may be attained if different weights are employed. Thus, the trajectory dissimilarity decision-making depends on reasonable and appropriate index weights to be assigned. Normally, the determination of weights can be done by domain experts in power systems according to their professional experience and knowledge. However, unreasonable weights could be assigned to the dissimilarity indexes if domain experts are not familiar with the given power system, and then the similarity of the trajectories cannot be correctly identified.

In this chapter, the CRITIC method, which can determine objective weights of relative importance for MCDM problems, is proposed to solve the trajectory dissimilarity. The CRITIC method could incorporate both contrast intensity and conflict which are contained in the structure of the decision-making problem [15]. In the CRITIC method, the entropy, which can measure the relative importance of each evaluating index, is utilized to measure the contrast intensity, and the Spearman's rank correlation coefficient, which is a nonparametric measure of the rank correlation, is used to measure the conflict (or correlation).

For determining the weights in (6.22) based on the CRITIC method, (6.22) should be converted to a multicriteria decision-making model which has $M(M-1)/2$ schemes (i.e., the coherency identification on all pairs of trajectories) and K indexes for evaluation. Thus, the converted decision-making matrix can be represented by

$$\mathbf{U}_{\mathbf{SD}}^{\mathbf{T}} = \left(u_{qk}\right)_{Q\times K} = \begin{bmatrix} u_{11} & u_{12} & \dots & u_{1K} \\ u_{21} & u_{22} & \dots & u_{2K} \\ \dots & \dots & \dots & \dots \\ u_{Q1} & u_{Q2} & \dots & u_{QK} \end{bmatrix} \tag{6.23}$$

where Q denotes the number of the schemes for evaluation, and its value is $M(M-1)/2$; $q \in \{1, 2, \dots, Q\}$; u_{qk} is the qth element of the upper triangular matrix of the kth trajectory dissimilarity matrix with diagonal elements removed.

6.2.2.1 Entropy for contrast intensity

In statistical mechanics, disorder could be measured by entropy. The entropy is also utilized to measure the information disorder in information theory [10]. In this chapter, the results attained include the values of ten trajectory dissimilarity indexes. Because there are disorder values for ten trajectory dissimilarity indexes, the entropy could be applied to compute the disorder degree among ten indexes. The entropy H_k for the kth index can be formulated as

$$H_k = -\kappa \sum_{q=1}^{Q} f_{qk} \ln f_{qk} \qquad k = 1, 2, \dots, K \tag{6.24}$$

where $f_{qk} = u_{qk}/\sum_{q=1}^{Q} u_{qk}$, and $\kappa = 1/\ln Q$. If $f_{qk} = 0$, $f_{qk} \ln f_{qk}$ is supposed to be zero.

The smaller the entropy is, the more different the values of a trajectory dissimilarity index are, and the more information the index provides. Thus, this trajectory dissimilarity index is supposed to be paid more attention. In this chapter, $1 - H_k$ is utilized to demonstrate the relative importance of each evaluating index.

6.2.2.2 Spearman's rank correlation coefficient for conflict

In statistics, Spearman's rank correlation coefficient is utilized to measure the correlation of the rank between two variables and to evaluate their monotonic relationships [16]. If there are no same values in any of the variables and one variable can be represented by a strictly monotonic function of the other one, the Spearman's rank correlation coefficient of these two variables could be equal to 1 or −1. In this chapter, Spearman's rank correlation coefficient is utilized to evaluate the correlation of any of two column vectors of the trajectory dissimilarity indexes. It is assumed that $\mathbf{U_X} = [u_{1X}, u_{2X}, \ldots, u_{QX}]^{\mathrm{T}}$ and $\mathbf{U_Y} = [u_{1Y}, u_{2Y}, \ldots, u_{QY}]^{\mathrm{T}}$ are two column vectors of $\mathbf{U}_{\mathbf{SD}}^{\mathbf{T}}$, where $X \in \{1, 2, \ldots, K\}$ and $Y \in \{1, 2, \ldots, K\}$; and $\mathbf{U}_{\mathbf{X}}^{\mathbf{R}} = [u_{1X}^R, u_{2X}^R, \ldots, u_{QX}^R]^{\mathrm{T}}$ and $\mathbf{U}_{\mathbf{Y}}^{\mathbf{R}} = [u_{1Y}^R, u_{2Y}^R, \ldots, u_{QY}^R]^{\mathrm{T}}$ are their rank column vectors, respectively. The Spearman's rank correlation coefficient of these two column vectors of the trajectory dissimilarity indexes is defined as

$$r_{XY} = \frac{\mathrm{cov}\left(\mathbf{U}_{\mathbf{X}}^{\mathbf{R}}, \mathbf{U}_{\mathbf{Y}}^{\mathbf{R}}\right)}{\sigma_{\mathbf{U}_{\mathbf{X}}^{\mathbf{R}}}\sigma_{\mathbf{U}_{\mathbf{Y}}^{\mathbf{R}}}} = \frac{\sum_{i=1}^{Q}\left(u_{iX}^R - \overline{u_X^R}\right)\left(u_{iY}^R - \overline{u_Y^R}\right)}{\sqrt{\sum_{i=1}^{Q}\left(u_{iX}^R - \overline{u_X^R}\right)^2 + \sum_{i=1}^{Q}\left(u_{iY}^R - \overline{u_Y^R}\right)^2}} \tag{6.25}$$

where $\mathrm{cov}\left(\mathbf{U}_{\mathbf{X}}^{\mathbf{R}}, \mathbf{U}_{\mathbf{Y}}^{\mathbf{R}}\right)$ is the covariance of the rank vectors $\mathbf{U}_{\mathbf{X}}^{\mathbf{R}}$ and $\mathbf{U}_{\mathbf{Y}}^{\mathbf{R}}$; $\sigma_{\mathbf{U}_{\mathbf{X}}^{\mathbf{R}}}$ and $\sigma_{\mathbf{U}_{\mathbf{Y}}^{\mathbf{R}}}$ are the standard deviations of the rank vectors $\mathbf{U}_{\mathbf{X}}^{\mathbf{R}}$ and $\mathbf{U}_{\mathbf{Y}}^{\mathbf{R}}$, respectively; and $\overline{u_X^R}$ and $\overline{u_Y^R}$ are the average value of the rank vectors $\mathbf{U}_{\mathbf{X}}^{\mathbf{R}}$ and $\mathbf{U}_{\mathbf{Y}}^{\mathbf{R}}$, respectively.

If there are no same values in each column vectors of the indexes $\mathbf{U_X}$ and $\mathbf{U_Y}$, all the elements of their rank vectors $\mathbf{U}_{\mathbf{X}}^{\mathbf{R}}$ and $\mathbf{U}_{\mathbf{Y}}^{\mathbf{R}}$ would be different integers, and then the defined Spearman's rank correlation coefficient in (6.25) could be rewritten as

$$r_{XY} = 1 - \frac{6\sum_{i=1}^{Q}\theta_i^2}{Q(Q^2 - 1)} \tag{6.26}$$

where $\theta_i = u_{iX}^R - u_{iY}^R$ is the difference between the ith elements of the rank vectors $\mathbf{U}_{\mathbf{X}}^{\mathbf{R}}$ and $\mathbf{U}_{\mathbf{Y}}^{\mathbf{R}}$.

6.2.2.3 Objective weight with the integration of entropy and Spearman's rank correlation coefficient

The objective weight is determined by the CRITIC method with entropy and Spearman's rank correlation coefficient integrated and is formulated as

$$C_k = (1 - H_k)\sum_{k'=1}^{K}(1 - r_{kk'}) \tag{6.27}$$

where $\sum_{k'=1}^{K}(1 - r_{kk'})$ denotes the correlation coefficient ρ_k^S of the kth index relative to the other $K - 1$ indexes.

Then, the objective weight should be normalized as

$$w_k = \frac{C_k}{\sum_{k'=1}^{K} C_{k'}} \tag{6.28}$$

It can be seen from (6.28) that the kth index weight w_k after normalization is within [0, 1] and $\sum_{k=1}^{K} w_k = 1$.

6.2.3 Steps of wide-area coherency identification based on the CRITIC method and AHC algorithm

As mentioned above, the generators' trajectory dissimilarity evaluation is described as an MCDM problem with the weights determined by the CRITIC method. After the evaluation results, the generators' trajectory dissimilarity is attained, the next step is to identify CGs. The AHC algorithm [13] is adopted to identify CGs in this chapter. Every synchronous generator in power systems is first divided into one cluster, and then the most coherent synchronous generators (i.e., these with the smallest dissimilarity or the highest similarity) are classified into one CG. Thus, the steps of the CRITIC and AHC-based methods for identifying all CGs are described in the following.

1. Set the initial CG number N_{CG} to zero.
2. Divide every generator into one cluster (i.e., one CG).
3. Input the wide-area measurement data associated with the synchronous generators (i.e., the generators' trajectories).
4. Convert the WAMS data to the COI-based rotor angles and rotor speeds of the trajectories.
5. Calculate ten COI-based trajectory dissimilarities shown in (6.6), (6.7), (6.10), (6.11), (6.13), (6.14), and (6.16)–(6.19), and then obtain ten dissimilarity matrices $\mathbf{S_k}$, where $k \in \{1, 2, \ldots, K\}$.
6. Normalize $\mathbf{S_k}$ to $\mathbf{S_k^*}$ for comparisons by using (6.20) and (6.21), respectively.
7. Convert ten trajectory dissimilarity matrices $\mathbf{S_k^*}$ ($k \in \{1, 2, \ldots, K\}$) to the MCDM model which has $M(M-1)/2$ schemes and K indexes for evaluating.
8. Use (6.24) to compute the entropy for each dissimilarity index, (6.25) or (6.26) to compute Spearman's rank correlation coefficient between the evaluating indexes, and (6.27)–(6.28) to compute the weights w_k ($k \in \{1, 2, \ldots, K\}$) of ten dissimilarity indexes.
9. Use (6.24) to compute the synthesized trajectory dissimilarity matrix which has all the dissimilarity values of any pair of the generators' trajectories.
10. Sort the trajectory dissimilarity synthesized in ascending order.
11. Pick out the pair of the synchronous generators with the smallest trajectory dissimilarity synthesized. Then, classify the generator in a single cluster (i.e., the cluster with one generator only) into the cluster where the other generator located if one of the synchronous generators is in a single cluster only; classify any one generator into the cluster where the other generator located if both of the generators are in two single clusters, and $N_{CG} = N_{CG} + 1$; and ignore this pair if both of the generators are not in a single cluster.
12. Delete the selected generators' pair with the smallest dissimilarity, and go to step 10 until all the pairs of the generators are selected.
13. Acquire the CG number N_{CG} classified and an AHC dendrogram.

The flowchart of the CRITIC method and AHC algorithm-based wide-area coherency identification of generators is shown in Figure 6.1.

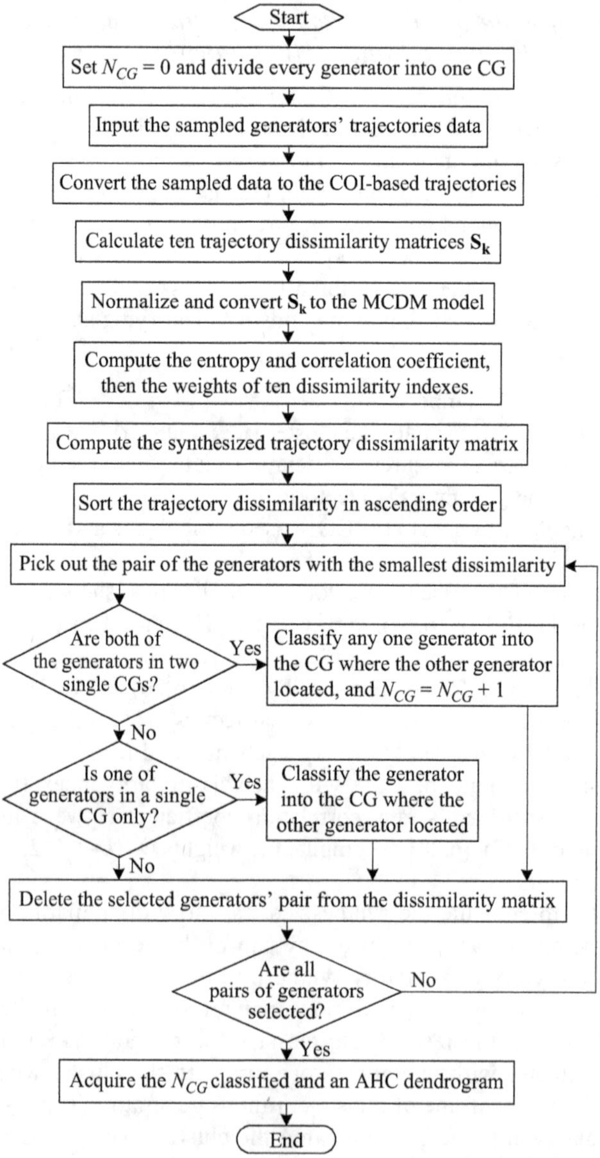

Figure 6.1 The flowchart of wide-area coherency identification of generators based on the CRITIC method and AHC algorithm

6.3 Case studies

Both the modified NENYPS with a large number of renewables [10,17], the simplified actual WECCPS [18], and the eastern part of Guangdong power system in China with a recorded oscillation event happened are utilized to demonstrate the

CRITIC and AHC-based wide-area generator coherency identification methodology proposed. It is assumed that each synchronous generator bus is equipped with a PMU. Many types of PMUs are available around the globe, such as the TR-2000 multifunction recorder manufactured by Rochestor [19], Model 1133A Power Sentinel by Arbiter Systems [20], CSD-361 PMU by Beijing Sifang Automation [21], and PCS-996 PMU by NR Electric [22,23]. Some types of PMUs (such as CSD-361 PMU and PCS-996 PMU) can attain rotor angles and rotor speeds by theoretical calculation based on electrical measurements and/or mechanical measurements employing key phase pulse signals, while the others (such as TR-2000 multifunction recorder and Model 1133A Power Sentinel) cannot measure rotor angles and rotor speeds. In this chapter, it is assumed that rotor angles and rotor speeds can be measured by PMUs. Hence WAMS could sample the generators' trajectories of rotor angles and rotor speeds directly. The sampling rate of an actual PMU is generally between 1.2 kHz and 4.8 kHz and up to 384 samples per cycle, but the data measured by a PMU cannot be transferred and recorded by WAMS at a high sampling rate [19–22]. Normally, there are several options for data transfer rates or data record rates, such as 25 Hz, 50 Hz, and 100 Hz for a 50 Hz power system (or 15 Hz and 30 Hz for a 60 Hz power system) [19–22]. Furthermore, the wide-area data are integrated with a white Gaussian noise in order to simulate the status of WAMS in a noise condition, and the wide-area data are prefiltered by using the median filter algorithm.

6.3.1 The modified NENYPS with a large number of renewables

NENYPS shown in Figure 6.2 is a 5-area power system with 68 buses and 16 synchronous generators, representing a reduced-order equivalent of the New England–New York interconnected power system. This interconnected power system consists of 5 areas, including New England area with nine generators G_1–G_9, New York area with four generators G_{10}–G_{13}, and the other three neighboring areas which are represented by equivalent generators G_{14}, G_{15}, and G_{16}, respectively. Each synchronous generator is represented by a fourth-order model and an IEEE Type 1 rotating excitation system model. The detailed topological and electrical parameters of NENYPS can be found in [2,17]. Furthermore, NENYPS is also extended to include a large number of renewables. In the modified NENYPS with a large number of renewables, an integrated 600 MVA DFIG-based wind farm (DFIG-WF) with variable speed wind turbines is assumed to be connected with bus 67. The Weibull distribution is used in wind speed models. The aggregated DFIG model can be found in [17].

6.3.1.1 Scenario 1 for demonstrating the CRITIC method and AHC algorithm

Suppose that a fault occurs in bus 16 and disappears after 160 ms, then the simulated generators' trajectories of *NENYPS with a DFIG-WF connected* at bus 67 are acquired. It is assumed that the recording rate of the generators' trajectories

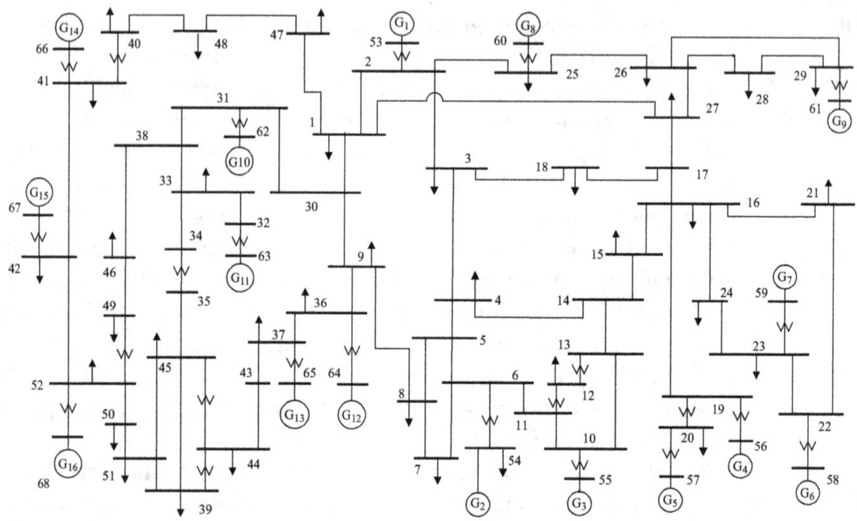

Figure 6.2 The New England–New York interconnected power system

measured by PMUs is 100 Hz. The generators' trajectories of rotor angles in the original and COI coordinates are shown in Figure 6.3(a) and (b), respectively. Comparing the trajectories in Figure 6.3(a) with these in Figure 6.3(b), it can be seen that the noise mingled with the trajectories in the original coordinate is filtered to a certain extent after the COI transformation. It can also be seen from Figure 6.3 (a) and (b) that it is very difficult to identify CGs through the trajectories' oscillation changes. Thus, the CRITIC and AHC-based wide-area coherency identification method proposed is utilized to identify CGs for the generators' trajectories. Then ten trajectory dissimilarity matrices are acquired, and the entropy, correlation coefficients, and synthesized weights of ten indexes are shown in Table 6.1, respectively.

It can be seen from Table 6.1 that the synthesized weights of correlation dissimilarity S_9 and corner deviation dissimilarity S_6 of the angle trajectories are equal to 0.2821 and 0.0013, which are the maximal and minimal weights, respectively. For S_9, its entropy is equal to 0.8789, which is the minimum of the entropy of all the indexes. It means that S_9 is greatly different from the other indexes, and more available information could be provided by this index in the MCDM. For S_6, it is with the biggest correlation coefficient and is equal to 0.1497, which means that S_6 is not great relative to the other indexes, and more available information could be provided by this index. However, it is with the biggest entropy and is equal to 0.9997, which means that S_6 is not greatly different from the other indexes, and few available information could be provided by this index. After the integration of the entropy and correlation coefficient, its synthesized weight is still the smallest. It can be also seen that the entropy weight of S_9 and S_6 are equal to 0.3057 and 0.0007, which are also the maximal and minimal weights, respectively. However,

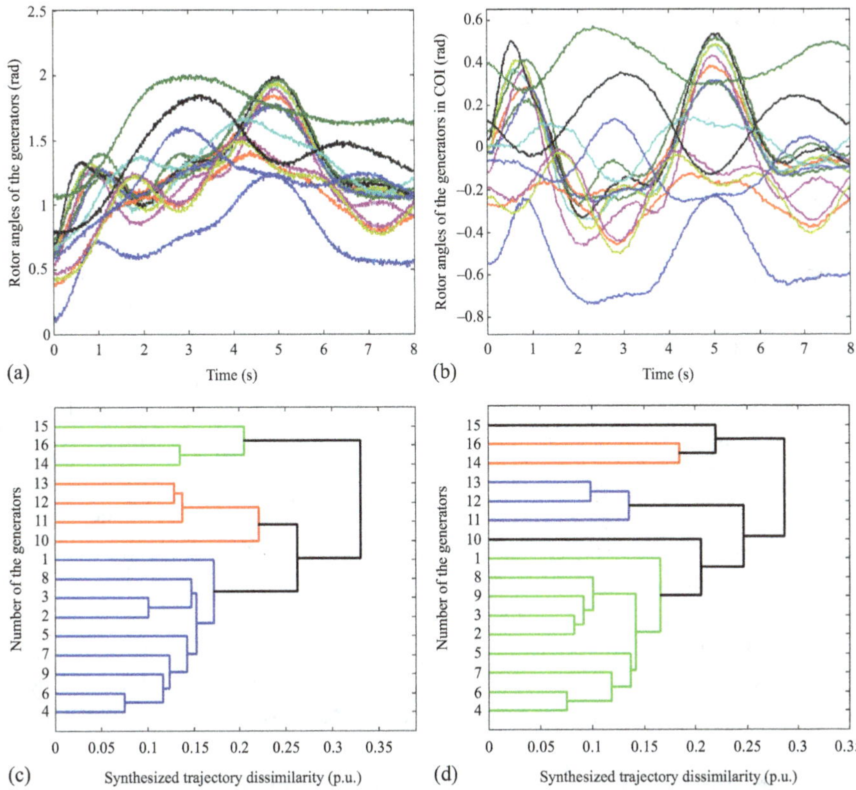

Figure 6.3 The swing trajectories of the generators' angles and the AHC dendrogram in Scenario 1 of NENYPS with a DFIG-WF connected: (a) trajectories in the original coordinate; (b) trajectories in the COI coordinate; (c) the dendrogram attained by the CRITIC and AHC-based method; and (d) the dendrogram attained by the entropy weight-based method

Table 6.1 The entropy, correlation coefficients, and synthesized weights of the trajectory dissimilarity indexes in Scenario 1

Indexes	S_1	S_2	S_3	S_4	S_5	S_6	S_7	S_8	S_9	S_{10}
Entropy	0.9495	0.9808	0.9598	0.9875	0.9991	0.9997	0.9508	0.9456	0.8789	0.9521
Correlation coefficient	0.0942	0.0842	0.0767	0.0727	0.1467	0.1497	0.0853	0.1248	0.0827	0.0830
Synthesized weight	0.1339	0.0456	0.0868	0.0256	0.0038	0.0013	0.1181	0.1911	0.2821	0.1118
Entropy weight [10]	0.1274	0.0485	0.1015	0.0315	0.0023	0.0007	0.1241	0.1372	0.3057	0.1208

the difference of the entropy weights of any two indexes is great, which means that the synthesized trajectory dissimilarity is determined by one or two indexes with great weights. After the integration of the entropy and correlation coefficient, the difference of the weights of any two indexes could be shortened to a certain extent, which could be seen from the synthesized weights in Table 6.1.

Then, the synthesized trajectory dissimilarity matrix is attained, and the AHC algorithm is utilized to identify CGs. After optimized by the CRITIC method and the entropy weight-based method [10], the AHC dendrograms are plotted and shown in Figure 6.3(c) and (d), respectively. It could be concluded from Figure 6.3(c) that all synchronous generators are classified into three CGs, that is, $\Omega_1^C = \{G_1, G_2, \ldots, G_9\}$, $\Omega_2^C = \{G_{10}, G_{11}, G_{12}, G_{13}\}$, and $\Omega_3^C = \{G_{14}, G_{15}, G_{16}\}$. The synchronous generators in a CG are listed together and their trajectory dissimilarity is smaller than that between the different coherent groups of generators. It can also be concluded from Figure 6.3(d) that all synchronous generators are classified into five CGs, that is, $\Omega_1^E = \{G_1, G_2, \ldots, G_9\}$, $\Omega_2^E = \{G_{10}\}$, $\Omega_3^E = \{G_{11}, G_{12}, G_{13}\}$, $\Omega_4^E = \{G_{14}, G_{16}\}$, and $\Omega_5^E = \{G_{15}\}$. It can be seen that there are two synchronous generators which are respectively classified into single CG (i.e., Ω_2^E and Ω_5^E) in the entropy weight-based method [10], and the trajectory dissimilarity of G_{15} is close to these of G_{14} and G_{16}, which means that the trajectories of G_{15}, G_{14}, and G_{16} are relatively similar to a certain extent. However, the trajectory dissimilarity of G_{10} is closer to these of Ω_1^E (not Ω_3^E), which is inconsistent to the results of traditional slow coherency. Hence, the CRITIC-based method proposed could identify CGs well.

6.3.1.2 Scenario 2 for demonstrating the impacts of the DFIG-WF and data sampling rate

Suppose that a fault occurs in bus 45 and disappears after 600 ms, then by simulations the generators' trajectories can be acquired. It is assumed that the sampling or recording rate of the generators' trajectories measured by PMUs is 100 Hz. The COI-based generators' trajectories of rotor angles in NENYPS with and without DFIG-WF at bus 67 are shown in Figure 6.4(a) and (b), respectively. The CRITIC and AHC-based wide-area coherency identification method proposed is utilized to identify CGs. After optimized by the CRITIC method proposed and the entropy weight-based method, the AHC dendrograms are plotted and shown in Figure 6.4 (c) and (d), respectively.

It could be concluded from Figure 6.4(c) that all synchronous generators are classified into four CGs, that is, $\Omega_1^W = \{G_1, G_2, \ldots, G_{13}\}$, $\Omega_2^W = \{G_{14}\}$, $\Omega_3^W = \{G_{15}\}$, and $\Omega_4^W = \{G_{16}\}$. It could also be concluded from Figure 6.4(d) that all synchronous generators are classified into two CGs, that is, $\Omega_1^O = \{G_1, G_2, \ldots, G_{13}\}$ and $\Omega_2^O = \{G_{14}, G_{15}, G_{16}\}$. Thus, it could be concluded that the connection of the integrated DFIG-WF in bus 67 could lead to the weakly damped oscillation, which leads to the oscillations between generators G_{14}, G_{15}, and G_{16}, and then the coherent group Ω_2^O in the power system without DFIG-WF is divided into three coherent groups, such as Ω_2^W, Ω_3^W, and Ω_4^W.

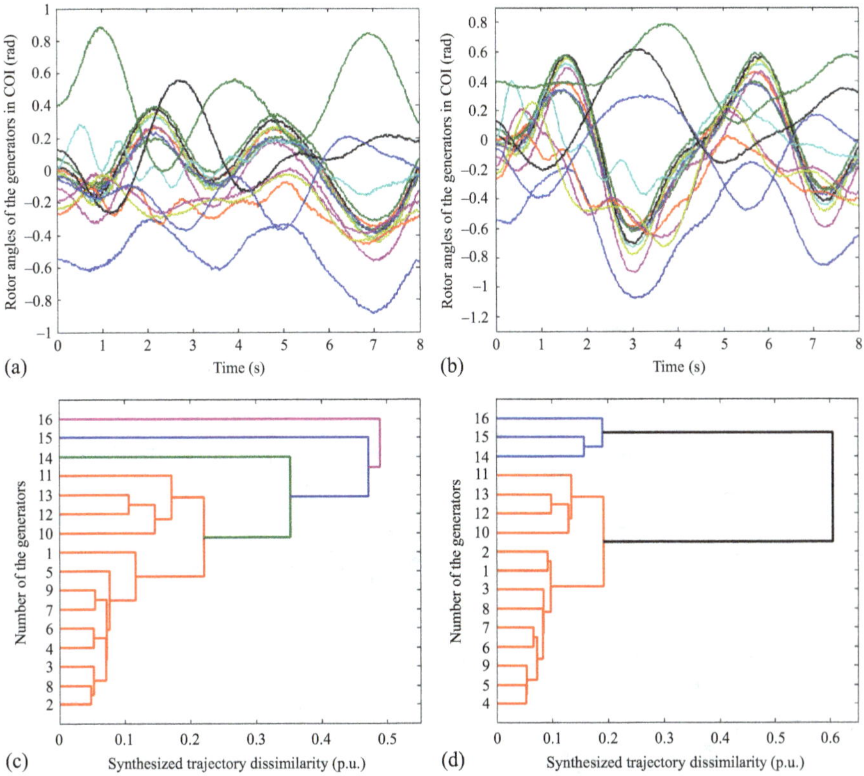

Figure 6.4 The swing trajectories of the generators' angles and the AHC dendrogram in Scenario 2: (a) the swing trajectories of the generators' angles in NENYPS with a DFIG-WF connected; (b) the swing trajectories of the generators' angles in NENYPS without a DFIG-WF connected; (c) the AHC dendrogram of NENYPS with a DFIG-WF connected at bus 67; and (d) the AHC dendrogram of NENYPS without a DFIG-WF connected

To demonstrate the impacts of sampling or recording rates of generators' trajectories on the identification of coherent generators, sensitivity analysis with respect to different recording rates (i.e., 25 Hz, 50 Hz, and 100 Hz) is carried out for Scenario 2. Figure 6.5(a)–(d) shows the agglomerative hierarchical clustering dendrograms attained by the proposed CRITIC method and the entropy weight-based method employing the generators' trajectories recorded with 50 Hz and 25 Hz sampling rates in NENYPS with and without a DFIG-WF connected at bus 67, respectively. It can be seen from Figures 6.4(c), 6.5(a), and 6.5(c) that all synchronous generators are classified into four coherent generators, that is, $\Omega_1^W = \{G_1, G_2, \ldots, G_{13}\}$, $\Omega_2^W = \{G_{14}\}$, $\Omega_3^W = \{G_{15}\}$, and $\Omega_4^W = \{G_{16}\}$ for all three groups of the generators' trajectories with 100 Hz, 50 Hz, and 25 Hz sampling

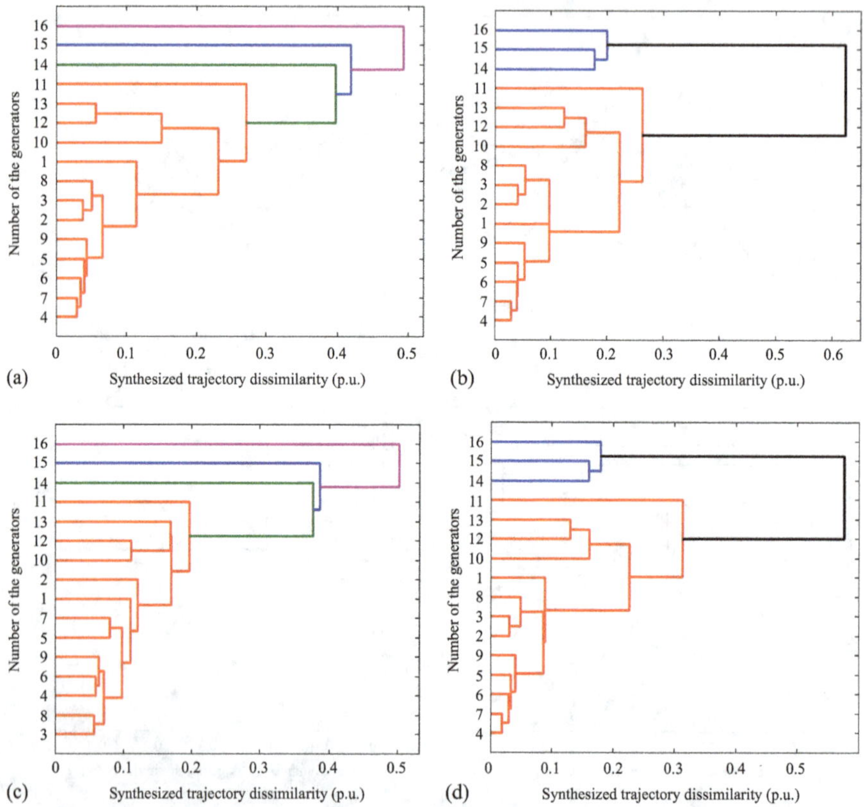

Figure 6.5 The impacts of trajectories' sampling rates on the AHC dendrogram and coherency identification in Scenario 2: (a) the 50 Hz trajectories' sampling rate in NENYPS with a DFIG-WF connected at bus 67; (b) the 50 Hz trajectories' sampling rate in NENYPS without a DFIG-WF connected; (c) the 25 Hz trajectories' sampling rate in NENYPS with a DFIG-WF connected at bus 67; and (d) the 25 Hz trajectories' sampling rate in NENYPS without a DFIG-WF connected

rates, respectively, in NENYPS with a DFIG-WF connected at bus 67, although there is a little difference in the synthesized trajectory dissimilarity of generator pairs. It can also be seen from Figures 6.4(d), 6.5(b), and 6.5(d) that all synchronous generators are classified into two CGs, that is, $\Omega_1^{\mathrm{O}} = \{G_1, G_2, \ldots, G_{13}\}$ and $\Omega_2^{\mathrm{O}} = \{G_{14}, G_{15}, G_{16}\}$ for all three groups of the generators' trajectories with 100 Hz, 50 Hz, and 25 Hz sampling rates respectively in NENYPS without a DFIG-WF connected, although there is a little difference in the synthesized trajectory dissimilarity of generator pairs. Hence, it can be concluded that the sampling or recording rates do not have significant impacts on the identification results of coherent generators for the studied scenarios in NENYPS with and without a DFIG-WF connected.

6.3.2 The simplified WECCPS

A simplified version of WECCPS as shown in Figure 6.6 with 179 buses and 29 synchronous generators is also utilized to demonstrate the CRITIC and AHC-based method. Each synchronous generator is represented by the classical second-order differential equations; the damping parameter of each synchronous generator is initially set to 4 and is adjusted in different cases for creating different

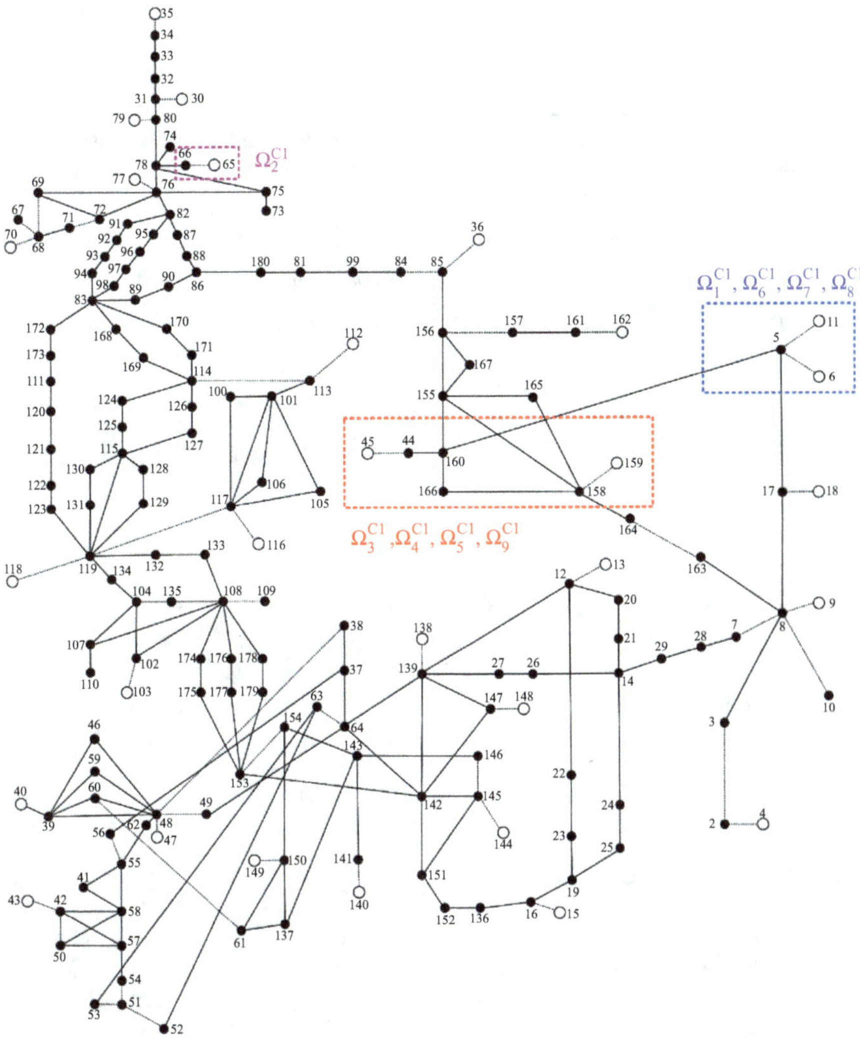

Figure 6.6 A simplified version of WECCPS

natural or forced oscillation modes. The detailed topological and electrical parameters of the WECCPS can be found in [18,24].

Nine poorly damped oscillation cases proposed in [18,24] and their measurement data are utilized to demonstrate the effectiveness of the proposed wide-area coherency identification method. It is assumed that the sampling or recording rate of the generators' trajectories measured by PMUs is 30 Hz. The results of the oscillation sources located in [18,24] and the coherent groups of the generators identified by the entropy weight-based method and the proposed method are shown in Table 6.2 for comparisons. In Table 6.2, $\Omega_{G,all}$ denotes the set of all the generators in the WECCPS; Ω_i^{C1} and Ω_i^{C2} (or $\Omega_{i,E}^{C1}$, $\Omega_{i,E}^{C2}$, and $\Omega_{i,E}^{C3}$) denote the sets of

Table 6.2 Comparisons among the oscillation source location method, the entropy weight-based method, and the proposed method

Case no.	Case name	Fault location	Oscillation sources in [18,24]	Coherent groups attained by the entropy weight-based method	Coherent groups attained by the proposed method
1	ND_1	G_{159}	G_{45}	$\Omega_{1,E}^{C1} = \{G_{45}, G_{159}\}$, $\Omega_{1,E}^{C2} = \Omega_{G,all} - \Omega_{1,E}^{C1}$	$\Omega_1^{C1} = \{G_{45}, G_{159}\}$, $\Omega_1^{C2} = \Omega_{G,all} - \Omega_1^{C1}$
2	ND_2	G_{79}	G_{65}	$\Omega_{2,E}^{C1} = \{G_{65}\}$, $\Omega_{2,E}^{C2} = \Omega_{G,all} - \Omega_{2,E}^{C1}$	$\Omega_2^{C1} = \{G_{65}\}$, $\Omega_2^{C2} = \Omega_{G,all} - \Omega_2^{C1}$
3	ND_3	G_{30}	G_{11}	$\Omega_{3,E}^{C1} = \{G_{30}\}$, $\Omega_{3,E}^{C2} = \{G_6, G_{11}\}$, $\Omega_{3,E}^{C3} = \Omega_{G,all} - \Omega_{3,E}^{C1} - \Omega_{3,E}^{C2}$	$\Omega_3^{C1} = \{G_6, G_{11}\}$, $\Omega_3^{C2} = \Omega_{G,all} - \Omega_3^{C1}$
4	ND_4	G_6	G_{11}	$\Omega_{4,E}^{C1} = \{G_6, G_{11}\}$, $\Omega_{4,E}^{C2} = \Omega_{G,all} - \Omega_{4,E}^{C1}$	$\Omega_4^{C1} = \{G_6, G_{11}\}$, $\Omega_4^{C2} = \Omega_{G,all} - \Omega_4^{C1}$
5	ND_5	G_{30}	G_{11}	$\Omega_{5,E}^{C1} = \{G_{30}\}$, $\Omega_{5,E}^{C2} = \{G_6, G_{11}\}$, $\Omega_{5,E}^{C3} = \Omega_{G,all} - \Omega_{5,E}^{C1} - \Omega_{5,E}^{C2}$	$\Omega_5^{C1} = \{G_6, G_{11}\}$, $\Omega_5^{C2} = \Omega_{G,all} - \Omega_5^{C1}$
6	ND_6	G_{159}	G_{45}, G_{159}	$\Omega_{6,E}^{C1} = \{G_{45}, G_{159}\}$, $\Omega_{6,E}^{C2} = \Omega_{G,all} - \Omega_{6,E}^{C1}$	$\Omega_6^{C1} = \{G_{45}, G_{159}\}$, $\Omega_6^{C2} = \Omega_{G,all} - \Omega_6^{C1}$
7	ND_7	G_{159}	G_{45}, G_{159}	$\Omega_{7,E}^{C1} = \{G_{45}, G_{159}\}$, $\Omega_{7,E}^{C2} = \Omega_{G,all} - \Omega_{7,E}^{C1}$	$\Omega_7^{C1} = \{G_{45}, G_{159}\}$, $\Omega_7^{C2} = \Omega_{G,all} - \Omega_7^{C1}$
8	ND_8	G_{159}	G_{36}, G_{45}	$\Omega_{8,E}^{C1} = \{G_{45}, G_{159}\}$, $\Omega_{8,E}^{C2} = \Omega_{G,all} - \Omega_{8,E}^{C1}$	$\Omega_8^{C1} = \{G_{45}, G_{159}\}$, $\Omega_8^{C2} = \Omega_{G,all} - \Omega_8^{C1}$
9	ND_9	G_{79}	G_{11}	$\Omega_{9,E}^{C1} = \{G_6, G_{11}\}$, $\Omega_{9,E}^{C2} = \Omega_{G,all} - \Omega_{9,E}^{C1}$	$\Omega_9^{C1} = \{G_6, G_{11}\}$, $\Omega_9^{C2} = \Omega_{G,all} - \Omega_9^{C1}$

two (or three) coherent groups of generators identified by the proposed method (or the entropy weight-based method) in the *i*th poorly damped oscillation cases, respectively. The first coherent group of generators Ω_i^{C1} identified by the proposed method is also marked in Figure 6.6.

It can be seen from Table 6.2 and Figure 6.6 that two coherent groups of generators are identified for all nine cases by the proposed method; the coherent groups of the generators of Cases 2, 6, and 7 identified by the proposed method are the same with the results attained by the oscillation source location method; for Cases 1, 3, 4, 5, and 9, the located oscillation source G_{11} is one of the generators (i.e., G_6 and G_{11}) in the first coherent group identified by the proposed method, which means that the effectiveness of the proposed method. Furthermore, both G_6 and G_{11} are directly connected with the bus 5 which can be seen in Figure 6.6. However, two located oscillation sources (i.e., G_{36} and G_{45}) in Case 8 are divided into different CGs by the method proposed. Hence, it can be concluded that the proposed method could identify CGs correctly in most of the test cases. It can also be seen from Table 6.2 that the coherent groups of generators in Cases 1, 2, 4, and 6–9 identified by the entropy weight-based method are the same with those by the proposed method, but the coherent groups of generators in Cases 3 and 5 identified by the entropy weight-based method are different from those by the proposed method. In Cases 3 and 5, all synchronous generators are divided into three coherent groups, that is, $\Omega_{3,E}^{C1} = \{G_{30}\}$, $\Omega_{3,E}^{C2} = \{G_6, G_{11}\}$, and $\Omega_{3,E}^{C3} = \Omega_{G,all} - \Omega_{3,E}^{C1} - \Omega_{3,E}^{C2}$ by the entropy weight-based method, but two coherent groups, that is, $\Omega_3^{C1} = \{G_6, G_{11}\}$ and $\Omega_3^{C2} = \Omega_{G,all} - \Omega_3^{C1}$ by the proposed method. According to the oscillation sources identified in [18,24], G_{30} is not included, thus the coherent group $\Omega_{3,E}^{C1}$ identified by the entropy weight-based method is not correct. Hence, it can be concluded that the proposed method is better than the entropy weight-based method.

To demonstrate the impacts of sampling or recording rates of generators' trajectories on the identification of coherent generators, sensitivity analysis with respect to different sampling rates (i.e., 15 Hz and 30 Hz) is carried out for the simplified version of WECCPS, and it is found that the results of the coherent groups of generators identified by the CRITIC method and the entropy weight-based method employing the generators' trajectories recorded with 15 Hz sampling rates are the same with those shown in Table 6.2 with the 30 Hz sampling rate. Hence, it can be concluded that the sampling or recording rates do not have significant impacts on the identification results of coherent generators for the studied scenarios in WECCPS.

6.3.3 The recorded oscillation event happened in the eastern part of Guangdong power system in China

In Guangdong power system, all 500 kV substations and power plants are equipped with PMUs and some 220 kV substations and power plants are also equipped with PMUs. So far, 77 substations and 70 power plants are equipped with PMUs in Guangdong power system. Due to data storage capability limit, only the data recorded by the PMUs close to an oscillation source are stored for post oscillation analysis, while those far from the oscillation source are not saved. On July 9, 2013, an oscillation event happed in the east part of Guangdong power system was detected and recorded by PMUs and WAMS in the time period from 18:05:00 to 18:09:59. Figure 6.7(a) shows the recorded rotor angle trajectories of four

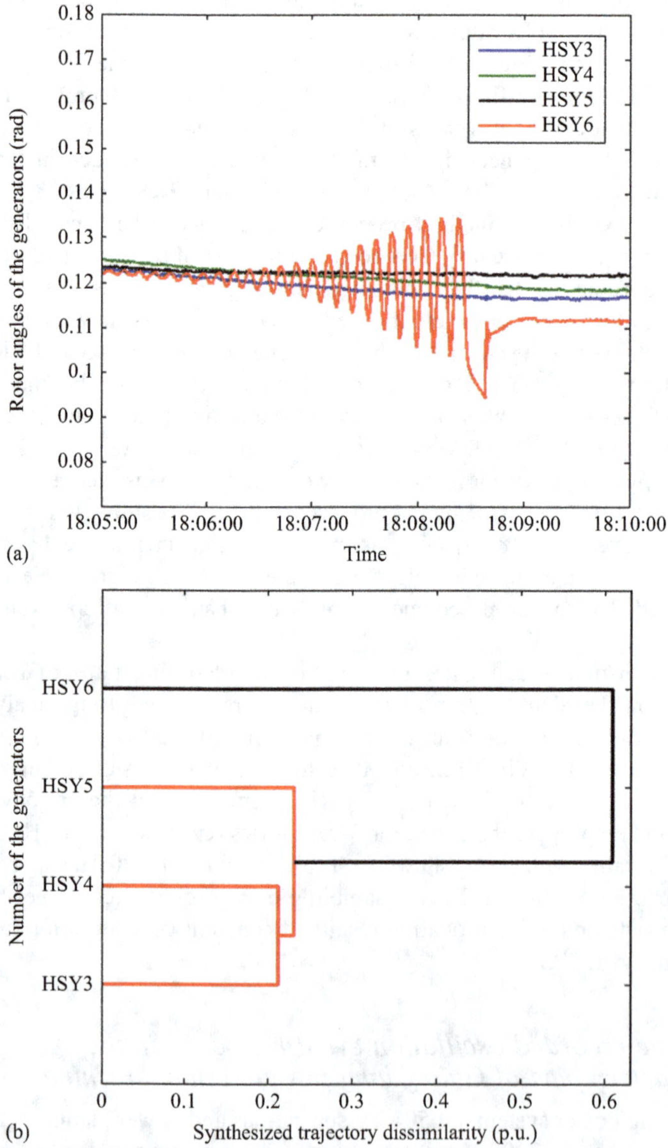

Figure 6.7 A recorded oscillation event happened in the eastern part of Guangdong power system in China: (a) the recorded swing trajectories of four generators' angles and (b) the dendrogram for coherency identification based on the proposed CRITIC and AHC method

generators. The actual recording rate of generators' trajectories by PMUs is 100 Hz. It can be seen from Figure 6.7(a) that generator HSY6 oscillates with the other three generators (i.e., HSY3, HSY4, and HSY5). This actual oscillation event can be well served for demonstrating the effectiveness of the proposed CRITIC and AHC-based coherency identification method. The coherency identification is repeatedly carried out on the generators' trajectories, and coherent groups of generators are correctly identified at 18:05:29. Figure 6.7(b) shows the AHC dendrogram of coherency identification at 18:05:29 for this oscillation event. It can be seen from Figure 6.7(b) that HSY6 is identified as one coherent group, while HSY3, HSY4, and HSY5 as the other one. Hence, it can be concluded that the proposed method could correctly identify the coherency of generators at the early stage of the oscillation event. The entropy weight-based method is also carried out for comparisons with the proposed one, and it can identify the coherent groups of generators correctly at 18:05:49. It can be seen that the proposed method can identify coherent generators faster than the entropy weight-based one. Furthermore, sensitivity analysis with respect to different sampling rates (i.e., 25 Hz and 50 Hz) is also carried out, and it is found that the coherent groups of generators can be correctly identified by the CRITIC method at 18:05:29. Hence, it can be concluded that the sampling or recording rates do not have significant impacts on the identification results of coherent generators for the recorded oscillation event that happened in the eastern part of Guangdong power system in China.

6.4 Concluding remarks

Ten trajectory similarity indexes are presented for the MCDM of the wide-area coherency identification of generators, and the CRITIC method utilized to integrate these indexes, while the differences and correlations among multiple indexes are considered by the entropy and the Spearman's rank correlation coefficient, respectively. Then, the AHC algorithm is proposed to cluster the generators based on the synthesized trajectory dissimilarity. Finally, case studies on a 16-generator 68-bus power system with a large number of renewables, a simplified version of the WECC Power System, and the eastern part of Guangdong power system in China with a recorded oscillation event happened show that the proposed method could well identify coherent generators. The integration of large-scale DFIG-WF could introduce weakly damped oscillations into the power system concerned.

References

[1] Liu, S.W., Li, G.Y., and Zhou, M. "Power system transient stability analysis with integration of DFIGs based on center of inertia." *CSEE Journal of Power and Energy Systems*. 2016, **2**(2), 20–29

[2] Jiang, T., Jia, H.J., Yuan, H.Y., *et al.* "Projection pursuit: a general methodology of wide-area coherency detection in bulk power grid." *IEEE Transactions on Power Systems*. 2015, **31**(4), 1–11

[3] Susuki, Y., and Mezic, I. "Nonlinear Koopman modes and coherency identification of coupled swing dynamics." *IEEE Transactions on Power Systems*. 2011, **26**(4), 1894–1904
[4] Vahidnia, A., Ledwich, G., Palmer, E., *et al.* "Generator coherency and area detection in large power systems." *IET Generation, Transmission and Distribution*. 2012, **6**(9), 874–883
[5] Ariff, M.A.M., and Pal, B.C. "Coherency identification in interconnected power system: an independent component analysis approach." *IEEE Transactions on Power Systems*. 2013, **28**(2), 1747–1755
[6] Avdakovic, S., Becirovic, E., Nuhanovic, A., *et al.* "Generator coherency using the wavelet phase difference approach." *IEEE Transactions on Power Systems*. 2014, **29**(1), 271–278
[7] Barocio, E., Pal, B.C., Thornhill, N.F., *et al.* "A dynamic mode decomposition framework for global power system oscillation analysis." *IEEE Transactions on Power Systems*. 2015, **30**(6), 2902–2912
[8] Gomez, O., and Rios, M.A. "Real time identification of coherent groups for controlled islanding based on graph theory." *IET Generation, Transmission & Distribution*. 2015, **9**(8), 748–758
[9] Sun, K., Hur, K., and Zhang, P. "A new unified scheme for controlled power system separation using synchronized phasor measurements." *IEEE Transactions on Power Systems*. 2011, **26**(3), 1544–1554
[10] Lin, Z.Z., Wen, F.S., Zhao, J.H., *et al.* "Controlled islanding schemes for interconnected power systems based on coherent generator group identification and wide-area measurements." *Journal of Modern Power Systems and Clean Energy*. 2016, **4**(3), 440–453
[11] Salimian, M.R., and Aghamohammadi, M.R. "Intelligent out of step predictor for inter area oscillations using speed-acceleration criterion as a time matching for controlled islanding." *IEEE Transactions on Smart Grid*. 2018, **9**(4), 2488–2497
[12] Zou, J.X., Peng, C., Xu, H.B., *et al.* "A fuzzy clustering algorithm-based dynamic equivalent modeling method for wind farm with DFIG." *IEEE Transactions on Energy Conversion*. 2015, **30**(4), 1329–1337
[13] Zhang, Y., Zhang, Y., Meng, G., *et al.* "A wide area information based clustering recognition method of coherent generators." *Power System Technology*. 2015, **39**(10), 2889–2893
[14] Lee, J.G., Han, J.W., and Whang, K.Y. "Trajectory clustering: a partition-and-group framework." *Proceedings of the 2007 ACM SIGMOD International Conference on Management of Data*, New York, NY, Association for Computing Machinery, June 2007, 1–12
[15] Diakoulaki, D., Mavrotas, G., and Papayannakis, L. "Determining objective weights in multiple criteria problems: the CRITIC method." *Computers & Operations Research*. 1995, **22**(7), 763–770
[16] Pirie, W. *Spearman Rank Correlation Coefficient, Encyclopedia of Statistical Sciences*. Hoboken, NJ, Wiley-Interscience, 2006

[17] Milano, F. "An open source power system analysis toolbox." *IEEE Transactions on Power Systems*. 2005, **20**(3), 1199–1206

[18] Maslennikov, S., Wang, B., Zhang, Q., *et al.* "A test cases library for methods locating the sources of sustained oscillations." *Proceedings of IEEE PES General Meeting*, Boston, MA, IEEE, July 2016, 1–5

[19] TR-2000 Multi-Function Recorder-Overview. http://www.ametekpower.com/Fault-Recorders/TR-2000-Multi-Function-Recorder.aspx, accessed February 25, 2017

[20] Model 1133A Power Sentinel. http://www.arbiter.com/catalog/product/model-1133a-power-sentinel.php#tabs-2, accessed February 25, 2017

[21] CSD-361 Phasor Measurement Unit. http://www.sifang-electric.com/index.php/en/products-solutions/generation/55-wide-area-measurement-system/220-csd-364-synchrophasor-measurement-unit, accessed February 25, 2017

[22] PCS-996 Phasor Measurement Unit. http://www.nrec.com/en/product/PCS-996.html, accessed February 25, 2017

[23] Lu, C., Shi, B.N., Wu, X.C., *et al.* "Advancing China's smart grid: phasor measurement units in a wide-area management system." *IEEE Power and Energy Magazine*. 2015, **13**(5), 60–71

[24] Test Cases Library of Power System Sustained Oscillations. http://web.eecs.utk.edu/~kaisun/Oscillation, accessed December 9, 2016

Chapter 7

A fast state estimator for systems including limited number of PMUs

Ting Wu[1,2], *Chi Yung Chung*[3] *and Innocent Kamwa*[4]

7.1 Introduction

Since its initial introduction by Fred Schweppe in 1970 [1–3], state estimation (SE) has become an essential tool for providing accurate system snapshots to several crucial applications in energy management system (EMS). In conventional power system SE [4], remote terminal units (RTUs) collect various types of measurements, such as bus voltage magnitude, line current magnitude, generator outputs, loads, active and reactive power flows, and power injections and transfer them to control center. As these measurements are nonlinear with respect to the power system states, SE is generally formulated as a nonlinear weighted least squares (WLS) problem and solved in an iterative manner. The accuracy and efficiency of SE can be enhanced from two aspects, that is, software and hardware. In former aspect, different SE algorithms have been proposed by modifying the estimation model and introducing other mature techniques to SE. The advent of phasor measurement units (PMUs) brought a hardware breakthrough in SE. Synchronized phasor measurements provided by PMUs can enhance SE in the following aspects: the faster reporting rate of PMUs can sharply increase the measurement redundancy level of SE [5]; PMUs can directly obtain the phase angle measurements, which can improve the accuracy of the phase angle estimates and then provide a more reliable estimator [6]; and the unique time-tags of measurements can be used to synchronize the results estimated from different systems [7]. When a power system is fully observable using only PMU measurements and rectangular coordinates are used for phasor measurements and state variables, the estimation problem becomes linear and can be noniteratively solved [8]. Despite the undisputed advantages of using PMUs in power system SE, three factors demonstrate that SE cannot be

[1]College of Information Engineering, Zhejiang University of Technology, Hangzhou, China
[2]Department of Electrical Engineering, The Hong Kong Polytechnic University, Hong Kong, China
[3]Department of Electrical and Computer Engineering, University of Saskatchewan, Saskatoon, Saskatchewan, Canada
[4]Hydro-Québec/IREQ, Power System and Mathematics, Varennes, Quebec, Canada

implemented using only PMUs. First, due to the financial constraints, the number of PMUs deployed in real systems worldwide is still insufficient to make the system fully observable [9]. Second, if there is a communication problem, SE should still provide state estimates for the whole system without PMU measurements. Lastly, the SCADA infrastructures were significantly invested in the past, which can provide valuable information and should not be discarded. Therefore, conventional and synchronized phasor measurements will coexist for several years to come, and the power system SE should effectively integrate both of these available data sources and provide a unified view [10,11]. To address the SE problem under incomplete PMU observability, two basic approaches, characterized structurally by the number of estimators used, are addressed: (1) mixing the SCADA and PMU measurements using a single estimator [9,12–16]; and (2) processing each set of measurements separately and using two or more estimators [8,10,17–20]. Although the former approach produces globally optimal estimates, it faces some serious implementation challenges. One such challenge is the significant difference between the refresh rates of SCADA and PMU measurements, which can be successfully circumvented by the latter approach.

Even though computing power is now orders of magnitude greater than four decades ago when power system SE was championed by Schweppe, the need for computationally more efficient techniques to integrate data from both the SCADA system and PMUs still remains due to expansion of the geographical scope of many SEs and the huge number of measurement points provided by the ongoing deployment of new digital devices [21,22]. A reduced-order two-stage method, proposed in [8] to perform SE using both SCADA and PMU measurements, featured lower computational complexity. To reduce the required execution time, the authors in [17] offered an alternative approach to incorporate PMU data into conventional SE through a postprocessing step. This scheme was modified in [23] by introducing supplementary current measurements in the postprocessing step to provide slightly higher precision. A two-stage state estimator was proposed in [24] to reduce computational effort. A distributed SE algorithm was presented in [19] and the simulation results demonstrated its efficacy in reducing the computational time. A distributed two-level state estimator function [25,26] used a linear SE algorithm at the control center level but moved the bad data identification and local topology processing to the substation level, making this processing very fast.

Bad data (BD) analysis is an essential attribute of any power system state estimator and a variety of methods for gross error detection and identification are available. Among existing methods, the largest normalized residual test [4] and the hypothesis testing identification (HTI) method [27] are widely used for SCADA-based SE. With the increasing penetration of PMUs in power system SE, BD analysis in PMU measurements has been a matter of interest for several years. A strategic PMU placement algorithm was proposed in [28] to improve the BD processing capability of state estimators. The authors in [13] applied the largest normalized residual test to address BD analysis in the context of hybrid SE. However, for multiple BD, updates on the measurements and states continue until all normalized residuals drop below a predefined detection threshold. Thus, the re-estimation step constitutes the main computational bottleneck of the BD analysis. To efficiently identify multiple gross errors and eliminate a BD smearing effect, an enhanced BD processing approach was developed in [11]. However, computational burden associated with the nonlinear re-estimation step of SCADA-based SE may lower its performance when there are multiple bad SCADA measurements in PMU-unobservable area. Therefore, there is a need to develop a new SE algorithm to improve computational efficiency which is crucial not only for the solution but also for the BD processing scheme.

This chapter describes the development of a new SE method to improve computational efficiency and maintain high estimation accuracy simultaneously, especially for large-scale networks with multiple BD. First, a three-stage SE method, composed of two WLS linear problems and a nonlinear transformation, is used to process the SCADA data. Then, the PMU observable states are calculated by a linear estimator, where pseudo-state values with large variances are assigned to PMU unobservable states. A fusion stage is used to combine the results obtained from the above two estimators. Bad SCADA measurements in PMU-observable area or bad PMU measurements are identified and processed all at once, which can dramatically reduce the implementation time. In addition, the linear re-estimation step, caused by bad SCADA measurements in PMU-unobservable area, can further improve computational efficiency.

7.2 A fast state estimator

This section describes a fast state estimator for integrating SCADA and PMU measurements. Consider a power system with n buses, l branches, and measured by m_s SCADA measurements and m_p PMU measurements [29].

7.2.1 First stage of SCADA-based SE

In the first stage, an auxiliary vector, y, is introduced in such a way that a linear measurement model is represented by [21]

$$z_{\text{scada}} = Ay + e_s \tag{7.1}$$

where z_{scada} is the SCADA measurement vector with dimension $(m_s \times 1)$; A is a constant Jacobian matrix; and e_s is the measurement error vector that features a normal distribution with zero mean and covariance matrix R_{SCADA}.

Specifically, for each bus i, the voltage magnitude V_i is squared to get the new variable U_i:

$$U_i = V_i^2 \tag{7.2}$$

Moreover, for each branch connecting buses i and j, the following two variables are constructed:

$$K_{ij} = V_i V_j \cos\theta_{ij} \tag{7.3}$$

$$L_{ij} = V_i V_j \sin \theta_{ij} \tag{7.4}$$

where θ_i and θ_j denote the voltage phase angles of buses i and j, respectively, and $\theta_{ij} = \theta_i - \theta_j$.

Then, the intermediate state vector, composed of $n + 2l$ variables, is given in block partitioned form:

$$y = \{U_i \quad K_{ij} \quad L_{ij}\} \tag{7.5}$$

Any SCADA measurement can be linearly expressed in terms of the intermediate state vector as follows [21].

- *Power flow measurements*

  For a branch connecting buses i and j, active and reactive power flow measurements taken at terminal bus i are

$$P_{ij} = (g_{sh,i} + g_{ij})U_i - g_{ij}K_{ij} - b_{ij}L_{ij} + \varepsilon_P \tag{7.6}$$

$$Q_{ij} = -(b_{sh,i} + b_{ij})U_i + b_{ij}K_{ij} - g_{ij}L_{ij} + \varepsilon_Q \tag{7.7}$$

  where g_{ij} and b_{ij} are the conductance and susceptance of the series branch connecting buses i and j, respectively, and $g_{sh,i}$ and $b_{sh,i}$ are the values of the shunt branch connected at bus i.

- *Power injection measurements at bus i*

  The active and reactive power injection measurements at bus i are

$$P_i = \sum_{j\in i} P_{ij} + \varepsilon_P \tag{7.8}$$

$$Q_i = \sum_{j\in i} Q_{ij} + \varepsilon_Q \tag{7.9}$$

- *Voltage magnitude measurement*

  Here, the voltage magnitude measurements should be squared to retain linearity in this stage:

$$V_i^2 = U_i + \varepsilon_U \tag{7.10}$$

  The standard deviation of the error associated with V_i^2 is calculated using the original measurement V_i and its standard deviation, as follows:

$$\sigma(V_i^2) = 2V_i\sigma(V_i) \tag{7.11}$$

The components of constant Jacobian matrix A in (7.1) are illustrated in Table 7.1 [21,22].

The WLS solution to the linear problem (7.1), $\widehat{y}^{(\text{scada})}$, and its error covariance matrix, $R_y^{(\text{scada})}$, can be directly obtained, respectively, as

$$\widehat{y}^{(\text{scada})} = G_a{}^{-1}A^{\text{T}}R_{\text{SCADA}}^{-1}z_{\text{scada}} \tag{7.12}$$

Table 7.1 Jacobian components in the first stage of SCADA-based SE

	$y = U_i$	$y = U_j$	$y = K_{ij}$	$y = L_{ij}$
$\partial V_i^2/\partial y$	1	0	0	0
$\partial P_{ij}/\partial y$	$g_{sh,i} + g_{ij}$	0	$-g_{ij}$	$-b_{ij}$
$\partial P_{ji}/\partial y$	0	$g_{sh,j} + g_{ij}$	$-g_{ij}$	$-b_{ij}$
$\partial Q_{ij}/\partial y$	$-(b_{sh,i} + b_{ij})$	0	b_{ij}	$-g_{ij}$
$\partial Q_{ji}/\partial y$	0	$-(b_{sh,j} + b_{ij})$	b_{ij}	g_{ij}

$$R_y^{(\text{scada})} = G_a{}^{-1} \tag{7.13}$$

where $G_a = A^{\mathrm{T}} R_{\text{SCADA}}^{-1} A$ is a gain matrix and the notation $(\cdot)^{\mathrm{T}}$ denotes transpose.

7.2.2 PMU-based SE

PMUs offer fast acquisition of voltage and current phasors in a synchronized manner. When rectangular coordinates are used for the phasor measurements and state variables, the SE problem becomes linear and states can be obtained by a noniterative algorithm. However, the phasor measurements derived from PMUs are generally stored and uploaded in polar form. The relationship between incremental representation in polar and rectangular coordinates is

$$\begin{aligned} &\left[z_{(1)\text{re}} \quad z_{(2)\text{re}} \quad \cdots \quad z_{(1)\text{im}} \quad z_{(2)\text{im}} \quad \cdots\right]^{\mathrm{T}} \\ &\quad = T_{\text{PMU}}\left[\Delta\theta_1 \quad \Delta\theta_2 \quad \cdots \quad \Delta|z_1| \quad \Delta|z_2| \quad \cdots\right]^{\mathrm{T}} \end{aligned} \tag{7.14}$$

where

$$T_{\text{PMU}} = \begin{bmatrix} -|z_1|\sin\theta_1 & 0 & \cdots & \cos\theta_1 & 0 & \cdots \\ 0 & -|z_2|\sin\theta_2 & \cdots & 0 & \cos\theta_2 & \cdots \\ \vdots & \vdots & \cdots & \vdots & \vdots & \cdots \\ |z_1|\cos\theta_1 & 0 & \cdots & \sin\theta_1 & 0 & \cdots \\ 0 & |z_2|\cos\theta_2 & \cdots & 0 & \sin\theta_2 & \cdots \\ \vdots & \vdots & \cdots & \vdots & \vdots & \cdots \end{bmatrix}$$

$z_{(1)\text{re}}$ and $z_{(1)\text{im}}$ represent the rectangular form of PMU measurements, and $|z_1|$ and θ_1 denote their polar form. The error covariance matrix of the measurements in polar coordinates, R_{PMU}^p, is thereby transformed by a rotation matrix T_{PMU} to obtain the error covariance matrix of measurements in rectangular coordinates, R_{PMU}^r, shown as follows:

$$R_{\text{PMU}}^r = T_{\text{PMU}} R_{\text{PMU}}^p (T_{\text{PMU}})^{\mathrm{T}} \tag{7.15}$$

After the above transformation, a set of positive sequence voltage and current phasors in rectangular coordinates, z_{pmu}, is obtained. PMU observable states are uniquely defined by the PMU placements and the type of PMU measurements [22].

We assume there are n_{ob} PMU observable states and n_{un} PMU unobservable states, which are denoted by x_{ob} and x_{un}, respectively. A permutation matrix Π is easily defined to reorder the state vector x such that the PMU observable states are stacked over the unobservable states, that is, $\Pi x = \begin{bmatrix} x_{ob}^{T} & x_{un}^{T} \end{bmatrix}^{T}$. Because $\Pi^{T}\Pi = I$, the PMU measurement model can be expressed as follows:

$$z_{pmu} = Bx + e_p = B\Pi^{T}\Pi x + e_p = [\,C_1 \quad C_2\,]\begin{bmatrix} x_{ob} \\ x_{un} \end{bmatrix} + e_p \tag{7.16}$$

where B is the PMU measurement Jacobian, with dimension $(m_p \times 2n)$, and $B\Pi^{T} = [\,C_1 \quad C_2\,]$. The dimensions of C_1 and C_2 are $m_p \times n_{ob}$ and $m_p \times n_{un}$, respectively. Because all entries in C_2 are zero, (7.16) can be rewritten as

$$z_{pmu} = C_1 x_{ob} + e_2 \tag{7.17}$$

Then, the estimate for the PMU observable state vector and its error covariance are given, respectively, by

$$\widehat{x}_{ob} = \left[C_1^{T}\left(R_{PMU}^{r}\right)^{-1}C_1\right]^{-1} C_1^{T}\left(R_{PMU}^{r}\right)^{-1} z_{pmu} \tag{7.18}$$

$$R_{ob} = \left[C_1^{T}\left(R_{PMU}^{r}\right)^{-1}C_1\right]^{-1} \tag{7.19}$$

In preparation for bad data processing and the fusion stage, standard "expected" values for bus complex voltages (real part = 1, imaginary part = 0) are assigned to the PMU-unobservable state vector $\widehat{x}_{un}$. Because these PMU-unobservable state values are inaccurate, their variances are usually some orders of magnitude larger than the tele-measurement variances. Let R_{un} ($n_{un} \times n_{un}$ diagonal matrix) be the error covariance matrix of $\widehat{x}_{un}$. Then, the estimate for the state vector $\widehat{x}$ and its error covariance matrix $R_x^{(pmu)}$ are calculated using

$$\widehat{x} = \Pi^{T}\begin{bmatrix} \widehat{x}_{ob} \\ \widehat{x}_{un} \end{bmatrix} \tag{7.20}$$

$$R_x^{(pmu)} = \Pi_1^{T} R_{ob} \Pi_1 + \Pi_2^{T} R_{un} \Pi_2 \tag{7.21}$$

where $\Pi = \begin{bmatrix} \Pi_1^{T} & \Pi_2^{T} \end{bmatrix}^{T}$, and the dimensions of Π_1 and Π_2 are $n_{ob} \times 2n$ and $n_{un} \times 2n$, respectively.

7.2.3 Bad data processing scheme

Sections 7.2.1 and 7.2.2 describe SCADA and PMU measurements independently and individually. When "scada" and "pmu" are used as superscripts, they denote in which estimator a vector/matrix is computed, whereas subscripts "scada" and "pmu" represent the type of measurements to which a vector/matrix is associated [11].

The state estimate $\widehat{x}^{(pmu)}$ is obtained together with its error covariance matrix $R_x^{(pmu)}$ from processing PMU measurements only. Then, using $\widehat{x}^{(pmu)}$, the estimated intermediate state vector $\widehat{y}^{(pmu)}$ is calculated as follows:

$$\widehat{y}^{(pmu)} = \left\{ \widehat{U}_i^{(pmu)} \quad \widehat{K}_{ij}^{(pmu)} \quad \widehat{L}_{ij}^{(pmu)} \right\} \tag{7.22}$$

Table 7.2 Jacobian components in the BD processing scheme

	$x = x_{(i)\mathrm{re}}$	$x = x_{(i)\mathrm{im}}$	$x = x_{(j)\mathrm{re}}$	$x = x_{(j)\mathrm{im}}$
$\partial U_i/\partial x$	$2x_{(i)\mathrm{re}}$	$2x_{(i)\mathrm{im}}$	0	0
$\partial K_{ij}/\partial x$	$x_{(j)\mathrm{re}}$	$x_{(j)\mathrm{im}}$	$x_{(i)\mathrm{re}}$	$x_{(i)\mathrm{im}}$
$\partial L_{ij}/\partial x$	$-x_{(j)\mathrm{im}}$	$x_{(j)\mathrm{re}}$	$x_{(i)\mathrm{im}}$	$-x_{(i)\mathrm{re}}$

where

$$\widehat{U}_i^{(\mathrm{pmu})} = \left(\widehat{x}_{(i)\mathrm{re}}^{(\mathrm{pmu})}\right)^2 + \left(\widehat{x}_{(i)\mathrm{im}}^{(\mathrm{pmu})}\right)^2 \tag{7.23}$$

$$\widehat{K}_{ij}^{(\mathrm{pmu})} = \widehat{x}_{(i)\mathrm{re}}^{(\mathrm{pmu})}\widehat{x}_{(j)\mathrm{re}}^{(\mathrm{pmu})} + \widehat{x}_{(i)\mathrm{im}}^{(\mathrm{pmu})}\widehat{x}_{(j)\mathrm{im}}^{(\mathrm{pmu})} \tag{7.24}$$

$$\widehat{L}_{ij}^{(\mathrm{pmu})} = \widehat{x}_{(i)\mathrm{im}}^{(\mathrm{pmu})}\widehat{x}_{(j)\mathrm{re}}^{(\mathrm{pmu})} - \widehat{x}_{(i)\mathrm{re}}^{(\mathrm{pmu})}\widehat{x}_{(j)\mathrm{im}}^{(\mathrm{pmu})} \tag{7.25}$$

where $\widehat{x}_{(i)\mathrm{re}}^{(\mathrm{pmu})}$ and $\widehat{x}_{(i)\mathrm{im}}^{(\mathrm{pmu})}$ are the real and imaginary parts of the estimated voltage phasor for bus i, respectively, which are obtained from (7.20). From the above relationships (7.23)–(7.25), the $(n + 2l) \times 2n$ Jacobian $F_y = \partial y/\partial x$ is evaluated at $x = \widehat{x}^{(\mathrm{pmu})}$, and its components are provided in Table 7.2.

Therefore, the error covariance matrix of $\widehat{y}^{(\mathrm{pmu})}$ is calculated as

$$R_y^{(\mathrm{pmu})} = F_y R_x^{(\mathrm{pmu})} F_y^{\mathrm{T}} \tag{7.26}$$

Using $\widehat{y}^{(\mathrm{pmu})}$, the estimated SCADA measurement vector $\widehat{z}_{\mathrm{scada}}^{(\mathrm{pmu})}$ and its error covariance matrix $M_{\mathrm{scada}}^{(\mathrm{pmu})}$ are given by

$$\widehat{z}_{\mathrm{scada}}^{(\mathrm{pmu})} = A\widehat{y}^{(\mathrm{pmu})} \tag{7.27}$$

$$M_{\mathrm{scada}}^{(\mathrm{pmu})} = AR_y^{(\mathrm{pmu})}A^{\mathrm{T}} \tag{7.28}$$

Thus, considering the SCADA measurement vector z_{scada}, the following difference vector and its covariance matrix are computed, respectively, as [11]

$$r_{\mathrm{scada}}^{(\mathrm{pmu})} = z_{\mathrm{scada}} - \widehat{z}_{\mathrm{scada}}^{(\mathrm{pmu})} \tag{7.29}$$

$$J_{\mathrm{scada}}^{(\mathrm{pmu})} = R_{\mathrm{SCADA}} + M_{\mathrm{scada}}^{(\mathrm{pmu})} \tag{7.30}$$

The components of $r_{\mathrm{scada}}^{(\mathrm{pmu})}$ are normalized as follows:

$$r_{N\mathrm{scada}}^{(\mathrm{pmu})}(i) = \frac{r_{\mathrm{scada}}^{(\mathrm{pmu})}(i)}{\sqrt{J_{\mathrm{scada}}^{(\mathrm{pmu})}(i,i)}} \tag{7.31}$$

A flag vector α with the same dimension as z_{scada} is defined as

$$\alpha(i) = \begin{cases} 0, & r_{N\text{scada}}^{(\text{pmu})}(i) \leq \text{threshold} \\ 1, & r_{N\text{scada}}^{(\text{pmu})}(i) > \text{threshold} \end{cases} \tag{7.32}$$

Similarly, $\hat{y}^{(\text{scada})}$ is used to calculate the estimated SCADA measurement vector $\hat{z}_{\text{scada}}^{(\text{scada})}$ and its error covariance matrix $M_{\text{scada}}^{(\text{scada})}$, respectively, using

$$\hat{z}_{\text{scada}}^{(\text{scada})} = A\hat{y}^{(\text{scada})} \tag{7.33}$$

$$M_{\text{scada}}^{(\text{scada})} = AR_y^{(\text{scada})}A^{\text{T}} \tag{7.34}$$

Then, the estimation residual vector and its covariance matrix are given by

$$r_{\text{scada}}^{(\text{scada})} = z_{\text{scada}} - \hat{z}_{\text{scada}}^{(\text{scada})} \tag{7.35}$$

$$J_{\text{scada}}^{(\text{scada})} = R_{\text{SCADA}} - M_{\text{scada}}^{(\text{scada})} \tag{7.36}$$

The components of $r_{\text{scada}}^{(\text{scada})}$ are normalized as

$$r_{N\text{scada}}^{(\text{scada})}(i) = \frac{r_{\text{scada}}^{(\text{scada})}(i)}{\sqrt{J_{\text{scada}}^{(\text{scade})}(i,i)}} \tag{7.37}$$

and a flag vector β with the same size of z_{scada} is defined as

$$\beta(i) = \begin{cases} 0, & r_{N\text{scada}}^{(\text{scada})}(i) \leq \text{threshold} \\ 1, & r_{N\text{scada}}^{(\text{scada})}(i) > \text{threshold} \end{cases} \tag{7.38}$$

Table 7.3 presents four scenarios of BD indication and corresponding diagnosis.

We assume that BD in SCADA and PMU measurements do not occur simultaneously. *Scenario 1:* If flag vectors α and β are zero vectors, that is, no threshold violations in either $r_{N\text{scada}}^{(\text{pmu})}$ or $r_{N\text{scada}}^{(\text{scada})}$, then there are no BD in the SCADA and PMU measurements. *Scenario 2:* In the case of threshold violations in $r_{N\text{scada}}^{(\text{scada})}$ indicated by the nonzero value of the sum of all components in flag vector β and no threshold violations in $r_{N\text{scada}}^{(\text{pmu})}$, bad SCADA measurements occur in the PMU-unobservable area. Then, the conventional BD processing method, that is, repeated application

Table 7.3 Diagnosis of bad data

	Bad data indication	**Diagnosis**
1	$\text{sum}(\alpha) = 0$ and $\text{sum}(\beta) = 0$	No BD
2	$\text{sum}(\alpha) = 0$ and $\text{sum}(\beta) \neq 0$	BD in z_{scada} (all of these BD are in the PMU-unobservable area)
3	$\text{sum}(\alpha) \neq 0$ and $\text{sum}(\beta) \neq 0$	BD in z_{scada} (some/all of these BD are in the PMU-observable area)
4	$\text{sum}(\alpha) \neq 0$ and $\text{sum}(\beta) = 0$	BD in PMU measurements

of BD detection, identification, correction, and re-estimation, is performed to eliminate the BD one-by-one. As shown in (7.11), if a bad voltage magnitude measurement exists in the PMU-unobservable area, the variance of its squared version is corrupted and cannot be used in the correction step. Thus, the bad voltage magnitude measurements in this scenario should be deleted, whereas other bad SCADA measurements, such as active/reactive power flows and power injections, are corrected as follows:

$$z_{\text{scada}}^{\text{new}}(i) = z_{\text{scada}}^{\text{bad}}(i) - \frac{R_{\text{SCADA}}(i,i)}{J_{\text{scada}}^{(\text{scada})}(i,i)} r_{\text{scada}}^{(\text{scada})}(i) \tag{7.39}$$

Scenario 3: Nonzero vectors α and β indicate the presence of gross errors in z_{scada}, and some or all of these BD occur within the network area observed by PMU measurements. First, the suspect SCADA measurements flagged by α can easily be replaced all at once by their corresponding estimates obtained from (7.27), weighted according to the diagonal elements of the matrix $M_{\text{scada}}^{(\text{pmu})}$ defined in (7.28). The intermediate state vector y is then re-estimated using the updated SCADA measurements. Lastly, the normalized estimation residual vector is recalculated to check whether there are still bad SCADA measurements in the PMU-unobservable area. If elements in $r_{N\text{scada}}^{(\text{scada})}$ exceed the predefined threshold, the bad SCADA measurements can only be eliminated one-by-one using a conventional BD processing scheme, as in Scenario 2. *Scenario 4:* If zero vector β validates SCADA measurements but elements of $r_{N\text{scada}}^{(\text{pmu})}$ exceed the detection threshold, one can conclude that the estimated measurement $\hat{z}_{\text{scada}}^{(\text{pmu})}$ obtained in (7.27) is contaminated by the presence of BD in PMU measurements. In this situation, $\hat{y}^{(\text{scada})}$ obtained in (7.12) is valid and can be used to obtain state vector $\hat{x}^{(\text{scada})}$ (in rectangular form) and its error covariance matrix $R_x^{(\text{scada})}$ as per Section 7.2.4. Then, the estimated PMU measurements $\hat{z}_{\text{pmu}}^{(\text{scada})}$ and the error covariance matrix $M_{\text{pmu}}^{(\text{scada})}$ are given by

$$\hat{z}_{\text{pmu}}^{(\text{scada})} = B\hat{x}^{(\text{scada})} \tag{7.40}$$

$$M_{\text{pmu}}^{(\text{scada})} = BR_x^{(\text{scada})}B^{\text{T}} \tag{7.41}$$

Similarly, the following differences, $r_{\text{pmu}}^{(\text{scada})}$, and its error covariance matrix, $J_{\text{pmu}}^{(\text{scada})}$, can be computed as follows:

$$r_{\text{pmu}}^{(\text{scada})} = z_{\text{pmu}} - \hat{z}_{\text{pmu}}^{(\text{scada})} \tag{7.42}$$

$$J_{\text{pmu}}^{(\text{scada})} = R_{\text{PMU}}^{r} + M_{\text{pmu}}^{(\text{scada})} \tag{7.43}$$

The components of $r_{\text{pmu}}^{(\text{scada})}$ are normalized and flagged as (7.44) and (7.45), respectively:

$$r_{N\text{pmu}}^{(\text{scada})}(i) = \frac{r_{\text{pmu}}^{(\text{scada})}(i)}{\sqrt{J_{\text{pmu}}^{(\text{scada})}(i,i)}} \tag{7.44}$$

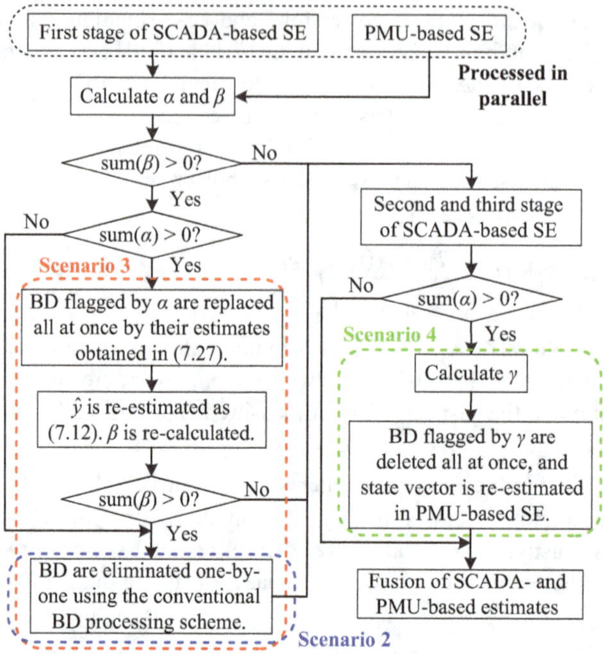

Figure 7.1 Architecture of proposed SE and BD processing scheme

$$\gamma(i) = \begin{cases} 0, & r_{N\text{pmu}}^{(\text{scada})}(i) \leq \text{threshold} \\ 1, & r_{N\text{pmu}}^{(\text{scada})}(i) > \text{threshold} \end{cases} \tag{7.45}$$

Compared to tele-measurement, the estimated PMU measurements $\hat{z}_{\text{pmu}}^{(\text{scada})}$ are not sufficiently accurate. Furthermore, as shown in (7.14) and (7.15), the covariance matrix R_{PMU}^{r} is contaminated if there are BD in PMU measurements and therefore cannot be used to correct the BD in the hybrid estimator [14] and the alternative estimator [17] mentioned in Sections 7.3 and 7.4. Thus, instead of replacing the BD by their estimates obtained in (7.40), the spurious PMU measurements flagged by the vector γ are deleted all at once. The block diagram depicted in Figure 7.1 represents the proposed SE and BD processing scheme. The PMU-based SE and the first stage of SCADA-based SE are processed in parallel.

7.2.4 Second and third stages of SCADA-based SE

So far, we can confirm no BD in the SCADA measurements. The state vector is computed through the second and third stages of SCADA-based SE as follows.

In the second stage, a new vector u, composed similar to y of $n + 2l$ variables, is obtained by performing an explicit nonlinear transformation [21]:

$$u = f_u(y) \tag{7.46}$$

The components of vector u,

$$u = \{\zeta_i \quad \zeta_{ij} \quad \theta_{ij}\} \tag{7.47}$$

can be explicitly related to those of vector y as follows:

$$\zeta_i = \ln U_i \tag{7.48}$$

$$\zeta_{ij} = \ln\left(K_{ij}^2 + L_{ij}^2\right) \tag{7.49}$$

$$\theta_{ij} = \arctan\left(L_{ij}/K_{ij}\right) \tag{7.50}$$

From the above explicit nonlinear relationships (7.48)–(7.50), the $(n+2l)\times(n+2l)$ Jacobian $F_u = \partial u/\partial y$ is evaluated at $y = \widehat{y}^{(\text{scada})}$, and its components are given as follows:

$$\partial\zeta_i/\partial U_i = 1/U_i \tag{7.51}$$

$$\begin{bmatrix} \partial\zeta_{ij}/\partial K_{ij} & \partial\zeta_{ij}/\partial L_{ij} \\ \partial\theta_{ij}/\partial K_{ij} & \partial\theta_{ij}/\partial L_{ij} \end{bmatrix} = \frac{1}{K_{ij}^2 + L_{ij}^2}\begin{bmatrix} 2K_{ij} & 2L_{ij} \\ -L_{ij} & K_{ij} \end{bmatrix} \tag{7.52}$$

Then, the error covariance matrix of u is calculated as follows:

$$R_u = F_u R_y^{(\text{scada})} F_u^{\mathrm{T}} \tag{7.53}$$

In the final stage, the state vector s is slightly modified in such a way that the new vector u is a linear function thereof. For this purpose, u and s in blocked form are expressed as

$$u = [\zeta^T \quad \zeta_b^T \quad \theta_b^T]^T; \qquad s = [\zeta^T \quad \theta^T]^T$$

where the subscript "b" denotes the set of branch variables, and logarithmic version ζ is chosen as the state rather than the ordinary voltage magnitude. Then, the linear measurement model in blocked form is given by

$$u = Ds + e_u \Rightarrow \begin{bmatrix} \zeta \\ \zeta_b \\ \theta_b \end{bmatrix} = \begin{bmatrix} I & 0 \\ |S^{\mathrm{T}}| & 0 \\ 0 & N^{\mathrm{T}} \end{bmatrix}\begin{bmatrix} \zeta \\ \theta \end{bmatrix} + e_u \tag{7.54}$$

where S is the branch-to-node incidence matrix and N is a reduced matrix obtained by omitting the reference phase angle in S. The WLS estimate $\widehat{s}$ and its error covariance matrix can be computed as follows:

$$\widehat{s} = G_d^{-1} D^{\mathrm{T}} R_u^{-1} \widehat{u} \tag{7.55}$$

$$R_s = G_d{}^{-1} \tag{7.56}$$

where $G_d = D^{\mathrm{T}} R_u{}^{-1} D$ is the gain matrix and $\widehat{u}$ is obtained from (7.47)–(7.50). Once ζ in $\widehat{s}$ is available, voltage magnitude $V = 0.5\exp(\zeta)$ can be readily obtained. The relationship between the incremental representation of the state vector in polar form $x_p = \left[V^{\mathrm{T}}\ \theta^{\mathrm{T}}\right]^T$ and the state vector in (7.54) is given by

$$\Delta x_p = F_x \Delta s \Rightarrow \begin{bmatrix} \Delta V \\ \Delta\theta \end{bmatrix} = \begin{bmatrix} 0.5\exp(\zeta) & \\ & I \end{bmatrix} \begin{bmatrix} \Delta\zeta \\ \Delta\theta \end{bmatrix} \tag{7.57}$$

where Jacobian $F_x = \partial x / \partial s$ is evaluated at $s = \widehat{s}$. The error covariance matrix of x_{p}, R^p_{SCADA}, is shown as follows:

$$R^p_{\mathrm{SCADA}} = F_x R_s F_x{}^{\mathrm{T}} \tag{7.58}$$

To formulate the fusion step, $\widehat{x}_p$ should be transformed to achieve the state vector in rectangular coordinates $\widehat{x}^{(\mathrm{scada})}$. Meanwhile, the error covariance matrix of $\widehat{x}^{(\mathrm{scada})}$ is given by

$$R_x^{(\mathrm{scada})} = T_{\mathrm{SCADA}} R^p_{\mathrm{SCADA}} (T_{\mathrm{SCADA}})^{\mathrm{T}} \tag{7.59}$$

where rotation matrix T_{SCADA} is similar to that in (7.15) but different in detail and size.

7.2.5 *Fusion of SCADA- and PMU-based estimates*

The final state estimate, $\widehat{x}^{(\mathrm{final})}$, can be computed as [10]

$$\widehat{x}^{(\mathrm{final})} = W_1 \widehat{x}^{(\mathrm{scada})} + W_2 \widehat{x}^{(\mathrm{pmu})} \tag{7.60}$$

where the weights in the above equation are given by

$$W_1 = R_x^{(\mathrm{pmu})} \left(R_x^{(\mathrm{scada})} + R_x^{(\mathrm{pmu})} \right)^{-1} \tag{7.61}$$

$$W_2 = R_x^{(\mathrm{scada})} \left(R_x^{(\mathrm{scada})} + R_x^{(\mathrm{pmu})} \right)^{-1} \tag{7.62}$$

As mentioned in Section 7.2.2, large variances are assigned to the PMU-unobservable states, so the estimates provided for those states in PMU-based SE receive very small weights. The corresponding SCADA-based estimates eventually prevail at the fusion step and the final estimates are not contaminated by inaccurate PMU-unobservable states.

7.3 Hybrid state estimator

To incorporate PMU measurements into conventional SE, this section develops a nonlinear WLS estimator with current phasor measurements either in rectangular or in polar coordinates [14].

7.3.1 State estimation formulation

Consider a power system having N buses. The SE measurement model is given as follows:

$$z = \begin{bmatrix} z_1 \\ z_2 \\ \vdots \\ z_m \end{bmatrix} = \begin{bmatrix} h_1(x_1, x_2, \ldots, x_n) \\ h_2(x_1, x_2, \ldots, x_n) \\ \vdots \\ h_m(x_1, x_2, \ldots, x_n) \end{bmatrix} + \begin{bmatrix} e_1 \\ e_2 \\ \vdots \\ e_m \end{bmatrix} = h(x) + e \tag{7.63}$$

where z is the $m \times 1$ measurement vector, which is composed of conventional measurements provided by SCADA system (active and reactive power flows, active and reactive power injections, and voltage magnitudes) and synchrophasor measurements obtained from PMUs (voltage and current phasors); x represents the system state vector with dimension ($n \times 1$); $h^{\mathrm{T}} = [h_1(x), h_2(x), \ldots, h_m(x)]$; $h_i(x)$ is the function relating measurement i to the state vector x; and $e^{\mathrm{T}} = [e_1, e_2, \ldots, e_m]$ denotes the vector of measurement errors. The errors are commonly assumed to be independent and uncorrelated, having a Gaussian distribution with zeros mean $E(e_i) = 0,\ i = 1, 2, \ldots, m$ and covariance $R = \mathrm{cov}(e) = diag\{\sigma_1^2, \sigma_2^2, \ldots, \sigma_m^2\}$, where σ_i represents the standard deviation of the ith measurement.

A typical two-bus system (Figure 7.2) is used to illustrate the mathematical model of the hybrid SE problem. The series admittance of branch $i - k$ is denoted by $y_{ik} = g_{ik} + jb_{ik}$, and the shunt admittance between bus i and the ground is represented by $y_{sh,i} = g_{sh,i} + jb_{sh,i}$. The voltage phasors at bus i and bus k are denoted by $V_i\angle\delta_i$ and $V_k\angle\delta_k$, respectively. The current phasor from bus i to bus k is represented by $I_{Ik}\angle\theta_{ik}$. The expressions for conventional measurements as functions of system state vector and branch parameters can be found in [4].

According to Kirchhoff's circuit law (KCL), the current phasor $I_{Ik}\angle\theta_{ik}$ can be expressed as

$$I_{ik}\angle\theta_{ik} = \left(g_{sh,i} + jb_{sh,i}\right)V_i\angle\delta_i + (g_{ik} + jb_{ik})(V_i\angle\delta_i - V_k\angle\delta_k) \tag{7.64}$$

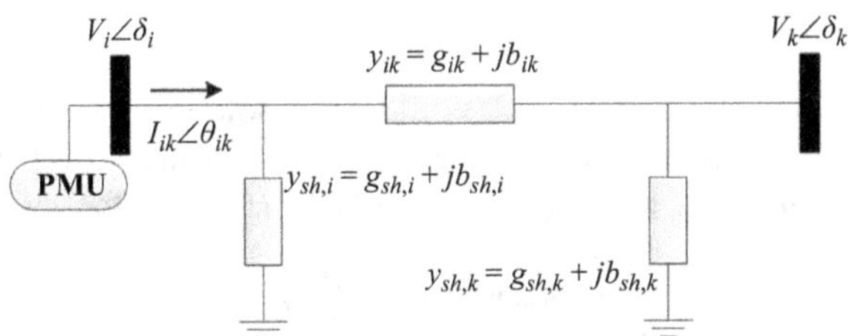

Figure 7.2 Two-bus system having a PMU at one end

If the current phasor is expressed in rectangular coordinates, the real and imaginary parts of the current phasor can be written as

$$I_{ik}^{r} = V_i\left[\left(g_{sh,i} + g_{ik}\right)\cos\delta_i - \left(b_{sh,i} + b_{ik}\right)\sin\delta_i\right] - V_k\left[g_{ik}\cos\delta_k - b_{ik}\sin\delta_k\right] \tag{7.65}$$

$$I_{ik}^{i} = V_i\left[\left(g_{sh,i} + g_{ik}\right)\sin\delta_i + \left(b_{sh,i} + b_{ik}\right)\cos\delta_i\right] - V_k\left[g_{ik}\sin\delta_k + b_{ik}\cos\delta_k\right] \tag{7.66}$$

where the superscripts "*r*" and "*i*" represent the real and imaginary parts of the current phasor, respectively.

If the current phasor is expressed in polar coordinates, its magnitude and phase angle can be written as

$$I_{ik} = \sqrt{\left(I_{ik}^{r}\right)^2 + \left(I_{ik}^{i}\right)^2} \tag{7.67}$$

$$\theta_{ik} = \arctan\left(\frac{I_{ik}^{i}}{I_{ik}^{r}}\right) \tag{7.68}$$

With reference to Figure 7.2, the power factor angle $\varphi_{ik} = \delta_i - \theta_{ik}$. From (7.65)–(7.68), we can obtain

$$I_{ik} = \sqrt{A_{ik}V_i^2 + B_{ik}V_k^2 + 2C_{ik}V_iV_k} \tag{7.69}$$

$$\theta_{ik} = \arctan\left(\frac{V_i\left[\left(g_{sh,i}+g_{ik}\right)\sin\delta_i + \left(b_{sh,i}+b_{ik}\right)\cos\delta_i\right] - V_k\left[g_{ik}\sin\delta_k + b_{ik}\cos\delta_k\right]}{V_i\left[\left(g_{sh,i}+g_{ik}\right)\cos\delta_i - \left(b_{sh,i}+b_{ik}\right)\sin\delta_i\right] - V_k\left[g_{ik}\cos\delta_k - b_{ik}\sin\delta_k\right]}\right) \tag{7.70}$$

where

$$A_{ik} = \left(g_{sh,i} + g_{ik}\right)^2 + \left(b_{sh,i} + b_{ik}\right)^2 \tag{7.71}$$

$$B_{ik} = g_{ik}^2 + b_{ik}^2 \tag{7.72}$$

$$C_{ik} = \left(g_{ik}b_{sh,i} - b_{ik}g_{sh,i}\right)\sin\left(\delta_i - \delta_k\right) - \left(g_{ik}^2 + b_{ik}^2 + g_{ik}g_{sh,i} + b_{ik}b_{sh,i}\right)\cos\left(\delta_i - \delta_k\right) \tag{7.73}$$

To obtain the optimal state estimates, the NE is formulated as follows:

$$G(x^{\lambda})\Delta x^{\lambda+1} = H^{\mathrm{T}}(x^{\lambda})R^{-1}\left[z - h(x^{\lambda})\right] \tag{7.74}$$

where λ represents the iteration index; x^{λ} is the solution vector at iteration λ; $\Delta x^{\lambda+1} = x^{\lambda+1} - x^{\lambda}$; $H(x) = \partial h(x)/\partial x$ denotes the measurement Jacobian matrix with dimension ($m \times n$); and $G(x^{\lambda}) = H^{\mathrm{T}}(x^{\lambda})R^{-1}H(x^{\lambda})$ is the gain matrix with dimension ($n \times n$).

The Jacobian matrix is denoted as H_p and H_r, when using polar coordinates and rectangular coordinates for branch current phasors, respectively (polar coordinates are used for bus voltage phasors in both cases). The structures of the two Jacobian matrices are shown as (7.75) and (7.76) in which P_{ik} and Q_{ik} represent the active and reactive power flows on branch i–k, respectively; P_i and Q_i are the active and reactive power injection at bus i, respectively.

$$H_p = \begin{bmatrix} \partial P_{ik}/\partial\delta & \partial P_{ik}/\partial V \\ \partial Q_{ik}/\partial\delta & \partial Q_{ik}/\partial V \\ \partial P_i/\partial\delta & \partial P_i/\partial V \\ \partial Q_i/\partial\delta & \partial Q_i/\partial V \\ \partial\delta_i/\partial\delta & \partial\delta_i/\partial V \\ \partial V_i/\partial\delta & \partial V_i/\partial V \\ \partial\theta_{ik}/\partial\delta & \partial\theta_{ik}/\partial V \\ \partial I_{ik}/\partial\delta & \partial I_{ik}/\partial V \end{bmatrix} \begin{matrix} \left.\begin{matrix} \text{Power flow} \\ \\ \text{Power injection} \end{matrix}\right\} \begin{matrix}\text{SCADA}\\ \text{measurements}\end{matrix} \\ \\ \left.\begin{matrix} \text{Voltage phasor} \\ \\ \text{Current phasor} \end{matrix}\right\} \begin{matrix}\text{SCADA}\\ \text{measurements}\end{matrix} \end{matrix} \tag{7.75}$$

$$H_r = \begin{bmatrix} \partial P_{ik}/\partial\delta & \partial P_{ik}/\partial V \\ \partial Q_{ik}/\partial\delta & \partial Q_{ik}/\partial V \\ \partial P_i/\partial\delta & \partial P_i/\partial V \\ \partial Q_i/\partial\delta & \partial Q_i/\partial V \\ \partial\delta_i/\partial\delta & \partial\delta_i/\partial V \\ \partial V_i/\partial\delta & \partial V_i/\partial V \\ \partial I_{ik}^r/\partial\delta & \partial I_{ik}^r/\partial V \\ \partial I_{ik}^i/\partial\delta & \partial I_{ik}^i/\partial V \end{bmatrix} \begin{matrix} \left.\begin{matrix} \text{Power flow} \\ \\ \text{Power injection} \end{matrix}\right\} \begin{matrix}\text{SCADA}\\ \text{measurements}\end{matrix} \\ \\ \left.\begin{matrix} \text{Voltage phasor} \\ \\ \text{Current phasor} \end{matrix}\right\} \begin{matrix}\text{PMU}\\ \text{measurements}\end{matrix} \end{matrix} \tag{7.76}$$

The expression of partial derivatives for the SCADA measurements as well as magnitude and phase angle of voltage phasor measurements with respect to state variables, that is, bus voltage magnitudes and phase angles, can be found in [4].

For H_p, the magnitude, I_{ik}, and phase angle, θ_{ik}, of current phasor measurement with respect to state variables at buses i and k, are given as follows:

$$\frac{\partial\theta_{ik}}{\partial\delta_i} = \frac{A_{ik}V_i^2 + C_{ik}V_iV_k}{D_{ik}} \qquad \frac{\partial\theta_{ik}}{\partial\delta_k} = \frac{B_{ik}V_k^2 + C_{ik}V_iV_k}{D_{ik}} \tag{7.77}$$

$$\frac{\partial\theta_{ik}}{\partial V_i} = -\frac{E_{ik}V_k}{D_{ik}} \qquad \frac{\partial\theta_{ik}}{\partial V_k} = \frac{E_{ik}V_i}{D_{ik}} \tag{7.78}$$

$$\frac{\partial I_{ik}}{\partial\delta_i} = \frac{E_{ik}V_iV_k}{\sqrt{D_{ik}}} \qquad \frac{\partial I_{ik}}{\partial\delta_k} = -\frac{E_{ik}V_iV_k}{\sqrt{D_{ik}}} \tag{7.79}$$

$$\frac{\partial I_{ik}}{\partial V_i} = \frac{A_{ik}V_i + C_{ik}V_k}{\sqrt{D_{ik}}} \qquad \frac{\partial I_{ik}}{\partial V_k} = \frac{B_{ik}V_k + C_{ik}V_i}{\sqrt{D_{ik}}} \tag{7.80}$$

where A_{ik}, B_{ik}, and C_{ik} are shown in (7.71), (7.72), and (7.73), respectively, and

$$D_{ik} = A_{ik}V_i^2 + B_{ik}V_k^2 + 2C_{ik}V_iV_k \tag{7.81}$$

$$\begin{aligned} E_{ik} = &\left(g_{ik}^2 + b_{ik}^2 + g_{ik}g_{sh,i} + b_{ik}b_{sh,i}\right)\sin(\delta_i - \delta_k) \\ &+ \left(g_{ik}b_{sh,i} - b_{ik}g_{sh,i}\right)\cos(\delta_i - \delta_k) \end{aligned} \tag{7.82}$$

As shown in (7.73) and (7.77)–(7.82), in the case of polar coordinates, the phase angle difference $\delta_i - \delta_k$ is needed when calculating the partial derivatives for branch current phasor measurements with respect to state variables.

The expressions of partial derivatives for the current phasor measurements in rectangular coordinates with respect to state variables at buses i and k are given as follows:

$$\frac{\partial I_{ik}^r}{\partial \delta_i} = -V_i\left[\left(g_{sh,i} + g_{ik}\right)\sin\delta_i + \left(b_{sh,i} + b_{ik}\right)\cos\delta_i\right] \tag{7.83}$$

$$\frac{\partial I_{ik}^r}{\partial \delta_k} = V_k\left[g_{ik}\sin\delta_k + b_{ik}\cos\delta_k\right] \tag{7.84}$$

$$\frac{\partial I_{ik}^r}{\partial V_i} = \left(g_{sh,i} + g_{ik}\right)\cos\delta_i - \left(b_{sh,i} + b_{ik}\right)\sin\delta_i \tag{7.85}$$

$$\frac{\partial I_{ik}^r}{\partial V_k} = b_{ik}\sin\delta_k - g_{ik}\cos\delta_k \tag{7.86}$$

$$\frac{\partial I_{ik}^i}{\partial \delta_i} = V_i\left[\left(g_{sh,i} + g_{ik}\right)\cos\delta_i - \left(b_{sh,i} + b_{ik}\right)\sin\delta_i\right] \tag{7.87}$$

$$\frac{\partial I_{ik}^i}{\partial \delta_k} = V_k\left[b_{ik}\sin\delta_k - g_{ik}\cos\delta_k\right] \tag{7.88}$$

$$\frac{\partial I_{ik}^i}{\partial V_i} = \left(g_{sh,i} + g_{ik}\right)\sin\delta_i + \left(b_{sh,i} + b_{ik}\right)\cos\delta_i \tag{7.89}$$

$$\frac{\partial I_{ik}^i}{\partial V_k} = -g_{ik}\sin\delta_k - b_{ik}\cos\delta_k \tag{7.90}$$

We can see that the absolute phase angles, that is, δ_i and δ_k, are used in (7.83)–(7.90) when using rectangular coordinates.

7.3.2 State estimation formulation

Using branch current phasor measurements in an un-appropriate way may lead to various numerical problems, and further deteriorate the SE performance. For a flat start ($V_i = V_k = 1$ and $\delta_i = \delta_k = 0$) or a lightly loaded branch i–k ($V_i \approx V_k$ and $\delta_i \approx \delta_k$), the Jacobian elements for the branch current phasor in polar coordinates are provided in Table 7.4.

As shown in Table 7.4, if the shunt conductance $g_{sh,i}$ and susceptance $b_{sh,i}$ of the PMU bus i almost equal to zero, such as a short line or a transformer, all the

Table 7.4 Jacobian elements of branch current phasor in polar coordinates for a flat start or a lightly loaded line

	Flat start	Lightly loaded line
$\frac{\partial \theta_{ik}}{\partial V_i}$	$\frac{b_{ik}g_{sh,i} - g_{ik}b_{sh,i}}{g_{sh,i}^2 + b_{sh,i}^2}$	$\frac{b_{ik}g_{sh,i} - g_{ik}b_{sh,i}}{\left(g_{sh,i}^2 + b_{sh,i}^2\right)V_i}$
$\frac{\partial \theta_{ik}}{\partial V_k}$	$\frac{g_{ik}b_{sh,i} - b_{ik}g_{sh,i}}{g_{sh,i}^2 + b_{sh,i}^2}$	$\frac{g_{ik}b_{sh,i} - b_{ik}g_{sh,i}}{\left(g_{sh,i}^2 + b_{sh,i}^2\right)V_i}$
$\frac{\partial I_{ik}}{\partial \delta_i}$	$\frac{g_{ik}b_{sh,i} - b_{ik}g_{sh,i}}{\sqrt{g_{sh,i}^2 + b_{sh,i}^2}}$	$\frac{g_{ik}b_{sh,i} - b_{ik}g_{sh,i}}{\sqrt{g_{sh,i}^2 + b_{sh,i}^2}}V_i$
$\frac{\partial I_{ik}}{\partial \delta_k}$	$\frac{b_{ik}g_{sh,i} - g_{ik}b_{sh,i}}{\sqrt{g_{sh,i}^2 + b_{sh,i}^2}}$	$\frac{b_{ik}g_{sh,i} - g_{ik}b_{sh,i}}{\sqrt{g_{sh,i}^2 + b_{sh,i}^2}}V_i$
$\frac{\partial \theta_{ik}}{\partial \delta_i}$	$1 + \frac{g_{ik}g_{sh,i} + b_{ik}b_{sh,i}}{g_{sh,i}^2 + b_{sh,i}^2}$	
$\frac{\partial \theta_{ik}}{\partial \delta_k}$	$-\frac{g_{ik}g_{sh,i} + b_{ik}b_{sh,i}}{g_{sh,i}^2 + b_{sh,i}^2}$	
$\frac{\partial I_{ik}}{\partial V_i}$	$\frac{g_{sh,i}^2 + b_{sh,i}^2 + g_{ik}g_{sh,i} + b_{ik}b_{sh,i}}{\sqrt{g_{sh,i}^2 + b_{sh,i}^2}}$	
$\frac{\partial I_{ik}}{\partial V_k}$	$-\frac{g_{ik}g_{sh,i} + b_{ik}b_{sh,i}}{\sqrt{g_{sh,i}^2 + b_{sh,i}^2}}$	

Table 7.5 Jacobian elements of branch current phasor in rectangular coordinates for a flat start or a lightly loaded line

	Flat start	Lightly loaded line
$\frac{\partial I_{ik}^r}{\partial \delta_i}$	$-b_{sh,i} - b_{ik}$	$-V_i\left[\left(g_{sh,i} + g_{ik}\right)\sin\delta_i + \left(b_{sh,i} + b_{ik}\right)\cos\delta_i\right]$
$\frac{\partial I_{ik}^r}{\partial \delta_k}$	b_{ik}	$V_i[g_{ik}\sin\delta_i + b_{ik}\cos\delta_i]$
$\frac{\partial I_{ik}^r}{\partial V_i}$	$g_{sh,i} + g_{ik}$	$\left(g_{sh,i} + g_{ik}\right)\cos\delta_i - \left(b_{sh,i} + b_{ik}\right)\sin\delta_i$
$\frac{\partial I_{ik}^r}{\partial V_k}$	$-g_{ik}$	$b_{ik}\sin\delta_i - g_{ik}\cos\delta_i$
$\frac{\partial I_{ik}^i}{\partial \delta_i}$	$g_{sh,i} + g_{ik}$	$V_i\left[\left(g_{sh,i} + g_{ik}\right)\cos\delta_i - \left(b_{sh,i} + b_{ik}\right)\sin\delta_i\right]$
$\frac{\partial I_{ik}^i}{\partial \delta_k}$	$-g_{ik}$	$V_i[b_{ik}\sin\delta_i - g_{ik}\cos\delta_i]$
$\frac{\partial I_{ik}^i}{\partial V_i}$	$b_{sh,i} + b_{ik}$	$\left(g_{sh,i} + g_{ik}\right)\sin\delta_i + \left(b_{sh,i} + b_{ik}\right)\cos\delta_i$
$\frac{\partial I_{ik}^i}{\partial V_k}$	$-b_{ik}$	$-g_{ik}\sin\delta_i - b_{ik}\cos\delta_i$

Jacobian elements will be undefined. The rectangular coordinates for branch current phasor measurements are used to circumvent this issue. Table 7.5 illustrates the Jacobian elements of current phasor measurements in rectangular coordinates for a flat start and a lightly loaded line.

7.3.3 Bad data processing scheme

The largest normalized residual test is adopted in the hybrid state estimator to identify the BD in conventional and phasor measurements.

After obtaining the state estimates vector $\widehat{x}$, the measurement residual of the ith measurement can be calculated as follows:

$$r_i = z_i - h_i(\widehat{x}) \tag{7.91}$$

Then, the normalized residual of the ith measurement is obtained:

$$r_i^N = |r_i|/\sqrt{\Omega_{ii}} \tag{7.92}$$

where $\Omega = R - HG^{-1}H^{\mathrm{T}}$ and Ω_{ii} is the ith diagonal entry of Ω. The measurement having the largest normalized residual $r_{\max}^N > 3$ is suspected as BD. Once the largest normalized residual is found, the corresponding measurement is corrected as follows:

$$z_i^{\mathrm{new}} = z_i^{\mathrm{bad}} - \frac{R_{ii}}{\Omega_{ii}} r_i^{\mathrm{bad}} \tag{7.93}$$

where R_{ii} is the ith diagonal entry of the error covariance matrix of the measurement. Based on the corrected measurements, the system states are re-estimated by using the hybrid estimator once the identified bad measurement is updated. The above process will be repeated until all normalized residuals drop below the detection threshold. The re-estimation process starts from the most recent estimates instead of flat start and hence takes fewer iterations to converge.

7.4 An alternative state estimator

An alternative state estimator is described in this section to incorporate PMU measurements and the results of traditional SE in a postprocessing linear estimator [17].

7.4.1 *Traditional state estimator*

Consider a set of conventional measurements z_s. The measurement function is given as follows:

$$z_s = h_s(x) + e_s \tag{7.94}$$

where $h_s(x)$ is composed of nonlinear functions of the state vector x expressed in polar coordinates and e_s is the measurement error vector with a covariance matrix R_s. If the state estimate vector at the kth iteration is denoted by x^k, according to [4], the next iterate x^{k+1} is expressed by

$$x^{k+1} = x^k + G_s(x^k)^{-1} H_s^{\mathrm{T}}(x^k) R_s^{-1} \left[z_s - h_s(x^k) \right] \tag{7.95}$$

where H_s is the Jacobian matrix obtained by taking partial derivatives of h_s with respect to x and $G_s(x^k) = H_s^{\mathrm{T}}(x^k) R_s^{-1} H_s(x^k)$ is the gain matrix. It is assumed that BD have been eliminated from the conventional measurements by using the largest normalized residual test. Let the estimates of the traditional state estimator be denoted by x_s^{polar}, which is expressed in polar rectangular coordinates. Then its error covariance matrix is given by

$$\operatorname{cov}(x_s^{\text{polar}}) = G_s^{-1} = \left[H_s^{\mathrm{T}} R_s^{-1} H_s\right]^{-1} \tag{7.96}$$

7.4.2 *Adding phasor measurements in a posting-processing way*

To formulate the postprocessing step in a linear way, the rectangular coordinates are used to express the state vector and the measurements vector.

First, the estimates of the traditional state estimator in polar coordinates, x_s^{polar}, are transformed to that in rectangular coordinates, denoted by x_s. Then, the new measurement vector, zz, is composed of the estimates of the traditional state estimator, x_s, and the PMU measurement vector, z_p, all expressed in rectangular coordinates. The error covariance matrix of x_s^{polar} is $\operatorname{cov}(x_s^{\text{polar}})$, shown in (7.96), should be modified to reflect the transformation of the state estimates from polar to rectangular coordinates:

$$\operatorname{cov}(x_s) = T_{\text{SCADA}} \operatorname{cov}(x_s^{\text{polar}}) (T_{\text{SCADA}})^{\mathrm{T}} = T_{\text{SCADA}} G_s^{-1} (T_{\text{SCADA}})^{\mathrm{T}} \tag{7.97}$$

where rotation matrix T_{SCADA} is same as that in (7.59).

The combined measurement equation is illustrated as

$$zz \equiv \begin{bmatrix} x_s \\ z_p \end{bmatrix} \equiv \begin{bmatrix} x_s^r \\ x_s^i \\ V^r \\ V^i \\ I^r \\ I^i \end{bmatrix} = \begin{bmatrix} I & 0 \\ 0 & I \\ I' & 0 \\ 0 & I' \\ C_1 & C_2 \\ C_3 & C_4 \end{bmatrix} \begin{bmatrix} x^r \\ x^i \end{bmatrix} \equiv A \begin{bmatrix} x^r \\ x^i \end{bmatrix} \tag{7.98}$$

where x_s^r and x_s^i denote the real and imaginary parts of state estimates of conventional state estimator, respectively; V^r and V^i represent the real and imaginary parts of bus voltage phasor measurements, respectively; I^r and I^i are the real and imaginary parts of branch current phasor measurements, respectively; x^r and x^i are the state variables in rectangular coordinates; I denotes a unit matrix while I' is a unit matrix with zeros on the diagonal where no voltage phasors have been measured; constant matrices C_1–C_4 are composed of line conductance and susceptance for those lines where current phasor measurements are available; and A denotes the Jacobian matrix.

The WLS solution to the linear problem (7.98), $\widehat{x}$, can be directly obtained as

$$\widehat{x} = \left[A^{\mathrm{T}} R^{-1} A\right]^{-1} A^{\mathrm{T}} R^{-1} zz \tag{7.99}$$

where the covariance matrix R is composed of error covariance submatrices of results of traditional state estimator and phasor measurements obtained from PMU, shown as follows:

$$R = \begin{bmatrix} \mathrm{cov}(x_s) & 0 \\ 0 & \mathrm{cov}(z_p) \end{bmatrix} \tag{7.100}$$

If the phasor measurements derived from PMUs are stored and uploaded in polar form, the error covariance matrix of phasor measurements in rectangular coordinates, $\mathrm{cov}(z_p)$, can be obtained using (7.15).

7.5 Discussion

In this section, the simulation results corresponding to the IEEE 30-bus, 57-bus, and 118-bus test systems are presented and discussed. Four distinct estimators are employed as references to evaluate the proposed fast SE method (FSE) in terms of solution accuracy and computational effort: a hybrid estimator (HSE) [14], a conventional SCADA-based estimator (SSE) [4], a bilinear SCADA-based estimator (BSE) [21], and an alternative estimator (ASE) developed in [17]. As described in Section 7.3.3, using branch current phasor measurements in polar coordinates may lead to various numerical problems, thus the rectangular coordinates for current phasor measurements are used in this study to construct the HSE. For SCADA measurements, it is assumed that voltage measurements and power injections are taken at all buses, and power flows are taken across all branches ("from" terminal only). A PMU placed at a bus is assumed to measure the voltage phasor at that bus as well as the current phasors on all lines incident to that bus. The PMU placement configurations adopted in this study are documented in [8] and illustrated in Table 7.6. Table 7.7 gives the measurement types and the corresponding standard

Table 7.6 PMU locations in the IEEE test systems

Test systems	PMU locations	Observability
IEEE 30-bus	6, 10, 15, 27	69.5%
IEEE 57-bus	4, 10, 15, 20, 24, 29, 32, 37, 41, 48, 54	76.99%
IEEE 118-bus	11, 19, 30, 32, 40, 49, 59, 69, 89, 96, 103	51%

Table 7.7 Standard deviations for SCADA and PMU measurements

SCADA measurements			PMU measurements		
$\lvert V\rvert$	Power flows	Power injections	$\lvert V\rvert$	$\lvert I\rvert$	Phase angle
1%	2%	2%	0.1%	0.1%	0.0017 rad

deviations used in this study. The variance of pseudo-states used in Section 7.2.2 is 1×10^8.

7.5.1 Assessment of estimation accuracy

Figures 7.3, 7.4, and 7.5 illustrate the absolute estimation errors (averaged over 100 statistical trials) of five estimators associated with the estimates of the voltage phase angle and magnitude at each bus for the IEEE 30-bus, 57-bus, and 118-bus test systems, respectively. The FSE plot clearly shows that the limited PMU measurements can greatly enhance the results of the bilinear SCADA-based estimator, labeled as BSE, through a fusion stage. HSE can produce global optimal estimates. As demonstrated in [17], ASE algorithm can provide the same results as the nonlinear HSE scheme. Moreover, FSE estimation errors are almost coincident with those of the optimal HSE, which demonstrates that the pseudo-states with large variances used in PMU-based SE are filtered out in the fusion procedure and do not affect the final estimation results.

The mean errors of voltage phase angle and magnitude calculated for all estimators depicted in Table 7.8 further confirm the accuracy of the FSE. The SSE and BSE only use the inaccurate SCADA measurements, so their mean estimation

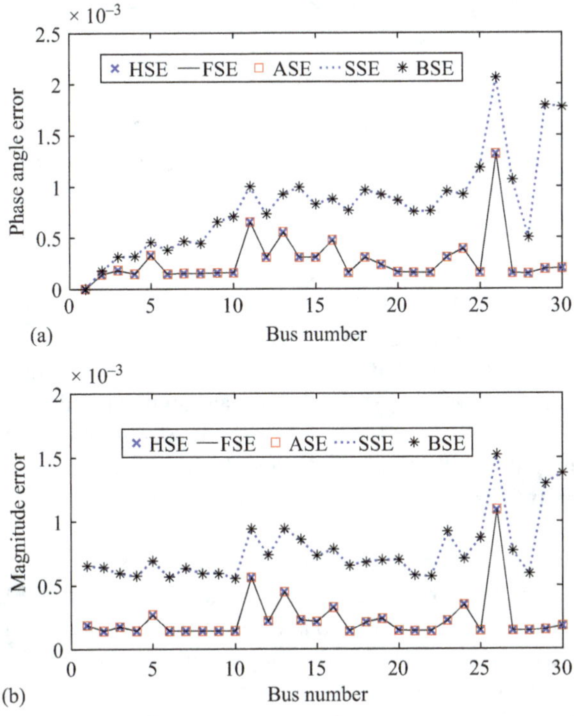

Figure 7.3 Absolute estimation errors for the IEEE 30-bus system without BD: (a) voltage phase angle error and (b) voltage magnitude error

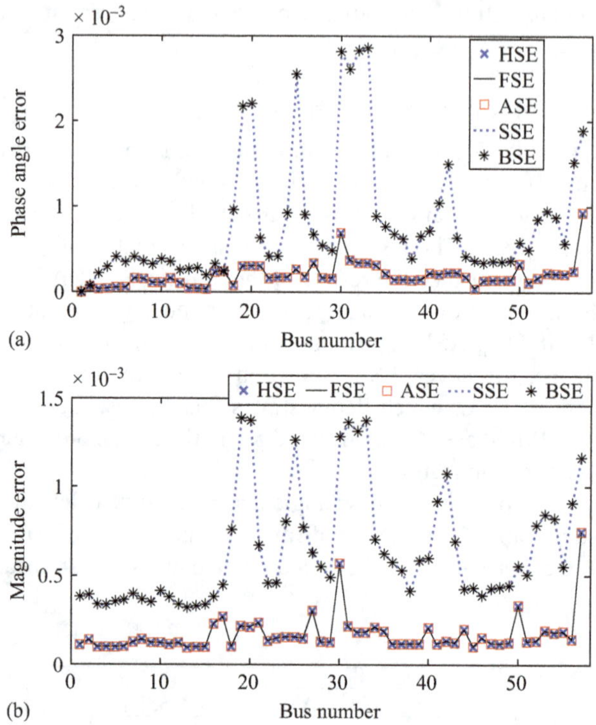

Figure 7.4 Absolute estimation errors for the IEEE 57-bus system without BD: (a) voltage phase angle error and (b) voltage magnitude error

errors are larger than those of the other three estimators. The SCADA measurement redundancy ratios of the IEEE 30-bus, 57-bus, and 118-bus test systems are equal to 2.92, 2.93, and 3.09, respectively. As expected, the estimation errors are lower for a higher redundancy. In addition, compared to SSE, the accuracy of the BSE deteriorates with the decrease in SCADA measurement redundancy level. As described in [17], the ASE can provide estimates having the same accuracy as those obtained by the optimal HSE strategy. Notice that the numerical indices computed for the three networks demonstrate the same accuracy as the FSE and HSE strategies.

The absolute estimation errors of the five estimators when four bad SCADA measurements in the PMU-observable area are introduced in the IEEE 30-bus test system are shown in Figure 7.6. Compared to the voltage phase angle and magnitude errors of SSE without BD, the accuracy of the SSE and BSE deteriorates at different levels due to the bad SCADA measurements. However, because these BD are located in the PMU-observable area, the results achieved by the HSE, FSE, and ASE strategies are barely distinguishable from the optimal one, that is, HSE without BD.

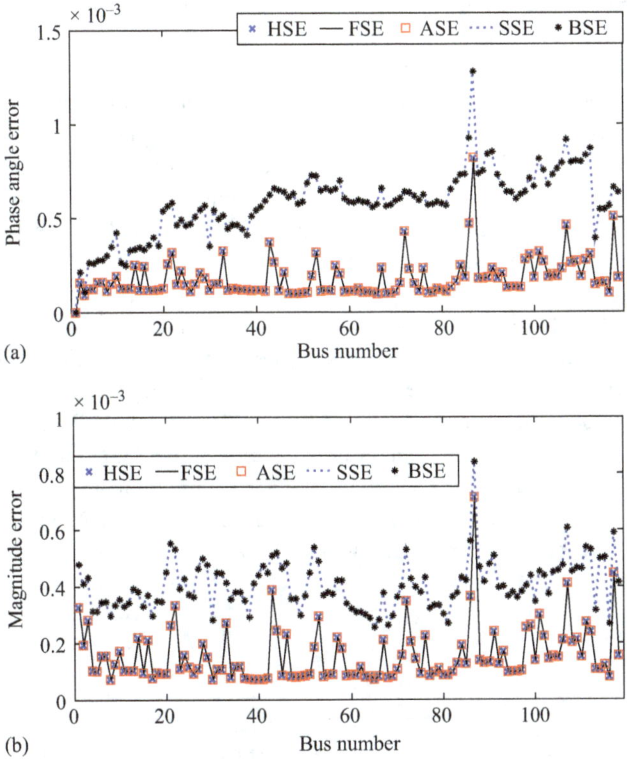

Figure 7.5 Absolute estimation errors for the IEEE 118-bus system without BD: (a) voltage phase angle error and (b) voltage magnitude error

Table 7.8 Accuracy performance of five estimators

Estimator	**IEEE 30-bus**		**IEEE 57-bus**		**IEEE 118-bus**	
	Error mean		**Error mean**		**Error mean**	
	$\|V\|$	θ	$\|V\|$	θ	$\|V\|$	θ
HSE	0.234	0.280	0.170	0.200	0.155	0.181
FSE	0.234	0.280	0.170	0.200	0.155	0.181
ASE	0.234	0.280	0.170	0.200	0.155	0.181
SSE	0.764	0.844	0.640	0.831	0.403	0.580
BSE	0.766	0.844	0.642	0.832	0.403	0.580

All results as factor of 10^{-3}.

Furthermore, the voltage phase angle and magnitude errors derived from five estimators when four bad SCADA measurements (voltage magnitudes V_1 and V_5, active power flow $P_{3\text{–}4}$, and active power injection P_{13}) in the PMU-unobservable area are introduced in the IEEE 30-bus system are depicted in Figure 7.7. Similar to what is observed in Figure 7.6, the SSE and BSE strategies cannot eliminate the detrimental effect of BD. As shown in Figure 7.7(a), the phase angle estimate of bus 13 computed from HSE, FSE, and ASE cannot achieve the same accuracy when no BD occurs because of the bad active power injection measurement P_{13} located in the PMU-unobservable area.

The absolute estimation errors of three estimators when four PMU measurements (both the phase angles and magnitudes of voltage phasor $\vec{V}_6$ and current phasor $\vec{I}_{6-8}$) are introduced in the 30-bus system are shown in Figure 7.8.

As shown in Figure 7.1, the bad PMU measurements are deleted rather than corrected after being identified. Thus, the originally observable PMU states, that is, the voltage phase angle and magnitude of bus 8, become unobservable after deleting $\vec{V}_6$ and $\vec{I}_{6-8}$. For this reason, the voltage phase angle and magnitude errors of bus 8 derived from the HSE, ASE, and FSE strategies are much larger than those obtained with no bad PMU measurements.

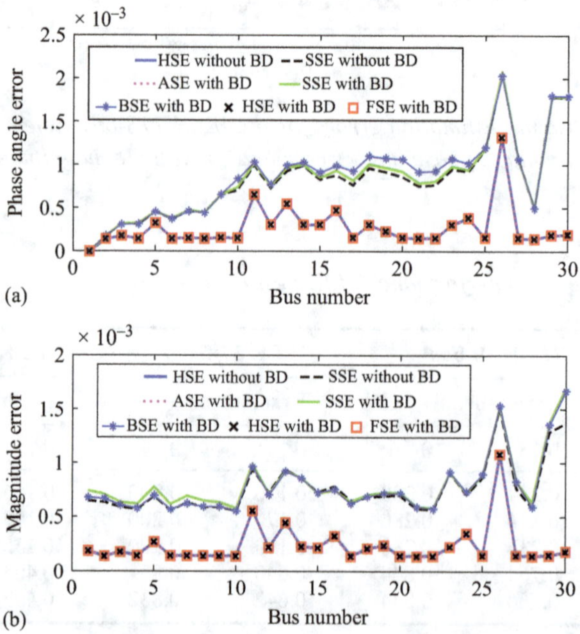

Figure 7.6 Absolute estimation errors for the IEEE 30-bus system with four bad SCADA measurements in the PMU-observable area: (a) voltage phase angle error and (b) voltage magnitude error

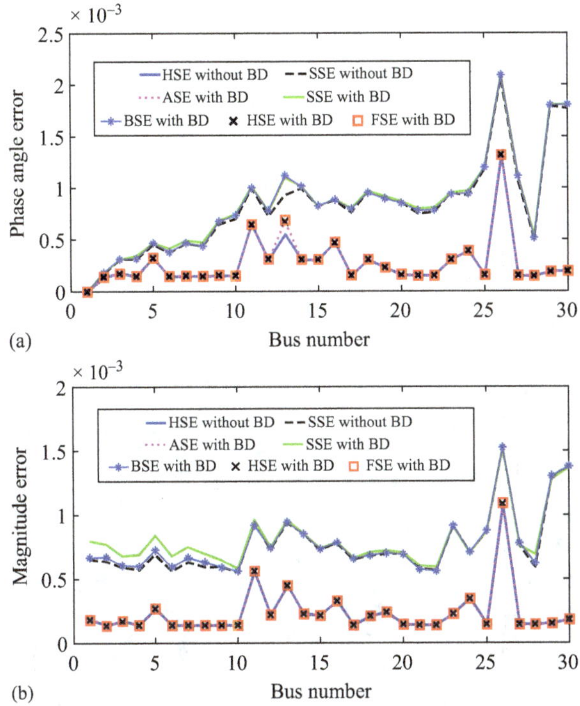

Figure 7.7 Absolute estimation errors for the IEEE 30-bus system with four bad SCADA measurements in the PMU-unobservable area: (a) voltage phase angle error and (b) voltage magnitude error

In summary, the results presented so far demonstrate that the proposed FSE strategy has almost the same estimation accuracy as HSE and ASE strategies for different BD scenarios.

7.5.2 Assessment of computational effort

All five estimators mentioned in Section 7.4.1 were coded in a MATLAB® R2013a environment and carried out on a 3.3-GHz desktop with 4 GB memory. Except for sparse matrix methods, no special effort was made to optimize the program codes. For each test system, the proposed FSE was assessed by comparing it with the other estimators in terms of computing time under five cases:

- Case 1: No BD;
- Case 2: Five BD in SCADA measurements (all of these BD are in the PMU-observable area);
- Case 3: Five BD in SCADA measurements (three BD are in the PMU-observable area and two BD are in the PMU-unobservable area);
- Case 4: Five BD in SCADA measurements (all of these BD are in the PMU-unobservable area);

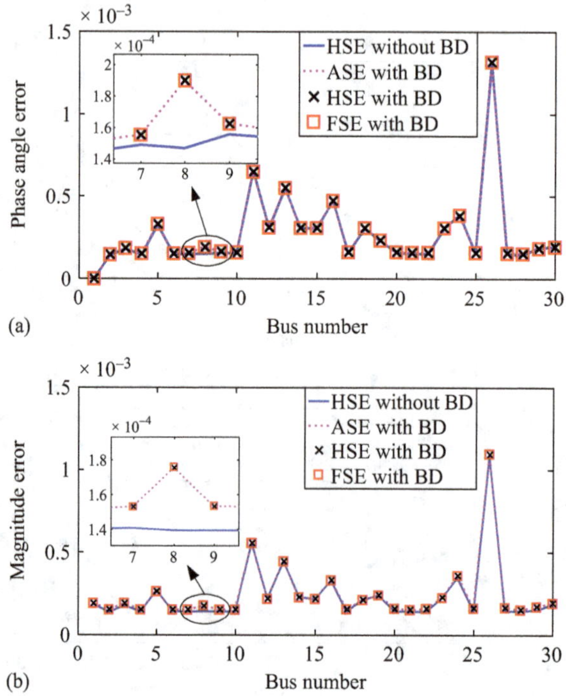

Figure 7.8 Absolute estimation errors for IEEE 30-bus system with four bad PMU measurements: (a) voltage phase angle error and (b) voltage magnitude error

- Case 5: Four BD, that is, two bad voltage/current phasor measurements (both their phase angle and magnitude) in PMU measurements.

In each case, 100 Monte Carlo simulations were executed with different Gaussian errors, and BD values were assigned to different measurements in each simulation run. The BD detection threshold adopted in this study for normalized residual tests is 3, that is, 99.7% confidence level. Because the PMU-based estimator and the first stage of the SCADA-based estimator constitute separate modules and process distinct measurement sets, it is reasonable to consider a parallelization of these two processes (Figure 7.1). Accordingly, if t_{PMU}, t_{fusion}, t_1, t_2, and t_3 denote the execution time of PMU-based SE, fusion stage, and the first, second, and third stages of SCADA-based SE, respectively, then the computing time for the FSE strategy, t_{FSE}, can be derived as follows:

$$t_{\mathrm{FSE}} = \max(t_1, t_{\mathrm{PMU}}) + t_2 + t_3 + t_{\mathrm{fusion}} \tag{7.101}$$

Table 7.9 provides the computing time for each individual stage of the FSE strategy as well as the total execution time for Case 1. For the IEEE 30-bus system, although the PMU-based SE and the first stage of SCADA-based SE are both

Table 7.9 Computing time (ms) for FSE strategy without BD

Test system	t_1	t_{PMU}	t_2	t_3	t_{fusion}	t_{FSE}
IEEE 30-bus	0.7	1.1	0.5	0.5	0.8	2.9
IEEE 57-bus	3.0	2.9	0.5	0.8	2.1	6.4
IEEE 118-bus	16.2	3.9	0.7	1.7	8.2	26.8

Table 7.10 Computational effort of five estimators under five cases

Test system	Case	Computing time (ms)				
		t_{HSE}	t_{FSE}	t_{ASE}	t_{SSE}	t_{BSE}
IEEE 30-bus	1	10.3	2.9	5.3	4.6	1.7
	2	42.3	5.2	32.1	30.5	8.3
	3	45.7	8.1	31.7	30.4	8.4
	4	48.3	9.9	32.0	30.5	8.4
	5	40.8	4.8	12.7	–	–
IEEE 57-bus	1	27.6	6.4	15.4	11.7	4.3
	2	113.1	14.1	88.5	84.8	23.2
	3	133.3	20.6	88.4	84.5	23.5
	4	143.6	25.3	88.4	84.7	23.3
	5	109.2	11.6	34.8	–	–
IEEE 118-bus	1	122.1	26.8	75.1	57.5	18.6
	2	526.1	60.4	389.8	373.0	78.8
	3	580.3	78.6	390.0	372.5	78.6
	4	622.5	86.8	389.5	372.4	78.7
	5	507.9	35.7	151.6	–	–

linear, and a larger number of measurements are processed in the latter, $t_{PMU} > t_1$ due to the additional calculation of the PMU measurement vector in rectangular coordinates as well as its error covariance matrix. With growing network size, t_1 increases considerably and even takes up a much larger proportion in t_{FSE}; this is expected due to the sharp rise in the number of SCADA measurements.

Table 7.10 presents the computing time for five estimators under five cases, in which t_{HSE}, t_{FSE}, t_{ASE}, t_{SSE}, and t_{BSE} represent the execution time of the HSE, FSE, ASE, SSE, and BSE strategies, respectively. The SSE and BSE use only SCADA measurements. When there is no BD (Case 1), the speedup achieved by BSE, defined as t_{SSE}/t_{BSE}, ranges from 2.7 to 3.1, with higher values obtained for a larger network with more redundant SCADA measurements. The speedup values, ranging from 3.6 to 4.7, are obtained when five BD occur in the SCADA measurements (Cases 2–4). This can be explained as follows. Although BD in these two estimators are identified and corrected one-by-one, the re-estimation step in BSE is linear and can significantly reduce the required computing time. The other three estimators, that is, HSE, FSE, and ASE, deal with both SCADA and PMU measurements. The FSE methodology has superior computational performance, resulting in up to

3.6–4.6 and 1.8–2.8 orders of implementation time reduction compared to HSE and ASE, respectively, for the three test systems under Case 1. For the cases with bad measurements, the resulting speedups, defined as t_{HSE}/t_{FSE} and t_{ASE}/t_{FSE}, are remarkably high (from 4.9 to 14.2 and 2.6 to 6.5, respectively). In Case 4, all five bad SCADA measurements are in the PMU-unobservable area, so these BD are identified and eliminated one-by-one in the FSE. On the contrary, the same amount of BD in Case 2 can be dealt with all at once and that is why the computing time of the FSE for Case 2 is much less than for Case 4. Before BD processing, the estimates derived from the HSE for Case 2 are more accurate than for Case 4 because the detrimental effect of bad SCADA measurements in the PMU-observable area is weakened by the accurate PMU measurements. Moreover, a hot-start execution is implemented in the re-estimation step of BD processing for the HSE solution. Thus, compared to Case 4, the average number of iterations in the re-estimation step in Case 2 can be significantly reduced, and its execution time decreases accordingly. However, for each test system, the ASE solution in Cases 2–4 takes almost the same execution time. In addition to dealing with a smaller BD set, HSE and FSE in Case 5 omit the BD correction step, which gives rise to a shorter implementation time than for Cases 2–4. It is worth mentioning that the execution time of ASE in Case 5 is about 2.5–fold less compared to Cases 2–4 for the three test systems because the re-estimation step in the former case is noniterative.

7.6 Summary

Table 7.9 provides the computing time for each individual stage of the FSE strategy as well as the total execution time for Case 1. For the IEEE 30-bus system, although the PMU-based SE and the first stage of SCADA-based SE are both linear, and a larger number of measurements are processed in the latter, $t_{PMU} > t_1$ due to the additional calculation of the PMU measurement vector in rectangular coordinates as well as its error covariance matrix. With growing network size, t_1 increases considerably and even takes up a much larger proportion in t_{FSE}; this is expected due to the sharp rise in the number of SCADA measurements.

References

[1] Schweppe F. C., and Wildes J. “Power system static-state estimation, part I: exact model.” *IEEE Transactions on Power Apparatus and Systems*. 1970, vol. PAS-89, no. 1, pp. 120–125.

[2] Schweppe F. C., and Rom D. B. “Power system static-state estimation, part II: approximate model.” *IEEE Transactions on Power Apparatus and Systems*. 1970, vol. PAS-89, no. 1, pp. 125–130.

[3] Schweppe F. C. “Power system static-state estimation, part III: implementation.” *IEEE Transactions on Power Apparatus and Systems*. 1970, vol. PAS-89, no. 1, pp. 130–135.

[4] Abur A., and Exposito A. G. *Power System State Estimation Theory and Implementation*. New York, NY: Marcel Dekker, 2004.

[5] Phadke A. G., and Thorp J. S. *Synchronized Phasor Measurements and Their Applications*. New York, NY: Springer, 2008.

[6] Thorp J. S., Phadke A. G., and Karimi K. J. "Real time voltage-phasor measurement for static state estimation." *IEEE Transactions on Power Apparatus and Systems*. 1985, vol. PAS-104, no. 11, pp. 3098–3106.

[7] Gomez-Exposito A., Abur A., Jaen A. D., and Gomez-Quiles C. "A multi-level state estimation paradigm for smart grids." *Proceedings of the IEEE*. 2011, vol. 99, no. 6, pp. 952–976.

[8] Kashyap N., Werner S., Huang Y. F., and Riihonen T. "Power system state estimation under incomplete PMU observability: a reduced-order approach." *IEEE Journal of Selected Topics in Signal Processing*. 2014, vol. 8, no. 6, pp. 1051–1062.

[9] Valverde G., Chakrabarti S., Kyriakides E., and Terzija V. "A constrained formulation for hybrid state estimation." *IEEE Transactions on Power Systems*. 2011, vol. 26, no. 3, pp. 1102–1109.

[10] Costa A. S., Albuquerque A., and Bez D. "An estimation fusion method for including phasor measurements into power system real-time modeling." *IEEE Transactions on Power Systems*. 2013, vol. 28, no. 2, pp. 1910–1920.

[11] Do Coutto M. B. Filho, de Souza J. C. S., and Guimaraens M. A. R. "Enhanced bad data processing by phasor-aided state estimation." *IEEE Transactions on Power Systems*. 2014, vol. 29, no. 5, pp. 2200–2209.

[12] Gol M., and Abur A. "A hybrid state estimator for systems with limited number of PMUs." *IEEE Transactions on Power Systems*. 2015, vol. 30, no. 3, pp. 1511–1517.

[13] Bi T. S., Qin X. H., and Yang Q. X. "A novel hybrid state estimator for including synchronized phasor measurements." *Electric Power Systems Research*. 2008, vol. 78, no. 8, pp. 1343–1352.

[14] Korres G. N., and Manousakis N. M. "State estimation and bad data processing for systems including PMU and SCADA measurements." *Electric Power Systems Research*. 2011, vol. 81, no. 7, pp. 1514–1524.

[15] Korres G. N., and Manousakis N. M. "A state estimator including conventional and synchronized phasor measurements." *Computers & Electrical Engineering*. 2012, vol. 38, no. 2, pp. 294–305.

[16] Chakrabarti S., Kyriakides E., Ledwich G., and Ghosh A. "Inclusion of PMU current phasor measurements in a power system state estimator." *IET Generation Transmission & Distribution*. 2010, vol. 4, no. 10, pp. 1104–1115.

[17] Zhou M., Centeno V. A., Thorp J. S., and Phadke A. G. "An alternative for including phasor measurements in state estimators." *IEEE Transactions on Power Systems*. 2006, vol. 21, no. 4, pp. 1930–1937.

[18] Zhao L., and Abur A. "Multiarea state estimation using synchronized phasor measurements." *IEEE Transactions on Power Systems*. 2005, vol. 20, no. 2, pp. 611–617.

[19] Jiang W. Q., Vittal V., and Heydt G. T. "A distributed state estimator utilizing synchronized phasor measurements." *IEEE Transactions on Power Systems*. 2007, vol. 22, no. 2, pp. 563–571.
[20] Jiang W. Q., Vittal V., and Heydt G. T. "Diakoptic state estimation using phasor measurement units." *IEEE Transactions on Power Systems*. 2008, vol. 23, no. 4, pp. 1580–1589.
[21] Gomez-Exposito A., Gomez-Quiles C., and Jaen A. D. "Bilinear power system state estimation." *IEEE Transactions on Power Systems*. 2012, vol. 27, no. 1, pp. 493–501.
[22] Gomez-Quiles C., Jaen A. D., and Gomez-Exposito A. "A factorized approach to WLS state estimation." *IEEE Transactions on Power Systems*. 2011, vol. 26, no. 3, pp. 1724–1732.
[23] Baltensperger R., Loosli A., Sauvain H., Zima M., Andersson G., and Nuqui R. "An implementation of two-stage hybrid state estimation with limited number of PMU." in *Proceedings of 10th IET International Conference on Developments in Power System Protection*, Manchester, UK, 2010, pp. 1–5.
[24] Manousakis N. M., Korres G. N., Aliprantis J. N., Vavourakis G. P., and Makrinas G. C. J. "A two-stage state estimator for power systems with PMU and SCADA measurements." in *Proceedings of IEEE Grenoble PowerTech (POWERTECH)*, Grenoble, France, IEEE, 2013, pp. 1–6.
[25] Yang T., Sun H. B., and Bose A. "Transition to a two-level linear state estimator - part I: architecture." *IEEE Transactions on Power Systems*. 2011, vol. 26, no. 1, pp. 46–53.
[26] Yang T., Sun H. B., and Bose A. "Transition to a two-level linear state estimator - part II: algorithm." *IEEE Transactions on Power Systems*. 2011, vol. 26, no. 1, pp. 54–62.
[27] Cutsem T. V., Ribbens-Pavella M., and Mili L. "Hypothesis testing identification: a new method for bad data analysis in power system state estimation." *IEEE Transactions on Power Apparatus and Systems*. 1984, vol. PAS-100, no. 11, pp. 3239–3252.
[28] Chen J., and Abur A. "Placement of PMUs to enable bad data detection in state estimation." *IEEE Transactions on Power Systems*. 2006, vol. 21, no. 4, pp. 1608–1615.
[29] Wu T., Chung C. Y., and Kamwa I. "A fast state estimator for systems including limited number of PMUs." *IEEE Transactions on Power Systems*. 2017, vol. 32, no. 6, pp. 4329–4339.

Chapter 8

Response-based event detection for one-shot wide-area stability controls

Steven M. Rovnyak[1], Daniel W. Longbottom[2], Diana C. Vasquez[3] and Maryam N. Nilchi[3]

8.1 Introduction

Phasor measurement units (PMU) calculate synchronized phase angle measurements which are useful for angle stability analysis [1–8]. Wide-area monitoring systems (WAMS) can communicate these measurements to a central location in nearly real time [9–12]. Wide-area control actions such as disconnection of loads and generators and fast power changes on DC interties can sometimes be used to prevent loss of synchronism [13]. Loss of synchronism between groups of synchronous generators in a power system is a form of instability that tends to produce large-scale blackouts when it occurs.

Renewable resources such as wind and solar generation provide intermittent power. In large scale power systems, the contribution of wind and solar generation depends on local properties of the weather over a large geographic area. Renewable resources are often distributed among a large number of smaller facilities which makes it difficult to measure their collective output compared to the output of a large utility-owned generating station.

The only way to observe the cumulative effect of intermittent renewable sources on a large scale power system in real time is indirectly through their effect on voltages and currents measured by PMU. Pattern recognition methods can create decision criteria for actuating wide-area stability controls based on PMU measurements in the presence of uncertainty. Pattern recognition tools such as decision trees (DT) create these rules by analyzing large amounts of time-domain transient simulation data from a variety of operating scenarios [13–22].

Training set simulations for a pattern recognition method can include a wide range of disturbances and operating conditions. Disturbances can include short circuit events with varying location, duration, and severity. Operating conditions

[1]Department of Electrical and Computer Engineering, Indiana University-Purdue University Indianapolis (IUPUI), Indianapolis, Indiana, USA
[2]Rolls Royce, Indianapolis, Indiana, USA
[3]Midcontinent Independent System Operator (MISO), Carmel, Indiana, USA

can include different amounts of customer load and renewable generation. The DTs in this chapter are only trained on a variety of disturbances but the extension to a variety of operating conditions would be straightforward. The decision criteria in this chapter for actuating wide-area stability controls are simple and robust which makes them good candidates for handling a variety of operating conditions.

Early methods for real time transient stability prediction and control assumed there would be some way of knowing that a fault had just occurred so that a short window of measurements collected immediately after the fault could be used to predict loss of synchronism [23,24]. The challenge in applying this method to a real power system is how to know when a fault has occurred somewhere in the network. One idea would be to monitor the status of circuit breakers all over the network in order to know that a fault had just been cleared. Such a methodology would not be response-based control because it could only respond to faults on a fixed set of transmission lines. Response-based control is supposed to infer that an event has occurred by monitoring voltage and current transducers that are not necessarily connected to the faulted or disconnected component [4,25,26].

Response-based prediction and control were accomplished in previous papers by turning every time sample from every simulation into an input–output pair for a pattern recognition tool training data set. References [15,23] give examples of response-based control using apparent resistance measured around the middle of the Pacific AC Intertie (PACI). Reference [13] extends this methodology to decision trees (DT) using wide-area phasor data for deciding whether to actuate a particular combination of one-shot controls. The controls in these papers consisted of step changes in the real power injections at pairs of AC buses. These actions approximate the effect of a fast power change on an HVDC intertie between the AC buses. They could also be approximately realized by disconnecting a generator at one bus and disconnecting load at another bus.

The DT training software has a parameter called relative misclassification cost that can be configured to make either the stable or the unstable region almost pure. For example, if the cost of misclassifying a stable case as unstable is set very large, then the unstable region is approximately the largest region in the feature space that is almost pure in terms of containing only unstable samples. The complement of this region is approximately the smallest region in feature space that encloses nearly all the portions of swings for which loss of synchronism is not imminent [13].

The relative misclassification cost can be used to influence whether the resultant DT is more or less trigger-happy in terms of predicting instability. For example, if the cost of misclassifying a stable case as unstable is set very large, then the DT tends to wait until loss of synchronism is almost imminent before predicting instability. If this relative misclassification cost is reduced, then the DT tends to predict instability earlier in the case of unstable events, but the DT also tends to misclassify a larger number of stable events as unstable. When used to order one-shot control, trigger-happy DTs are successful in stabilizing a larger number of events at the cost of an increased number of unnecessary control applications.

A closer examination of the performance of trigger-happy DTs revealed that they were frequently predicting instability during the fault-on period. Furthermore, these early predictions turned out to be when many of the prediction errors were occurring. The methodology in the present work was developed in an attempt to preserve the earliness of prediction of trigger-happy DTs while reducing the number of errors. This goal was accomplished by developing response-based methods for event detection so that a short window of measurements collected immediately after the event could be used to predict loss of synchronism.

Previous work proposed using the variance of generator frequency measurements to detect dynamic events and estimate their location [25]. For the present results, we use bus voltage phasor measurements instead of generator rotor frequency measurements because bus voltage phasors are easier to measure. We also aggregate the information from several bus phasor measurements instead of identifying the single generator with the largest variance in frequency. Our first efforts in this direction involved calculating an integral square bus angle (ISBA) index [27]. The variance of ISBA over a short window turned out to be useful for detecting dynamic events although it had a relatively slow response and it did not provide a clear indication of fault clearing time.

Visual examination revealed that the ISBA had sharp corners at the beginning and end of each fault-on period. We developed a variety of methods to detect these corners. One approach involved training a DT to detect the onset of a dynamic event using the ISBA and its time derivative [28]. We also applied wavelet analysis to a sliding window of ISBA values and calculated the signal energy in one level of detail coefficients in order to infer whether the window contains a dynamic event. Another method we developed for dynamic event detection involves calculating the error of a second-order polynomial fit to successive windows of integral square generator angle (ISGA) values [29].

The combination of separate event detection and control DTs for transient stability control is new and results in a higher rate of success in stabilizing events using one-shot control. Control actions may include disconnection of generation and load which are very expensive except compared to a blackout. The proposed control will be more acceptable if it is never activated unless there is a major transient event. The control scheme proposed in this chapter satisfies this requirement better than previous methods of using DTs for one-shot control.

8.2 Simulated measurements and control

A 29-machine, 176-bus simplified model of the Western Electricity Coordinating Council (WECC) system is employed to illustrate and evaluate the decision-tree-based wide-area control system (WACS). Transmission lines retained in the 176-bus model are shown in Figure 8.1. The two DC interties are modeled as loads, positive and negative, at the terminal buses. The dynamic data contains detailed synchronous machine models with exciters and power system stabilizers. Load models are a combination of constant impedance, constant current, and constant power loads. The nominal operating frequency is 60 Hz. DSA PowerTools from Powertech Labs was used to simulate the events with trapezoidal integration and a step size of 0.5 cycles.

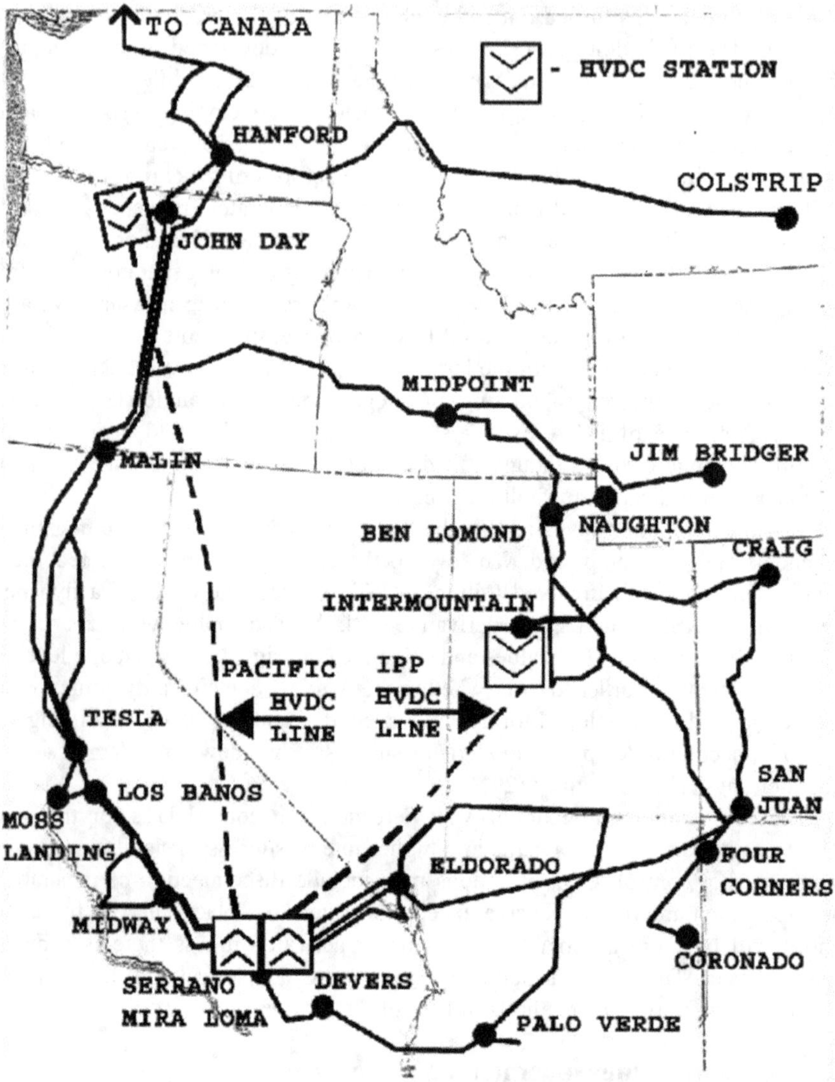

Figure 8.1 Transmission lines retained in the 176-bus model of the WECC. Source: [2011] IEEE. Reprinted, with permission, from [13]

8.2.1 Phasor measurements

PMUs in this study were assumed to be installed at 17 buses, in which 15 buses are 500 kV substations. Voltage phasor measurements are recorded at all the PMU buses at a rate of 30 samples per second. Unlike generator rotor angles, the bus angles do not go outside –180° to 180°. They wrap around 180° as shown in Figure 8.2. A solution to this problem consists of unwrapping the angle trajectories

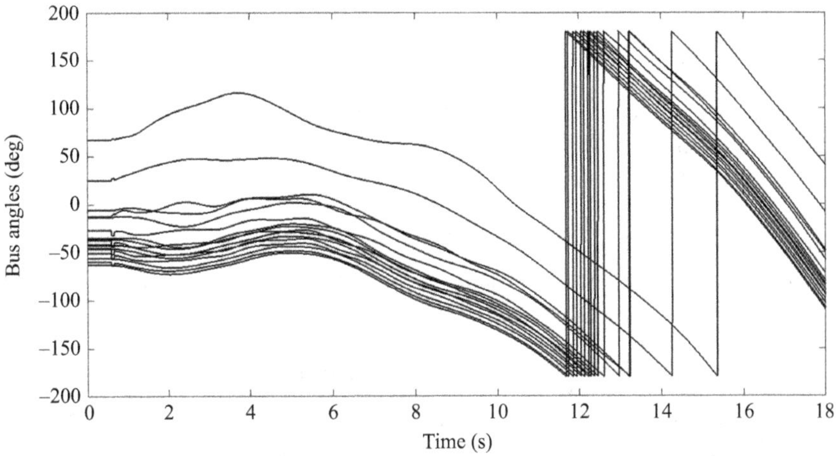

Figure 8.2 Discontinuity of bus angles

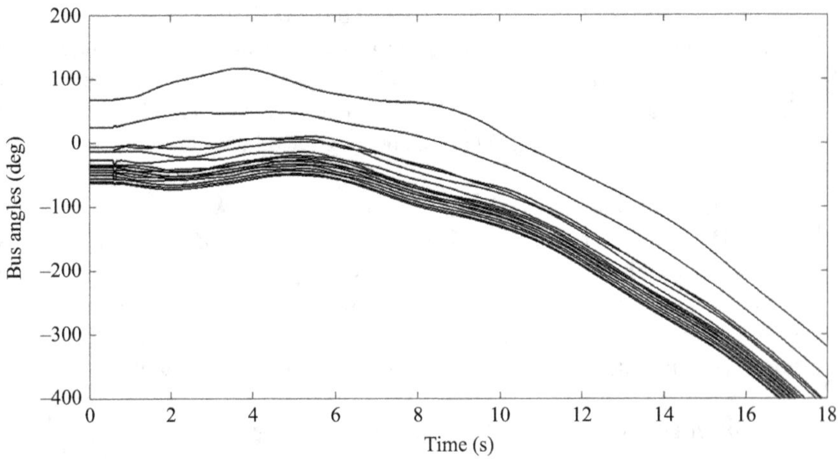

Figure 8.3 Reconstruction of continuous trajectories from the bus angles

as shown in Figure 8.3 using a procedure similar to the unwrap function in MATLAB®.

After reconstructing continuous trajectories, the bus phasor angle of Westwing is subtracted from the angle of each phasor. Westwing is located in the southern part of the network near Palo Verde. This choice reflects the fact that the southern area has relatively strong transmission connections to other areas in this model whereas the AC transmission connections between the northern ends of the DC interties are less strong. The square bus angle index is calculated in center of angle coordinates, so it is not affected by the choice of Westwing as the reference angle.

Let the relative phasor measurements for time sample k be denoted as

$$V_i[k] = |V_i[k]|\angle\theta_i[k], i = 1,\ldots,17$$

Bus 17 is the reference angle and it is subtracted from all angles so $\theta_{17}[k] = 0$. Derivatives of the angles are calculated as point to point differences between samples 30 times per second.

$$\theta dot_i[k] = 30(\theta_i[k] - \theta_i[k-1]), i = 1,\ldots,17$$

Decision trees use these, and other derived indices described in Section 8.3 to order one-shot stabilizing control.

8.2.2 Wide-area one-shot control combination

A single combination of one-shot controls is selected for its ability to reduce phase angle differences for the operating condition which has large power flows from north to south. Decision trees are used to determine whether to actuate this combination of controls at each time step. The one-shot controls can only be actuated once during a simulation.

The control actions consist of step changes in the real power injections at pairs of AC buses. This control approximates the effect of a fast power change on an HVDC intertie between the AC buses. One 500 MW fast load increase is applied to Colstrip in Montana and another 500 MW fast load increase is applied to a bus in Canada. Corresponding step load decreases occur at two buses around the southern ends of the two HVDC lines.

These control combinations would be similar to 500 MW fast power increases on both HVDC lines if they extended all the way to Canada and Montana. The effect of these controls could also be approximated by disconnecting generation in Canada and Montana and shedding load in southern California. Figure 8.4 shows the results of simulating the one-shot control combination during steady-state conditions. The subsequent equilibrium has smaller steady-state angle differences.

8.2.3 Simulations for training and test data sets

The training set contains data from 385 six-second simulations of the 176-bus model. It is the same set of simulations used for training in [13]. We found about 30 transmission lines in this model where long, three-phase faults could cause angle instability. We added ten more lines from the Pacific AC Intertie (PACI) to this list. The final list of 40 lines contained about 15 transmission line segments on the PACI between Tesla and John Day. The training set simulations included four single outage contingencies for each of the 40 test lines, 210 double outage contingencies each involving two lines on the PACI, plus 15 more single contingency events on the PACI. The four single outage contingencies for each line were (1) line removal after a momentary fault, (2) one-cycle three-phase fault, (3) four-cycle three-phase fault, and (4) six-cycle one phase fault. The double contingencies had a four-cycle, single line to ground fault ending at 40 cycles followed by a different four-cycle, single line to ground fault 2 s later.

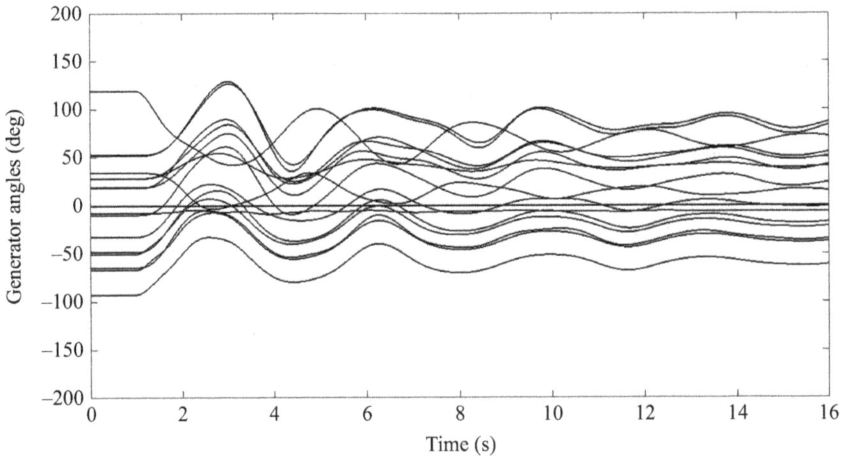

Figure 8.4 Generator angles for wide-area one-shot control at 1.0 s

Table 8.1 Test set information

Fault type	**1 phase**	**3 phase**
Simulations in test set	480	480
Simulation length	6 s	6 s
Fault duration	4–15 cycles	0–11 cycles
Stable simulations	276	250
Unstable simulations	204	230

The test set includes 960 simulations of three-phase short circuit faults and single line to ground faults. For each of the 40 lines, there were 12 simulations for three-phase faults ranging from 0 to 11 cycles in length. There were 12 additional simulations for single line to ground faults ranging from 4 to 15 cycles in length. The simulations used for testing are summarized in Table 8.1.

The fault durations were selected so the event was often stable for the shortest faults and unstable for the longest faults. With this test set, the number of simulations stabilized can be used for estimating the average increase in critical clearing time (CCT). For example, increasing the CCT of both fault types by one cycle would result in roughly 80 more of the 960 simulations being stable. Stabilizing 40 of the 960 simulations corresponds to an average CCT increase of 0.0083 s for the 40 lines.

8.3 Indices for event detection and one-shot control

This paper proposes and compares three indices for response-based control. The SBA index measures the overall stress of the network in terms of angle differences. It is based on the ISGA index [26], which is described first.

8.3.1 Integral square generator angle index

The ISGA index [26] is suitable for assessing stability margins in simulations. The index can be used to judge the severity of both stable and unstable transient events. The ISGA can be useful for finding combinations of one-shot controls to stabilize severely unstable transient events [26]:

$$J = \int_0^T \sum_i M_i(\delta_i(t) - \delta_{coi}(t))^2 dt \Big/ \sum_i M_i$$

Terms $\delta_i(t)$ represent the generator angles as functions of time. Constants M_i are machine inertias and δ_{coi} is the center of inertia (COI):

$$\delta_{coi}(t) = \sum_i M_i\delta_i(t) \Big/ \sum_i M_i$$

Generators farther from the COI are penalized more by the ISGA. Any disconnected generators are removed from the calculation of COI and ISGA. The ISGA is suitable for simulation data and it can easily include rotor angle data from all generators.

The ISGA is a coherency-based index similar to one of the indices proposed for dynamic security assessment (DSA) in [30,31]. Here and in [30,31], the term coherency-based means that the index reflects the degree of coherency. Specifically, the ISGA is correlated with the degree of incoherency. The ISGA measures the aggregate differences of generator angles during both transient and equilibrium conditions. This index is not a form of kinetic energy for the network.

8.3.2 Square bus angle index

If $\theta_i(t)$ represent the continuous trajectories reconstructed from the bus angles and constants M_i have been chosen to weight the angles from different locations, then the SBA index at time sample k is

$$SBA[k] = \sum_i M_i(\theta_i[k] - \theta_{coa}[k])^2 \Big/ \sum_i M_i$$

$$\theta_{coa}[k] = \sum_i M_i\theta_i[k] \Big/ \sum_i M_i$$

The bus angles used in the ISBA calculation should come from major buses all around the network but the sampling of buses to measure does not need to be exhaustive. Whereas the ISGA uses machine inertias for weights, it is possible to assign equal weights to all the monitored buses for the ISBA. Another possibility would be to weight buses in terms of voltage levels. The derivative of the SBA is calculated from point to point differences 30 times per second.

$$SBAdot[k] = 30(SBA[k] - SBA[k-1])$$

The use of a squared angle index for deciding whether to actuate control in real time is new compared to its previous use which was off-line.

8.3.3 Bus magnitude indices

Similar indices can be calculated from the bus voltage magnitudes. The average and variance of bus magnitudes are:

$$BMavg[k] = \sum_i |V_i[k]|/17$$

$$BMvar[k] = \sum_i (|V_i[k]| - BMavg[k])^2/17$$

Derivatives of these quantities are calculated from point to point differences 30 times per second:

$$BMavgdot[k] = 30(BMavg[k] - BMavg[k-1])$$

$$BMvardot[k] = 30(BMvar[k] - BMvar[k-1])$$

The values used for *SBA* and *BMvar* in the proposed control schemes are actually square roots of the above quantities, so they are really standard deviations.

8.3.4 Illustration of indices during and after a fault

The index values for a simulation with a short circuit fault from 0.58 s to 0.67 s are shown in Figure 8.5. The change in bus voltage magnitudes during a fault was expected to be useful for detecting the end of a fault. This phenomenon is observed in the large negative and positive values of *BMavgdot* during and after the fault. The bus voltage angles increase during a fault as reflected in the peak and dip of the *SBAdot* during and after the fault. The *BMvardot* was added for completeness but it was not anticipated to be a highly useful index.

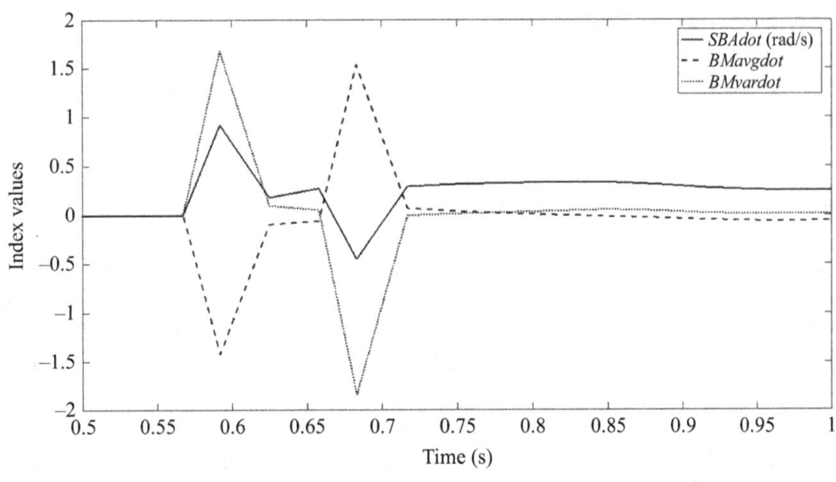

Figure 8.5 Indices calculated from a simulation with a fault from 0.58 s to 0.67 s

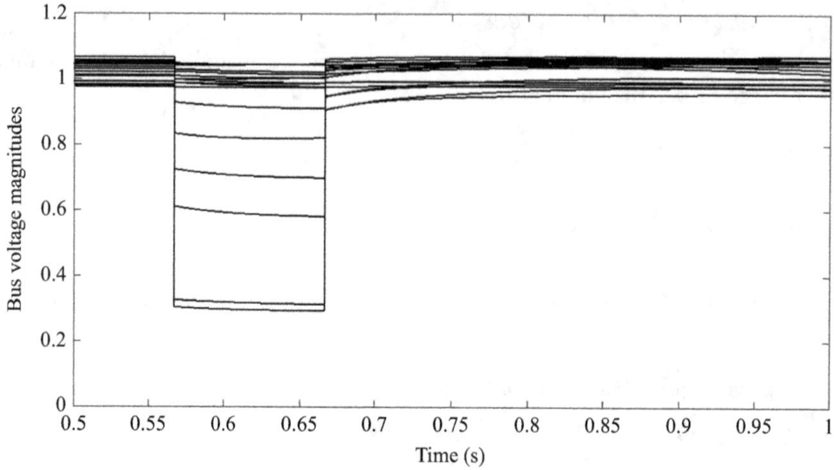

Figure 8.6 Bus voltage magnitudes from a simulation with a fault from 0.57 s to 0.67 s

Simulation results in Section 8.6 show that variance among bus voltage magnitudes is in fact one of the most reliable indicators of short circuit faults. A severe short circuit fault affects the voltage magnitudes over a wide area which greatly increases the variance over a geographic region. Figure 8.5 shows how the change in variance of bus voltage magnitudes during and after a fault contains a spike followed by a dip.

In retrospect, it is obvious why variance among bus voltage magnitudes over a geographic region is useful for detecting short circuit faults. Figure 8.6 shows the 17 PMU bus voltage magnitudes during a simulation with a short circuit fault that lasts 0.1 s. The familiar pattern illustrates how the variation among the bus voltage magnitudes increases during the fault-on period and subsequently decreases when the voltages substantially recover in the same way that the set of numbers {0.3, 0.7, 1.0} has a larger variance than the set of numbers {0.95, 1.0, 1.05}.

8.4 DTs for response-based control

DTs learn from examples and require a training set of data consisting of input–output pairs. DTs in this paper are trained using the rpart function in Rattle. Rattle is a graphical interface for R [32]. The rpart function follows closely the methodology of the CART algorithm which uses recursive partitioning to grow a DT [33].

There is an input–output vector for every time sample of the PMU measurements, which is 30 times per second. The input vector can include magnitudes and angles from the 17 PMU buses as well as indices calculated from them. The target refers to the correct value of the output for a given input vector.

8.4.1 Training DT for one-shot control

If a simulation is stable, then the target for every time sample from that simulation is set to "No Control." If a simulation is unstable then the target for every time sample from that simulation is set to "Control." The trained DT processes a new input vector for every measurement time sample and control is actuated the first time the output is "Control."

In order to refrain from actuating control, the DT must output "No Control" for every time sample during a simulation. For this reason, it is necessary to specify during DT training that it is, for example, 100 times the cost to misclassify a "No Control" sample compared to the cost of misclassifying a "Control" sample. The relative misclassification costs can be set to different values to increase or decrease the number of simulations having control action.

The complexity cost can be used to control the size of the DT that is generated. It is sometimes possible to adjust the complexity parameter until the DT consists of only a threshold test on a single variable. One of the DTs trained this way outputs "Control" for any sample that has $SBAdot[k] > 24.18$ degrees per second. Threshold values are rounded to the nearest integer if they are in degrees or degrees per second. Thresholds involving per unit voltages are rounded to the nearest hundredth.

The DT and supervisor reset logic are shown in Figure 8.7. As mentioned earlier the $SBAdot[k]$ is actually the square root of the squared bus angle index and the units are degrees per second. For comparison the units of $SBAdot[k]$ in Figure 8.5 are radians per second. The addition of event detection to the one-shot control scheme as shown in Figure 8.8 is new. The DT that determines whether to

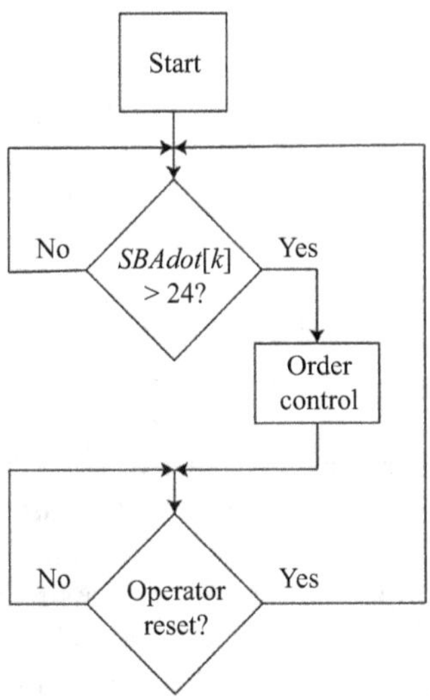

Figure 8.7 DT and supervisor reset logic for response-based one-shot control

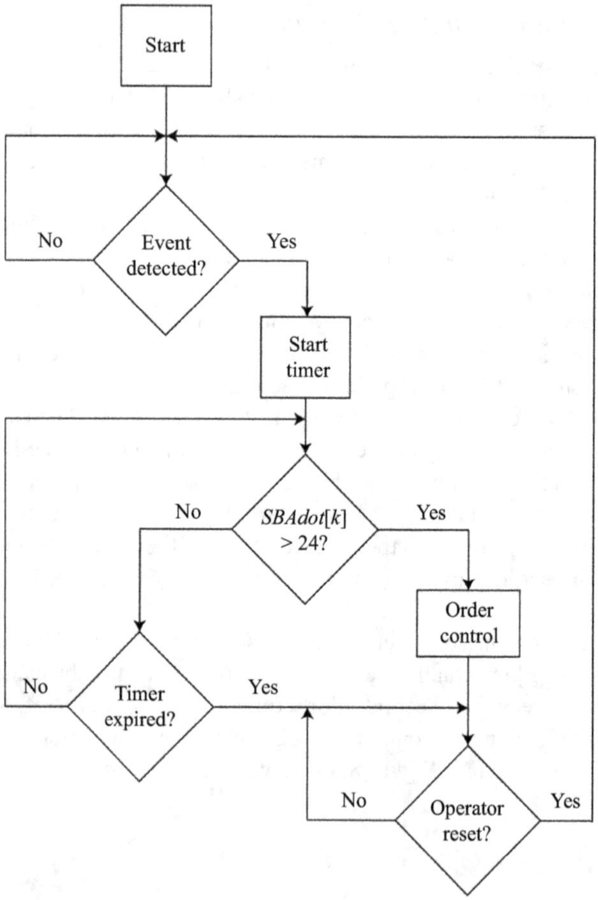

Figure 8.8 Flow chart for response-based control that uses event detection

order one-shot control is only enabled for a few samples after an event is detected. This enhancement reduces the number of unnecessary control actuations.

8.4.2 Training DT for event detection

All the simulations in the training set have a short circuit to ground fault that is cleared at 0.67 s and the fault beginning time is varied to obtain different fault durations. Data from each time sample is included in the input vector and the target is set to "Postfault" if the sample time is between 0.67 s and 0.72 s. The target is "Not Postfault" for all other samples. Typically, there is one sample per simulation with the target set to "Postfault." It is necessary to set the cost of misclassifying a "Postfault" sample to something like 10 or 100 times the cost of misclassifying a "Not Postfault" sample.

A DT that is trained to use only *BMvardot* for event detection outputs "Postfault" for any sample that has *BMvardot*[*k*] < –0.03. This one-node DT which is shown in Figure 8.9 can be used for the event detection function in Figure 8.8.

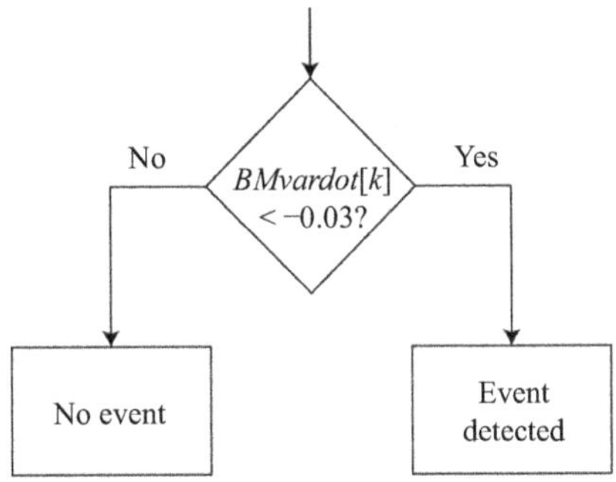

Figure 8.9 Flow chart for one-node DT that performs event detection

8.4.3 Assessing performance of DTs for one-shot control

Performance is assessed by using the DT logic to decide whether to actuate control for each simulation in the test set. The whole simulation is run once without control and again with one-shot control at the time ordered by the DT. Statistics are compiled by calculating the following quantities.

The number of simulations controlled includes all the simulations during which control was actuated. The number of simulations stabilized includes events that are stable with the control scheme and unstable without the proposed control scheme. The number of unnecessary controls includes stable events that have one-shot control actuation and are also stable without control. The success rate of a control configuration is calculated as the number of simulations stabilized divided by the number of simulations controlled.

The number of simulations destabilized is zero for all the test set simulations in this chapter. This means there are no events that are stable without control and unstable with control. One-shot controls that destress the network by reducing overall angle differences rarely cause instability. The success rates of several control configurations in this paper are greater than 0.1 or 10%. The cost of one large blackout could be more than a thousand times the cost of actuating the control. The large cost of blackouts resulting from transient instability would likely justify using the proposed control scheme.

8.5 Simulations without event detection

We tested the DT without event detection using each variable by itself as the only input and raising the complexity cost during training to result in a DT with only one or two nodes. The variables included bus angles, bus velocities, index values, and their derivatives as described in Section 8.3. The DT was trained on the training set of 385 simulations and tested on the test set of 480 single line to ground (SLG) faults. The simulations we generated using SLG faults are generally more challenging to stabilize because of their long duration. The three-phase faults are

Table 8.2 Training the control DT

Variable	cp	Relative cost	Control criteria
θ_9	0.1	10:1	$\theta_9 > 193$
θdot_9	0.1	10:1	$\theta dot_9 > 71$
SBA	0.1	10:1	$SBA > 59$
SBAdot	0.1	10:1	$SBAdot > 24$
BMavg	0.1	10:1	$BMavg < 0.85$
BMavgdot	0.035	10:1	$-2.84 < BMavgdot < -0.20$
BMvar	0.035	10:1	$BMvar > 0.13$
BMvardot	0.050	10:1	$0.10 < BMvardot < 3.37$

Table 8.3 Performance of DT control without event detection

Variable	Controlled	Unnecessary	Stabilized	T_{avg}	Success rate
θ_9	173	0	6	2.66	0.035
θdot_9	207	32	29	1.73	0.140
SBA	171	0	4	2.34	0.023
SBAdot	345	166	29	1.08	0.084
BMavg	139	12	0	2.47	0.000
BMavgdot	323	119	33	0.85	0.102
BMvar	332	166	16	0.88	0.048
BMvardot	398	194	42	0.72	0.106

analogous to experiencing a short, hard shove toward the edge of a cliff while standing a few meters away from the edge. The SLG faults are analogous to being pushed more nearly to the edge of a cliff, albeit more slowly.

The results for response-based control schemes that do not use event detection are shown in Tables 8.2 and 8.3. For each variable listed in the left column, a DT is created which uses only that variable. Training data for the first DT contained the 17 bus angle measurements in the DT input vector and the complexity parameter *cp* was set to 0.1 which resulted in a DT with a single threshold test on θ_9. Training data for the second DT contained the 17 bus angle velocities in the DT input vector and *cp* was set to 0.1 which resulted in a DT with a single threshold test on θdot_9. Training data for each of the remaining DTs contained only the one variable that is listed in the input vector. Setting *cp* to 0.1 resulted in no DT being created for some variables. This parameter was decreased until Rattle produced a DT with one or two nodes. Units for the angle variables are in degrees whereas the bus magnitude values are per unit.

The quantity "Controlled" in Table 8.3 is the number of events in the test set of 480 SLG fault scenarios that have one-shot control action. "Stabilized" is the number of simulations stabilized by the one-shot control. "Unnecessary" is the number of unnecessary controls. "T_{avg}" is the average control actuation time and "Success rate" equals "Stabilized" divided by "Controlled." A large number for

"Unnecessary" indicates that the number of control actuations is excessive. Trigger happy DTs would be difficult to justify if the control actions are expensive as they are with disconnection of generation and load.

DTs that use derivatives have better performance in both the number of simulations stabilized and the success rate. Variable θ_9 represents the angle of Colstrip which is one of four buses that have fast power changes in the one-shot control action. There are several control configurations that have success rates around 0.1 or higher and their use could likely be justified due to the enormous costs of blackouts.

The DT that uses θdot_9 is the only one among the DTs that use derivatives that has a reasonable frequency of control application. Normally the relative misclassification cost mentioned earlier would be set to 100:1 instead of 10:1 but the larger misclassification cost makes it more difficult to produce a one or two node DT for some of the variables.

8.6 Simulations with event detection

The DT that orders control with event detection is trained the same as without event detection except that the relative misclassification cost is set to 1:1 instead of 10:1. This even ratio increases the number of control actuations and makes it more suitable for use with event detection. When using event detection, the DT for control makes a decision for a much smaller subset of the measurements from a simulation. The length of the timer in Figure 8.8 was set to four samples (4/30 s).

Each of the derivative variables is evaluated for use by the event detection DT. The relative misclassification costs for the detection trees are set to 1:100 because the number of samples containing immediate postfault measurements is so small. For each of these tests the DT that decides on control actuation uses θdot_9 and was trained using a relative misclassification cost set to 1:1. The resulting control criterion is $\theta dot_9 > 65$. Results of using different variables for the event detection DT are shown in Tables 8.4 and 8.5.

The best performance is obtained from DT controllers that use indices based on bus voltage magnitudes for event detection. Their success ratios are larger than any of the DTs that do not use event detection.

When the control DT consists of a single threshold test then it is possible to vary the threshold to learn the effect of the parameter on the success rate of the

Table 8.4 Training the event detection DT

Variable	*cp*	Relative cost	Detection criteria
θdot_9	0.1	1:100	$\theta dot_9 < -26$
SBAdot	0.1	1:100	$SBAdot < -18$
BMavgdot	0.1	1:100	$BMavgdot > 0.07$
BMvardot	0.1	1:100	$BMvardot < -0.03$

Table 8.5 Performance of DT control with event detection

Variable	Controlled	Unnecessary	Stabilized	T_{avg}	Success rate
θdot_9	89	0	7	3.37	0.077
SBAdot	104	0	10	2.06	0.096
BMavgdot	72	0	13	0.77	0.181
BMvardot	72	0	13	0.76	0.181

Table 8.6 Performance of one-shot control with event detection as the control DT is adjusted

Control DT	Controlled	Unnecessary	Stabilized	T_{avg}	Success rate
$\theta dot_9 > 65$	72	0	13	0.76	0.181
$\theta dot_9 > 60$	76	0	16	0.76	0.211
$\theta dot_9 > 50$	96	6	22	0.76	0.229
$\theta dot_9 > 40$	126	18	22	0.76	0.175
$\theta dot_9 > 30$	155	44	22	0.77	0.142

Table 8.7 Performance of one-shot control with event detection as the detection DT is adjusted

Detection DT	Controlled	Not detected	Stabilized	T_{avg}	Success rate
BMvardot < -0.03	96	0	22	0.76	0.229
BMvardot < -0.50	96	12	21	0.77	0.219
BMvardot < -1.00	95	16	21	0.91	0.221
BMvardot < -1.50	97	57	21	1.00	0.216

proposed control. Table 8.6 shows the simulations stabilized by the control scheme with event detection using different values for the threshold in the control DT. There is a peak in the success rate around the parameter value of 50. Using $\theta dot_9 > 50$ for the control threshold, Table 8.7 shows how the performance varies as the detection threshold is varied.

Simulations are normally in perfect steady state until an event like a short circuit occurs. A detection criterion that is triggered by the slightest movement in any of the variables would be successful in these simulations. In reality such a sensitive criterion would detect an event for minor events or just real-time variations.

The "Not detected" column contains the number of simulated events that do not have a fault detected by the criteria. In the application of real-time control if an event is too insignificant to be worth controlling then it does not matter if the event is not detected. The detection criterion will be much less likely to declare an event for something insignificant if the threshold is changed from –0.03 to –0.5. Decreasing the threshold to –1.0 increases the control actuation time, which is undesirable. The final configuration for control with event detection uses *BMvardot* < –0.5 for event detection and $\theta dot_9 > 50$ for control actuation.

To obtain a fair comparison between control with and without event detection we perform similar experiments adjusting the threshold for the DT without event detection. Table 8.8 shows the simulations stabilized without event detection using different values for the threshold in the control DT. The final configuration for control without event detection uses $\theta dot_9 > 80$ for control actuation because larger thresholds significantly decrease the success rate.

Table 8.8 illustrates how it can be difficult to reduce the number of unnecessary control operations by changing threshold values in a control scheme that does not use event detection. As the threshold value is increased the number of successful operations goes to zero before the number of unnecessary operations goes to zero. Unnecessary operations often occur during and immediately after the fault-on period when measured values experience large sudden changes. The purpose of adding event detection is to prevent control operation until excessive swings associated with fault-on conditions have subsided.

There is precedent for preventing the operation of response-based control during large swings in measured quantities that are associated with fault-on conditions. Engineers who designed and tested the R-Rdot relay for out-of-step swing detection and mitigation in the western United States had similar concerns about improper operation associated with short circuit conditions [34,35]. Consequently, they designed the R-Rdot relay to avoid operation until large swings caused by fault inception and clearing subsided.

Tables 8.9 and 8.10 show the performance of the final configurations without event detection and with event detection. The control scheme with event detection has a higher success rate and fewer unnecessary controls. For a very similar test set of 960 simulations, the best method in [13] stabilizes 72 out of the 507 events having control action for a success rate of 0.142. The success rate of the new method with event detection is higher than the success rate in [13].

Table 8.8 Performance of one-shot control without event detection as the control DT is adjusted

Control DT	Controlled	Unnecessary	Stabilized	T_{avg}	Success rate
$\theta dot_9 > 160$	183	12	1	2.51	0.005
$\theta dot_9 > 140$	184	12	9	2.37	0.049
$\theta dot_9 > 120$	185	12	17	2.26	0.092
$\theta dot_9 > 100$	186	13	18	2.13	0.097
$\theta dot_9 > 90$	186	13	21	2.03	0.113
$\theta dot_9 > 80$	196	22	27	1.86	0.138
$\theta dot_9 > 70$	207	32	29	1.70	0.140
$\theta dot_9 > 60$	218	37	30	1.58	0.138

Table 8.9 Performance of DT control without event detection for one phase and three phase faults

Test set	Controlled	Unnecessary	Stabilized	T_{avg}	Success rate
One phase	196	22	27	1.86	0.138
Three phase	248	41	41	1.63	0.165
Total	444	63	68	1.74	0.153

Table 8.10 Performance of DT control with event detection for one phase and three phase faults

Test set	Controlled	Unnecessary	Stabilized	T_{avg}	Success rate
One phase	96	6	21	0.77	0.219
Three phase	112	13	28	0.91	0.250
Total	208	19	49	0.85	0.236

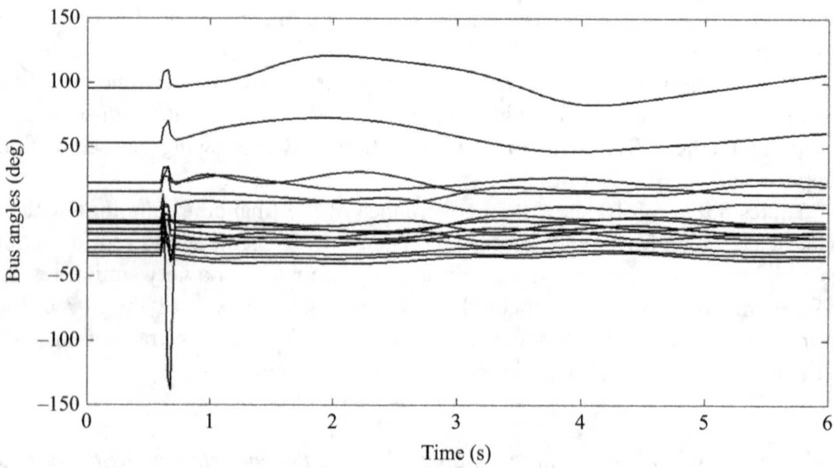

Figure 8.10 Bus angles for a four-cycle single line to ground fault near Malin

Figure 8.10 shows the 17 PMU bus angles for a four-cycle SLG fault on a transmission line connected to Malin substation. The faulted line is removed at 0.67 s when the fault is cleared. Bus angles in this section have been processed to remove any discontinuities at ±180° and to calculate center of angle coordinates.

For the final configuration without event detection the decision to order control occurs at 0.625 s followed by 0.1 s actuation delay. Control action occurs at 0.725 s and the bus angles with control are shown in Figure 8.11. This control action is classified as unnecessary because the event is stable with and without control.

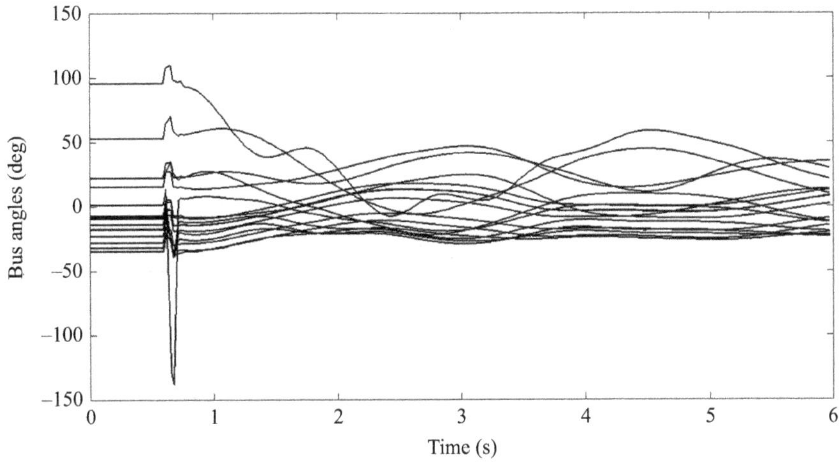

Figure 8.11 Unnecessary control action ordered by the scheme without event detection

Figures 8.10 and 8.11 illustrate how the one-shot control reduces phase angle differences. These angle differences drive current flows which sometimes cause protective relays to trip on overload. Many cascading outages involve relays tripping on overload. An event that would be stable without any circuit breakers opening could lead to cascading outages if relays disconnect transmission lines and other components.

The simulations in this chapter do not model the action of over current relays and other protective devices. Such a detailed study could potentially justify a decision to order control such as the one shown in Figure 8.11. For example, researchers at the Bonneville Power Administration reported in [36] that a 500 MW fast power increase on the IPP HVDC link could have prevented the December 14, 1994 blackout in the Western United States by reducing phase angle differences in the AC network and preventing relays from tripping on overload.

Figure 8.12 shows the 17 PMU bus angles for a four-cycle SLG fault on a transmission line connected to Hanford. The faulted line is removed at 0.67 s when the fault is cleared. For the final configuration with event detection the decision to order control occurs at 0.75 s followed by 0.1 s actuation delay. Control action occurs at 0.85 s and the resulting swing is stable as shown in Figure 8.13.

8.7 Conclusions

Renewable generation resources such as wind and solar are intermittent and their power output has significant real-time variations. Renewable resources are often distributed geographically which makes it difficult to monitor their output from a central location. PMU measurements of voltage and current on the transmission grid provide a means to observe at least indirectly the effects of renewable power

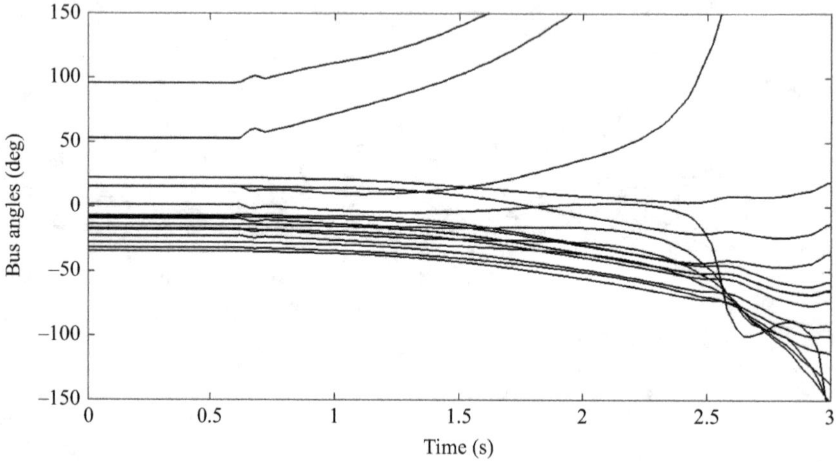

Figure 8.12 Bus angles for a four-cycle single line to ground fault near Hanford

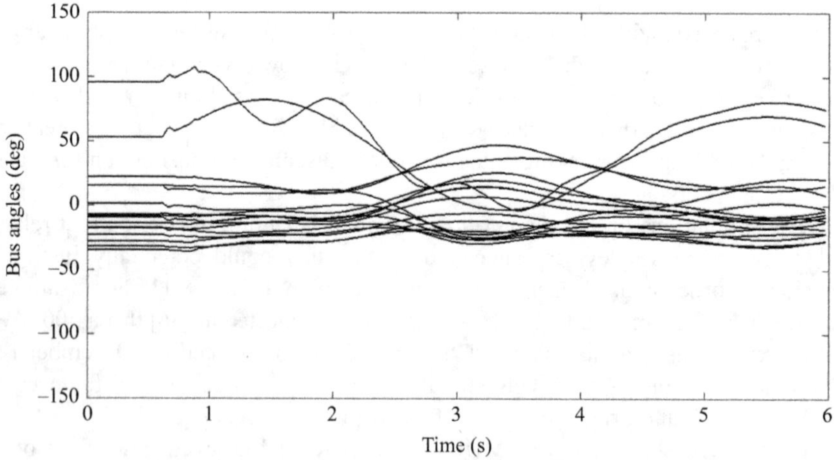

Figure 8.13 Event in Figure 8.12 with one-shot control using event detection

variations. This chapter describes improved methods for using pattern recognition tools such as DT to design one-shot controls that use PMU measurements to decrease the likelihood of blackouts.

Earlier work on pattern recognition for transient stability prediction and control assumed there would be some way to detect that an event had just occurred. Then the post-event phasor measurements could be processed by a DT to predict future instability and order control. This chapter shows how to use phasor measurements to perform this event detection.

The rate of change of bus voltage magnitudes worked well for event detection as expected but so did the rate of change in variance of the magnitudes. The authors chose to evaluate the variance of magnitudes index because it has a nice symmetry with the SBA index. The results illustrate how the variance of bus voltage magnitude reflects the severity of a short circuit.

The combination of separate event detection and control DTs for transient stability control is new and results in a higher rate of success in stabilizing events using one-shot control. Control actions may include disconnection of generation and load which are very expensive except compared to a blackout. The proposed control will be more acceptable if it is never activated unless there is a major transient event. The control proposed in this chapter satisfies this objective better than the previous method of using DTs for one-shot control.

The chapter evaluates some new and old indices for both the event detection and control actuation functions. Indices are compared by training a DT to order control using only one variable in the input vector. This chapter also uses the complexity cost to produce DTs with one or two nodes that are easy to interpret and modify. A line search for the best performance is done for each of the thresholds to get better performance.

Our results show that performance remains acceptable as decision thresholds are varied as much as 10° or 20°. This robustness to changes in the decision thresholds suggests that the methods will be robust to changes in operating conditions from intermittent renewable sources such as solar and wind. The only modification to the methodology required to handle different operating conditions will be to include simulations from different operating conditions in the training and test data as in [24].

Acknowledgments

This work was supported by the National Science Foundation Grant No. ECCS-1711521.

References

[1] Sun K., Likhate S., Vittal V., Kolluri V.S., and Mandal S. "An online dynamic security assessment scheme using phasor measurements and decision trees." *IEEE Transactions on Power Systems*. 2007;22(4): 1935–1943

[2] Diao R., Vittal V., and Logic N. "Design of a real-time security assessment tool for situational awareness enhancement in modern power systems." *IEEE Transactions on Power Systems*. 2010;25(2):957–965

[3] Diao R., Sun K., Vittal V., *et al.* "Decision tree-based online voltage security assessment using PMU measurements." *IEEE Transactions on Power Systems*. 2009;24(2):832–839

[4] Kamwa I., Samantaray S.R., and Joos G. "Development of rule-based classifiers for rapid stability assessment of wide-area post-disturbance records." *IEEE Transactions on Power Systems*. 2009;24(1):258–270

[5] Kundu P., and Pradhan A.K. "Wide area measurement based protection support during power swing." *International Journal of Electrical Power & Energy Systems*. 2014;63(Dec):546–554

[6] Amraee T., and Ranjbar S. "Transient instability prediction using decision tree technique." *IEEE Transactions on Power Systems*. 2013;28(3):3028–3037

[7] Cao J., Du W., Wang H., Chen Z., and Li H.F. "A novel emergency damping control to suppress power system inter-area oscillations." *IEEE Transactions on Power Systems*. 2013;28(3):3165–3173

[8] Beiraghi M., and Ranjbar A.M. "Online voltage security assessment based on wide-area measurements." *IEEE Transactions on Power Delivery*. 2013; 28(2):989–997

[9] Chaudhuri B., Majumder R., and Pal B.C. "Wide-area measurement based stabilizing control of power system considering signal transmission delay." *IEEE Transactions on Power Systems*. 2004;19(4):1971–1979

[10] Nuqui R.F., and Phadke A.G. "Phasor measurement unit placement techniques for complete and incomplete observability." *IEEE Transactions on Power Delivery*. 2005;20(4):2381–2388

[11] De La Ree J., Centeno V., Thorp J.S., and Phadke A.G. "Synchronized phasor measurement applications in power systems." *IEEE Transactions on Smart Grid*. 2010;1(1):20–27

[12] Bernabeu E.E., Thorp J.S., and Centeno V. "Methodology for a security/dependability adaptive protection scheme based on data mining." *IEEE Transactions on Power Delivery*. 2012;27(1):104–111

[13] Gao Q., and Rovnyak S.M. "Decision trees using synchronized phasor measurements for wide-area response-based control." *IEEE Transactions on Power Systems*. 2011;26(2):855–861

[14] Xu Y., Dong Z.Y., Guan L., Zhang R., Wong K.P., and Luo F. "Preventive dynamic security control of power systems based on pattern discovery technique." *IEEE Transactions on Power Systems*. 2012;27(3):1236–1244

[15] Mei K., and Rovnyak S.M. "Response-based decision trees to trigger one-shot stabilizing control." *IEEE Transactions on Power Systems*. 2004;19(1): 531–537

[16] Guo T., and Milanovic J.V. "Probabilistic framework for assessing the accuracy of data mining tool for online prediction of transient stability." *IEEE Transactions on Power Systems*. 2014;29(1):377–385

[17] Senroy N., and Heydt G.T. "A conceptual framework for the controlled islanding of interconnected power systems." *IEEE Transactions on Power Systems*. 2006;21(2):1005–1006

[18] Kamwa I., Pradhan A.K., Joos G., and Samantaray S.R. "Fuzzy partitioning of a real power system for dynamic vulnerability assessment." *IEEE Transactions on Power Systems*. 2009;24(3):1356–1365

[19] Samantaray S.R., Kamwa I., and Joos G. "Ensemble decision trees for phasor measurement unit-based wide-area security assessment in the operations time frame." *IET Generation, Transmission & Distribution*. 2010; 4(12):1334–1348

[20] Gomez F.R., Rajapakse A.D., Annakkage U.D., and Fernando I.T. "Support vector machine-based algorithm for post-fault transient stability status prediction using synchronized measurements." *IEEE Transactions on Power Systems*. 2011;26(3):1474–1483

[21] Ma J., Makarov Y.V., Diao R., Etingov P.V., Dagle J.E., and De Tuglie E. "The characteristic ellipsoid methodology and its application in power systems." *IEEE Transactions on Power Delivery*. 2012;27(4):2206–2214

[22] Guo T., and Milanovic J.V. "Online identification of power system dynamic signature using PMU measurements and data mining." *IEEE Transactions on Power Systems*. 2016;31(3):1760–1768

[23] Rovnyak S.M., Li G., and Mei K. "One-shot controls for preventing loss of synchronism." *Proceedings of the IEEE PES General Meeting*, Toronto, CA, USA, IEEE, July 2003. pp. 2045–2052

[24] Rovnyak S.M., Kretsinger S.E., Thorp J.S., and Brown D.E. "Decision trees for real-time transient stability prediction." *IEEE Transactions on Power Systems*. 1994;9(3):1417–1426

[25] Mei K., Rovnyak S.M., and Ong C.M. "Clustering-based dynamic event location using wide-area phasor measurements." *IEEE Transactions on Power Systems*. 2008;23(2):673–679

[26] Li G., and Rovnyak S.M. "Integral square generator angle index for stability ranking and control." *IEEE Transactions on Power Systems*. 2005; 20(2):926–934

[27] Rovnyak S.M., Nilchi M.N., Longbottom D.W., and Vasquez D.C. "Angle stability predictive indices." *Proceedings of the IEEE PES General Meeting*, San Diego, USA, IEEE, July 2012

[28] Nilchi M.N. *Electric utility planning methods for the design of one-shot stability controls*. MS thesis, Purdue University Indianapolis, 2012

[29] Longbottom D.W. *Polynomial curve fitting indices for dynamic event detection in wide-area measurement systems*. MS thesis, Purdue University Indianapolis, 2012

[30] Brandwajn V., Kumar A.B.R., Ipakchi A., Bose A., and Kuo S.D. "Severity indices for contingency screening in dynamic security assessment." *IEEE Transactions on Power Systems*. 1997;12(3):1136–1142

[31] Fu C., and Bose A. "Contingency ranking based on severity indices in dynamic security analysis." *IEEE Transactions on Power Systems*. 1999;14 (3):980–986

[32] R Development Core Team. R: A language and environment for statistical computing. R Foundation for Statistical Computing, Vienna, Austria [online]. 2011. Available from http://www.R-project.org/

[33] Breiman L., Friedman J., Stone C.J., and Olshen R.A. *Classification and Regression Trees*. Belmont, CA: Wadsworth, 1984.

[34] Taylor C.W., Haner J.M., Hill L.A., Mittelstadt W.A., and Cresap R.L. "A new out-of-step relay with rate of change of apparent resistance augmentation." *IEEE Transactions on Power Apparatus and Systems*. 1983;102 (3):631–639

[35] Haner J.M., Laughlin T.D., and Taylor C.W. "Experience with the R-Rdot out-of-step relay." *IEEE Transactions on Power Delivery*. 1986;1(2):35–39
[36] Rovnyak S.M., Taylor C.W., Mechenbier J.R., and Thorp J.S. "Plans to demonstrate decision tree control using phasor measurements for HVDC fast power changes" [online]. *Proceedings of the Fault and Disturbance Analysis & Precise Measurements in Power Systems Conference*, Arlington, USA, November 1995. Available from http://www.engr.iupui.edu/~srovnyak/pubs/PMconf95.pdf

Chapter 9

Online short-term voltage stability monitoring based on wide-area trajectory data analytics

Lipeng Zhu[1] *and Chao Lu*[2]

9.1 Introduction

With a continuous increase of load demands yet a mismatched expansion of transmission lines, modern power systems are delivering a large amount of power near their stability limits, which may lead to high risk of instability after encountering large disturbances. Meanwhile, due to the unprecedented penetration of renewable energy sources, some unfavorable features such as volatility, intermittency, and low inertia are introduced into power system operation and control. In urban areas, with the increasing integration of dynamic loads such as air conditioners and industrial motors, the short-term voltage stability (SVS) problem driven by postcontingency dynamics of load demand restoration is becoming more and more pronounced [1,2]. Recent decades have witnessed several SVS related blackouts, for example, Athens [3] and Brazil blackouts [4]. With timely SVS assessment performed during online monitoring, such failures could have been remedied by corrective actions, thereby avoiding unnecessary power outage and economic loss.

However, owing to the complicated driving mechanisms of various dynamic loads, it is difficult to analytically establish unified SVS assessment criteria for all sorts of power systems. Although some attempts such as energy function based methods [5] and transient P-V plane based stability boundary analysis [6] have been made for SVS, they are only competent in small test systems and become invalid when applied to large-scale power grids. Rooted from ergodic theory, a time series (TS) Lyapunov exponents (LE) approach is developed in [7] to assess SVS via the signs of LE values of TS; yet it may cost a relatively long time to ascertain the stability status in case that turbulent LEs linger around 0. In [8], by estimating Thevenin equivalent parameters in real time, an SVS predictor is put forward, whereas the estimated time-varying parameters could be sensitive to PMU noises. On the whole, today's power system operation and online monitoring are still in urgent need of reliable and efficient SVS assessment solutions.

[1]Department of Electrical and Electronic Engineering, The University of Hong Kong, Hong Kong, China
[2]Department of Electrical Engineering, Tsinghua University, Beijing, China

With the rapid development of wide-area measurement systems (WAMS) in smart grids, the availability of huge volumes of synchronized data acquired by phasor measurement units (PMUs) provides immense opportunities for data-driven dynamic stability/security assessment (DSA). In the research community, substantial attempts have been made to fulfill such tasks via machine learning (ML) techniques, including pattern discovery-based fuzzy classifiers [9], artificial neural network (ANN)-based extreme learning machines (ELMs) [10,11], support vector machines (SVMs) [12], decision trees (DTs), fuzzy rules [13], etc. Most of the existing methods extract features from individual snapshots of system dynamics. In fact, as the general time scale of SVS is 0~10 s, instead of focusing on snapshots, more attention should be paid to sequential TS for feature extraction.

Moreover, these representative ML-based efforts assume that their learning datasets can be generated by system simulations as many as needed. As precise simulation and modeling, especially load modeling [1,2], are recognized as great challenges in power systems, errors are inevitably introduced into learning datasets. Preferably, learning datasets can be acquired from practical PMU records, which would significantly improve the quality and reliability of the knowledge base. Following this idea, however, a learning machine is likely to encounter a serious class imbalance problem. In a realistic power grid, the system remains stable after most disturbances, while becoming unstable only in a few situations. If not treated properly, such an imbalance would dramatically deteriorate the learning machine's performances, with the minority class overlooked and thus mistaken. Besides, another crucial issue is often inherently tied to class imbalance: misdetection of power system instability would lead to irreversible voltage collapse or catastrophic outage, whereas mislabeling a stable case as unstable is usually remediable at a much lower expense. This implies that the imbalanced dataset is cost-sensitive as well, that is, unstable/unsecure cases are quite scarce yet rather expensive.

Focusing on tackling the above challenging issues, this chapter aims to develop a new framework for online SVS monitoring from the perspective of wide-area data analytics. Considering ML techniques' significant advantage in exploring discriminative features and discovering new knowledge from numerous complicated scenarios, the whole chapter will emphasize on the interpretability of the data analytics framework, which is expected to help system operators better comprehend SVS characteristics and patterns in complex power grids. With special attention to temporal correlations between successive snapshots of system trajectories, a novel TS learning approach called time-series shapelet classification (TSSC) [14–16] will be introduced into the framework to implement sequential SVS feature learning. In particular, faced with the practical class imbalance and cost-sensitive issues, a TSSC based imbalance learning machine will be systematically developed to construct a reliable and explainable SVS assessment model. Note that, as it is almost impossible for initial offline learning to cover all the potential conditions in future, an incremental learning strategy will be embedded into the learning machine, so as to enhance its adaptability and robustness during online monitoring.

The remainder of this chapter is structured as follows. First, the TSSC method for wide-area TS trajectory data analytic is described in detail. Then, the comprehensive

TSSC based imbalance learning machine for online SVS assessment is elaborately presented. Based on the proposed learning machine, numerical case studies are carried out on the Nordic test system and the real-world China Southern Power Grid (CSG) to demonstrate its efficacy. Afterward, further discussions on relevant future work are provided. Finally, the chapter ends with some concluding remarks.

9.2 Time series trajectory data analytics

As mentioned previously, during SVS dynamics, critical features could exist in TS of transient trajectories in a few seconds instead of several single snapshots. Therefore, learning TS trajectories directly rather than mining individual snapshots may result in more reliable SVS assessment. In terms of TS classification methods, the simple k-nearest neighbor algorithm seems to be accurate and robust, but its classification results are difficult to be explained. TSSC is a novel methodology recently proposed in the computer science community [14] for TS feature learning, and it can not only hold a high accuracy of classification but also provide a desirable interpretation for the classification results. Considering such appealing advantages in tackling TS data analytics issues, the TSSC method is introduced and employed as the key technique for SVS feature learning in this chapter. Technical details about TSSC and relevant techniques for improving its learning efficiency are presented in what follows.

9.2.1 Brief overview of TSSC

TSSC, first proposed in [14], is a promising TS learning approach in the field of ML. Here the term *shapelet* means a critical "sub-shape," more specifically, a TS subsequence with the highest performance of discriminating one class from another. For a multidimensional TS dataset, each dimension is expected to produce a shapelet via TSSC, and each shapelet in turn serves as an attribute, measuring the dissimilarities between TS samples by calculating the Euclidean distances between it and corresponding TS. In this way, the original TS dataset is transformed into a distance dataset, which is also called *shapelet transformation* [17], and conventional classification approaches such as DT, SVM, and ANN are all readily available for subsequent classification learning. For the sake of retaining of shapelets' interpretability, the DT algorithm is recommended here for its superiority in knowledge representation and comprehension. For a deeper understanding of TSSC, more descriptions about the shapelet transformation procedure and related concepts are presented below.

Given a d-dimensional dataset $\boldsymbol{S}$ with n TS samples of the same length m (number of data points), all the TS subsequences in the kth ($1 \leq k \leq d$) dimension constitute the kth set of shapelet candidates, where the kth shapelet will be finally picked out. Assuming the maximum and minimum lengths (number of sampling points) of these candidates are $L_{\min}$ and $L_{\max}$, respectively, the kth candidate set is

$$\boldsymbol{SC}_k = \left\{\boldsymbol{S}_{k,L_{\min}}, \boldsymbol{S}_{k,L_{\min}+1}, \ldots, \boldsymbol{S}_{k,i}, \ldots, \boldsymbol{S}_{k,L_{\max}}\right\} \tag{9.1}$$

where $\boldsymbol{S}_{k,i}$ is the collection of all the i-point TS subsequences in the kth dimension, for $L_{\min} \leq i \leq L_{\max}$. For a certain candidate X_i with length i in $\boldsymbol{SC}_k$, the distance between it and a full-length TS in the kth dimension, denoted by Y, is defined as

$$D(X_i, Y) = \min_{Y_{i,j} \in \boldsymbol{W}_i} \left(D(X_i, Y_{i,j}) \right) \tag{9.2}$$

where $\boldsymbol{W}_i$ is the set of all i-point subsequences of Y, $Y_{i,j}$ denotes a subsequence randomly chosen from $\boldsymbol{W}_i$, and $D(X_i, Y_{i,j})$ represents the Euclidean distance between X_i and $Y_{i,j}$. As shown in Figure 9.1, the minimum criterion in (9.2) indicates the subsequence of Y that best matches X_i. In this regard, the dissimilarities between the candidate X_i and all the TS are quantified by an array of distances, denoted as $\boldsymbol{D}_i = \{D_{1,i}, D_{2,i}, \ldots, D_{n,i}\}$. Given $\boldsymbol{D}_i$ and all the n samples' class labels, the potential of X_i in classification is measured by information gain (IG); the larger the IG, the greater the capability of X_i in discriminating different classes. Hence, by IG computations and comparisons, the optimal subsequence with the largest IG, that is, the shapelet, is selected from the candidate set. More detailed techniques about shapelet searching can be found in [14,15].

As for TSSC-based SVS assessment, in a specific power system, TS of each bus's *V/I/P/Q*, that is, voltage, current, active power, and reactive power injected into the bus, are acquired from PMU measurements, and TS from all kinds of cases are then gathered as a TS dataset. With shapelet transformation, such a TS dataset is transformed into a distance dataset, based on which a DT model can be built for online SVS assessment. A DT is essentially a group of *if-then* rules organized in a flowchart form. From top to bottom, by if-then comparisons of one sample's attribute values with a series of splitting values (thresholds) on the tree, a specific path consisting of the root node, a number of internal nodes, a terminal node, and relevant branches is determined. In this manner, the class label stored at the terminal node finally predicts whether the system can keep stable or not. Based on

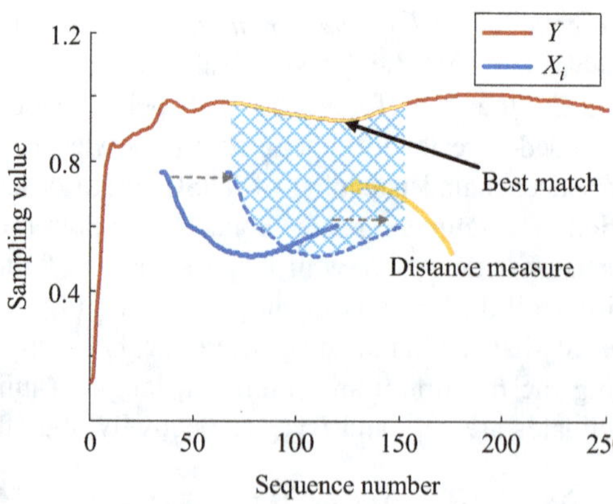

Figure 9.1 Illustration of the distance between X_i and Y

different node splitting criteria, concrete DT induction algorithms are categorized as ID3, C4.5, CART, etc. [18].

9.2.2 Learning efficiency improvement

The original TSSC methodology has a severe weakness that its shapelet searching procedure is almost brute-force, and it is computationally prohibitive for real-world problems. Although [14,19] have provided some techniques to improve the search efficiency, when those techniques are applied to SVS assessment, the efficiency still remains unsatisfactory. For a receiving-end region, suppose each bus is placed with a PMU for SVS monitoring. If there exist h buses in the region, the number of dimensions of the multivariate TS, denoted as d, would be $d = 4h$. Empirical tests show that it costs tens of hours to finish the shapelet searching procedure for a five-bus region using the improved method in [14]. Although the searching procedure can be carried out offline before the system is commissioned, or in an incremental way after some disturbances, its low efficiency would lead to a reduction of the reliability for online application. This is because online SVS assessment needs to run for 24*7 h based on the latest searching and learning results, while the searching procedure with a too heavy computational burden cannot timely update the latest learning results. In order to speed up shapelet searching and make the TSSC methodology more practical, two acceleration techniques proposed by the authors [15] are presented as follows.

9.2.2.1 Sampling search

In the original methodology, exhaustive search is performed to find the optimal shapelet. If the number of cases in the dataset is too large, substantial shapelet candidates will be generated and compared one by one. The total times of Euclidean norm computations and comparisons for determining the distances between a candidate and TS will increase significantly as well. With a lot of IGs needed to be calculated and compared, the computational efficiency would be even lower. If the total number of TS samples is cut down, these problems may be alleviated. Hence, the strategy of random sampling is taken into consideration. The main idea of sampling search is to cut the big dataset into small subsets, search shapelets in each subset (one shapelet per subset) and select the optimal shapelet with the largest IG from these subsets. Its main procedure is summarized in Algorithm 9.1

Algorithm 9.1: Sampling search

Input: A d-dimensional TS dataset $\boldsymbol{S} = \{\boldsymbol{TS}_1, \boldsymbol{TS}_2, \ldots, \boldsymbol{TS}_n\}$, $\boldsymbol{S} \in \mathbb{R}^{m*d*n}$.
Number of sampling times k.

Output: A set of k shapelets $\{Shp_1, Shp_2, \ldots, Shp_d\}$

(1) **for** $i = 1$ to d **do**

(2) **for** $j = 1$ to k **do**

(3) create subset $\boldsymbol{S}_j$ by arbitrarily choosing $[n/k]$ (round toward 0) objects from $\boldsymbol{S}$;

(4) find the shapelet shp_j in $\boldsymbol{S}_j$ by shapelet searching;
(5) **endfor**
(6) select the optimal shapelet Shp_i from $\{shp_1, shp_2, \ldots, shp_k\}$ by IG comparison;
(7) **endfor**

Given the number of sampling times k, the search procedure needs k iterations, while the time consumptions of computations and comparisons decrease significantly, resulting in a total speed-up. In particular, this simple method produces k times speed-up. In this respect, the value setting of k seems very important. If k is too large, the number of samples in each subset could be too small to find a high-performance shapelet; otherwise, the improvement of computational efficiency would be trivial. Experimental test results show that setting k among the range of 5~10 would be an appropriate trade-off.

9.2.2.2 Adaptive variable-step search

Suppose the TS is divided into a number of pieces by an appropriate piecewise linear fitting method. Since the evolution of individual TS often differs from each other, corresponding piece sizes would change automatically. Each piece is taken as the step for shapelet searching. Instead of tediously searching point by point, here searching is performed piece by piece, which speeds up the search procedure dramatically. For example, if a certain TS with the length of 100 is divided into 10 pieces by the piecewise linear fitting approach, only 10 steps are needed to finish the search, while, in contrast, 100 steps are needed in the original method (step size set to 1). Although increasing the fixed step size in the latter method can also improve the computational efficiency, some critical characteristics or information may disappear. On the contrary, as an approximate representation of TS, piecewise linear fitting can retain the evolution trends and critical characteristics of the original TS. Even if the real optimal shapelets could get lost sometimes, the obtained sub-optimal shapelets can still achieve excellent classification performances.

As for TS fitting, the fitting error, denoted as *ERR*, is an important metric of the fitting precision. Given the original TS $Y = \{y_1, y_2, y_3, \ldots, y_m\}$ and the fitted TS $Y' = \{y_1', y_2', y_3', \ldots, y_m'\}$, *ERR* is calculated as

$$ERR(Y, Y') = \frac{1}{m}\sum_{i=1}^{m}(y_i - y_i')^2 \tag{9.3}$$

To make the fitting error as small as possible, the least square method (LSM) is employed in linear fitting. Besides, the number of objective segments should not be too large in consideration of both fitting precision and computational efficiency. Otherwise, too many pieces in the TS need to be merged. Based on a piecewise segmenting and merging approach named the bottom-up algorithm [20], its improved version is developed here. Taking the kth dimension in $\boldsymbol{S}$ for example, the implementation of this improved method is detailed in Algorithm 9.2.

Algorithm 9.2: Improved bottom-up segmentation

Input: A TS dataset $\boldsymbol{S}_k = \{\text{TS}_1, \text{TS}_2, \ldots, \text{TS}_n\}$, $\boldsymbol{S}_k \in \boldsymbol{S}$
Initial number of segments K_0
Objective number of segments K
Output: Piecewise information of all objects in $\boldsymbol{S}_k$
(1) **for** $i = 1$ to n **do**
(2) cut TS_i into K_0 pieces uniformly, and cut them into more small pieces by extreme points;
(3) denote the number of new pieces as K_1 and fit all pieces with LSM;
(4) **repeat**
(5) calculate fitting errors for two adjacent pieces in turn and choose the minimum for merging (with LSM);
(6) update the number of pieces as $K_1 = K_1 - 1$;
(7) **until** $K = K_1$
(8) collect piecewise information for TS_i;
(9) **endfor**

9.3 TSSC-based imbalance learning for SVS monitoring

Faced with class imbalance problems, the ML community has made considerable efforts to deal with them [21–24]. Broadly speaking, synthetic sampling is the most commonly used method to rebalance class distributions. However, it cannot be directly applied to SVS assessment, because datasets created by naïve replication or linear interpolation can hardly be related to actual power system dynamics. In other words, these synthesized samples may not exist in practice and thus make no physical sense. For power system operators, they usually favor transparent and interpretable learning machines during both offline analysis and online decision making [25]. Hence, for the sake of comprehending concepts and patterns learned from knowledge bases, it is indispensable to customize specific synthetic methods based on actual SVS dynamics. In addition to sampling related techniques, some cost-sensitive tricks have been proposed to construct cost-sensitive classifiers, such as cost-sensitive ANN [21] and cost-sensitive DT [26]. With preset costs appended to different classes, such techniques manage to reinforce minority class learning via raising more concerns on minority cases. In this way, those cost-sensitive classifiers are able to make an appropriate trade-off regarding the classification precisions of majority and minority classes.

Taking the above points into consideration, an imbalance learning machine based on TSSC is systematically designed, as illustrated in Figure 9.2. The whole learning machine consists of two critical techniques, that is, forecasting-based nonlinear synthetic minority oversampling technique (FN-SMOTE) and cost-sensitive learning. In the following, before presenting the two techniques, several

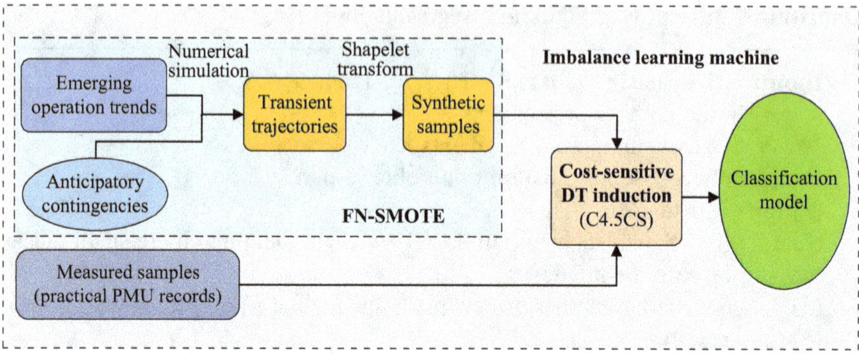

Figure 9.2 Architecture of the proposed imbalance learning machine

metrics for measuring the learning machine's performances in the context of class imbalance are first introduced.

9.3.1 Classification performance metrics

Suppose a certain SVS dataset is composed of 4% unstable instances and 96% stable instances, respectively. If a simple classifier labels all the instances as stable, it will achieve a total accuracy of 96%, which suggests that the conventional metric *accuracy* cannot be solely relied on to assess the performance of imbalanced classification. Hence, several other metrics for performance evaluation of class imbalance problems are introduced here. First, a confusion matrix is illustrated in Table 9.1. According to predicted and actual classes, all the instances are assigned to four groups, that is, N_{TS}, N_{FS}, N_{FU}, and N_{TU}, and related indices [13,18] are defined as follows.

$$Accuracy: Accu = (N_{TU} + N_{TS})/(N_U + N_S) \tag{9.4}$$

$$Misdetection: Mis = N_{FS}/(N_U + N_S) \tag{9.5}$$

$$False-alarm: Fal = N_{FU}/(N_U + N_S) \tag{9.6}$$

$$G-mean: G = \sqrt{\frac{N_{TU}}{N_{TU} + N_{FS}} \cdot \frac{N_{TS}}{N_{TS} + N_{FU}}} \tag{9.7}$$

Among them, *Accuracy* (*Accu*) is widely used without attention to class imbalance. *Misdetection* (*Mis*) stands for the ratio of falsely dismissed unstable cases, while *False-alarm* (*Fal*) represents what percentage of stable samples is falsely alarmed. *G-mean* (*G*) acts as a geometric mean of the true stable rate and the true unstable rate, evaluating the overall performance of imbalanced learning.

As mentioned above, mistaking unstable cases as stable would cause disastrous consequences, while the opposite one leads to much less loss. A cost matrix is shown in Table 9.2. The nondiagonal elements are denoted as C_1 and C_2 ($C_1 > C_2$), which correspond to the costs of misdetection and false alarm, respectively, while the remaining ones of correct classification are set to 0.

Table 9.1 Confusion matrix

Confusion matrix	Stable (actual)	Unstable (actual)
Stable (predicted)	N_{TS} (true stable)	N_{FS} (false stable)
Unstable (predicted)	N_{FU} (false unstable)	N_{TU} (true unstable)
Total	N_S (stable)	N_U (unstable)

Table 9.2 Cost matrix

Cost matrix	Stable (actual)	Unstable (actual)
Stable (predicted)	0	$C_1 = C(S, U)$
Unstable (predicted)	$C_2 = C(U, S)$	0

9.3.2 Class-imbalanced learning

9.3.2.1 Forecasting-based nonlinear SMOTE (FN-SMOTE)

This technique aims at generating reasonable samples according to the current and predicted impending operating points in a specific power system. Load forecasting data and generation schedule data are collected for operation forecasting. In particular, by integrating scheduling/prediction data under certain operation constraints, power flow analysis is conducted to predict the impending operation states of the system.

Mathematically, power system dynamics can be formulated as

$$\begin{cases} \dot{\mathbf{x}}_i = f(\mathbf{x}_i, \mathbf{y}_i, \mathbf{p}_i) \\ 0 = g(\mathbf{x}_i, \mathbf{y}_i, \mathbf{p}_i) \end{cases} \tag{9.8}$$

where $\mathbf{x}_i$, $\mathbf{y}_i$, and $\mathbf{p}_i$ are the vectors of dynamic variables (e.g., TS of *V/I/P/Q*), algebraic variables (e.g., steady-state voltages and power), and parameter variables (e.g., generator controller parameters), respectively. Given the predicted operating trend, $\mathbf{y}_i$ and part of $\mathbf{p}_i$ are specified, and the remaining part of $\mathbf{p}_i$ and $\mathbf{x}_i$ are determined by specific SVS dynamics following a certain contingency. With $\mathbf{x}_i$, $\mathbf{y}_i$, and $\mathbf{p}_i$ determined, they comprehensively shape the postcontingency trajectory $\mathcal{T}_i$. Taking into account all credible contingencies, power system numerical simulations are performed to obtain a series of trajectories, from which TS of *V/I/P/Q* are extracted. With the help of TSSC, these TS from trajectory $\mathcal{T}_i$ are transformed into

$$\{\boldsymbol{D}^i_V, \boldsymbol{D}^i_I, \boldsymbol{D}^i_P, \boldsymbol{D}^i_Q\} = \mathbb{ST}(\boldsymbol{TS}^i_V, \boldsymbol{TS}^i_I, \boldsymbol{TS}^i_P, \boldsymbol{TS}^i_Q) \tag{9.9}$$

where $\boldsymbol{TS}^i_V$, $\boldsymbol{TS}^i_I$, $\boldsymbol{TS}^i_P$, and $\boldsymbol{TS}^i_Q$ are TS of *V/I/P/Q* obtained from trajectory $\mathcal{T}_i$, $\mathbb{ST}(*)$ denotes shapelet transformation, and $\boldsymbol{D}^i_V$, $\boldsymbol{D}^i_I$, $\boldsymbol{D}^i_P$, and $\boldsymbol{D}^i_Q$ are transformed *V/I/P/Q* distance data. $\boldsymbol{D}^i_V$, $\boldsymbol{D}^i_I$, $\boldsymbol{D}^i_P$, and $\boldsymbol{D}^i_Q$ are then integrated as a synthetic instance in the feature space. Since (9.8) and (9.9) are nonlinear, the synthetic instance is not limited to the intervals of linear segments between the original minority instances. Compared with the widely used linear interpolation based synthetic sampling [21], this relief helps to depict the class boundaries more naturally and reasonably.

Since synthesis aims to mitigate class imbalance to some degree, it is suggested to generate more unstable cases than stable ones. A feasible way is to duly adjust simulation parameters, such as setting more severe faults and imposing more dynamic loads. However, even the state-of-the-art load forecasting techniques [27] unavoidably introduce errors to predicted data. Thus, it is not intended to achieve an absolute balance by this technique. For this reason, the number of synthetic samples should not be too large in case of poorer classification performances. In fact, according to the authors' empirical tests with a certain level of load forecasting errors [27] and load modeling errors, the synthetic/measured proportion being 15%~30% would be a satisfactory trade-off.

Due to possible errors caused by forecasting and simulation, there may exist outliers or noises among synthetic SVS datasets, and a data cleaning method called Tomek link detection [21,28] is adopted here to remove such disqualified samples. Let the original measured dataset be $\boldsymbol{S} = \boldsymbol{S}_{\mathrm{S}} \cup \boldsymbol{S}_{\mathrm{U}}$, where $\boldsymbol{S}_{\mathrm{S}}$ and $\boldsymbol{S}_{\mathrm{U}}$ denote the stable and unstable subsets, respectively. Given a synthetic sample $\boldsymbol{x}$ which is labeled as unstable (stable), a measured stable (unstable) sample $\boldsymbol{y}$ is randomly chosen from $\boldsymbol{S}_{\mathrm{S}}$ ($\boldsymbol{S}_{\mathrm{U}}$), and they constitute a detection pair ($\boldsymbol{x}$, $\boldsymbol{y}$). The Euclidian distance between $\boldsymbol{x}$ and $\boldsymbol{y}$ is denoted as $D(\boldsymbol{x}, \boldsymbol{y})$, and the pair ($\boldsymbol{x}$, $\boldsymbol{y}$) presents a Tomek link if there exists no sample $\boldsymbol{z}$ in $\boldsymbol{S}$ subject to the condition $D(\boldsymbol{x}, \boldsymbol{y}) > D(\boldsymbol{x}, \boldsymbol{z})$ or $D(\boldsymbol{x}, \boldsymbol{y}) > D(\boldsymbol{y}, \boldsymbol{z})$. Such a link implies that $\boldsymbol{x}$ is a disqualified sample, which will be deleted for its poor quality. In this way, all the synthetic samples are examined one by one to filter out possible outliers.

9.3.2.2 Cost-sensitive learning

By setting distinct costs, this technique imposes bias or weights toward the minority unstable class, aiming to achieve a totally minimal misclassifying cost. Based on the C4.5 DT algorithm employed in [14] for classification, an improved version named C4.5CS [26] is utilized for cost-sensitive learning here. Supposing $\boldsymbol{S} = \boldsymbol{S}_{\mathrm{S}} \cup \boldsymbol{S}_{\mathrm{U}}$ contains N_1 unstable samples and N_2 stable ones, the proportion of objects in class i ($1 \leq i \leq 2$, hereafter class 1 and 2 represent the unstable class and the stable class, respectively) is

$$p_i = N_i/N \qquad (N_1 + N_2 = N) \tag{9.10}$$

Without additional settings, C4.5 specifies each object's weight as 1 by default. On the contrary, C4.5CS modifies all the samples' weights, making them be proportional to respective misclassifying costs:

$$w_i = C_i N \Bigg/ \left(\sum_{j=1}^{2} C_j N_j \right) \tag{9.11}$$

where C_1 and C_2 denote the aforementioned costs of misdetection and false alarm, respectively. Thus, the weighted proportion of class i is

$$p_{wi} = w_i N_i \bigg/ \left(\sum_{j=1}^{2} w_j N_j \right) = w_i N_i / N \tag{9.12}$$

With the weighted proportion, new splitting criteria devoted to shapelet performance evaluation of TSSC and to DT induction are constructed in the following. First, the weighted entropy/expected information for classifying an object in $\boldsymbol{S}$ is computed by

$$Info_w(\boldsymbol{S}) = -\sum_{i=1}^{2} p_{wi} \log_2(p_{wi}) \tag{9.13}$$

Given an attribute A_k and the distance measure of a shapelet, whose value ranges among $[d_1, d_2]$, by partitioning on A_k with a splitting point sp_i ($d_1 < sp_i < d_2$), $\boldsymbol{S}$ is cut into two subsets $\boldsymbol{S}_1$ and $\boldsymbol{S}_2$. As a new metric of shapelets' performances, the weighted information gain (WIG) is calculated as

$$WIG(A_k) = Info_w(\boldsymbol{S}) - \sum_{i=1}^{2} p_{w\boldsymbol{S}_i} Info_w(\boldsymbol{S}_i) \tag{9.14}$$

Here $p_{w\boldsymbol{S}_i}$ represents the weighted proportion of objects in $\boldsymbol{S}_i$:

$$p_{w\boldsymbol{S}_i} = \sum_{j=1}^{2} w_j N_{\boldsymbol{S}_{ij}} \bigg/ \left(\sum_{i=1}^{2} w_i N_i \right) = \sum_{j=1}^{2} w_j N_{\boldsymbol{S}_{ij}} / N \tag{9.15}$$

where $N_{\boldsymbol{S}ij}$ denotes the number of class j objects in $\boldsymbol{S}_i$. Analogous to (9.13), the potential weighted information generated by the split of sp_i is

$$SplitInfo_w(\boldsymbol{S}) = -\sum_{i=1}^{2} p_{w\boldsymbol{S}_i} \text{ log2 } p_{w\boldsymbol{S}_i} \tag{9.16}$$

With $SplitInfo_w(\boldsymbol{S})$, WIG can be normalized as weighted information gain ratio (WIGR), which is written as

$$WIGR(A_k) = \frac{WIG(A_k)}{SplitInfo_w(\boldsymbol{S})} \tag{9.17}$$

For each attribute, its maximum WIGR can be obtained by scanning all the possible splitting values. Generally the attribute with the largest WIGR indicates by far the best partition pattern, which not only derives fairly pure subsets but also minimizes misdetection of high-cost unstable objects. Hence, WIGR is designated as the splitting criterion for cost-sensitive DT construction, during which top-down splits are recursively conducted by searching the maximum WIGR at each node.

9.3.3 SVS assessment scheme

The above two learning techniques are systematically combined to form the imbalance learning machine (see Figure 9.2), and it is further exploited to construct an SVS assessment scheme for online monitoring. The implementation of the scheme mainly includes three phases, as presented in the following.

9.3.3.1 Initial learning

First of all, measured samples are mainly collected from historical PMU records of system operation. Months of operation data are inquired, and TS are extracted from PMU data to form representative measured samples for diverse conditions and scenarios. As a matter of fact, this dataset is typically class-imbalanced. With the proposed FN-SMOTE, a number of synthetic samples are generated in the light of the current and ongoing trends of the system. Then the 2 groups of samples are integrated as the initial TS dataset, and shapelet transformation is performed to extract shapelets from it and further transform it into a distance dataset. With this transformed dataset, C4.5CS cost-sensitive learning is executed to build an initial DT model.

9.3.3.2 Incremental update

Based on initial offline learning results, this phase runs continually in parallel with online assessment. By periodically inspecting and learning various new situations, it is devised to strengthen the reliability and adaptability of the DT classifier and to keep the classifier tightly tied to the newest system conditions. Since significant disturbances and contingencies in practice take place infrequently, the incremental update period is set to 72 h, during which operating conditions involving various disturbances are picked out and respective PMU data makes up measured samples. Similar to initial learning, a reasonable number of synthetic samples can be also generated and combined with measured ones, constituting the incremental dataset.

Incremental learning is then performed to update shapelets and the dataset, and to update the cost-sensitive DT classifier, as presented in Figure 9.3. Instead of searching shapelets in a large dataset, this incremental strategy searches new shapelets only in the incremental dataset and compares these new shapelets with the old ones extracted from the initial dataset, determining preferable ones by WIG comparisons. Meanwhile, the dataset is updated in a rolling manner: on the basis of timeline, the newest incremental samples are injected into the initial dataset and meanwhile an identical number of the oldest samples are ejected. Note that although old samples are replaced along with time, discriminative shapelets derived from them can still be reserved unless their new competitors from incremental datasets are shown to have higher discriminability. Subsequently, the updated datasets and shapelets would in turn serve as the initial database for the next update period. As a result, this incremental strategy saves large volumes of computation time and avoids the heavy burden of storing accumulated large-size databases, with only a slight reduction of the classification performance.

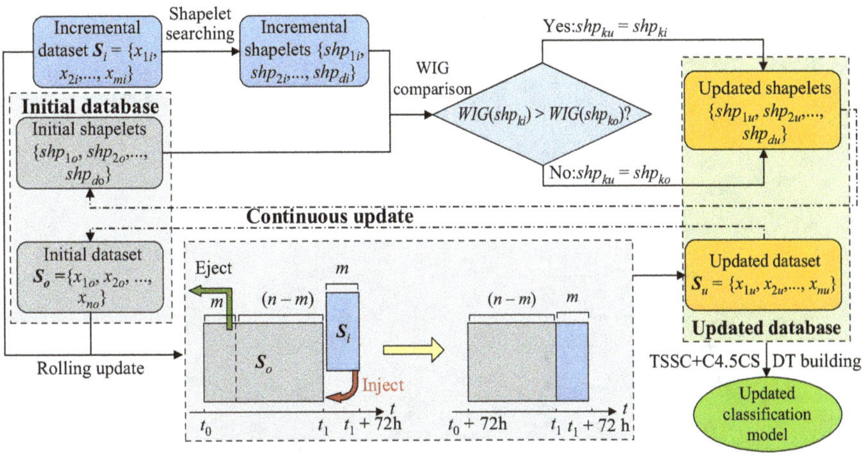

Figure 9.3 Schematic diagram of incremental learning

9.3.3.3 Online monitoring

This phase works in real time, and once contingencies occur, TS acquisitions are conducted by collecting postcontingency PMU measurements of *V/I/P/Q* at each bus. Distances between TS and those shapelets are computed and compared with the corresponding thresholds on the DT, tracing the path from the root node to a certain terminal node. Ultimately, the class label at the terminal node predicts whether the system can maintain stability. If so, the system will keep on monitoring. Otherwise, alerting signals will be sent out to alarm that remedial actions be taken as soon as possible. As both stable and unstable cases are precious measured samples, they will be added into the periodical collection of newly incremental cases for the next incremental update.

9.4 Numerical case studies

The proposed imbalance learning machine was extensively tested on the Nordic test system and the realistic CSG. Numerical tests on the two systems focus on demonstrating the effectiveness of the learning machine and illustrating how it contributes to SVS knowledge discovery and analysis in practice, respectively.

9.4.1 Test on Nordic test system

The Nordic test system is a simplified version of the real-world Swedish and Nordic power grid [29,30], as shown in Figure 9.4. The system includes four areas, that is, North, South, Equiv., and Central, and the last one is the load center with large amounts of load demands. The receiving-end region in Central consisting of five transmission buses (bus 1041~bus 1045) and their respective distribution sides was selected for SVS assessment demonstration.

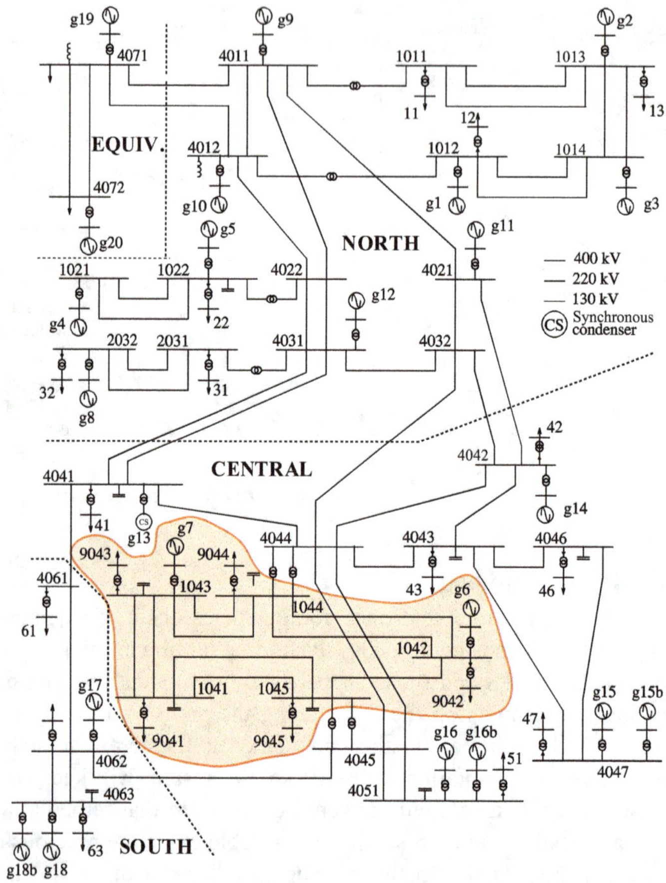

Figure 9.4 One-line diagram of the Nordic test system

9.4.1.1 Simulation settings and assumptions

As for system modeling, dynamic components, including synchronous machines, exciters, automatic voltage regulators, and power system stabilizers, were modeled in accordance with [29] except for load models. Instead, a composite load model with an induction motor and static ZIP load in parallel was applied to all the loads, so as to follow the primal concern of SVS in this chapter. The operating point B in [29] was specified as the base operating condition. Based on it, various operating points, proportions of dynamic load, fault types, fault locations, and fault clearing time were set to mimic diverse historical conditions and contingencies with respect to SVS. Detailed settings were listed in Table 9.3. With these settings, time-domain numerical simulations were performed and 900 original samples were generated, within which 720 ones (7.6% being unstable) were randomly chosen as "measured" training samples and the rest as testing samples. In order to sufficiently capture the

Table 9.3 Condition and contingency settings

Category	Shifts and specifications
Operating points	80%, 100%, 120% (power flow adjustment)
Ratios of dynamic load	60%, 75%, 90%
Fault types	Three-phase short circuit
Fault location	0, 20%, 40%, 60%, 80% (of the whole line length)
Fault line	1041–1043, 1041–1045, 1042–1044, 1042–1045, 1043–1044, 4031–4041, 1043–4042, 4032–4044, 4041–4061, 4045–4062 (bus *m*–bus *n*)
Fault clearing time	0.05/0.25 s for near ends; 0.15/0.3 s for remote ends (primary/backup protection)

critical shapelets behind system instability, the TS length of PMU data acquisition was set to 2.5 s with a 100-Hz sampling rate.

9.4.1.2 Reliability tests on imbalance learning machine

1. *Verification of FN-SMOTE:* Let the operating point A [29] be the forecasted impending operating trend of the system. Since the average load forecasting error generally ranges among 0.9%~8.27% [27], taking load modeling errors into additional account, an average load demand deviation of 10% was set to mimic the total load modeling and forecasting errors. FN-SMOTE was performed and 180 synthetic samples were generated. After Tomek link-based data cleaning, 168 eligible instances were left, including 127 unstable ones; only these unstable cases were chosen for imbalance mitigation. Then measured and synthetic cases formed the initial dataset, containing 665 stable cases (78.5%) and 182 unstable cases (21.5%). Compared with the previous proportion of 7.6%, the imbalance degree was relatively alleviated.

   Besides FN-SMOTE, other popular methods for tackling class imbalance issues in the ML community, including ROS [31], SMOTE [32], B-SMOTE [33], and ADASYN [34], were also carried out for comparative study. With no concern about forecasting errors, the latter four methods synthesized unstable samples until their total proportion reached 50%, that is, an absolute balance. In order to test their capability of tackling class skewness on the same base, the original C4.5 DT algorithm without cost-sensitive treatment was utilized for classification. Five DT classifiers were built on respective rebalanced datasets. To make a fair comparison, all these models were tested on the same 720 training samples and 180 testing samples, respectively. Furthermore, the 720 samples without any sampling treatment were also fed into the DT algorithm for classification (called *untreated* below). Learning performances of these classifiers are summarized in Table 9.4.

   As can be seen, although the untreated one achieves the highest accuracy (98.19%) on the training dataset, it has the poorest generalization capability for detecting the unseen testing cases. Among the five sampling methods, ROS and B-SMOTE exhibit moderate performances, especially failing to detect a significant proportion of minority unstable cases in the testing phase. Intuitively, ROS simply replicates unstable samples in the original dataset, which makes quite a number of instances overlap with each other, thus the induced rules become too specific, leading to overfitting problems. For B-SMOTE, since it pays special attention to unstable instances close to stable/unstable boundaries in the feature space, the induced DT becomes too detailed and trivial. To some degree, such a trivialness is likely to deteriorate its generalization capability. On the other hand, FN-SMOTE, SMOTE, and ADASYN appear amazingly competent and comparable in tackling the skewed class distribution, with extremely low rates of misdetection and false alarm for both training and testing datasets. In fact, it is because all the three approaches manage to interpolate data points in the feature space in a more general and adaptive way that they result in such desirable generalization performances.

Note that, however, if the forecasting and simulation errors are further reduced, FN-SMOTE would achieve an even better performance. Besides, in contrast to other existing SMOTE methods producing fictive SVS samples, FN-SMOTE generates samples which can be reasonably related to realistic system dynamics. This is of great importance for subsequent comprehension of critical instability patterns from sub-shapes of TS samples. Hence, FN-SMOTE would be preferred in practical contexts of SVS assessment.

2. *Cost-sensitive learning and cost ratio tuning:* In Table 9.4, it is found that among the 180 testing samples, misdetection happens more often than false alarm for all the methods, thus cost-sensitive learning is necessary to further enhance the detection of unstable cases. As for cost-sensitive DT, the parameter of misclassifying cost ratio C_1/C_2 needs to be prescribed. Without quantitative prior information about cost distinctions, C_1/C_2 was separately set to 5, 10, 15, and 20 based on general qualitative cognition. Given these ratios, a series of DT classifiers were built, and then tested on the same 720 training samples and 180 testing ones, with corresponding results summarized in Table 9.5. Note that $C_1/C_2 = 1$ indicates an ordinary DT classifier without cost-sensitive concern.

Table 9.4 Comparison of sampling techniques for offline learning

Method	**720 training samples**				**180 testing samples**			
	***Mis* (%)**	***Fal* (%)**	***G***	***Accu* (%)**	***Mis* (%)**	***Fal* (%)**	***G***	***Accu* (%)**
Untreated	0.97	0.83	0.930	98.19	19.44	0	0.879	80.56
FN-SMOTE	1.39	0.56	0.902	98.06	1.67	0.56	0.971	97.78
ROS	0.56	2.64	0.949	96.81	5.56	0	0.967	94.44
SMOTE	0.28	2.64	0.968	97.08	2.22	0.56	0.968	97.22
B-SMOTE	0.14	2.92	0.975	96.94	5.00	0	0.967	95.00
ADASYN	0.97	1.67	0.926	97.36	1.67	0.56	0.971	97.78

Table 9.5 Comparison of DT performances with different cost ratios

Cost ratio C_1/C_2	**720 training samples**				**180 testing samples**			
	***Mis* (%)**	***Fal* (%)**	***G***	***Accu* (%)**	***Mis* (%)**	***Fal* (%)**	***G***	***Accu* (%)**
1	1.39	0.56	0.902	98.06	1.67	0.56	0.971	97.78
5	0.69	1.39	0.946	97.92	1.11	1.11	0.955	97.78
10	0.42	1.67	0.964	97.92	0.56	2.22	0.917	97.22
15	0.14	2.22	0.979	97.64	0	3.89	0.855	96.11
20	0.14	2.50	0.977	96.36	0	5.00	0.809	95.00

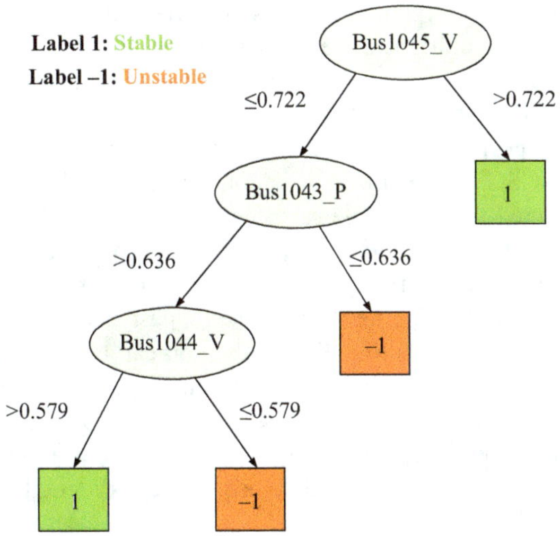

Figure 9.5 Offline learned SVS assessment model (Nordic)

From an overall perspective, with the increase of the cost ratio, C4.5CS successfully recalls more unstable cases. On the other hand, the reduction of the number of accurately identified stable cases is also seen, while the DTs still remain a high accuracy (above 95%) on the whole. When C_1/C_2 is set above 15, the classifiers do not gain significant benefits of reducing misdetection any more, whereas mistaking more stable cases. Hence, as a compromise between misdetection and false alarm, the ratio C_1/C_2 is set to 10. With such a cost ratio, the imbalance learning machine is expected to accurately identify both unstable and stable cases to some extent during online monitoring.

3. *Analysis of SVS assessment model:* Here the DT model with $C_1/C_2 = 10$, as illustrated in Figure 9.5, is taken as an example for further analysis of the DT based online SVS assessment model.

   The variables at internal nodes of the DT in Figure 9.5 represent critical features that are highly associated with system stability/instability. For example, the root node (Bus1045_V) stands for the feature that measures the dissimilarity (distance) between the shapelet extracted from voltage sequences of bus 1045 and the corresponding TS. If the distance value is larger than the splitting value 0.722, the system is deemed to be stable; otherwise, it needs to be further checked by other features, for example, Bus1043_P and Bus1044_V. By such recursive distance comparisons, a specific path from the root node to leaf nodes (blocks with 1/−1 marks indicating stability/instability) is traced. In essence, each path of the DT represents a specific instability or stability pattern of the system. As can be seen, there exist four main patterns on the DT. To further analyze these patterns, the three shapelets corresponding to three nodes on the DT and their best-matched subsequences from TS of individual instances are depicted in Figure 9.6.

From Figures 9.5 and 9.6, it can be summarized that low levels and drastic variations of *V/P* evolutions at heavy loaded buses (1043~1045) are generally dominant indicators of voltage instability in the specific region. In fact, such statistical profiles can further help to derive an intuitive SVS assessment criterion for the specific system, which will be demonstrated in the subsequent case study of CSG. Furthermore, it is noticed that the three critical shapelets were all extracted from unstable samples. This suggests that with the help of the imbalance learning machine, TSSC is able to capture the critical trends and shapes of voltage instability regardless of the scarceness of unstable cases.

It is also observed in Figure 9.6 that the longest shapelet's length is 0.87 s, indicating each round of online SVS evaluation can be completed in less than 1 s. Let the maximum length of shapelets be $T_{\max}$, by setting a sliding time window with the same length, SVS assessment results can be issued periodically (with a period of $T_{\max}$) during continuous online monitoring. If faster online SVS assessment is needed, those shapelets' lengths can be further shortened.

9.4.1.3 Adaptability tests of online application

1. *Validation in typical scenarios:* Suppose the operating point A in [29] was the starting point of online assessment. Three typical scenarios of operation changes (corresponding to three incremental learning periods) were considered here for adaptability tests of online application.

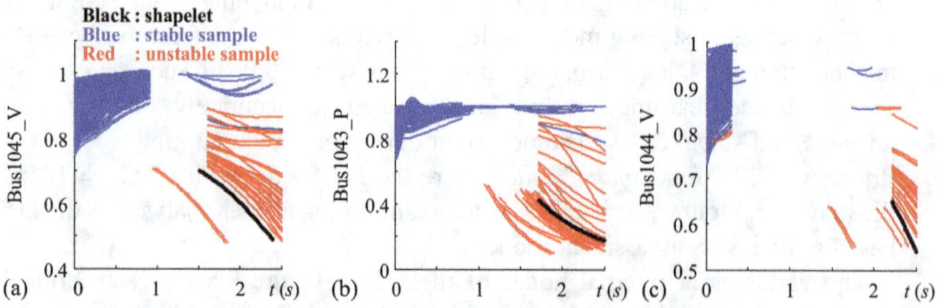

Figure 9.6 Initial shapelets and their best-matched subsequences (Nordic)

Scenario 1 (1~3 days): The total load level in the system rose by 30% and the proportion of dynamic motor loads increased by 25% because of imminent midsummer. Under such a serious condition, for the sake of enhancing system security, the system operating point was adjusted by starting five reserve generating units in parallel with g14, g15, g16, g17, and g18, respectively (similar to shifting the operating point from A to B in [29]).

Scenario 2 (4~6 days): The generators g5 and g7 in Central area were out of service for maintenance. Accordingly, the deficiency of active power and reactive power is supplemented by dispatching and rescheduling other generators and by adjusting nearby shunt compensations, respectively.

Scenario 3 (7~9 days): The tie lines bus 4031~bus 4041 and bus 4044~bus 4045 tripped off due to accidental failures. With such a topology change, all the running generators and other equipment were accordingly regulated to meet new balances of power transmission.

Given these scenarios, the proposed learning machine continually executed incremental update and online SVS assessment. Clearly, system conditions under these scenarios are significantly different from the starting point A. In each scenario, based on the current and ongoing conditions as well as credible contingencies, 180 synthetic samples were generated and fed into incremental learning. On the other hand, these new samples were exploited to test online assessment performances of the old DT model obtained at the end of last period and of the new DT model induced from incremental learning. Classification results of those old/new DTs under the three scenarios are summarized in the middle column of Table 9.6. Besides, with rolling update, classification performances were further validated on the updated datasets (667 old samples + 180 new samples), as shown in the right column of Table 9.6.

Note that the old DT in scenario ($n + 1$) is also the new DT in scenario n ($n = 1, 2$). By comparing all the three old DTs, it can be observed that the misdetection and false alarm rates are generally reduced, resulting in the improvement of the overall *Accuracy* and *G-mean*. This indicates that the SVS assessment model's adaptability to totally unforeseen scenarios and conditions is generally enhanced step by step. For each incremental learning period, after learning information from newly synthetic samples, these new DTs

Table 9.6 SVS assessment performances in incremental learning scenarios (Nordic)

Scenario	N_{FS}	N_{FU}	N_{TU}	N_{TS}	***Mis* (%)**	***Fal* (%)**	***G***	***Accu* (%)**
1 (old DT)	6	9	43	122	2.95	4.37	0.903	92.68
1 (new DT)	3	5	46	125	0.71	2.24	0.969	97.05
2 (old DT)	4	8	43	125	2.83	3.90	0.909	93.27
2 (new DT)	2	4	45	129	0.59	2.01	0.974	97.40
3 (old DT)	4	6	48	122	2.60	4.01	0.913	93.39
3 (new DT)	2	5	50	123	0.59	2.13	0.973	97.28

significantly achieve better classification performances than those old DTs do, with the overall classification accuracy improved by 4%~4.5%. From a long-run point of view, owing to such a consecutive enhancement, the constantly upgraded DT assessment model would gain persistent adaptability to a wide range of unknown conditions and contingencies.

2. *Discussion on incremental learning:* A major concern of incremental learning is its computational efficiency. Without incremental learning, shapelet searching and the subsequent learning would have to follow the same procedure as initial offline learning mode, called the *original mode*. In the above three scenarios, both modes of learning were tested on the same computing platform (Intel Core i7 CPU/2.50 GHz + 4.0 GB RAM), and their learning efficiencies are summarized in Table 9.7. Obviously, the computational efficiency is remarkably improved by the incremental mode, due to avoiding laborious shapelet searching in large-size datasets mixed by old and new samples. In general, it costs less than 25 minutes to fulfil the update work for each period, with almost no influence on the SVS assessment model's real-time performance.

   To further exemplify the benefits of incremental learning, taking Scenario 3 for example, three representative shapelets and their best matched subsequences before and after incremental updates are depicted in Figures 9.7 and 9.8, respectively. Compared to the old shapelets in Figure 9.7, their upgraded counterparts in Figure 9.8 are able to distinguish unstable profiles from stable ones more discriminatively, with fewer inter-class subsequences overlapped. Besides, by quantitative WIG comparison, as shown in Table 9.8, it also suggests that the updated shapelets outperform the old ones with respect to

Table 9.7 Computation time of shapelet searching

Mode	Scenario 1	Scenario 2	Scenario 3
Original (time/s)	6,613.5	6,708.9	6,671.8
Incremental (time/s)	1,336.4	1,269.7	1,288.2

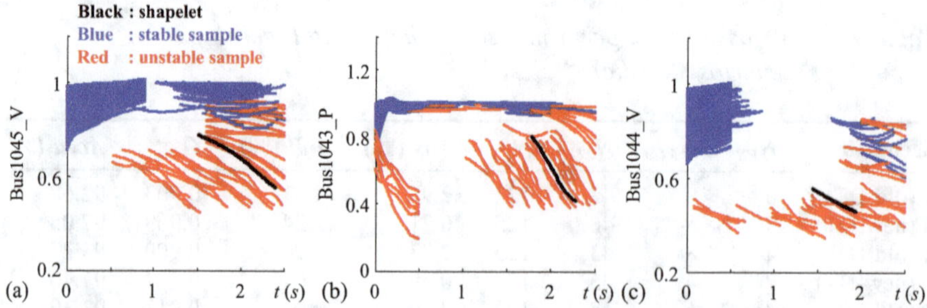

Figure 9.7 Old shapelets and their best-matched subsequences (Scenario 3)

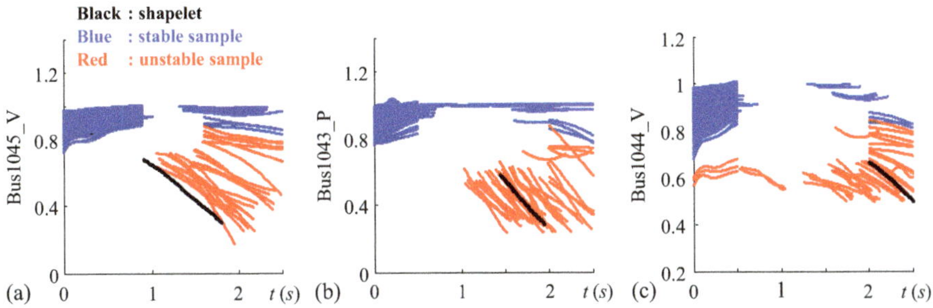

Figure 9.8 New shapelets and their best-matched subsequences (Scenario 3)

Table 9.8 Comparison of shapelets' WIG (Scenario 3)

Shapelets' WIG	Bus1045_V	Bus1043_P	Bus1044_V
Old (unit-free)	0.3912	0.4087	0.4179
New (unit-free)	0.4707	0.4478	0.4642

classification capability (improved by 10%~20%). In this regard, new shapelets are capable to capture the critical features and trends of system instability more proficiently. Actually, this upgrade benefits from new concepts and knowledge learned from those incremental instances. With the newest instances continually learned along with time, incremental updates would facilitate everlasting maintenance and enhancement of the online SVS assessment model's performance.

9.4.2 Test on China Southern Power Grid

The proposed learning machine was further tested on CSG, which is one of the two major power grids in China. This large-scale system has 12,995 buses, with a total capacity of 270 GW. The main receiving-end area in CSG is Guangdong, whose power consumption accounts for 56% of the whole grid. In Guangdong, Guangzhou is the largest city which supplies power to nearly 20 million people. While substantial industrial motors and residential air conditioners are heavily loaded in Guangzhou, its power is mainly imported from power plants thousands of miles away. Such a factor makes the area extremely vulnerable to the SVS problem. Hence, as shown in Figure 9.9, the Guangzhou subsystem falls into the main scope of online SVS monitoring. Note that although the case study below mainly focuses on the Guangzhou subsystem, the complete CSG model without any simplification was utilized for numerical simulations, so that system SVS dynamics can be reproduced with a high fidelity. As for SVS monitoring, eight 500-kV buses in Guangzhou, that is, {HD, CH, MM, BJ, ZC, SY, GN, SD}, were configured with PMUs for sequential synchrophasor data acquisition.

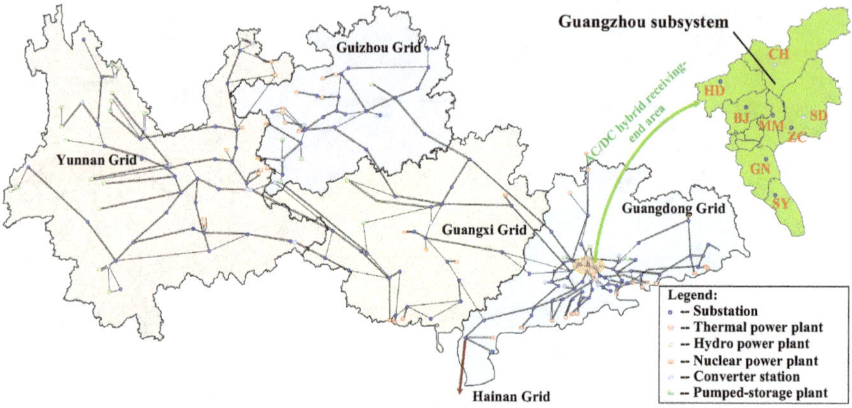

Figure 9.9 Structure of China Southern Power Grid

9.4.2.1 Simulation setting and case generation

Given the SVS concern in CSG, a composite load model with an induction motor (60%) and ZIP load (40%) in parallel was applied to all the loads in the system. Taking four typical operating points in CSG from January 2015 to June 2015 as the basic operating conditions, a wide variety of representative operating scenarios and contingencies were considered for SVS case generation: (1) for each basic operating point, the overall load demands were set to 80%, 90%, 100%, 110%, and 120% of the base level, respectively, and the power supplied by individual power plants was accordingly adjusted to meet new power balances; (2) five different 500-kV transmission lines near Guangzhou were separately disconnected to simulate typical topological changes in the receiving-end area; (3) 24 representative $N-1$~$N-3$ events were carefully selected from the credible contingency list (provided by system operators in CSG) to mimic typical transient contingencies in the first half of 2015. According to the guideline on security and stability analysis for CSG [35], the fault clearing time was specified as 0.1 s.

Based on the above simulation settings, 2,400 SVS cases were generated in batch by time-domain numerical simulations. Among them, 2,000 cases (8.8% being unstable) were randomly chosen as "measured" training cases and the rest as testing cases. Following the data acquisition setting in the Nordic test system, here the simulated PMU measurement data of 0~2.5 s after fault clearance were sequentially acquired with a sampling rate of 100 Hz.

9.4.2.2 SVS assessment in class imbalance contexts

Suppose the primary operating point prevalent in July 2015 was the impending operating condition in the system. Based on that, 10% of stochastic load demand deviation was set to simulate load modeling and forecasting errors. In order to make FN-SMOTE generate more unstable cases, the proportion of dynamic loads in composite load models was increased to 75%~90%, and $N-3$~$N-4$ contingencies were mainly considered to mimic severe SVS scenarios. In this way, the

FN-SMOTE technique produced 480 synthetic cases, within which 12 disqualified cases were removed by Tomek link-based data cleaning. For the sake of alleviating the class imbalance issue, only these unstable synthetic cases were integrated with the previous "measured" training cases. The two groups of cases formed an initial SVS dataset, with stable and unstable cases accounting for 75.9% (1,824 cases) and 24.1% (578 cases), respectively. Obviously, the imbalance degree of the unstable cases was significantly mitigated.

Taking the cost-sensitive issue into additional account, the cost ratio was specified as $R = 10$ for class imbalance learning. To observe the effect of FN-SMOTE and cost-sensitive learning on tackling the class imbalance issue, here a comparative study was carried out by learning with the class imbalance problem untreated (simple TSSC), learning with FN-SMOTE and ordinary DT induction, and learning with FN-SMOTE and cost-sensitive DT induction (proposed), respectively. To assess the three methods' learning performances on the same base, the initial 2,000 training cases and 400 testing cases were used to evaluate their classification performances, as summarized in Table 9.9.

Comparatively speaking, although the simple TSSC method without any treatment achieves the highest classification accuracy on the training cases, it has the worst extrapolation performances on the testing cases, with an extremely high misdetection rate. After the introduction of FN-SMOTE, more attention is paid to the scarce unstable cases, resulting in the improvement of extrapolation on the testing cases. Moreover, with the incorporation of cost-sensitive learning, a desirable trade-off between false alarm and misdetection is achieved, which leads the proposed learning machine to outperform the other two methods regarding generalization performances on the testing cases.

Taking the SVS assessment model derived by the proposed learning machine as the initial model, three incremental learning scenarios in July 2015 were considered below to perform incremental update for the SVS assessment model.

Scenario 1 (1~3 days): Due to extremely hot weather in summer, the total load demands in CSG rose by 20%. With substantial air conditioners put into use, the proportion of dynamic loads increased by 30%. Accordingly, a number of hydro and thermal power plants in CSG were rescheduled to meet the increasing power demands.

Table 9.9 SVS assessment performances of different learning approaches

Method	2,000 training cases				400 testing cases			
	Mis (%)	*Fal* (%)	*G*	*Accu* (%)	*Mis* (%)	*Fal* (%)	*G*	*Accu* (%)
Untreated	1.30	0.65	0.920	98.35	3.75	3.00	0.814	93.25
FN-SMOTE + DT	0.95	1.05	0.939	98.00	1.00	3.00	0.971	96.00
Proposed	**0.80**	**1.25**	**0.947**	**97.95**	**0.75**	**1.50**	**0.958**	**97.75**

Scenario 2 (4~6 days): To ensure the whole system can smoothly go through future dry seasons, thermal power plants in four provinces (Guangdong, Guangxi, Guizhou, and Yunnan) with a total capacity of 12 GW were arranged to shut down for examination and maintenance. The deficiency of power was supplemented by dispatching and rescheduling other hydropower plants in the system.

Scenario 3 (7~9 days): With an extremely bad weather condition (typhoon and thunderstorm), the double-circuit transmission line between two 500-kV buses SD and ZC abnormally tripped off. After this sudden change, the power plants and reactive power sources in Guangdong were regulated to maintain normal operation.

Considering the above three scenarios, the SVS assessment model was periodically updated with the incremental learning strategy. Specifically, for each scenario, by taking into account its current and ongoing conditions and presumed/actual contingencies, 400 synthetic cases were produced by FN-SMOTE. Such cases were further gathered as incremental learning datasets for incremental update of the SVS assessment model. Similar to the tests in the Nordic test system, these incrementally generated cases were used to evaluate both the old DT model's (obtained at the end of last period) and the new DT model's (newly induced from incremental learning) SVS assessment performances. The assessment results are summarized in the second column of Table 9.10. In addition, by periodically updating the whole dataset, the overall SVS assessment performances were further tested on the updated datasets (see the third column in Table 9.10).

As can be seen, with the help of incremental update, the SVS assessment model is capable of maintaining both the misdetection and false alarm rates at relatively low levels, which thus resulting in an overall high classification accuracy of more than 97%. On the other hand, with the assessment model periodically updated, the differences of classification performances between the old and new DTs are gradually reduced. This indicates that the SVS assessment model's adaptability to unknown scenarios and events is progressively enhanced. Hence, by incrementally assimilating new information and stability knowledge from new scenarios, the proposed learning machine can keep the SVS assessment model timely updated, thereby ensuring its adaptability and reliability along with time.

Table 9.10 SVS assessment performances in incremental learning scenarios (CSG)

Scenario	**Incremental dataset (400 cases)**				**Updated datasets (2,002 + 400 cases)**			
	N_{FS}	N_{FU}	N_{TU}	N_{TS}	***Mis* (%)**	***Fal* (%)**	***G***	***Accu* (%)**
1 (old DT)	6	13	92	302	1.71	2.83	0.946	95.38
1 (new DT)	**3**	**8**	**95**	**294**	**0.83**	**1.33**	**0.974**	**97.42**
2 (old DT)	6	9	95	290	1.42	2.58	0.953	96.00
2 (new DT)	**5**	**7**	**96**	**292**	**0.87**	**1.37**	**0.973**	**97.34**
3 (old DT)	5	9	94	292	1.21	2.50	0.959	96.29
3 (new DT)	**4**	**8**	**95**	**293**	**0.92**	**1.62**	**0.970**	**97.46**

9.4.2.3 SVS pattern analysis and comprehension

Taking the SVS assessment model updated after incrementally learning from *Scenario 3* for example, how the proposed learning machine can help to analyze and comprehend typical SVS patterns in practical power grids is illustrated in the following. As depicted in Figure 9.10, the DT-based SVS assessment model involves three critical variables, that is, the voltages of buses ZC and MM and the reactive power injected into bus MM. Based on the transient evolution of these three variables, the whole feature space is separated into four typical stable/unstable patterns, which can be further expressed by the following combinational *if-then* rules:

Rule 1 (Stable Pattern 1): If $D_{ZC\_V}(Shp_k, y) > 0.8761$, then the system will keep stable. ($D_{ZC\_V}(Shp_k, y) \rightarrow$ the distance between bus ZC's voltage TS and the corresponding shapelet of bus ZC's voltage, the same below.)

Rule 2 (Unstable Pattern 1): If $[D_{ZC\_V}(Shp_k, y) \leq 0.8761] \bigwedge [D_{MM\_V}(Shp_k, y) \leq 0.6796]$, then the system will be unstable.

Rule 3 (Stable Pattern 2): If $[D_{ZC\_V}(Shp_k, y) \leq 0.8761] \bigwedge [D_{MM\_V}(Shp_k, y) > 0.6796] \bigwedge [D_{MM\_Q}(Shp_k, y) \leq 0.5984]$, then the system will keep stable.

Rule 4 (Unstable Pattern 2): If $[D_{ZC\_V}(Shp_k, y) \leq 0.8761] \bigwedge [D_{MM\_V}(Shp_k, y) > 0.6796] \bigwedge [D_{MM\_Q}(Shp_k, y) > 0.5984]$, then the system will be unstable.

n fact, the above four rules constitute a data-driven SVS assessment criterion in the feature space. This criterion comprehensively combines the sequential trajectories of different quantities of buses ZC and MM to implement online SVS assessment. The thresholds and quantities of the *if-then* rules in the criterion are automatically derived from substantial SVS scenarios by the imbalance learning machine. On the

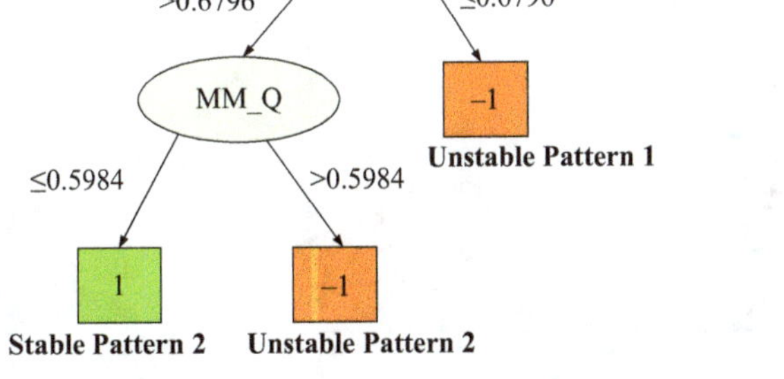

Figure 9.10 SVS assessment model after incremental update (CSG)

other hand, it is found that the two critical buses ZC and MM have extremely heavy load demands in Guangzhou, with both of them responsible for a large number of industrial, commercial, and residential loads in central areas of Guangzhou. The fact that these two heavily loaded buses' electrical quantities were selected as the key variables to construct the data-driven SVS assessment criterion implies that the proposed learning machine manages to capture the critical SVS evolution trends and features. As a result, such a data-driven criterion can achieve reliable online SVS assessment in practical power grids.

Furthermore, the three shapelets corresponding to the critical variables (nodes) on the DT model and their best matched subsequences from individual SVS cases were gathered, as presented in Figure 9.11. The black curves stand for shapelets, while the blue and red ones represent TS subsequences of stable and unstable cases, respectively. By statistically observing the evolution trends of these profiles, the above-mentioned *if-then* rules can be converted into more intuitive and practical SVS assessment rules:

Rule 1 (Stable Pattern 1): If the voltage of bus ZC recovers from 0.7 pu to more than 1.0 pu with a recovery speed of 0.2~0.3 pu/s, then the system will keep stable.

Rule 2 (Unstable Pattern 1): If Rule 1 is not satisfied and the voltage of bus MM drops below 0.8 pu with a speed of 0.1~0.3 pu/s, then the system will be unstable.

Rule 3 (Stable Pattern 2): If Rules 1 and 2 are not satisfied and the reactive power injected into bus MM falls below 3.5 pu with a falling speed of 2~2.5 pu/s, then the system will keep stable.

Rule 4 (Unstable Pattern 2): If Rules 1 and 2 are not satisfied and the reactive power injected into bus MM increases from 3~12 pu to 6~25 pu with an increasing speed of more than 2 pu/s, then the system will be unstable.

The above combinational rules base their SVS assessment on the observations of the magnitudes and velocities of bus voltages and reactive power during SVS dynamics. Compared with the previous counterpart based on distance measures of sub-trajectories, this group of rules directly focusing on the trajectories themselves are more explicit and comprehensible for system operators. Besides helping to achieve reliable online SVS assessment, such intuitive rules can be further

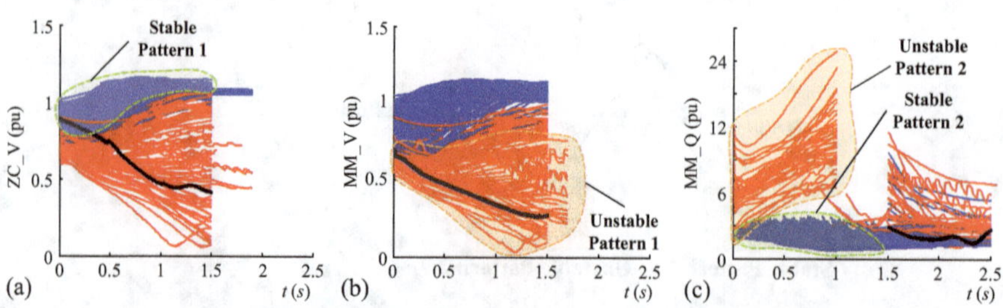

Figure 9.11 Critical shapelets and their best matched subsequences (Scenario 3)

leveraged to assist system operators to explore and understand SVS characteristics in specific power grids during offline analysis.

9.4.2.4 Performance in the presence of PMU measurement errors

Considering undesirable monitoring contexts in practical power grids, the proposed learning machine's performances were further tested in the presence of large PMU measurement errors. Taking the previous three incremental learning scenarios for instance, PMU measurement errors were added to all the *V/I/P/Q* TS instances. In particular, errors were simulated by adding white noises following the Gaussian distribution $N(0, \sigma^2)$ [30] to sequential PMU measurements, with the stand deviation set to $\sigma = 0.04$ pu. Obviously, such error settings are much more rigorous than what is required by the IEEE standard [36]. In order to smoothly discover sequential features hidden behind measurement noises, the simple moving average algorithm was utilized to preprocess these noisy *V/I/P/Q* TS.

With filtered TS datasets, imbalance learning was performed under the three incremental learning scenarios, and the corresponding learning results are summarized in Table 9.11. Compared with the PMU error-free assessment results in Table 9.10 (third column), here the SVS assessment performances are just a little weaker, with the total classification accuracy keeping above 95.5%. Such results imply that the SVS assessment model remains robust to fairly noisy environments. In fact, this appealing merit is due to the nature of TSSC that extracts features from subsequences of TS rather than from individual snapshots. Even if the acquired TS are contaminated by noises, dominant sequential evolution trends are still preserved beneath them. Concentrating on sequential feature discovery, TSSC is thus able to uncover crucial shapelets from scarce yet noisy TS of unstable cases, enhancing the SVS assessment model's robustness to noisy PMU measurements.

9.5 Further discussion

This chapter mainly focuses on learning temporal correlations from TS trajectories of multiple load buses for SVS assessment, with special attention to the practical class imbalance problem. However, considering the implementation of data-driven SVS monitoring in realistic power grids, more research aspects need to be investigated to enhance the data-driven SVS assessment solution's applicability in practice. As demonstrated in the authors' preliminary work in [37], by extending

Table 9.11 Learning performances after incremental update (with PMU errors)

Scenario	*Mis* (%)	*Fal* (%)	*G*	*Accu* (%)
1	1.33	2.87	0.954	95.80
2	1.08	3.00	0.958	96.00
3	1.12	2.71	0.959	96.17

the idea of learning temporal correlations to a wider perspective of exploring spatial-temporal correlations, the reliability of online SVS assessment can be further improved. In relevant future work, more systematic approaches for characterizing and learning spatial-temporal correlations remain to be studied. Besides, since recent researches have shown the great potential of deep learning in transient stability assessment [38], they may also be fully exploited to cope with the SVS problem, so as to achieve higher reliability and efficiency for online SVS assessment. Yet it should be pointed out that as ANN-based deep learning methods generally derive "black-box" learning models, they have little interpretability for the DSA learning results. In some occasions, when the SVS assessment model is expected to be transparent and explainable, interpretable learning approaches such as the TSSC-based learning machine developed by the authors would be favored in practice. Note that most of the existing ML-based SVS assessment schemes emphasize on learning from the data-driven perspective, that is, inferring the mapping relationship between system steady-state or dynamic measurement data and the eventual system stability status. In fact, some refined information derived from the model-based perspective, for example, stability indices obtained by theoretical analysis, may be assimilated as potential features or prior knowledge to facilitate a more reasonable and strategic ML procedure. In this regard, data-based ML approaches can be tactfully combined with model-based theoretical analysis to yield a more powerful SVS assessment model. Last but not least, on the basis of accurate prediction of the onset of impending instability, more efforts should be devoted to stability margin (SM) estimation. With the help of such quantitative SM information, data-driven online decision-making and countermeasures such as corrective control can be carried out to efficiently save the system from instability.

9.6 Conclusion

In this chapter, a TS data analytics-based imbalance learning machine for power system online SVS monitoring is systematically developed. It is dedicated to addressing the challenging class imbalance problem in practice, which is induced by the assumption that learning samples are mainly collected from historical and/or online PMU records. To deal with class skewness from both data-preprocessing and algorithm perspectives, the proposed learning machine tactfully integrates two critical techniques, that is, FN-SMOTE and cost-sensitive learning. Based on this learning machine, an online SVS assessment scheme is designed by further introducing an incremental learning strategy, which enhances the scheme's adaptability and reliability during online monitoring. Numerical test results on the Nordic test system and the real-word CSG illustrate that the proposed learning machine achieves excellent performances even if it is exposed to severe class imbalance. In addition to its reliability and adaptability during online monitoring, the learning machine exhibits desirable interpretability for SVS pattern discovery and comprehension in practical power grids. In relevant future work, an array of aspects, e.g., spatial-temporal correlation learning, advanced deep learning, and model/data mixed learning, can be explored and investigated to further enhance the data-driven SVS assessment solution's applicability in practice.

Acknowledgments

Lipeng Zhu's work was supported by the Research Grants Council of Hong Kong Special Administrative Region under the General Research Fund (GRF) through Project No. 17208817 and under the Theme-based Research Scheme (TRS) through Project No. T23-701/14-N. Chao Lu's work was supported by National Natural Science Foundation of China under Grants 51677097 and U1766214.

References

[1] Kundur P. *Power System Stability and Control.* New York, NY: McGraw-Hill, 1993.
[2] Van Cutsem T., and Vournas C. *Voltage Stability of Electric Power Systems.* Norwell, MA: Kluwer, 1998.
[3] Vournas C. D., Nikolaidis V. C., and Tassoulis A. A. "Postmortem analysis and data validation in the wake of the 2004 Athens blackout." *IEEE Trans. Power Syst.* 2006, vol. 21(3), pp. 1331–1339.
[4] Conti J. P. "The day the samba stopped." *Eng. Technol.*, 2010, vol. 5(4), pp. 46–47.
[5] Praprost K. L., and Loparo K. A. "An energy function method for determining voltage collapse during a power system transient." *IEEE Trans. Circuits Syst. I, Reg. Pap.* 1994, vol. 41(10), pp. 635–651.
[6] Kawabe K., and Tanaka K. "Analytical method for short-term voltage stability using the stability boundary in the PV plane." *IEEE Trans. Power Syst.* 2014, vol. 29(6), pp. 3041–3047.
[7] Dasgupta S., Paramasivam M., Vaidya U., *et al.* "Real-time monitoring of short-term voltage stability using PMU data." *IEEE Trans. Power Syst.* 2013, vol. 28(4), pp. 3702–3711.
[8] Glavic M., Novosel D., Heredia E., *et al.* "See it fast to keep calm: Real-time voltage control under stressed conditions." *IEEE Power Energy Mag.* 2012, vol. 10(4), pp. 43–55.
[9] Luo F., Dong Z. Y., Chen G., *et al.* "Advanced pattern discovery-based fuzzy classification method for power system dynamic security assessment." *IEEE Trans. Ind. Inform.* 2015, vol. 11(2), pp. 416–426.
[10] Xu Y., Dong Z. Y., Zhao J. H., *et al.* "A reliable intelligent system for real-time dynamic security assessment of power systems." *IEEE Trans. Power Syst.* 2012, vol. 27(3), pp. 1253–1263.
[11] Zhang Y., Xu Y., Dong Z. Y., and Zhang R. "A hierarchical self-adaptive data-analytics method for real-time power system short-term voltage stability assessment." *IEEE Trans. Ind. Inform.* 2018, vol. 15(1), pp. 74–84.

[12] Gomez F. R., Rajapakse A. D., Annakkage U. D., *et al.* "Support vector machine-based algorithm for post-fault transient stability status prediction using synchronized measurements." *IEEE Trans. Power Syst.* 2011, vol. 26(3), pp. 1474–1483.

[13] Kamwa I., Samantaray S. R., and Joos G. "Development of rule-based classifiers for rapid stability assessment of wide-area post-disturbance records." *IEEE Trans. Power Syst.* 2009, vol. 24(1), pp. 258–270.

[14] Ye L., and Keogh E. "Time series shapelets: a new primitive for data mining." *Proceedings of 15th ACM SIGKDD International Conference on Knowledge Discovery and Data Mining*, Paris, France, June–July 2009, pp. 947–955.

[15] Zhu L., Lu C., and Sun Y. "Time series shapelet classification based online short-term voltage stability assessment." *IEEE Trans. Power Syst.* 2016, vol. 31(2), pp. 1430–1439.

[16] Zhu L., Lu C., Dong Z. Y., *et al.* "Imbalance learning machine-based power system short-term voltage stability assessment." *IEEE Trans. Ind. Informat.* 2017, vol. 13(5), pp. 2533–2543.

[17] Hills J., Lines J., Baranauskas E., Mapp J., and Bagnall A. "Classification of time series by shapelet transformation." *Data Mining and Knowl. Discovery*, 2014, vol. 28(4), pp. 851–881.

[18] Han J., Kamber M., and Pei J. *Data Mining: Concepts and Techniques*. 3rd ed. San Francisco, CA: Morgan Kaufmann, 2011.

[19] Mueen A., Keogh E., and Young N. "Logical-shapelets: an expressive primitive for time series classification." *Proceedings of 17th ACM SIGKDD International Conference on Knowledge Discovery and Data Mining*, San Diego, CA, August 2011, pp. 1154–1162.

[20] Keogh E., Chu S., Hart D., and Pazzani M. "An online algorithm for segmenting time series." *Proceedings of IEEE International Conference on Data Mining*, San Jose, CA, November–December 2001, pp. 289–296.

[21] He H., and Garcia E. A. "Learning from imbalanced data." *IEEE Trans. Knowl. Data Eng.* 2009, vol. 21(9), pp. 1263–1284.

[22] Galar M., Fernandez A., Barrenechea E., *et al.* "A review on ensembles for the class imbalance problem: bagging-, boosting-, and hybrid-based approaches." *IEEE Trans. Syst. Man Cybernet. Part C: Appl. Rev.* 2012, vol. 42(4), pp. 463–484.

[23] Khoshgoftaar T. M., Hulse J. V., and Napolitano A. "Comparing boosting and bagging techniques with noisy and imbalanced data." *IEEE Trans. Syst. Man Cybern. A: Syst. Humans.* 2011, vol. 41(3), pp. 552–568.

[24] Prati R. C., Batista G. E., and Silva D. F. "Class imbalance revisited: a new experimental setup to assess the performance of treatment methods." *Knowl. Inform. Syst.* 2015, vol. 45(1), pp. 247–270.

[25] Kamwa I., Samantaray S. R., and Joos G. "On the accuracy versus transparency trade-off of data-mining models for fast response PMU based catastrophe predictors." *IEEE Trans. Smart Grid.* 2012, vol. 3(1), pp. 152–161.

[26] Ting K. M. "An instance-weighting method to induce cost-sensitive trees." *IEEE Trans. Knowl. Data Eng.* 2002, vol. 14(13), pp. 659–665.

[27] Khan A. R., Mahmood A., Safdar A., *et al.* "Load forecasting, dynamic pricing and DSM in smart grid: a review." *Renew. Sustain. Energy Rev.* 2016, vol. 54, pp. 1311–1322.

[28] Tomek I. "Two modifications of CNN." *IEEE Trans. Syst. Man Cybernet.* 1976, vol. 6(11), pp. 769–772.

[29] Van Cutsem T., and Papangelis L. "Description, modeling and simulation results of a test system for voltage stability analysis." 2013. [Online]. Available: http://orbi.ulg.ac.be/handle/2268/141234

[30] Glavic M., and Van Cutsem T. "Wide-area detection of voltage instability from synchronized phasor measurement. Part II: Simulation results." *IEEE Trans. Power Syst.* 2009, vol. 24(3), pp. 1417–1425.

[31] Batista G. E., Prati R. C., and Monard M. C. "A study of the behavior of several methods for balancing machine learning training data." *ACM SIGKDD Explor. Newslett.* 2004, vol. 6(1), pp. 20–29.

[32] Chawla N. V., Bowyer K. W., Hall L. O., and Kegelmeyer W. P. "SMOTE: synthetic minority over-sampling technique." *J. Artif. Intell. Res.* 2002, vol. 16, pp. 321–357.

[33] Han H., Wang W. Y., and Mao B. H. "Borderline-smote: a new over-sampling method in imbalanced data sets learning." *Proceedings of International Conference on Advances in Intelligent Computing.* August 2005, Hefei, China, pp. 878–887.

[34] He H., Yai B., Garcia E., and Li S. "Adasyn: adaptive synthetic sampling approach for imbalanced learning." *IEEE International Joint Conference on Neural Networks*, June 2008, Hong Kong, China, pp. 1322–1328.

[35] Guide on Security and Stability Analysis for CSG, China Southern Power Grid, Q/CSG 11004-2009, 2009 (in Chinese).

[36] IEEE Standard for Synchrophasor Measurements for Power Systems. IEEE Standard C37.118.1-2011.

[37] Zhu L., Lu C., Kamwa I., and Zeng H. "Spatial-temporal feature learning in smart grids: a case study on short-term voltage stability assessment." *IEEE Transactions on Industrial Informatics*, 2020, vol. 16(3), pp. 1470–1482.

[38] Gupta A., Gurrala G., and Sastry P. "An online power system stability monitoring system using convolutional neural networks." *IEEE Trans. Power Syst.* 2019, vol. 32(2), pp. 864–872.

Chapter 10

A hierarchical data-driven method for short-term voltage stability assessment of power systems

Yan Xu[1], Zhao Yang Dong[2] and Rui Zhang[2]

In the Smart Grid paradigm, growing integration of large-scale intermittent renewable energies has introduced significant uncertainties to the operations of an electric power system. This makes real-time dynamic security assessment (DSA) a necessity to enable enhanced situational-awareness against the risk of blackouts. Conventional DSA methods are mainly based on time-domain simulation, which are insufficiently fast and knowledge-poor. In recent years, the intelligent system (IS) strategy has been identified as a promising approach to facilitate real-time DSA. While previous works mainly concentrate on the rotor angle stability, this chapter focuses another yet increasingly important dynamic insecurity phenomenon—short-term voltage instability, which involves fast and complex load dynamics. The problem is modeled as a classification subproblem for transient voltage collapse and a prediction subproblem for unacceptable dynamic voltage deviation. A hierarchical IS is developed to address the two subproblems sequentially. The IS is based on ensemble learning of random-weights neural networks and is implemented in an off-line training, real-time application, and on-line updating pattern. Simulation results on the New England 39-bus system verify its superiority in both learning speed and accuracy over some state-of-the-art learning algorithms.

List of abbreviations

ANN	artificial neural network
AVR	automatic voltage regulator
DAE	differential-algebraic equation

[1]School of Electrical and Electronic Engineering, Nanyang Technological University, Singapore
[2]School of Electrical Engineering and Telecommunication, University of New South Wales, Sydney, Australia

DSA	dynamic security assessment
DT	decision tree
LM	large motor
LVRT	low-voltage-ride-through
MAPE	mean absolute percentage error
NNRW	neural networks with random weights
OPF	optimal power flow
PMU	phasor measurement unit
SLFN	single-hidden layer feed-forward networks
SVM	support vector machine
SM	small motor
ST-VSA	short-term voltage stability assessment
TDS	time-domain simulation
TVCI	transient voltage collapse index
TVSI	transient voltage severity index
UVLS	under-voltage load shedding

10.1 Introduction

The fundamental goal of an electric power system is to maintain a secure and economic process for electricity generation, transmission, and distribution. However, in the practical operation phase, a power system is always exposed to various disturbances such as sudden load changes, unexpected element outages, short-circuit faults, etc. Once occurred, the disturbance can harm the power system security and may result in catastrophic consequences such as wide-spread blackouts which affect the society and the economy significantly.

As one essential security criterion, dynamic security refers to the ability of a power system to survive the dynamic transition from an initial operating equilibrium to a viable new equilibrium with most limits bounded after being subject to a disturbance [1]. To protect a power system against the risk of blackouts, dynamic security assessment (DSA) needs to be performed to examine various credible disturbances in advance. Once the disturbance is identified as insecure, appropriate security control actions need to be implemented to protect the power system if the contingency really occurs. Mathematically, the dynamics of a power system can be modeled as a large set of nonlinear differential-algebraic equations (DAEs) which cannot be analytically solved. Conventional DSA is therefore based on time-domain simulation (TDS) which solves the DAEs via step-by-step integrations. Although accurate, the TDS method is very time-consuming and difficult to provide useful information about the system's dynamic security characteristics.

With the continuous growth of load and unparalleled infrastructure investment, modern power system is forced to operate near the dynamic security limit.

At the same time, with the development of smart grid paradigm [2], today's power system has been integrated with renewable energy (such as wind and solar energy) with increasing capacity, which are usually random and difficult to predict for the operation of power system. Therefore, system operators expect faster DSA tools to monitor the dynamic security of power system in real time.

In recent years, the intelligent system (IS) strategy has been regarded as a promising technique to solve such problem [3]. The nonlinear relationship between the power system operating parameters (input) and the corresponding security index (output) can be extracted and reformulated in an IS though learning from a dynamic security database. During the on-line application phase, the system's stability can be assessed when the input is available. The IS-based DSA not only provides much faster on-line decision-making speed but also outperforms conventional methods for its less data requirement, stronger generalization capacity, and extensibility [3], making it an ideal candidate for future Smart Grid applications [4].

In the literature, a variety of ISs based on learning algorithms like artificial neural network (ANN) [5,6], support vector machine (SVM) [7], decision tree (DT) [8], and data-mining technique [9] have been reported. However, most of them are proposed for transient stability assessment, which focuses on the generator rotor angle performance [5–10]. However, very few copes with other dynamic security criteria such as voltage stability [6]. Besides, it can be seen that the reported models usually suffer from extensive learning/tuning burden and unsatisfactory accuracy. A comprehensive review and classification on previous works can be found in [11].

This chapter focuses on another yet increasingly important dynamic security concern—short-term voltage stability [12–14] and develops a hierarchical IS for its real-time assessment. Unlike rotor angle stability, STVS involves fast and complex load dynamics, which is even more difficult to capture. The instability can propagate very rapidly and trigger significant load shedding or even widespread blackouts. Nowadays, due to the increased penetration of wind power, the short-term voltage stability is becoming a critical concern in terms of low-voltage-ride-through (LVRT) capability for wind turbines. To address the short-term voltage stability assessment (ST-VSA) problem, we employ a feedforward neural network with random weights (NNRW) [15] and employ the ensemble learning strategy to design a hierarchical IS. We model the ST-VSA problem as a transient voltage collapse subproblem which corresponds to a classification process and an unacceptable dynamic voltage deviation subproblem which corresponds to a regression process. The two subproblems are sequentially handled by two ensemble-based classifier and predictor models. The proposed IS is numerically tested on the New England 39-bus system, and compared with other popular learning algorithms. Simulation results have verified its effectiveness and superiority in accuracy and learning speed over the compared algorithms.

10.2 Problem descriptions

10.2.1 Short-term voltage instability: mechanism and phenomenon

The IEEE and CIGRE joint task force define the voltage stability as the ability of a power system to maintain steady voltages at all buses in the system after being subjected to a disturbance from a given initial operating condition [1]. As a major threat to power system security, the loss of voltage stability can lead to progressive voltage drop or rapid voltage collapse, which can trigger significant load shedding or even wide-spread blackouts. Over past decades, many major blackouts around the world have been found directly associated with this phenomenon [12–14].

In general, voltage instability stems from the attempt of load dynamics to restore the power consumption beyond the capability of the combined transmission and generation system [13]. According to [1], the voltage stability can be divided into long-term and short-term phenomena. The long-term voltage stability involves slower acting equipment such as tap-changing transformers, thermostatically controlled loads, and generator current limiters [1]. It can be modeled as algebraic equations and solved by power flow-based methods [13]. Therefore, its computation speed can be fast enough to enable (near) real-time implementation. By contrast, the short-term voltage stability phenomenon involves complex dynamics of load components such as induction motors tending to restore their consumed power in a very short time-frame (say, several seconds) [14]. Following a large disturbance, the induction motors decelerate dramatically by the voltage dip or may stall if the electrical torque cannot overcome the mechanical load. This in turn draws very high reactive current which affects adversely the voltage magnitudes. As a consequence, unacceptable transient voltage performance (e.g., delayed voltage recovery) may be experienced and/or transient voltage collapse may occur. With increased penetration of induction motor loads (e.g., air-conditioners), today's power systems tend to be more vulnerable to short-term voltage instability [14].

10.2.2 Dynamic load model

As mentioned above, the key driving force of voltage instability is the tendency of dynamic loads to restore their consumed power within a very short time frame. Therefore, modeling dynamic loads is a necessity for short-term voltage stability study.

We employ a composite load model "CLOD" defined in the commercial power system simulation package PSS®E to represent various dynamic load components. As illustrated in Figure 10.1, this load model is an aggregation of large motors (LM), small motors (SM), discharge lighting, transformer exciting current, and constant power load, as well as the voltage-dependent load, all of which are fed from many real-world substations. However, other dynamic load models can also be used if necessary.

The mathematical model of induction motors is also represented by DAEs. Hence, the whole mathematical model of the power system for short-term voltage

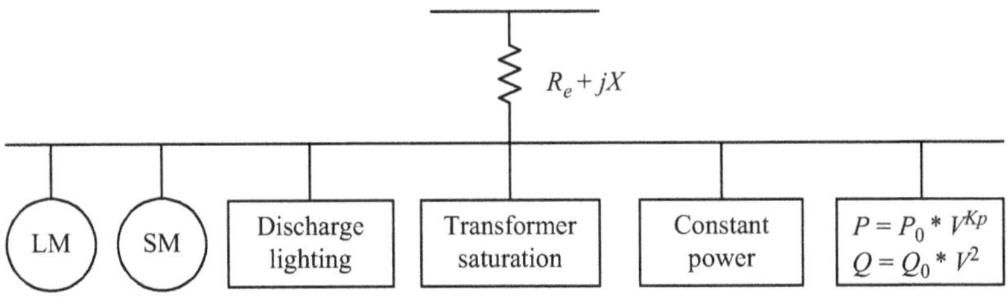

Figure 10.1 Structure of the composite load model "CLOD"

stability study composes a very large set of nonlinear DAEs which can be compactly formulated as follows:

$$\frac{d\mathbf{x}}{dt} = \mathbf{f}(\mathbf{x}, \mathbf{y}, \mathbf{p}, \boldsymbol{\lambda}) \tag{10.1}$$

$$0 = \mathbf{g}(\mathbf{x}, \mathbf{y}, \mathbf{p}, \boldsymbol{\lambda}) \tag{10.2}$$

where set (10.1) corresponds to the differential equations of the system components, including generators, motors, dynamic loads, as well as their control systems; and set (10.2) corresponds to the algebraic equations of the network and static loads. Vector $\mathbf{x}$ denotes the state variable (e.g., generator angles and dynamic bus voltages); vector $\mathbf{y}$ denotes algebraic variables (e.g., static load voltages and angles); vector $\mathbf{p}$ stands for controllable parameters [e.g., automatic voltage regulator (AVR) setpoints]; vector $\boldsymbol{\lambda}$ represents uncontrollable parameters (e.g., load levels).

The DAEs cannot be analytically solved. Rather, one can solve it via step-by-step integrations. In practice, the ST-VSA is realized based on TDS, which is however very computationally burdensome.

10.2.3 Simulation study

Using the composite load model "CLOD", we examine the impact of load dynamics on the short-term voltage performance of a power system. Figures 10.2 and 10.3 illustrate the dynamic voltage response in the New England 39-bus system following short-circuit disturbances with different severity degrees [16].

The voltage trajectories are obtained by TDS [16]. For each figure, the upper and lower windows respectively show the voltage response without the load dynamics (i.e., the dynamic load model is not used, and a constant power load is used), and with load dynamics (i.e., the dynamic load model is used). For each figure, it can be seen that without load dynamics, the bus voltages recover very fast to the normal level and the system is stable. In Figure 10.2, a very severe disturbance is applied, and the lower window shows that transient voltage collapse occurs around 1.6 s. The voltage collapse is an unacceptable situation for a practical power system since it can lead to more serious consequences such as cascading failures and/or blackouts. For Figure 10.3, a less severe disturbance is applied, and the lower window shows that some buses (in particular bus 20 and bus 34) experience a prolonged voltage depression and delayed voltage recovery.

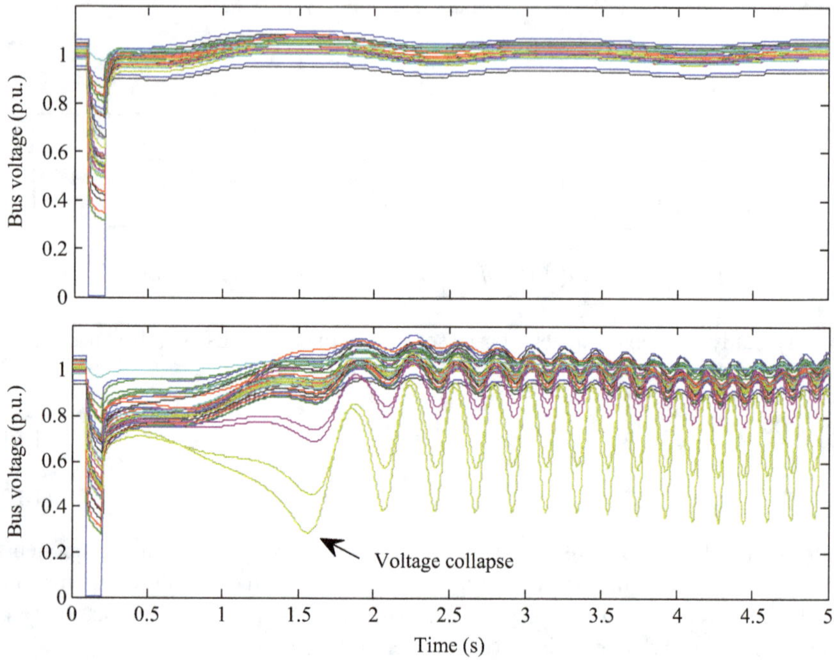

Figure 10.2 Dynamic voltage responses without (upper window) and with (lower window) load dynamics—a very severe fault

In practice, the delayed voltage recovery can trigger under-voltage load shedding (UVLS) devices that curtail a significant amount of loads.

10.2.4 Short-term voltage stability indices

As illustrated in Figures 10.2 and 10.3, short-term voltage stability study mainly focuses on (1) fast voltage collapse—see Figure 10.2 or (2) unacceptable post-disturbance voltage deviation—see Figure 10.3.

To evaluate the short-term voltage performance, some industrial criteria such as NERC criteria have been proposed [17]. However, they only define the unacceptable voltage-dip magnitude and the duration time, and can only provide a binary answer to the stability status, namely, stable or unstable.

We model the ST-VSA problem as a transient voltage collapse subproblem and an unacceptable dynamic voltage deviation subproblem. The first subproblem checks the existence of a short-term equilibrium of the postcontingency system and the second subproblem quantifies the degree of stability.

10.2.4.1 Transient voltage collapse index

The evaluation of transient voltage collapse is a binary question, that is, whether or not the system loses the short-term equilibrium following a disturbance. Consequently, the transient voltage collapse index (TVCI) can be a binary index, namely, 1 or 0.

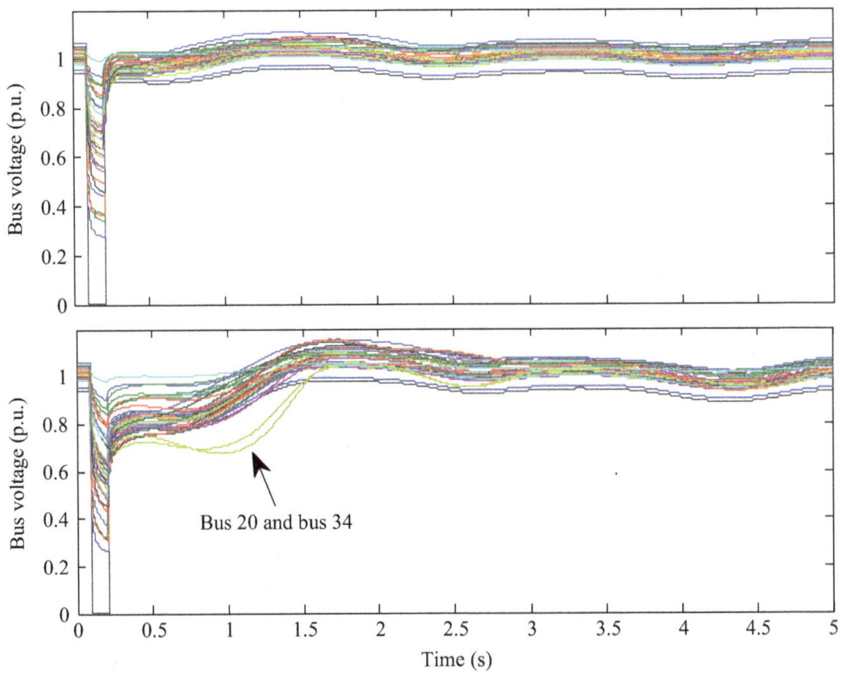

Figure 10.3 Dynamic voltage responses without (upper window) and with (lower window) load dynamics—a less severe fault

10.2.4.2 Transient voltage severity index for unacceptable dynamic voltage deviation

After the system withholds a short-term equilibrium, the system may still experience unacceptable voltage deviation such as prolonged voltage depression and delayed voltage recovery. Actions of motor protective devices can trip a significant amount of loads. In contrast to transient voltage collapse, the severity of unacceptable dynamic voltage deviation can be varying depending on the severity of a disturbance, such rule-based industrial criteria show great differences among the operating regions, countries, the utilities, and lack quantitative unified standard on voltage recovery behavior. In practice, the degree of severity in voltage deviation can be reflected by the magnitude of the voltage violation and the associated duration time. This chapter employs a straightforward index called *transient voltage severity index* (TVSI) proposed in our previous work [16] to quantify this phenomenon:

$$\mathrm{TVSI} = \frac{\sum\limits_{i=1}^{N} \sum\limits_{t=T_c}^{T} \mathrm{TVDI}_{i,t}}{N \times (T - T_c)} \tag{10.3}$$

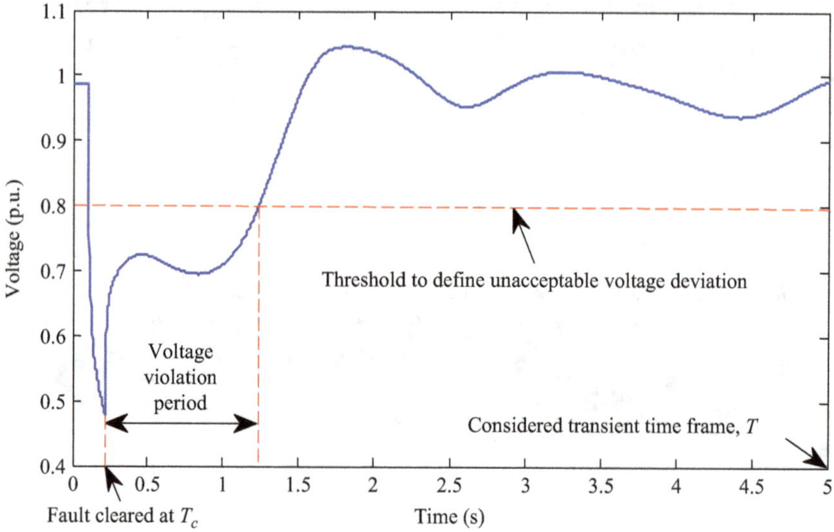

Figure 10.4 Illustration of the concept of TVSI

where N is the total number of buses in the system, T is the considered transient time frame, T_c is the fault clearing time, TVDI is the transient voltage deviation index, calculated by

$$\mathrm{TVDI}_{i,t} = \begin{cases} \dfrac{|V_{i,t} - V_{i,0}|}{V_{i,0}}, & \text{if } \dfrac{|V_{i,t} - V_{i,0}|}{V_{i,0}} \geq \mu \\ 0, & \text{otherwise} \end{cases} \quad \forall t \in [T_c, T] \tag{10.4}$$

where $V_{i,t}$ denotes the voltage magnitude of bus i at time t, which is obtained from time-domain simulation, and μ is the threshold to define unacceptable voltage deviation level, which can be set according to the industrial criteria, for example, 20%.

As illustrated in Figure 10.4, TVSI only accounts for the buses with unacceptable voltage violation during the transient period, and it measures not only the voltage violation magnitude but also the associated duration time in the whole system level. TVSI only considers the buses with incredible voltage violation in the transient period. In practice, the magnitude of the voltage violation and the associated duration time have the impact on the degree of severity in voltage deviation [10]. TVSI can measure both the voltage violation magnitude and the associated duration time in the whole system level. A smaller TVSI value indicates lower FIDVR severity, meaning a better voltage recovery performance.

Both of the indices can be calculated via performing TDS to obtain the voltage trajectories, while the former is a binary value and the latter is a continuous value. Taking the simulation results, for example, the TVCI is 0 for Figure 10.2 (lower window), and the TVSI is 0.37 for Figure 10.3 (lower window).

10.3 Proposed intelligent system

The ST-VSA can be converted as a classification subproblem for transient voltage collapse and a regression subproblem for unacceptable dynamic voltage deviation. The latter depends on the result of the former. In this chapter, we develop a hierarchical IS for real-time ST-VSA. The IS is based on the ensemble learning strategy where neural networks with random weights (NNRWs) are employed.

10.3.1 Neural networks with random weights (NNRWs)

NNRWs were originally proposed in [15] which suggested a fast implementation of learning for single-hidden layer feedforward networks (SLFN)—see Figure 10.5.

Given a training data set with N instances in total, $\aleph_N = \{(\mathbf{x}_j, \mathbf{t}_j) | \mathbf{x}_j \in R^n, \mathbf{t}_j \in R^m\}_{j=1}^{N}$, where $\mathbf{x}_j$ is the $n \times 1$ input vector and $\mathbf{t}_j$ is a $m \times 1$ target vector, the output function of the SLFN with $\widetilde{N}$ hidden nodes is

$$f_{\widetilde{N}}(\mathbf{x}_j) = \sum_{i=1}^{\widetilde{N}} \beta_i \cdot \vartheta(\mathbf{w}_i \cdot \mathbf{x}_j + b_i) = \mathbf{t}_j, \; j = 1, 2, \ldots, N \tag{10.5}$$

where ϑ is the activation function, $\mathbf{w}_i$ is the weight vector connecting the ith hidden node and the input nodes, β_i is the weight vector connecting the ith hidden node and the output nodes, b_i is the bias of the ith hidden node, and $\mathbf{w}_i \cdot \mathbf{x}_j$ denotes the inner product of $\mathbf{w}_i$ and $\mathbf{x}_j$.

NNRW is totally different from traditional learning algorithms since it learns though randomly choosing the input weights and biases for hidden nodes $\mathbf{w}$ and b, and analytically determining the output weights β via direct matrix calculations [15]. Equation (10.5) can be formulated as

$$\mathbf{H}\beta = \mathbf{T} \tag{10.6}$$

where $\mathbf{H}$ represents the hidden layer output matrix of the network.

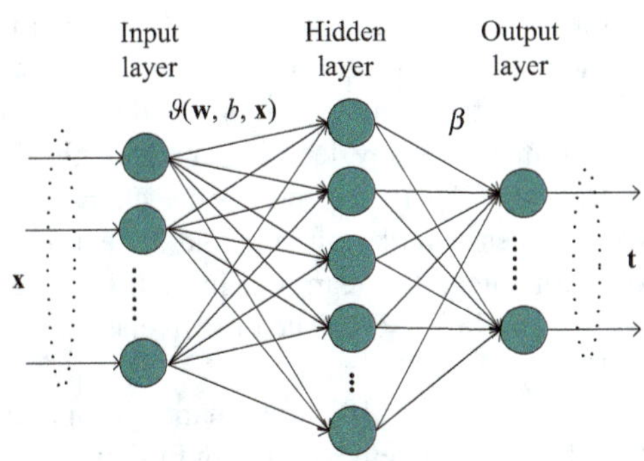

Figure 10.5 Structure of an SLFN

Based on a training database, given the activation function and hidden node number, the NNRW learning process can be briefly summarized as three major steps:

Step 1. Randomly generate the input weights $\mathbf{w}_i$ and b_i, for $i = 1, \ldots, \widetilde{N}$;
Step 2. Calculate the hidden layer output matrix $\mathbf{H}$; and
Step 3. Calculate output weights matrix $\beta = \mathbf{H}^{\dagger}\mathbf{T}$

where $\mathbf{H}^{\dagger}$ can be *Moor–Penrose (MP) generalized inverse* of $\mathbf{H}$.

NNRW can be employed for both regression and classification problems, involving real-time requirements for domain applications. Compared with the traditional learning algorithm, NNRW can train the learner model by solving the standard linear regression problem, so its learning speed can be thousands of times faster, and only needs a little computational memory. Another important advantage of NNRW is its effective adjustment mechanism given the activation function, only the number of hidden neuron nodes needs to be adjusted, which can be effectively realized by the linear verification program. In addition, NNRW can avoid the difficulties of traditional algorithms, such as stop criteria, learning rate setting, learning period, and local minimum.

In the literature, NNRW has been comprehensively verified for many benchmark and practical problems from the field of classification and regression, showing the feasibility of real-time data modeling. However, the operation of random weight neural network is not stable, because each result model can be regarded as an example of random variables. Some monitoring mechanisms should be studied so that the selection of random weights used to initialize NN can achieve better modeling performance (at least better than the mean of random models).

10.3.2 Neural network ensemble

The simplicity and efficiency properties of the NNRW make it a promising candidate for tackling large-scale learning tasks. However, as an undesired side effect, the randomly chose input parameters of the NNRW may result in stability issues.

As widely studied and proved in the literature, *ensemble learning* schemes (such as bagging and upgrading) [18] are effective strategies to improve the accuracy of individual learners. By combining a group of different individual learners, they can compensate each other, and the whole model can reduce the variance and deviation of aggregation, so they tend to improve the accuracy of individuals [18–20]. As a general strategy for any learning method, the feasibility and efficiency of integrated learning largely depend on the characteristics of individual learners. Because NNRW is essentially random, and its learning speed is very fast, it is an ideal candidate for integrated learning. In addition, the whole model can improve the robustness of single NNRW. The effectiveness of building for NNRW has been verified in [10].

Based on the idea proposed in [10], this chapter proposes a new ensemble model of NNRWs for ST-VSA problem. As shown in Figure 10.6, in the ensemble model, M single NNRWs are ensembled in parallel, all of which have common

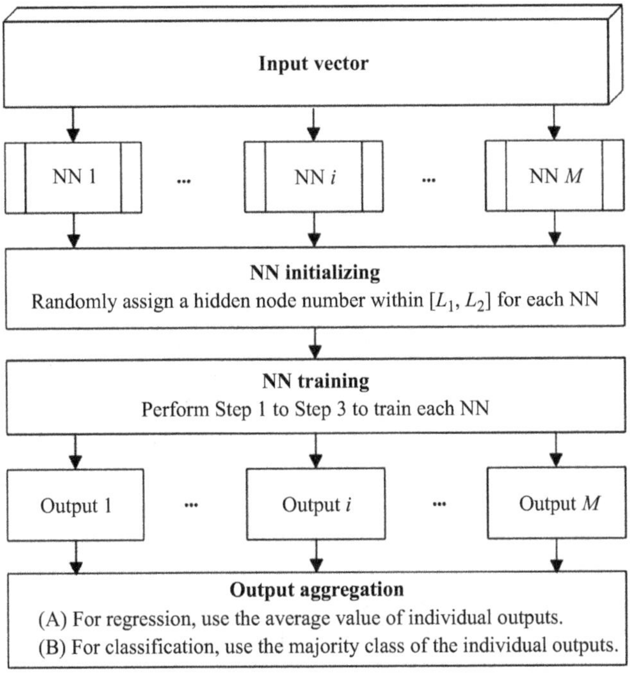

Figure 10.6 Proposed ensemble model of NNRWs

input variables. In the initialization stage, the ensemble model randomly assigns a hidden node number to each NNRW within an optimal range, $[L_1, L_2]$, which is determined from a pretuning procedure. In the training stage, the single NNRWs are trained according to Step 1 to Step 3 and export their individual outputs. Given the stochastic nature of the NNRW, the output of each NNRW can be diverse. In result aggregation stage, for regression problem, the average value of the individual outputs is used, and for classification problem, the majority class of the individual outputs is adopted as the final classification result.

In addition to the inherent selection of random input layer parameters for a single NNRW, the integration model also assigns random hidden node numbers to NNRW individuals. This aspect aims to expand the internal diversity of the integration model. Besides, it is inspired by the observation that NNRW usually has an optimal range, rather than a fixed number of hidden nodes (because the optimal number of hidden nodes can be changed under different sets of random input parameters) [10]. In this way, the randomness can be generalized, thereby enhancing the robustness.

10.3.3 Application for real-time ST-VSA

In the off-line process, a voltage stability database is prepared by performing comprehensive TDS considering various scenarios and/or collecting fault records.

Then, the IS can be trained by the database. In the real-time application process, once the input is available, the STVS can be assessed by the IS rapidly and the results are used to support control decision-makings. Besides, in the on-line process, with updated system information (e.g., network structure, load composition, etc.), a new stability database can be generated to update/enrich the IS. Note that this on-line updating aims to keep the IS adaptive to unexpected system changes.

The proposed hierarchical IS for real-time ST-VSA is conceptually described in Figure 10.7.

The IS includes an ensemble classifier and an ensemble predictor based on the proposed NNRW ensemble. They are pretrained in off-line stage with forecasted operating conditions (e.g., day ahead) and corresponding voltage collapse status and TVSI. For on-line application, the real-time system measurements (including generator outputs, bus loads, bus voltage magnitudes, bus angles, etc.) are collected by SCADA or PMU devices. Once the input measurements have been obtained,

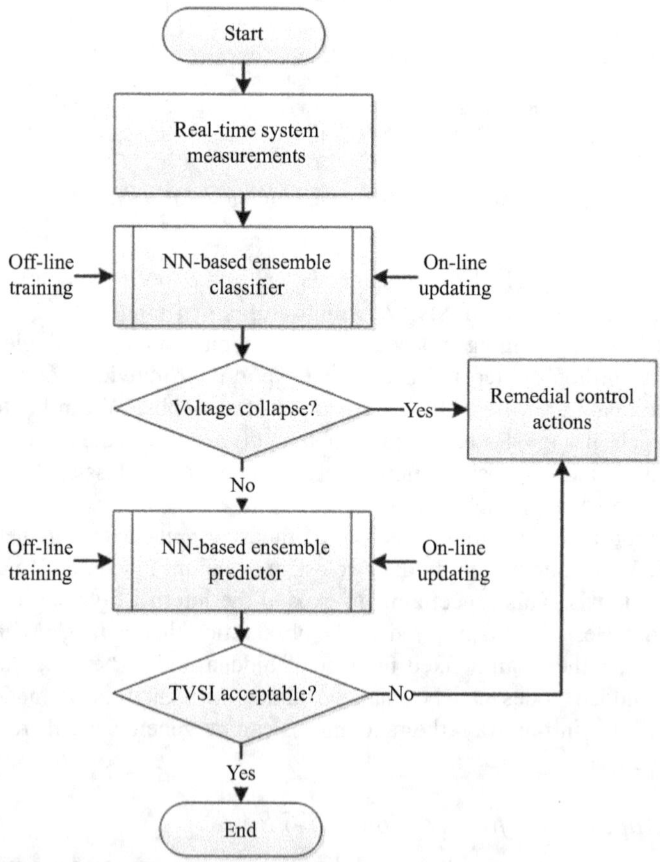

Figure 10.7 Proposed hierarchical IS for ST-VSA

they will be first dropped to the classifier which determines the transient voltage collapse condition of an unknown operating condition with respect to a certain disturbance, if voltage collapse is ascertained, remedial control actions should be taken to protect the power system from the voltage collapse should the disturbance really occurs; otherwise, the measurements are then dropped to the predictor to evaluate the TVSI of the operating condition. If the predicted TVSI is unacceptable, for example, less than a certain threshold, remedial actions will be activated. During the on-line application phase, the classifier or predictor can also be updated via the latest on-line operating information.

Since the training speed of NNRW is very fast, the IS-based model can be updated real-time. It can be updated regularly according to actual needs, such as every hour or half an hour every day. The updated data can be obtained from the TDS results on the real-time system operating point. The simulation results can be saved as new training data. Such new training data can be used as a new set, or can be attached to the initial training database to retrain the IS.

10.3.4 Automatic feature selection

The purpose of automatic feature selection is to select the most important subset from the entire input set to train the model. Therefore, irrelevant/redundant features can be removed, and the total size of the training data can be significantly reduced, thereby improving training speed and accuracy. For DSA applications in power systems, automatic feature selection can identify important operating variables that best describe the system's dynamic safety characteristics, thereby providing useful information for safety control [21]. In addition, using only important features as inputs to the DSA model can reduce the requirements for parameter measurement.

In [21], the author reviews the feature selection technology of intelligent DSA and introduces a new algorithm RELIEF. Unlike other feature selection algorithms, the idea of RELIEF is to evaluate the quality of a single feature based on the degree to which the value of the single feature is distinguished between instances next to each other. It not only considers the differences between feature values and classes but also considers the distance between instances, so important features can collect similar instances, separate different instances, and make a qualitative estimation of the importance of the features. The RELIEF has been successfully utilized in our previous work with satisfactory results [11,21]. It also applies RELIEF to the proposed ST-VSA model for automatic feature selection.

The basic RELIEF is to iteratively update the weight of each feature as follows [22]:

$$W[X]^{i+1} = W[X]^{i} - diff(X, R_i, H)/m + diff(X, R_i, M)/m, i \leq m \tag{10.7}$$

where X denotes a feature, R_i is the instance sampled in the ith iteration, H is the nearest instance from the same class as R_i (called *nearest hit*) while M is the nearest instance from the different class with R_i (called *nearest miss*), and m is the number of sampled instances guaranteeing the weights between [–1, 1]. Function $diff(X,R,R')$

calculates the difference between the values of feature X for two instances R and R':

$$diff(X, R, R') = \frac{|value(X, R) - value(X, R')|}{\max(X) - \min(X)} \tag{10.8}$$

From the statistics point of view, the weight of feature X is an approximation of the difference of probabilities

$$\begin{aligned} W[X] = {} & P(\text{diff. value of } X \mid \text{nearest inst. from diff. class}) \\ & - P(\text{diff. value of } X \mid \text{nearest inst. from same class}) \end{aligned} \tag{10.9}$$

To deal with noises, incomplete data, and multiclass problems, *RELIEF* was extended with the following weight updating equation [22]:

$$\begin{aligned} W[X]^{i+1} = {} & W[X]^{i} - \sum_{j=1}^{k} diff(X, R_i, H_j)/(m \cdot k) \\ & + \sum_{C \neq class(R_i)}^{k} \left[\frac{P(C)}{1 - P(class(R_i))} \cdot \sum_{j=1}^{k} diff(X, R_i, M_j(C)) \right] \Big/ (m \cdot k) \end{aligned} \tag{10.10}$$

where C is a class label, $P(C)$ is the prior probability of class C, and k is an user-defined parameter.

Instead of finding the nearest one hit and miss, (10.10) finds k nearest hits and misses to average their contribution in updating the weight, thereby reducing the risk of miss estimation; the introduction of $P(C)$ aims to estimate the ability of separating each pair of classes.

To identify a set of significant features for training the NNRW ensemble, the features with higher W values can be selected since they can better differentiate instances among classes. As supplementary benefits, the selected features can provide insight into the critical variables that impact the short-term voltage most, which is valuable for designing preventive controls [15].

10.4 Simulation results

The developed hierarchical IS is tested on a New England 10-machine 39-bus system (see Figure 10.8), which is a common test system for stability studies [23].

The numerical simulation is conducted on a 64-bit PC with 3.1 GHz CPU and 4.0 GB RAM. TDS is performed using commercial software PSS®E [23] and the proposed algorithms are realized in the MATLAB® platform. For the test system, the dynamic load parameter is assumed as LM—20%, SM—30%, discharge lighting—5%, transformer exciting current—5%, constant power load—30%, and voltage-dependent load—10%.

10.4.1 Database generation

To train the proposed IS model, a comprehensive stability database is necessary. In practice, this can be obtained by performing simulations on various scenarios

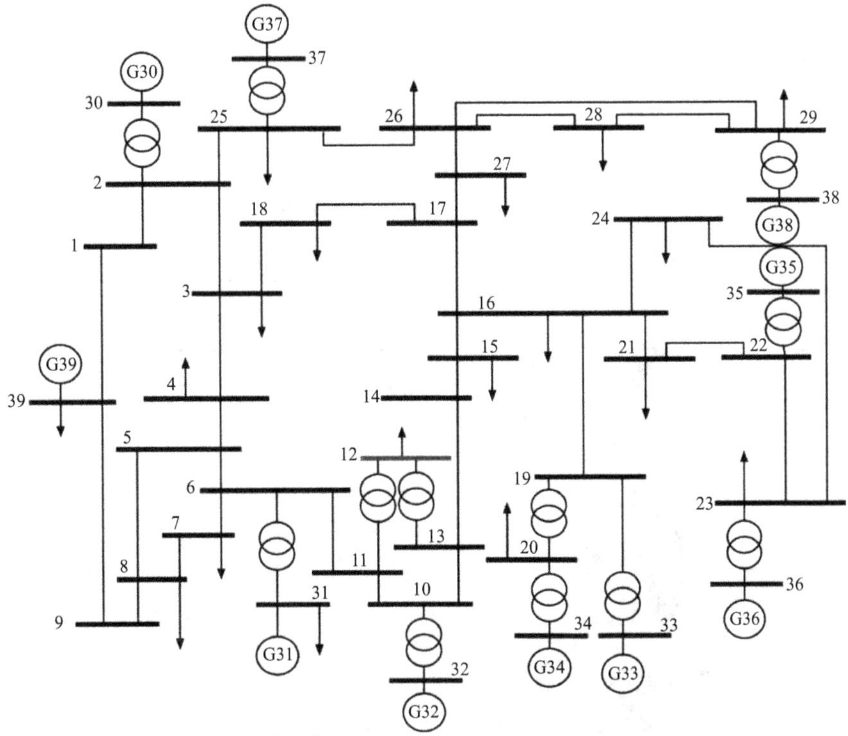

Figure 10.8 One-line diagram of the New England 10-machine 39-bus system

and/or fault recordings. In this study, we artificially generate such a database through optimal power flow (OPF) calculation and TDS [11]. A total of 700 different operating states covering a variety of different load/generation patterns ranging from 80% to 120% of the base loading level are generated through OPF calculation. A three-phase short-circuit fault is considered first (denoted as Fault #1). The fault is applied to bus 15 with a duration time of 0.2 s. For each operating condition, the following candidate features are selected as the input vector: P, Q load, voltage magnitude, and angle of each load bus; P, Q generation of each generator; and total P, Q load and generation of the whole system. For each operating condition, there are a total of 136 candidate features.

TDS is then performed on each generated operating condition to determine their TVCI and TVSI. The total simulation time is 5 s and the simulation step is 0.01 s. The illustration of the dynamic response of this system under the assumed fault is already given in the lower window of Figure 10.3. In calculating the TVSI, μ is set to 20%. For the 700 generated samples, 256 of them (36.6%) experience voltage collapse and 444 (63.4%) do not experience voltage collapse.

The structure of the short-term voltage stability database is illustrated in Table 10.1.

Table 10.1 Structure of the stability database

Sample ID	**Input vector**				**Output vector**	
	F_1	F_2	...	F_n	**TVCI**	**TVSI**
1	0.0	0.0	0.0	0.0	1	n/a
2	0.22	0.13	0.22	0.64	0	2.6
...	...	...	...	...	...	...
N	1.0	1.0	1.0	1.0	0	3.8

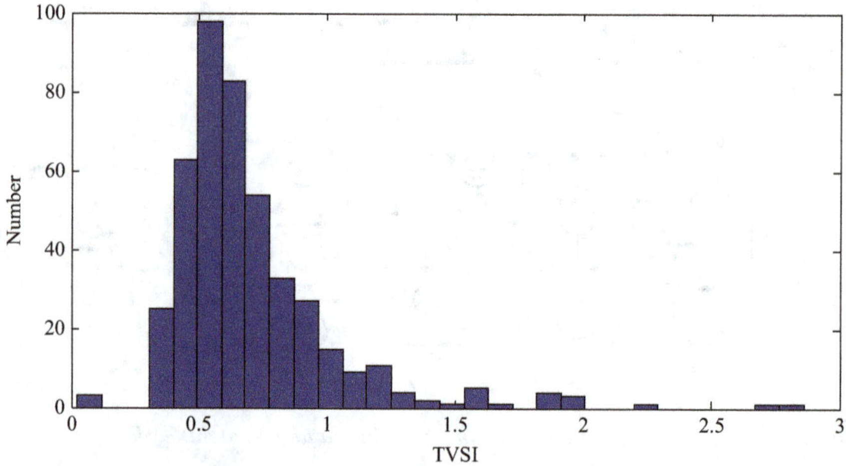

Figure 10.9 Histogram plot of the TVSI of nonvoltage collapse samples

Note that the input features are normalized into the range of [0, 1]. For TVCI, 1 represents the voltage collapse occurs while 0 not. Only the samples with 0 TVCI have a TVSI. The histogram plot of the TVSI of the samples is given in Figure 10.9, where it can be seen that the range [0.3, 1.5] occupies the major proportion.

Note that during the training phase, the classifier is trained by all of the samples and the TVCI is used as the training target, while the predictor is only trained by the samples whose TVCI is 0 and only the TVSI is used as the training target.

10.4.2 Feature selection

We apply the RELIEF algorithm to choose the important features. The ranked values of the 136 features are shown in Figure 10.10, where it is clear that different features can vary capabilities to distinguish the operating conditions in terms of the status of STVS. A large RELIEF value indicates a stronger capability. The top three features are the voltage angle of bus 24, Q load of bus 27, and voltage angle of bus 26, all of them are voltage related variables of the system. In the next test, we utilize the top 20 features as the model inputs.

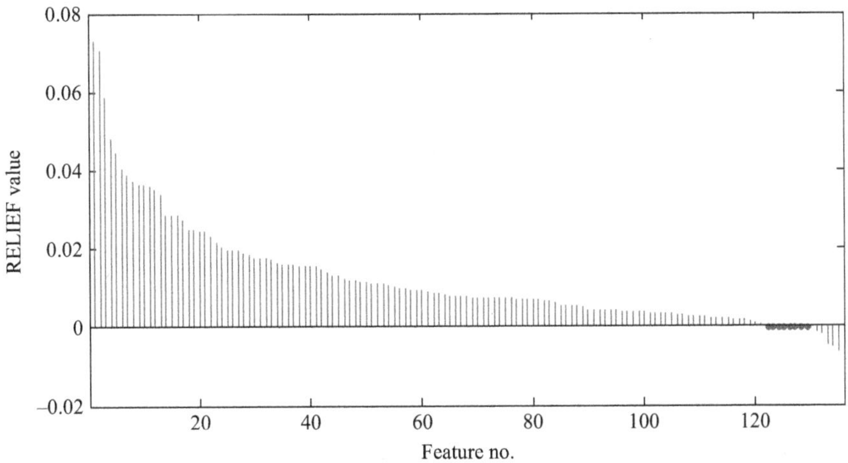

Figure 10.10 RELIEF value for each candidate feature

10.4.3 Model tuning

According to the NNRW theory, a certain hidden node number should be predetermined before training. This corresponds to the optimal hidden node range $[L_1, L_2]$ for the NNRW ensemble. Since the hidden node has a direct impact on the accuracy of a single NNRW, the range $[L_1, L_2]$ needs to be well tuned in advance. For this purpose, the training data set is randomly divided into two nonoverlapped sets, one for training and the other for validation. $[L_1, L_2]$ can be determined at the range where the lowest validation error is obtained. To measure the performance of the proposed IS, percentage classification accuracy is used for the classifier, and the mean absolute average percentage error (MAPE) is used for the predictor:

$$MAPE = \frac{\sum_{i=1}^{d} |y_i - \widehat{y}_i|}{|y_i| \cdot d} \tag{10.11}$$

where y_i and $\widehat{y}_i$ are the actual and predicted TVSI value, respectively, and d is the total number of the testing samples.

The tuning profile is shown in Figures 10.11 and 10.12. It can be seen that for the classifier, with the hidden nodes increasing, the classification accuracy increases sharply, but then gradually decreases. When using 30–60 hidden nodes, the classification accuracy can achieve the highest. For the predictor, a similar trend is also observed (in the case of continuously adding hidden nodes, the prediction error decreases sharply and then gradually increase). When using 15–50 hidden nodes, the obtained MAPE is in the lowest range. Therefore, these hidden nodes are utilized to train the integrated model.

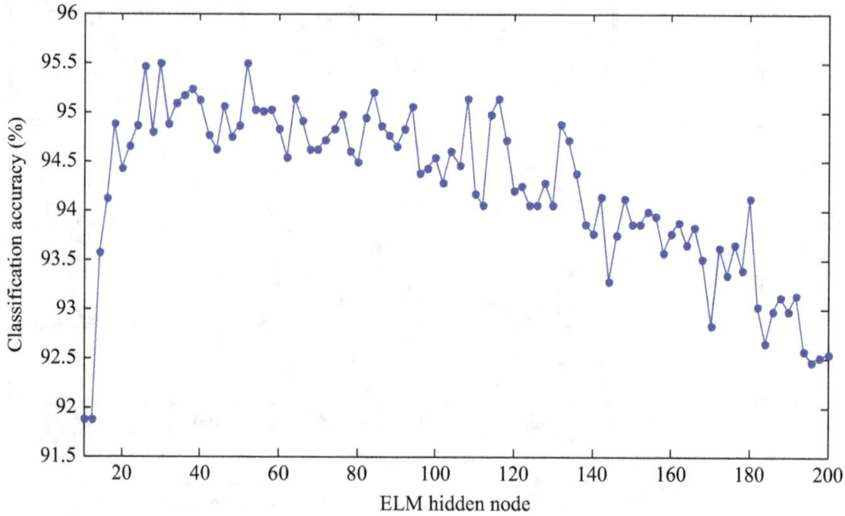

Figure 10.11 Tuning profile for the classifier

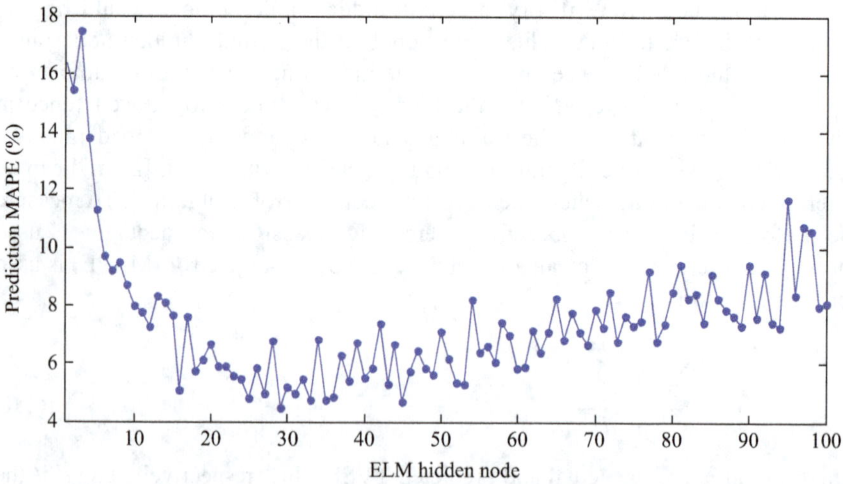

Figure 10.12 Tuning profile for the predictor

Moreover, in Figures 10.11 and 10.12, it can be seen from the tuning process that a single NNRW can achieve maximum 95.5% classification accuracy and minimum 4.56% prediction MAPE. Besides, it is worth mentioning that the training of a single NNRW is very fast, it only costs 0.01 s to learn the tuning data set.

10.4.4 Performance test

In this test, we use 100 individual NNRWs to establish the ensemble model. Once the model has been well trained, it will be applied on-line. As soon as the input variables (significant features) are available (e.g., measured by PMUs), the short-term voltage stability can be determined instantaneously.

The testing set is used to validate the trained IS model. For comparison purpose, the single NNRW is also tested. Besides, some state-of-the-art single learning algorithms including RBFNN, SVM, and DT are also tested with the same training and testing data sets. The performance of the algorithms is presented in Tables 10.2 and 10.3, respectively.

As Table 10.2 shows, the proposed IS outperforms the others in both classification and prediction performance. Importantly, it should be noted that the proposed IS significantly increases the performance over the single NNRW, achieving very high classification accuracy and very low prediction error.

As Table 10.3 shows, the training efficiency of the proposed IS is very high, although 100 NNRWs are used, the overall training time is 28 s, depending on the number of hidden nodes, one single NNRW only cost around 0.3 s for training, this owes to the unique learning process of the NNRW—randomly selecting the input weights and analytically determine the output weights. Compared with other learning models, DT is fastest since its learning procedure does not require time-consuming iterative process. SVM and RBFNN are generally slower due to the time-consuming optimization process during the model training. Given the very high training speed, the on-line updating of the model can be readily achieved, for example, periodically and half-daily. Once more TDS results are available, they

Table 10.2 Performance in terms of accuracy (Fault #1)

Model	**Classification**	**Prediction MAPE**
Proposed IS	98.6%	1.82%
Single NNRW	95.3%	3.83%
SVM	95.8%	3.96%
DT	96.2%	4.68%
RBFNN	94.1%	2.96%

Table 10.3 Performance in training efficiency (Fault #1)

Model	**Training time (s)**
Proposed IS	28
SVM	78
DT	16
RBFNN	62

can be attached to the original training data to re-train the IS model. For all of the models, the classification/prediction process is very fast, which costs around one millisecond for one single test sample, indicating a real-time decision-making speed for ST-VSA.

10.4.5 Extended accuracy test

In the above test, only one fault contingency is considered. In practice, it is usually needed to assess more fault contingencies depending on their likelihood to occur. For a comprehensive test of the accuracy, another two contingencies are tested here. The two contingencies have previously been studied in [16] and their details are given in Table 10.4. Given the 700 generated operating states, the two contingencies are simulated yielding to another two databases. The databases are then used to train and test the proposed IS model and compared it with state-of-the-art algorithms through the same procedure of the above test. The classification and prediction performance are shown in Tables 10.5 and 10.6, respectively.

Table 10.4 Fault contingencies for extended test

Fault no.	**Fault location**	**Fault duration**
#2	Bus 8	0.11 s
#3	Bus 21	0.11 s

Table 10.5 Voltage collapse classification accuracy

Model	**Fault #2**	**Fault #3**
Proposed IS	99.2%	98.8%
Single NNRW	96.7%	94.9%
SVM	97.2%	96.2%
DT	98.9%	97.4%
RBFNN	96.1%	95.3%

Table 10.6 TVSI prediction MAPE

Model	**Fault #2**	**Fault #3**
Proposed IS	1.66%	1.78%
Single NNRW	3.03%	3.77%
SVM	2.68%	3.85%
DT	3.78%	4.43%
RBFNN	2.83%	2.69%

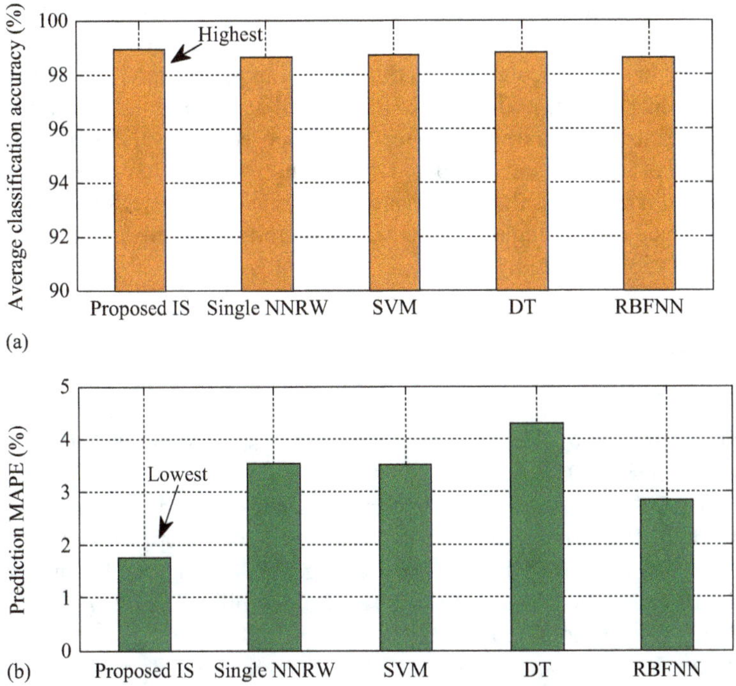

Figure 10.13 Average performance on the three contingencies: (a) classification for voltage collapse and (b) prediction for TVSI

To summarize the overall performance, the average performance metrics on the three contingencies for each model are compared in Figure 10.13. On average, the proposed IS model still outperforms the others in both classification and prediction subproblems, owing to the ensemble effect.

10.5 Conclusion

In recent years, power systems are combined with growing renewable energy capabilities such as wind and solar power. These renewable energy sources are intermittent in nature, so significant uncertainty and complexity can be introduced to ensure the safe and reliable operation of power systems. In order to protect the power system from dynamic unsafe factors that could cause a large power outage, the control center needs real-time digital subtraction. IS strategy is a promising method for real-time digital subtraction. Although there is extensive research in the literature, most of the research has focused on the problem of rotor angle stability. This chapter concentrates on another yet increasingly important dynamic insecurity phenomenon—short-term voltage instability, which involves fast and complex load dynamics. A so-called TVSI is used to quantify the unacceptable dynamic voltage deviation. Then, a hierarchical IS is developed to address the two subproblems sequentially. The IS is based on the ensemble learning strategy and NNRW algorithm. It can be implemented in an off-line training, real-time application, and on-line updating pattern. The RELIEFF algorithm is used to select the most significant features as the training input. The proposed IS is validated on the New England 39-bus system and compared with some other state-of-the-art learning algorithms. Simulation results show that the hierarchical IS can effectively increase the performance over a single NNRW and outperform the compared algorithms in both classification and prediction.

References

[1] IEEE/CIGRE Joint Task Force on Stability Terms and Definitions. "Definition and classification of power system stability." *IEEE Trans. Power Syst.* 2004, vol. 19, pp. 1387–1401.

[2] Li F., Qiao W., Sun H., *et al.* "Smart transmission grid: vision and framework." *IEEE Trans. Smart Grid.* 2010, vol. 1, no. 2, pp. 168–177.

[3] Dong Z.Y., Xu Y., Zhang P., and Wong K.P. "Using intelligent system to assess an electric power system's real-time stability." *IEEE Intelligent Syst.* 2013, vol. 28, no. 4, pp. 60–66,.

[4] Zhang P., Li F., and Bhatt N. "Next-generation monitoring, analysis, and control for the future smart control centre." *IEEE Trans. Smart Grid.* 2010, vol. 1, no. 2, pp. 186–192.

[5] Mansour Y., Vaahedi E., and El-Sharkawi A. "Dynamic security contingency screening and ranking using neural networks." *IEEE Trans. Neural Networks.* 1997, vol. 8, pp. 942–950.

[6] Zhou D.Q., Annakkage U.D., and Rajapakse A.D. "Online monitoring of voltage stability margin using an artificial neural network." *IEEE Trans. Power Syst.* 2010, vol. 25, pp. 1566–1574.

[7] Moulin L.S., Silva A.P., EI-Sharkawi M.A., and Marks R.J. "Support vector machines for transient stability analysis of large-scale power systems." *IEEE Trans. Power Syst.* 2004, vol. 19, pp. 818–825.

[8] Sun K., Likhate S., Vittal V., Kolluri S., and Mandal S. "An online dynamic security assessment scheme using phasor measurements and decision trees." *IEEE Trans. Power Syst.* 2007, vol. 22, pp. 1935–1943.

[9] Xu Y., Guan L., Dong Z.Y., and Wong K.P. "Transient stability assessment on China Southern Power Grid System with an improved pattern discovery-based method", in *Proceedings of International Conference on Power System Technology*, Hangzhou, China, October 2010.

[10] Xu Y., Dong Z.Y., Zhao J.H., Zhang P., and Wong K.P. "A reliable intelligent system for real-time dynamic security assessment of power systems." *IEEE Trans. Power Syst.* 2012, vol. 27, no. 3, pp. 1236–1244.

[11] Zhang R., Xu Y., Dong Z.Y., Meng K., and Xu Z. "Intelligent systems for power system dynamic security assessment: review and classification." 2011

4th International Conference on Electric Utility Deregulation and Restructuring and Power Technologies (DRPT). Weihai, China, July 2011, pp. 1–6.

[12] Taylor C.W. *Power System Voltage Stability*. New York, NY: McGraw-Hill, 1994.

[13] Van Cutsem T., and Vournas C. *Voltage Stability of Electric Power Systems*. Norwell, MA: Kluwer, 1998.

[14] Diaz de Leon J.A., and Taylor C.W. "Understanding and solving short term voltage stability problems." *Proceedings of IEEE PES Summer Meeting*, Chicago, IL, July 2002, pp. 745–752.

[15] Schmidt W., Kraaijveld M., and Duin R. "Feedforward neural networks with random weights." *Proceedings of 11th IAPR International Conference on Pattern Recognition Methodology and Systems*, The Hague, Netherlands, 1992, pp. 1–4.

[16] Xu Y., Dong Z.Y., Meng K., Yao W.F., Zhang R., and Wong K.P. "Multi-objective dynamic VAR planning against short-term voltage instability using a decomposition-based evolutionary algorithm." *IEEE Trans. Power Syst.* 2014, vol. 29, no. 6, pp. 2813–2822.

[17] Shoup D.J., Paserba J.J., and Taylor C.W. "A survey of current practices for transient voltage dip/sag criteria related to power system stability." *Proceedings of IEEE PES Power Systems Conference and Exposition*, New York, NY, 2004, pp. 1–6.

[18] Hansen L.K., and Salamon P. "Neural network ensemble." *IEEE Trans. Pattern Anal. Mach. Intell.* 1990, vol. 12, no. 10, pp. 993–1001.

[19] Alhamdoosh M., and Wang D.H. "Fast decorrelated neural network ensembles with random weights." *Inf. Sci.* 2014, vol. 264, pp. 104–117.

[20] Meng K., Dong Z.Y., Wong K.P., Xu Y., and Luo F. "Speed-up the computing efficiency of PSS/E-based power system transient stability simulations." *IET Gen. Trans. Dist.* 2010, vol. 4, pp. 652–661.

[21] Zhang R., Xu Y., Dong Z.Y., and Hill D.J. "Feature selection for intelligent stability assessment of power systems." *Proceedings of IEEE PES General Meeting*, San Diego, CA, July 2012, pp. 1–6.

[22] Kononenko I. "Estimating attributes: analysis and extension of RELIEF." *Proceedings of European Conference on Machine Learning*, Catania, Italy, 1994, pp. 171–182.

[23] Xu Y., Dong Z.Y., Guan L., Zhang R., Wong K.P., and Luo F. "Preventive dynamic security control of power systems based on pattern discovery technique." *IEEE Trans. Power Syst.* 2012, vol. 27, no. 3, pp. 1236–1244.

Part III

Wide-area control and decision making

Chapter 11

Impact of data quality on synchrophasor-based control to minimize wind curtailment[A]

Anurag Srivastava[1], Vignesh Venkata Gopala Krishnan[1], Shyam Gopal[1], Shikhar Pandey[1] and David Bakken[1]

11.1 Introduction

The increased penetration of wind power in the power grid enables energy security and sustainable grid. However, the intermittent and uncertain nature of power generation from wind farm may cause frequent violations of the ampacity ratings of the low voltage transmission lines. The storage of excess wind energy during insufficient transmission capacity could be a solution; however, an expensive solution given existing cost of storage [5]. The overload condition caused by excess wind generation sometimes leads to the power outages. Since load shedding is the last resort to rescue the system in this scenario, it could be safely addressed by curtailing some of the wind generations. The curtailment of wind energy can also limit the loading of the transmission lines without interrupting services to the consumers.

The objective of the proposed automatic control schemes or remedial action schemes (RASs) is to detect the overload in transmission lines and trip minimum number of wind farms or minimum wind energy curtailment such that the current in the transmission lines is well within the limits.

RAS is an automatic control mechanism designed to detect the abnormal system conditions and take fast control action for maintaining the system reliability [6]. Most of the existing RASs are usually hard coded, which do not adapt to changing system conditions. Redesign of the RASs may be required if power system network and/or operating conditions change. With the development of the smart grid technology, there is a significant increase in the deployments of the phasor measurement units (PMUs). Recent development in synchrophasor measurement technology allows high-speed time-synchronized phasor data available. Given high reporting rates, this technology enables faster control schemes to be implemented.

[A]Contents of this chapter are partially based on the work done by the authors in [1–4].

[1]Department of Electrical and Computer Sciences, Washington State University, Pullman, Washington, USA

Authors in [7] have analyzed multiple RAS schemes and a system-wide RAS arming (RASA) system. The automatic real-time process, from the RASA decision (or recommendation) to actual RAS actions in response to disturbances of the overall system is discussed. In this study, the authors have clearly given a contrast between an automated RAS and a nonautomated RAS. The nonautomatic RAS system requires operators to verify, validate, and manually download the RASA patterns to RAS schemes at arming stations even though the arming pattern determination may be automated. On the other hand, the automatic RAS system automates the entire process of determining, verifying, validating, and downloading the arming patterns to all RAS schemes within a balancing authority.

In [8], an optimal generation shedding for dynamic RAS is proposed. A generation-shedding cost function has been developed by minimizing the pre-RAS and post-RAS costs. The proposed method in this work updates control action schemes online based on the system operating conditions and minimal-cost objective. The criteria for selecting generators to trip is based on the minimum cost rather than minimum amount of generation to shed. Authors in [9] have proposed a method that in real-time determines remedial actions, which restores secure operation with respect to aperiodic small-signal rotor angle stability (ASSRAS) when insecure or unstable operation has been detected. A method to assess an ASSRAS event is used to monitor the stability boundary for each generator in a real-time manner. In [10], the study proposes a novel event-based RAS to overcome the frequency and voltage instability caused by super component contingencies (SCCs) through optimal load shedding.

In this chapter, PMU-based RAS is developed to minimize the wind power curtailment and protect the transmission lines from overload/congestion problems. Mathematical model of the RAS is described and its real-time validation using a testbed comprising of real-time simulators, PMUs, and Controllers is discussed. Sources of anomalies in PMU data, their detection, and the effect of data anomaly on RAS operation are also outlined in this chapter.

11.2 Wind curtailment: current practices

Minimizing wind curtailment is one of the problems that system operators and wind power producers need to solve with the transmission overload problems caused by the increasing wind penetration. Wind curtailment is a reduction in the output of wind generators from otherwise produced given available resources, typically on an involuntary basis [11,12]. Owing to technical and economic constraints [13], excessive portions of the wind generation are curtailed and hence wasted [11]. However, curtailment in wind generation may result in several undesirable consequences. Primarily, wasting freely available energy leading to economic loss. Second, revenue of the wind industry can be reduced significantly, which puts wind projects at risk of not meeting their debt obligations or reducing the amount of debt that the project's cash flows can support. Hence, there is a need to minimize wind curtailment by optimization techniques and not violating the line capacity limits.

System operators do the curtailment typically because of transmission congestion or lack of transmission access, but it can also occur for a variety of other reasons, such as excess generation during low load periods, voltage problems, or interconnection issues.

Transmission constraints have been the most common reason for wind curtailments in the United States [12]. Most curtailments have occurred when the construction of necessary transmission lines lags behind the pace of wind farm development, resulting in infrastructure that is insufficient for the amount of wind generation online. For instance, in ERCOT and SPP, substantial amounts of wind generation were brought online before the necessary transmission could be built leading to massive wind curtailments. ERCOT's reported wind curtailments rose from about 1% in 2007 to a peak of more than 17% in 2009 [12]. With the adoption of new transmission operating procedures that increased the transfer capacity on existing transmission, curtailments fell to 4% in 2012 and further to 1.6% in 2013.

In many cases, curtailments are implemented manually whereby grid operators call wind facilities to issue curtailment directives. Manual curtailment instructions are used by some utilities. For example, APS, Salt River Project, Tucson Electric, ISO-NE, PacifiCorp, SPP, and MISO [12]. As an example, PacifiCorp's generation center can manually control wind output levels for wind facilities they own.

Majority of the utilities, including ERCOT have realized the importance of the automated curtailment, rather than the manual curtailments. Manual curtailment processes for wind have been found to extend curtailment periods because of the time required to implement curtailment and hesitancy to release units from curtailment orders. Automatic communication procedures can speed the implementation of curtailment orders and reduce overall curtailment time. Moreover, automated wind curtailments can also relieve the system stress faster, thereby ensuring better reliability of the power supply.

Based on the limited literature survey, the following are a few issues that are identified in the existing wind curtailment approaches:

- Almost all of them are centralized approaches. The curtailment decisions are taken at a single control node. Failure of the central node may prove dangerous to the system operations.
- Most of the RAS schemes are hard coded. The RAS designed for one operating condition, may not be suitable for the other. Hence it is required to design a dynamic RAS to adapt to changing conditions and take near real-time control decisions.

11.3 Synchrophasor-based automated remedial action schemes

This section presents an overview of the architecture of the RAS. The architecture primarily consists of (a) data sensors, (b) data aggregator, (c) substation local area network (LAN), (d) wide area network (WAN), (e) computing nodes, and (f) actuators. The architecture is shown in Figure 11.1. Sections 11.3.1–11.3.5 describe these components in brief.

11.3.1 Data sensors

The RAS being discussed in this chapter makes use of local and remote measurements of voltage and current obtained at each substation. The RAS logic also requires the temperature and wind speed. The current network topology is obtained based on the circuit breaker status. The primary source of voltage and current phasors are field sensors such as PMUs, digital fault recorders (DFR), and remote terminal units (RTU).

The PMUs and DFRs report the measurements typically at one phasor value per cycle. The RTUs provide the voltage magnitude and currents for power flow measurements. The weather measurements can be periodically updated every 10 min. The data form the PMUs and DFRs are time stamped using global positioning system (GPS) at the source while the measurements from the RTUs are time stamped by the data acquisition module. PMU and DFR measurements use C37.118 over TCP layer and that of RTUs and weather sensors use DNP3.

11.3.2 Data aggregator

The measurements are sent to the data aggregator through a substation local area network. This module consists of a phasor data concentrator (PDC). It time aligns the data and the processed data is sent to the computing nodes.

11.3.3 Wide area network

Data from all the substations are communicated to the computing nodes through the wide area network. The control signals computed is also sent to the respective substations through this network.

11.3.4 Computing nodes

The data from the substations are obtained at the computing nodes. The line capacities are obtained by the dynamic line rating (DLR) module based on the weather data obtained through the weather sensors. The curtailment signals are computed by the RAS module based on the line capacities estimated and the voltage and current phasors.

A backup computing node is also present with all the modules in that of the main node. It receives all the phasor data and the weather data simultaneously with the main node. The backup node acts only when the main node fails or crashes and computes the control signal. Failures are detected by the backup node using the "Heartbeat" liveliness module.

11.3.5 Actuators

The actuators receive the control commands (curtailment commands) from the computing nodes through the WAN and issue these commands to the wind farm controls through the substation LAN.

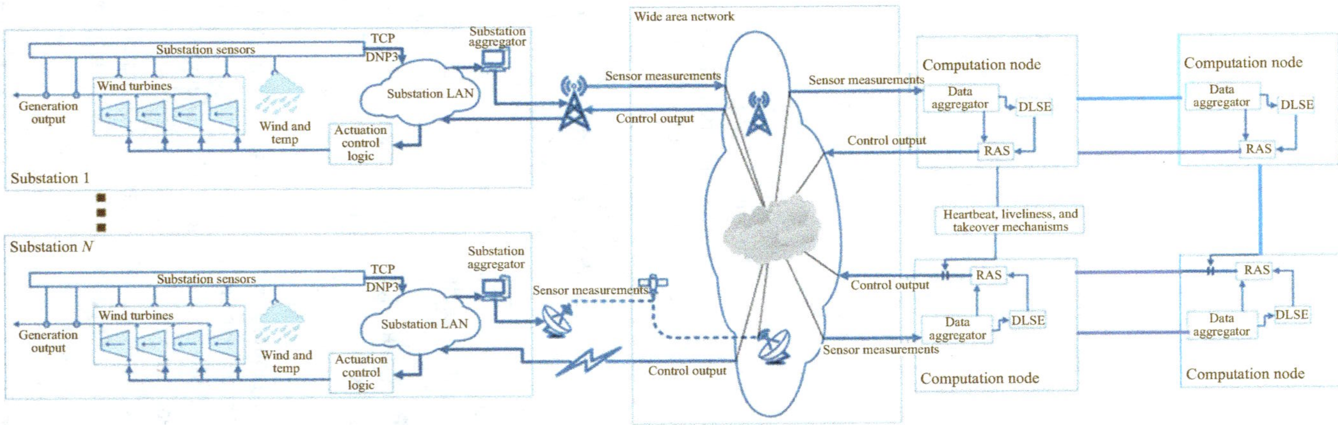

Figure 11.1 High-level architecture for the proposed cyber-resilient RAS

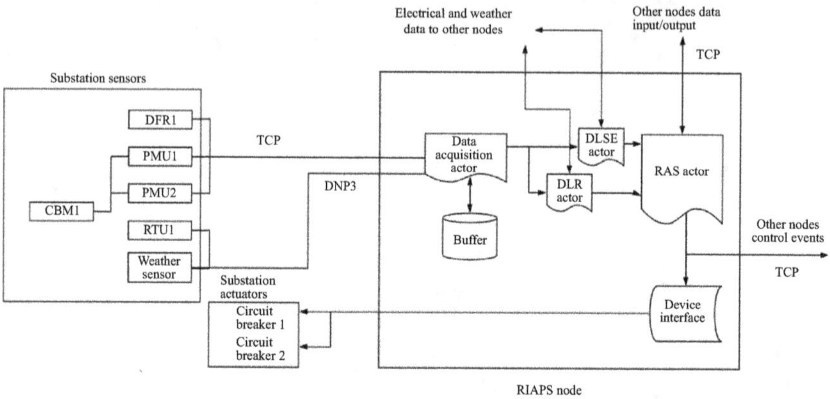

Figure 11.2 Actors in a RAS node

11.4 Architecture of PMU-based RAS

Figure 11.2 shows a typical configuration of actors required for the proposed RAS. The substation sensors consist of PMUs and DFRs which use TCP protocol. The electrical measurements from RTUs and circuit breaker monitors (CBMs) use DNP3 protocol. The weather information is obtained using various meteorological sensors which also use DNP3 protocol. The electrical and weather measurements for the boundary buses are collected from other nodes using TCP protocol.

The decentralized linear state estimation (DLSE) actor performs the state estimation using acquired data while DLR actor computes the dynamic line ratings using the weather information. The results are transmitted to the RAS actor which performs the optimal wind curtailment application. The required control actions are generated by the RAS actor and transmitted to the appropriate devices in the network.

The detailed configuration of the RAS actor for the leader node is shown in Figure 11.3. The DLSE and DLR actors have server ports and sender ports. Server ports send the data to rest of the nodes and the sender ports send data within the node. Based on the electrical states and weather information obtained from DLSE and DLR actors, respectively, the RAS generates the initial conditions required for solving the wind curtailment problem. The leader breaks the optimization problem into subproblems and sends data to other nodes to solve these subproblems. One of the subproblems is also solved by the leader node. The results of subproblem from other nodes are aggregated at the leader node and the convergence criteria are checked. If the criteria are met, the leader node generates the control actions and issues control commands to the local devices using the device interface service and to other appropriate nodes using TCP. If the convergence criteria are not met, the leader node rebuilds the initial condition matrix and repeats the process or issues a timeout signal when the convergence is not met within specified time.

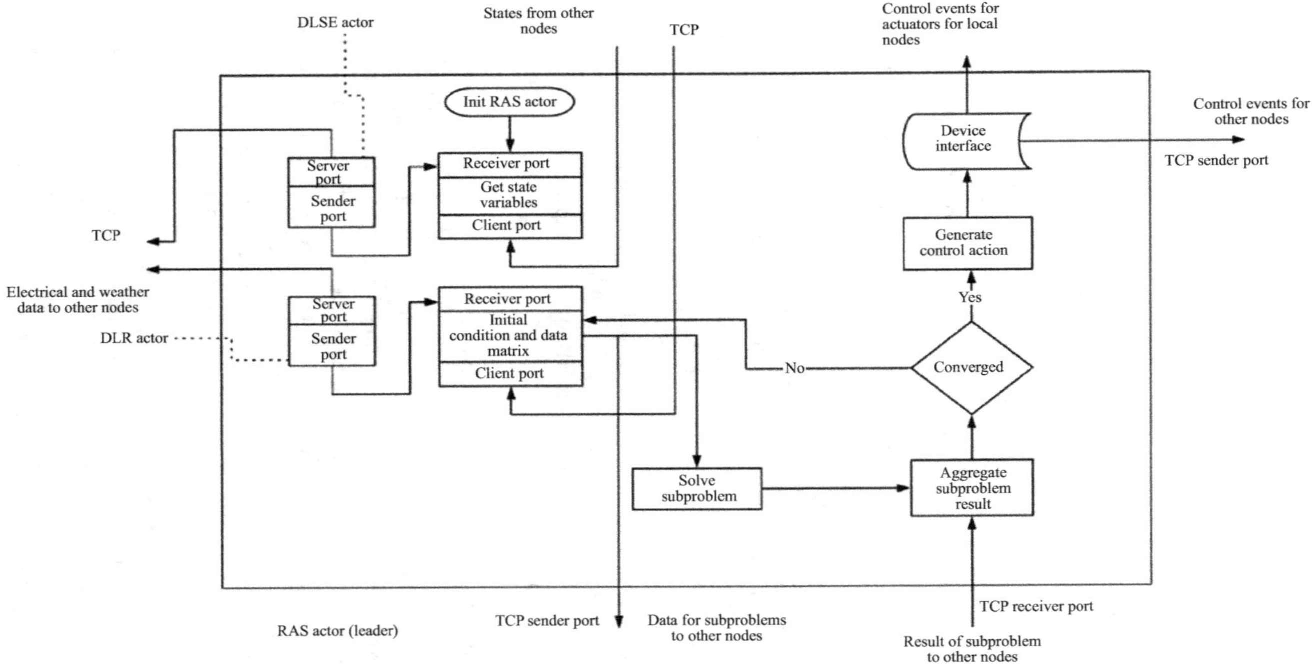

Figure 11.3 RAS actor in a leader node

11.4.1 Actors and components

This section describes the various modules in the proposed RAS. These modules are also known as actors. Each actor has a number of submodules and component to achieve various functionalities. The specification of actors and components related to RAS for wind curtailment are presented below:

- Data acquisition actor: The proposed RAS uses local and remote measurements of voltage and current phasors. The temperature and wind speed are also required for the proposed RAS. The topology processing is carried out based on CBM status. The primary source of voltage and current phasors are field sensors such as PMUs, DFRs, and RTUs. The PMUs and DFRs transmit phasor data in a periodic fashion at the rate of 1 phasor per second. The RTUs transmit the voltage magnitude and power flow measurements at the rate of 1 set of measurements per second. The weather measurements are periodically updated every 10 min. The data form the PMUs and DFRs are time stamped using GPS at the source while the measurements from the RTUs are time stamped by the data acquisition actor. The physical interface for data inputs from PMUs, DFRs, RTU, CBM, and weather measurements are Ethernet ports. The protocol for input data for PMU and DFR measurements is C37.118 over TCP layer and that of RTUs and weather sensors is DNP3. The portability among different protocols is carried out by the protocol conversion component of the Data acquisition actor. The CBMs transmit the circuit breaker status data to the PMUs and RTUs for periodic updates or it can transmit directly to the RAS node as the circuit breaker event occurs. The data acquisition actor calls time stamping component for the RTU measurement, whereas the time alignment component aligns the PMU data based on GPS time stamps.

  The specifications of the components used by Data acquisition actor are as follows:

  1. Protocol conversion component: The diverse category of sensors installed in the power system network uses different kinds of protocols like IEEE C37.118 and DNP3. The protocol conversion component extracts measurements from the data packets by decoding these protocols. The protocol conversion component uses the device interface service to collect the data from different sensors available with the node. The protocol conversion component does not use time synchronization service. It uses fault management service to detect errors in the data packets based on parity bits and error-detecting codes like cyclic redundancy check (CRC). This component discards the data packet upon detection of such errors.
  2. Periodic input data component: This component gets the measurements from the sensors upon the overflow of the timer. This component uses the time synchronization service for detecting the latency of the data arrival from the instant it is requested. The delayed data are rejected for the proposed RAS application.
  3. Event data input component: The event data input component receives data from the sensors when sudden changes in the system occur.

4. Time stamping component: The time-stamping component uses the best available time in the node for time stamping the incoming data packets.
5. Buffer input data component: This component buffers the measurement data for a few seconds in an array which is required by the proposed RAS. The maximum length of the buffer required for the proposed RAS is chosen to be ten instances for each measurement.
6. Time aligning component: The buffered data are time aligned based on the time stamps of the data packet using this component.

- DLSE actor: The DLSE provides the best estimate of the voltage and current phasors for the system considered. The DLSE uses redundant measurement values from a combination of sensors like PMUs, DFRs, and RTUs. The DLSE accounts for the network uncertainty by filtering bad data resulted from drop, delay, and corrupt data packets. The status from the CBMs is also used to identify and filter the bad data. The virtual and pseudo measurements are used by DLSE to replace the bad data which are previously identified. The DLSE actor invokes the topology processing component and bad data detection before using weighted least square to estimate the state of the system. More details of DLSE algorithm can be found in Section 11.5. The specifications by DLSE actor are as follows:
  1. Noise filtering component: The measurement data from different sensors are associated with different noise levels due to hardware variations and different estimation algorithms. This component filters the noise in the measurement data based on the prior information of the sensors from which the data are received.
  2. Bad data detection component: This component detects any bad data in the measurement using statistical techniques. The possible replacement of the bad data using virtual measurements and pseudo measurements are also carried out by this component.
  3. Topology processing component: The CBM data are used by this component for generating the network topology and validate them with the bad data detection component.
  4. Weighted least square component: The best estimation of the power system states is obtained by this component using weighted least square method.
- DLR actor: The dynamic line ratings specify the value of the current without overloading the transmission lines at a given temperature and wind speed. The increase in the ambient temperature reduces the dynamic line ratings whereas increase in wind speed increases the dynamic line ratings. The DLR actor uses weather data from the data acquisition actor which are required to estimate the DLR based on IEEE Std 738-2006. The DLR actor calls estimate DLR component to obtain the dynamic line ratings.
- RAS actor: The RAS actor receives the electrical and weather data from the DLSE and DLR actor. It formulates the optimization problem, solves it and

aggregates the results and generates the control action. The specifications of the components used by RAS actor are as follows:

1. Initialization component: This component resets all the iteration counters, variables, and timers.
2. Get state variables component: The RAS actor requests electrical states of the system from the DLSE actor using this component. The data are received from the local node of the DLSE actor at the receiver port and from the other nodes at the client port.
3. Initial condition component: This component forms the relational matrix and the constraint vector using the electrical measurements from DLSE actor and line ratings from the DLR actor for the optimization problem. The line ratings from the local node DLR are received at the receiver port and that from the other nodes are received at the client port.
4. Optimization time out component: The optimization logic of the RAS can be centralized [3] or distributed as in [4]. The distributed architecture is to ensure better tolerance against computational node and communication failures. The convergence of the optimization problem in a distributed environment is not guaranteed. The divergence of the solution after few iterations can be prevented by terminating the optimization using this component. This component is also used by the RAS to terminate the optimization for slow change in the solution after each iteration.
5. Optimization component: This component generates the solution of the optimization problem. The minimum wind generation curtailment such that the transmission line loadings are not violated, is the result of this component.
6. Generate control action component: The successful termination of the optimization will result in vector solution variables. This component translates the solution into required control action and uses the device interface service for execution to the local actuator. This component also transmits the control action to the other nodes relevant to the RAS output.

11.5 Distributed linear state estimation

The proposed RAS scheme uses the power generated by the wind farms and the power consumed by the load as input parameters. The estimated values of the load and generation are provided by the state estimation (SE) component. The traditional centralized SE is updated every 2–15 min, which is insufficient to meet the requirement of the proposed RAS. In this respect, a new DLSE is required in order to provide the accurate input data to the RAS.

With the increasing size and more data flow in the power system, traditional centralized state estimation (TCSE) needs to handle an increased quantity of state variables. TCSE will take more time to compute the system status or may not be able to converge. In order to address this challenge, power industry needs a new computational architecture and efficient algorithms for state estimation.

There are several variations of state estimation algorithms being developed by researchers. In [14], authors propose a star-like hierarchical two-level state estimation. In this method, the first level is the local state estimation. The second level is the centralized state estimation.

In [15] and [16], authors show a two-level linear state estimator. With this two-level processing, the bad data and topology errors can be removed in the substation level, so it provides the more accurate input to the system level state estimation. In [17], authors show an efficient way to calculate the difference of reference angle between different areas and utilize the PMU data in this DSE. In [18], a multiarea DSE is presented. Authors have used the local measurement and boundary information to do the local state estimation and calculate system-wide states. A fully DSE method for wide-area monitoring system is presented in [19]. Both DC state estimation and AC state estimation are done in [19]. Authors show that this method is implementable under more relaxed assumption. In [20], authors proposed a linear state estimation model using PMU. This method combines the TCSE results and PMU measurements together.

The main idea of DSE is to divide the power system computational data into set of groups, and process in distributed manner to reduce the computational burden [1,2]. Additionally, with the development of the smart grid technology, increasing number of power system intelligent electronic devices (IED) are deployed in many substations. These IEDs also have the computational ability, which can be used for decentralized applications. PMU is a special kind of IEDs and can directly measure the synchronized and accurate phasor values, which also provide enough information for the linear state estimation. In this study, we assume PMUs at all the buses which may be a case, as there is an increased installation of PMUs in the current grid.

For the DLSE, the first step is formation of initial grouping to generate number of groups. Note that set of groups formed initially may change dynamically to meet the system requirements. Figure 11.4 shows the example of the initial grouping algorithms. The given power system is divided into several groups based on the following constraints such as electrical distance, bus connectivity, and computational abilities [1]. Once the initial grouping is done, the computational node leader in each group needs to be selected to run the DLSE algorithm to monitor the system status.

Based on these requirements, a multiobjective graph partitioning algorithm is utilized for initial grouping [21]. The whole power system is considered as a graph. Each bus is considered as a vertex and each transmission line is considered as an edge. The weight of the edge is defined in two different ways: (1) electrical distance and (2) the connectivity of two buses, which are on the two sides of the edge. First, a k-way graph partitioning algorithm in METIS [22] is utilized to compute the partitioning for each of the two different weights. Second, based on the multiobjective graph partitioning algorithm, each edge is assigned for a combined weight. Third, METIS [22] is utilized to compute the final partitioning based on the new combined edge weight.

For developing a DLSE, the system needs to be modeled first. Figure 11.5 shows the Pi model of the transmission line, where, V_i and V_j are the voltage value on bus i and j. Y_{i0} and Y_{j0} are the transmission line charging admittance. Y_{ij} is the transmission line admittance between bus i and bus j. Based on only the KCL, the

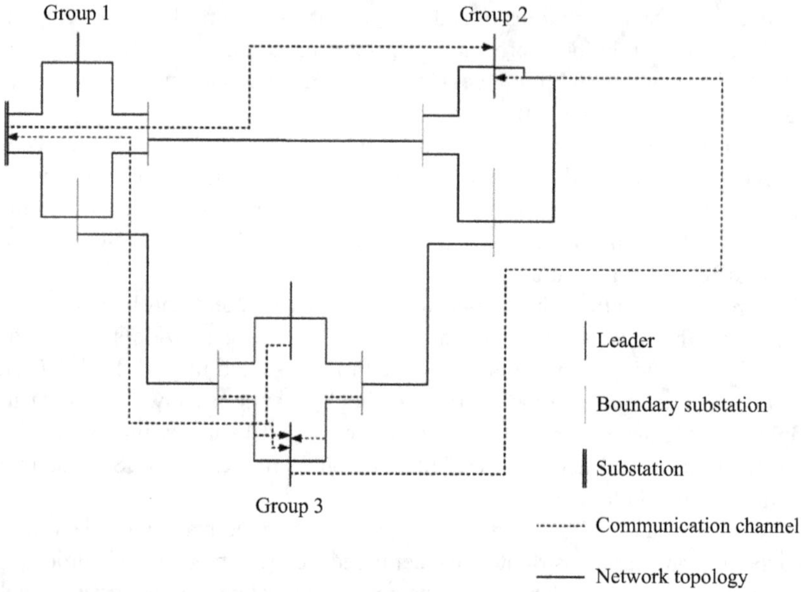

Figure 11.4 Power system topology and grouping example

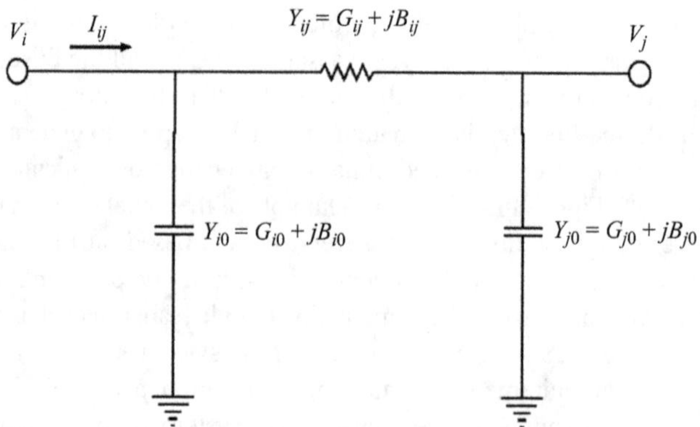

Figure 11.5 Pi model of the transmission line

current I_{ij} from bus i to bus j can be easily calculated by the following (11.1):

$$\bar{I}_{ij} = \bar{V}_i Y_{i0} + (\bar{V}_i - \bar{V}_j) Y_{ij} \tag{11.1}$$

where I_{ij_R} and I_{ij_I} are the real and imaginary part of current I_{ij}. V_{i_R} and V_{i_I} are the real and imaginary part of voltage value on bus i. V_{j_R} and V_{j_I} are the real and imaginary part of voltage value on bus j. Since the PMU can measure the voltage and current phasor values, the real part and imaginary part of current and voltage values can be converted from the phasor values. Therefore, (11.1) can be rewritten as the complex number format:

$$\begin{aligned} I_{ij\_R} + jI_{ij\_I} &= (V_{i\_R} + jV_{i\_I})(G_{i0} + jB_{i0}) \\ &\quad + (V_{i\_R} + jV_{i\_I} - V_{j\_R} - jV_{j\_I})(G_{ij} + jB_{ij}) \end{aligned} \tag{11.2}$$

Then the real part and imaginary part can be split using (11.2) as given in (11.3) and (11.4):

$$\begin{aligned} I_{ij\_R} &= (G_{i0} + G_{ij})V_{i\_R} - (B_{i0} + B_{ij})V_{i\_I} \\ &\quad - G_{ij}V_{j\_R} + B_{ij}V_{j\_I} \end{aligned} \tag{11.3}$$

$$\begin{aligned} I_{ij\_I} &= (B_{i0} + B_{ij})V_{i\_R} + (G_{i0} + G_{ij})V_{i\_I} \\ &\quad - B_{ij}V_{j\_R} - G_{ij}V_{j\_I} \end{aligned} \tag{11.4}$$

Assuming there are M substations and N measurements (including nearby boundary bus measurements) in a specific group. The general system measurement model can be represented as

$$z = Hx + e \tag{11.5}$$

where z is the N-dimensional vector of measurements. H is the $N \times 2M$ dimensional Jacobian matrix. x is the $2M$ dimensional system status which is the real and imaginary part of voltage values on each substation. e is the N-dimensional noise vector of the measurements.

Then, (11.3) and (11.4) can be substituted into (11.5) as

$$\begin{bmatrix} \begin{bmatrix} V_R \\ V_I \end{bmatrix}_{\text{PMU\_V}} \\ \begin{bmatrix} I_R \\ I_I \end{bmatrix}_{\text{PMU\_I}} \end{bmatrix} = \begin{bmatrix} H_{11} & H_{12} \\ H_{21} & H_{22} \\ H_{31} & H_{32} \\ H_{41} & H_{42} \end{bmatrix} \begin{bmatrix} V_R \\ V_I \end{bmatrix} + \begin{bmatrix} e_1 \\ e_2 \\ e_3 \\ e_4 \end{bmatrix} \tag{11.6}$$

where H_{11} and H_{22} are the unit matrix; H_{12} and H_{21} are zero matrix; and H_{31} and H_{42} consist of the real part of the transmission line admittance. H_{32} and H_{41} consist of the imaginary part of the transmission line admittance. V_R and V_I are the real and imaginary parts of voltage values which are the estimation targets.

In order to estimate the optimal value, weighted least squares (WLS) method is used to solve the linear state estimation model. The objective function of WLS is

$$J(x) = [z - Hx]^T R^{-1} [z - Hx] \tag{11.7}$$

where R is the variance matrix of measurements and the element of this matrix is the accuracy of the corresponding measurement. The optimal solution of WLS is

$$\widehat{x} = \left[H^T R^{-1} H\right]^{-1} H^T R^{-1} z \tag{11.8}$$

where $\widehat{x}$ is the estimated system status. When DLSE is utilized in each group, boundary bus voltage values from adjacent groups are also estimated. However, the estimated boundary bus voltage is more accurate when computed within the group. The estimation values for these boundary buses from adjacent groups are ignored and only the values computed within a group have been considered. Since the GPS signal is used in all the PMUs, the reference angle for the phasor data in all the groups is synchronized by the GPS signal.

The DLSE not only reduces the size and the complexity of the power system state estimation computation but also uses the noniterative method to solve the optimal system status. Both of these two features can significantly reduce the computational time of state estimation.

11.6 Mathematical modeling of PMU-based RAS for minimum wind curtailment

The objective of the proposed scheme is to minimize the amount of wind energy curtailed, which can avoid transmission line overloads. This problem can be mathematically formulated as

$$\max f(x) = \sum_{i=1}^{N} x_i P_i^{\text{Gen}} \tag{11.9}$$

In this equation, the power generation for the ith wind farm is represented by P_i^{Gen}. The number of wind farms installed in the whole power system is represented by N. The wind farm connection status for the ith wind farm is represented by x_i, $i \in Z(1, N)$.

DC power flow model is utilized to solve this wind power curtailment problem as following:

$$P_{\text{Gen}} - P_{\text{Load}} = B\theta \tag{11.10}$$

$$\theta = B^{-1}(P_{\text{Gen}} - P_{\text{Load}}) \tag{11.11}$$

$$P_{ij} = B_{ij}(\theta_i - \theta_j) \le LR_{ij} \tag{11.12}$$

In order to solve this problem by a standard linear programming (LP) solver, (11.10) and (11.12) are written as follows:

$$P_{\text{Gen}} - P_{\text{Load}} = Hx = \begin{bmatrix} P_1^{\text{Gen}} & 0 & \cdots & 0 & -P_1^{\text{Load}} \\ 0 & \ddots & \ddots & \vdots & \vdots \\ \vdots & \ddots & \ddots & 0 & \vdots \\ 0 & \cdots & 0 & P_N^{\text{Gen}} & -P_N^{\text{Load}} \\ 0 & \cdots & \cdots & 0 & -P_{N+1} \\ \vdots & & & \vdots & \vdots \\ 0 & \cdots & \cdots & 0 & -P_{N+K} \end{bmatrix} \begin{bmatrix} x_1 \\ \vdots \\ x_{N+1} \end{bmatrix} \tag{11.13}$$

$$T\theta \leq LR \Leftrightarrow \begin{bmatrix} 0 & B_{ij-1} & -B_{ij-1} & 0 \\ \vdots & \vdots & \vdots & \vdots \\ B_{ij\_M} & 0 & 0 & -B_{ij\_M} \end{bmatrix} \begin{bmatrix} \theta_1 \\ \vdots \\ \theta_{N+K} \end{bmatrix} \leq m \begin{bmatrix} LR_1 \\ \vdots \\ LR_M \end{bmatrix} \tag{11.14}$$

This can be expressed in the standard format of the linear programming problem as shown below [1]

$$\max f(x) \\ \text{s.t.} \begin{cases} Ax \leq b \\ lb \leq x \leq ub \end{cases} \tag{11.15}$$

$$A = TB^{-1}H \tag{11.16}$$

$$b = [LR_1 \quad \cdots \quad LR_M]^T \tag{11.17}$$

$$lb = [0 \quad \cdots \quad 0 \quad 1]^T \tag{11.18}$$

$$ub = [1 \quad \cdots \quad 1 \quad 1]^T \tag{11.19}$$

In this power grid, there are $N+K$ buses, in which N buses are installed with wind farms and no wind farm installed in the rest of K buses. The power injections on each bus are presented by H matrix with $N+K$ by $N+1$. T matrix, with the size of M by $N+K$, is for calculating the power flow on each transmission line. B is a $N+K$ by $N+K$ susceptance matrix of the power system. ub and lb are the upper and lower boundaries of the RAS variables. The line rating on the ith transmission line is represented by LR_i.

11.7 Synchrophasor anomalies and detection

PMUs, also known as synchrophasors, have opened numerous possibilities for power system operation and its control. The PMUs have the ability to make measurements with high accuracy and have high reporting rates. The reporting rates can be within the range of 10 frames per second to 60 frames per second. Whereas in the traditional system like SCADA, the typical reporting rate is 1 frame per 4 s. One of the other advantages the PMUs have over the traditional device is the time-synchronized measurements using GPS. This enables the possibilities for wide-area management and control applications. There are two standards that define the functioning of a PMU device and vendors have to comply to the standards to ensure interoperability. The first standard defines the requirements of PMUs as a measuring device [23]. This also has all the compliance test that each PMUs have to adhere to. The second standard is the communication standard [24] which gives a way to communicate, fairly encrypt, and decrypt the PMU measurements. Using this communication standard the timestamped measurements can be sent to a local

PDC which would accumulate the data for a few local PMUs. These local PDCs can then transfer all the data to a central PDC, further communicating it to control center-based applications. There is a long process before the data can be used for an application. This causes the possibilities of data anomalies which are discussed in Sections 11.7.1 and 11.7.2.

11.7.1 PMU data flow

As shown in Figure 11.6, the measurement starts with current transformers and potential transformers and makes a long journey before it is received at control centers for applications. This is discussed in detail in the document released by The North American Synchrophasor Initiative (NASPI), PMU Applications Requirements Task Force (PARTF) Framework [25].

The document provides a detailed overview of the data quality issues and different attributes of data and its definitions. To summarize the document there are potentially four points where the data could get anomalous.

- Within the PMU: CT and PT bias is a well-known phenomenon and it could cause an error in measurement, also the different classes of CT and PT utilized (measurement or protection), differences or errors within the PMU and its filtering and algorithms, or erroneous or missing timestamps associated with the measured grid conditions. Individual data points will be affected if the timestamps are missing but the whole data set can be affected if there are biases in CTs and PTs.
- At data aggregators: As seen in Figure 11.6, the data points are aggregated at at-least two phasor data concentrators before it gets to control center. This can induce error due to misalignment, erroneous compression, creation of duplicate data, or loss of data due to late deliveries. There can also be mislabeling or data losses at aggregators' end.
- Communications networks: The loss of communications network nodes can cause the loss of data packets, excess congestion can cause latency-delaying delivery beyond the acceptable time window of the phasor data concentrators, or data corruption can introduce erroneous values.

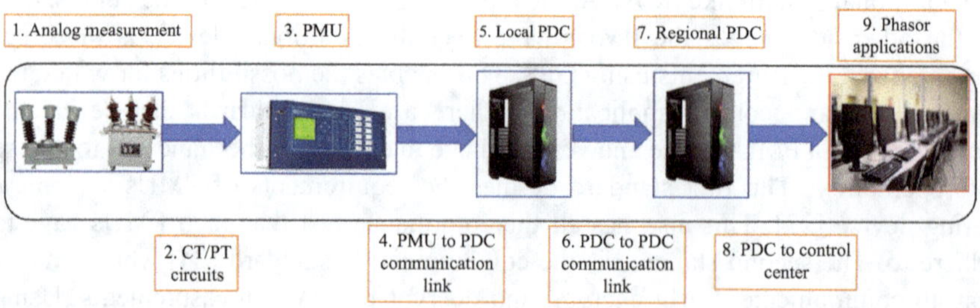

Figure 11.6 Synchrophasor data flow to application

- At application usage point: It is important that PMU-based applications are trained using enough historical data to make them efficiently produce accurate results. The data storage at the point of application can also cause bad results for the application if not done properly.

These issues might corrupt the PMU data making it unfit for use in a PMU application. Therefore a data preprocessing is required before it is used in an application.

11.7.2 Synchrophasor anomaly detection

The different kind of anomalies post different challenges for algorithms. A real-time algorithm like linear state estimator which takes into consideration each data point will give anomalous results for a missing data or outliers. Whereas an algorithm which takes a window of 30 or 60 samples as input might give acceptable result if there are a few outliers in the window. The PARTF framework discusses the need of algorithm to be trained on different kind of anomalies. It is very important while developing an algorithm to recognize the kind of anomalies that the algorithm can be resilient against. It is also necessary to preprocess the PMU data, identify the anomaly points, and flag the anomalies so that different applications can handle it accordingly.

Here, we present the architecture of synchrophasor anomaly detection (SyncAD) tool based on maximum likelihood estimator-based (MLE) unsupervised machine learning algorithm which uses statistical, clustering, and Prony analysis algorithms. This tool is used to detect the anomalies in the synchrophasor measurements. The architecture of the SyncAD tool is shown in Figure 11.7.

In the first step of the SyncAD algorithm, a set of base detectors linear regression, Chebyshev, and DBSCAN algorithm with loosely set thresholds analyze the PMU data window for anomalies. They make an assessment based on their functioning and compute an outlier score for each detected outlier [26]. Each base detector's outlier score has to be analyzed using the same frame of reference, this enables to compare the results of independent base detectors and increases the possibility of adding more base detectors in future if needed. The outlier scores are normalized using the expectation maximization (EM) algorithm. These normalized outlier scores are then passed through the MLE-Ensemble to compute the weight of each anomaly point. This learning of weights for each kind of base detector is the training process of the MLE-Ensemble. Finally, the inference algorithm uses the weights obtained from the MLE-Ensemble and flags the data anomalies in the new data set.

Since we are using statistical and clustering techniques we would come across a lot of event data points which may appear to be an outlier but are actually transients points between two steady states. The MLE-Estimator could have flagged these points as anomalies and to increase the precision of the algorithm and minimize false detection, we do a Prony analysis [27] on the data window to determine the transients. If the flagged anomalies lie during a transient window detected by the Prony analysis, the flagged anomaly is unflagged and said to be a good data.

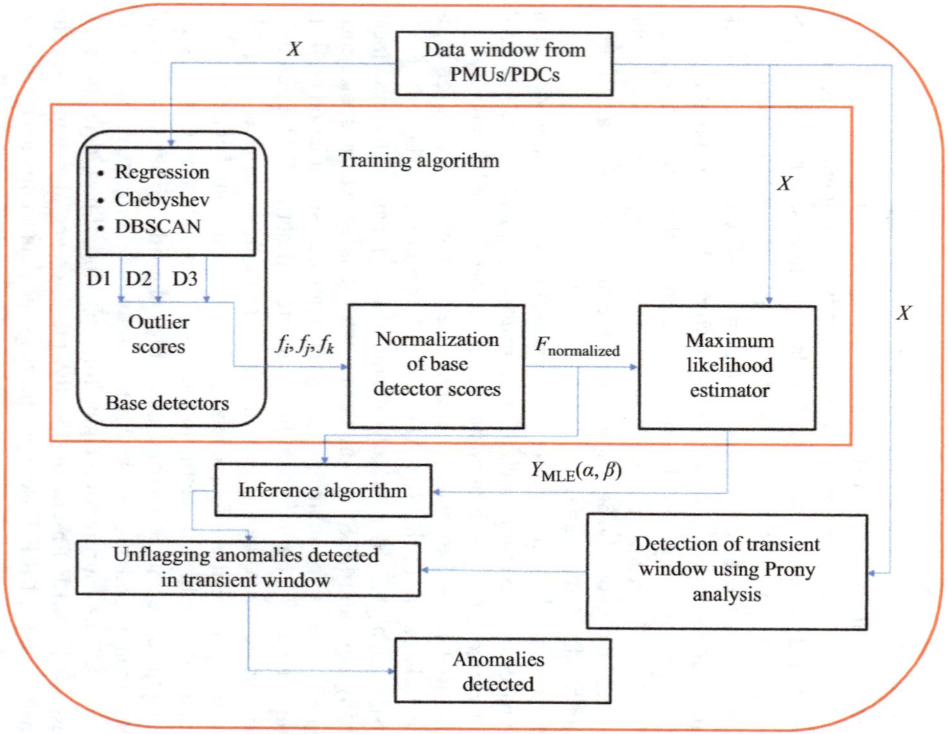

Figure 11.7 Anomaly detection architecture (Reproduced from [26])

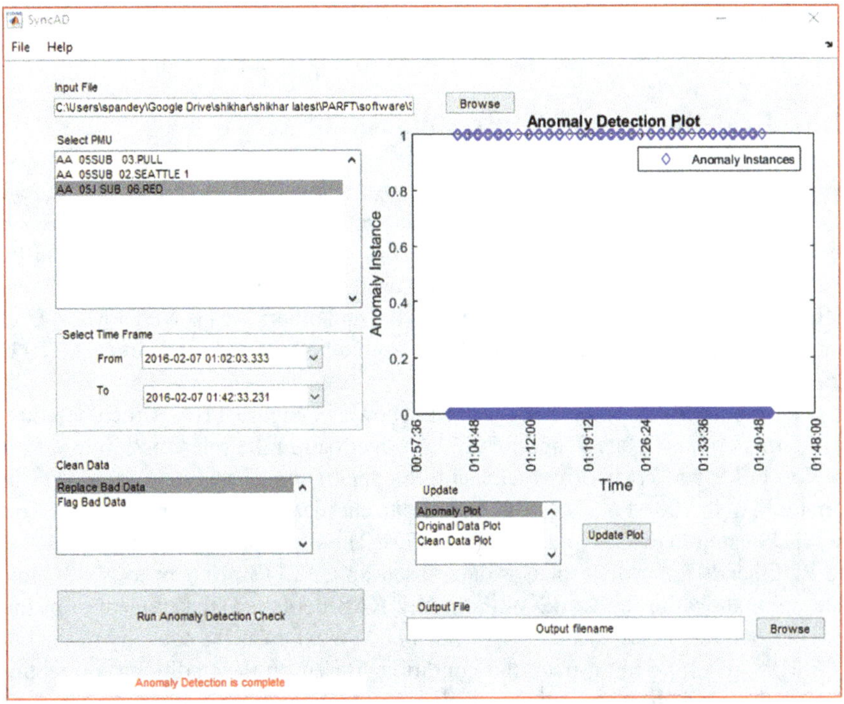

Figure 11.8 Synchrophasor anomaly detection (Sync-AD) tool (Reproduced from [26])

A screenshot of a test file detecting anomalies using SyncAD tool is shown in Figure 11.8. User can import a PMU file which can feature data from a number of PMUs. In the select PMU tab, the PMUs can be selected on which the user wishes to perform the anomaly detection test. The time frame can be selected for the analysis and if nothing is selected the tool takes the full time in the file by default. User also has the option to replace the selected bad data by imputing the average of the before and after data or they can simply flag the bad data instance. User can also plot the anomaly plot, original data plot, and clean data plot and finally they can save the output file with the desired name. Once the anomalies have been dealt with, the applications can use this clean data or flagged data file for analysis.

The anomaly detection tool can also be used to detect certain kind of cyber anomalies being injected in the PMU data. If the hacker decides to swap the PMU data of say, two PMUs, the tool will give an anomaly at the point of swapping, as there will be sudden change in the operating state of the PMU. This will alarm the operator of an event happening in the power system which can be cross-checked. If no such event is there the operator can check for cyber intrusion. If the attacker decides to inject certain outliers or tries to manipulate the data so that it looks like an event, such as opening of a line, faults, etc. This will also get detected as an anomaly for the particular PMU and if the no other PMUs in the vicinity detect the same, it will again alarm the operator of a possible cyber intrusion.

11.8 Real-time simulation results

The algorithm was validated through extensive testing with both offline and real-time simulations carried out on IEEE 14-bus system. It is assumed that the buses in the test system have PMUs, which are used as sensors, to read data and send the measurements to the controllers. Testbed used in this work has a total of nine PMUs, one of which is an SEL hardware PMU, and others are GTNET PMUs. Each controller connects to three of these PMUs from which it reads and parses C37.118 packets.

The modified IEEE 14 bus system is shown in Figure 11.9. It includes three wind farms on bus 2, bus 9, and bus 11. An overload in the transmission line from bus 7 to bus 9 was created by increasing the generation of the wind farm at bus 9 from 60 MW to 100 MW. The voltage and current measurements are obtained from the RTDS using hardware and GTNET PMUs. The PMU data are sent to SEL PDC and PDC sends the synchrophasor data based on C37.118 protocol to NS-3 simulated communication network. Without any RAS deployed, the transmission line from bus 6 to bus 7 carries a real power of 111 MW whereas the capacity of the line is 95 MW, resulting in an overload condition. Based on the traditional protection method, the wind farm generation, which is 180 MW, on bus 6 will be shed in order to solve this overload problem. It can protect the transmission line from the overload condition, but it sheds a large amount of renewable energy. In order to maximize the usage of wind power, the proposed RAS is implemented.

Cisco Fog is an infrastructure to connect applications at the edges of a network to the cloud in data centers. Fog uses a RESTful approach to describe and then manage this gap between the edges and the data center: network links, routers/switches, data, and computational nodes. This project successively evaluates Fog for suitability with advanced power infrastructures that can be very dynamic and include in-network processing (e.g., in a Fog coprocessor in a router).

The implementation of the controller is carried out using Cisco Fog Routers. The Fog routers are connected to RTDS and the PMUs using Ethernet cables. IEEE C37.118 protocol is used to communicate with the PMUs. Control commands are sent back to RTDS as text strings over its *ListenOnPort* interface using standard TCP Sockets. The network topology of the testbed is shown in Figure 11.10. The architecture of the testbed is shown in Figure 11.11.

A cyberattack on the system is simulated, as shown in Figure 11.12. The attack assumes that there is a malicious actor that has bypassed application security and is able to execute code and deploy applications within the same network or even on the same hardware as the application. It is assumed that this malicious actor can receive and read messages that are sent between application actors, and can also publish messages which can be received and read by other actors in the application. Typically, the RAS application detects overloads and calculates curtailment values

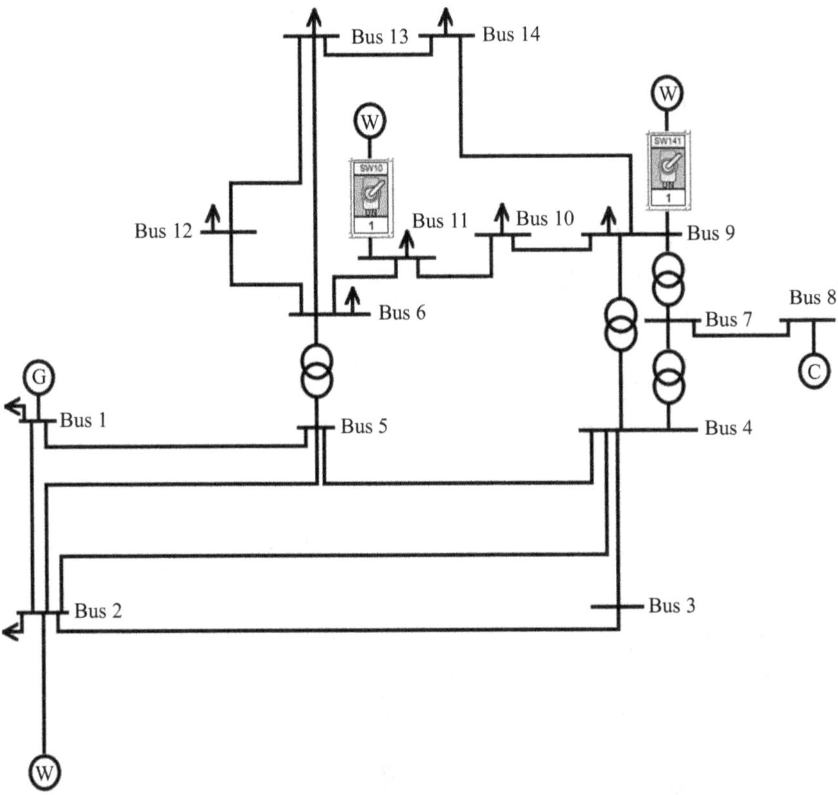

Figure 11.9 IEEE 14 bus system modeled in RTDS for real-time validation of the cyber-resilient RAS

by reading power generation values that are published by the DLSE component. The malicious actor carries out an attack by intercepting a message from the DLSE component, tampering with the data and publishing false values to the RAS component of the application. In case one of the wind power generators is producing excess power, the malicious actor can carry out an attack by artificially decreasing this excess value and publishing a value of power generation which is within operational limits. The RAS component interprets these malicious messages as authentic readings of power generation values and is then forced to perform the overload detection and curtailment calculation on these false values. This could lead to the application being unable to detect overloads or find an optimal curtailment value.

The effect of cyberattacks on the RAS application is also evaluated. A data injection attack on one of the nodes was simulated, and its effects on the different RAS algorithms are observed. This attack is simulated using a malicious actor that tampers with the input to the RAS application by changing the power generation values of one of the wind power generators. In this case, the true power generation

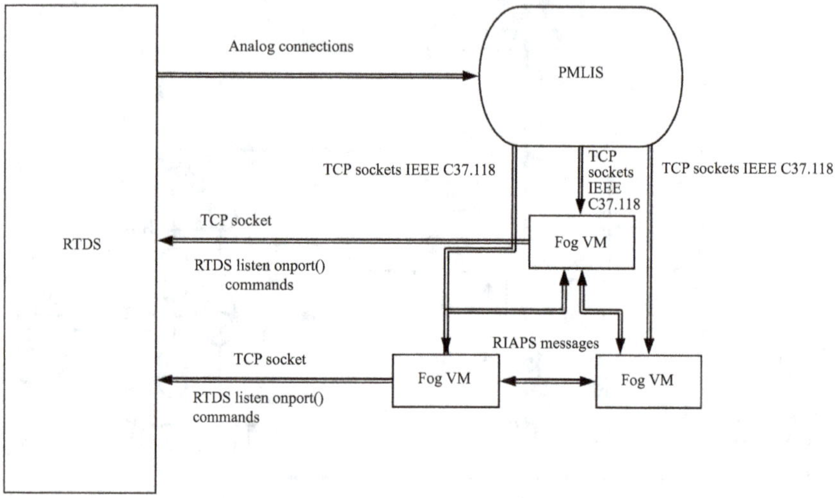

Figure 11.10 Network topology

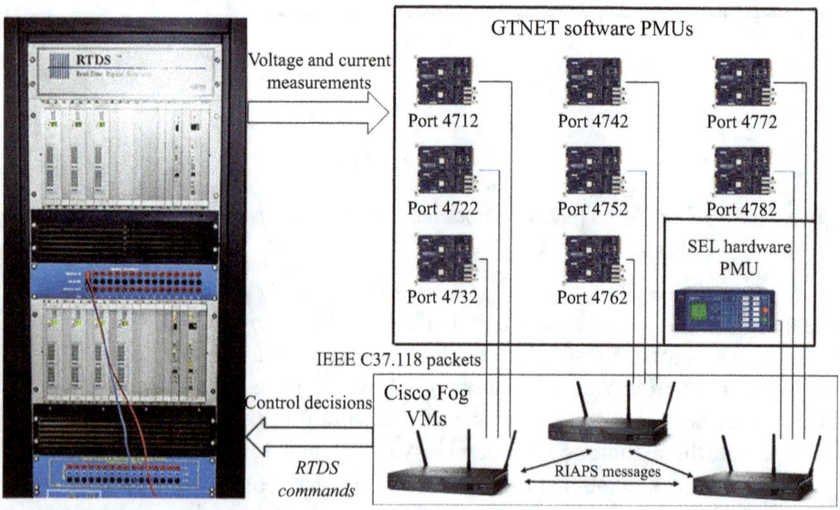

Figure 11.11 Realtime testbed architecture

value is 97.78 MW, which would be indicative of an overload. The malicious attacker intercepts this value and changes it to 60.0 MW, which would be well within the operational range for this system. These malicious values are sent as input to the RAS algorithm. The results of this experiment are shown in Table 11.1.

The results from Table 11.1 show that the centralized RAS algorithm is not able to detect an overload when the malicious attacker injects false values. The

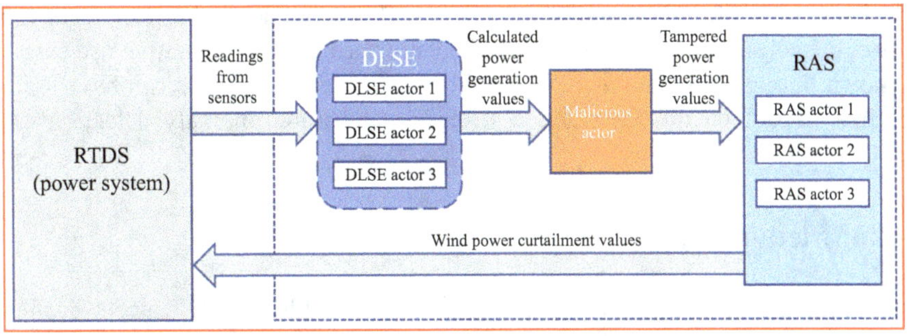

Figure 11.12 Simulation of a cyberattack on the RAS scheme

Table 11.1 Curtailment values generated by RAS algorithm

Case	Input power generation values in p.u.	Wind generation curtailment
Normal overload case	[1.4913, 0.3752, 0.9778, 0.7172]	[0.3133, 0.2477, 0.9432]
Attack simulation	[1.4913, 0.3752, 0.6000, 0.7172]	N/A

centralized RAS algorithm does not generate a curtailment value in this case and no action is taken. This would lead to tripping of line 6-11.

11.9 Summary

This chapter discusses the design, modeling, and implementation of PMU-based RAS for minimizing wind curtailment. Existing curtailment process is usually slow and hard coded. Owing to faster refresh rates of PMU, the proposed RAS is automatic and adapts to changing system conditions. The various modules and components of the automatic RAS are discussed in detail. The prosed PMU based RAS is modeled as an optimization problem with PMU data as the inputs. The algorithm issues control commands to curtail the amount of wind energy, thereby reducing the congestion on the transmission lines. Such a scheme ensures minimum curtailment of the wind, in addition to being fast and scalable. The RAS algorithm is fully implemented and validated on the proposed testbed consisting of real-time digital simulator (RTDS), PMUs, and Cisco Fog routers. The effectiveness of the proposed approach is validated through online simulations on IEEE 14-bus test system.

This chapter also outlines the sources of PMU data anomalies. The architecture of SyncAD tool based on MLE-based unsupervised machine learning algorithm which uses statistical, clustering, and Prony analysis algorithms is presented. This tool is used to detect the anomalies in the synchrophasor measurements. The effect of cyberattack on RAS scheme is investigated. PMU data anomalies can negatively impact the control algorithm, in which, they are used. As an illustrative use case, it is shown through simulations that, the false data injection attacks can have a detrimental impact on the PMU-based RAS operation and anomaly detection tool can help.

Acknowledgments

The authors thank US Department of Energy Advanced Research Projects Agency-Energy (ARPA-E), Power System Engineering Research Center (PSERC), and RTE-France for financially supporting this research work. They are also thankful to Dr Y. Wu, Dr R. Liu, and M. Zhou for their contributions toward this work.

References

[1] Liu R, Srivastava A, Askerman A, *et al.* Decentralized state estimation and remedial control action for minimum wind curtailment using distributed computing platform. In: *2016 IEEE Industry Applications Society Annual Meeting*, Portland, OR; 2016, pp. 1–9.

[2] Liu R, Srivastava AK, Bakken DE, *et al.* Decentralized state estimation and remedial control action for minimum wind curtailment using distributed computing platform. *IEEE Transactions on Industry Applications*. 2017; 53(6):5915–5926.

[3] Krishnan VVG, Liu R, Askerman A, *et al.* Resilient cyber infrastructure for the minimum wind curtailment remedial control scheme. In: *2017 IEEE Industry Applications Society Annual Meeting*, Cincinnati, OH; 2017, pp. 1–7.

[4] Krishnan VVG, Gopal S, Liu R, *et al.* Resilient cyber infrastructure for the minimum wind curtailment remedial control scheme. In: *IEEE Transactions on Industry Applications*; 2019;55(1):943–953

[5] Matevosyan J. Wind power integration in power systems with transmission bottlenecks. In: *IEEE Power Engineering Society General Meeting, 2007*, IEEE, Tampa, FL; 2007, pp. 1–7.

[6] NERC. *Remedial Action Scheme Definition Development*. North American Reliability Corporation, Atlanta, GA; 2014.

[7] Yao Z, Vinnakota VR, Zhu Q, *et al.* Forewarned is forearmed: an automated system for remedial action schemes. *IEEE Power and Energy Magazine*. 2014;12(3):77–86.

[8] Shrestha A, Cecchi V, and Cox RW. Optimal power flow-based generation shedding for dynamic remedial action scheme. In: *2016 IEEE/PES Transmission and Distribution Conference and Exposition (TD)*, Dallas, TX; 2016, pp. 1–5.

[9] Weckesser T, Jóhannsson H, and Ostergaard J. Real-time remedial action against aperiodic small signal rotor angle instability. *IEEE Transactions on Power Systems*. 2016;31(1):387–396.

[10] Maram MD, and Amjady N. Event-based remedial action scheme against super-component contingencies to avert frequency and voltage instabilities. *IET Generation, Transmission & Distribution*. 2014;8(9):1591–1603.

[11] Bird L, Cochran J, and Wang X. Wind and solar energy curtailment: experience and practices in the United States. US National Renewable Energy Laboratory, NREL/TP-6A20-60983. 2014; p. 3.

[12] Bird L, Cochran J, and Wang X. *Wind and Solar Energy Curtailment: Experience and Practices in the United States*. Golden, CO: NREL; 2014.

[13] Gu Y, and Xie L. Fast sensitivity analysis approach to assessing congestion induced wind curtailment. *IEEE Transactions on Power Systems*. 2014; 29(1):101–110.

[14] Cutsem TV, Horward JL, and Ribbens-Pavella M. A two-level static state estimator for electric power systems. *IEEE Transactions on Power Apparatus and Systems*. 1981; PAS-100(8):3722–3732.

[15] Yang T, Sun H, and Bose A. Transition to a two-level linear state estimator - Part I: architecture. *IEEE Transactions on Power Systems*. 201;26(1):46–53.

[16] Yang T, Sun H, and Bose A. Transition to a two-level linear state estimator - Part II: algorithm. *IEEE Transactions on Power Systems*. 2011;26(1):54–62.

[17] Huang Q, Schulz NN, Srivastava AK, *et al.* Distributed state estimation with PMU using grid computing. In: *IEEE Power Energy Society General Meeting*, Calgary, AB; 2009, pp. 1–7.

[18] Korres GN. A distributed multiarea state estimation. *IEEE Transactions on Power Systems*. 2011;26(1):73–84.

[19] Xie L, Choi DH, Kar S, *et al.* Fully distributed state estimation for wide-area monitoring systems. *IEEE Transactions on Smart Grid*. 2012;3(3):1154–1169.

[20] Soni S, Bhil S, Mehta D, *et al.* Linear state estimation model using phasor measurement unit (PMU) technology. In: *2012 9th International Conference on Electrical Engineering, Computing Science and Automatic Control (CCE)*, Mexico City; 2012, pp. 1–6.

[21] Schloegel K, Karypis G, and Kumar V. A new algorithm for multi-objective graph partitioning. In: *European Conference on Parallel Processing*. Springer, Berlin, Heidelberg; 1999, pp. 322–331.

[22] Karypis G, and Kumar V. A fast and high quality multilevel scheme for partitioning irregular graphs. *SIAM Journal on Scientific Computing*. 1998; 20(1):359–392.

[23] IEEE Standard for Synchrophasor Measurements for Power Systems. IEEE Std C371181-2011 (Revision of IEEE Std C37118-2005). December 2011.

[24] Khan R, McLaughlin K, Laverty D, *et al.* IEEE C37. 118-2 synchrophasor communication framework-overview, cyber vulnerabilities analysis and performance evaluation. In: *International Conference on Information Systems Security and Privacy (ICISSP)*; 2016, pp. 167–178.

[25] Methodology for examining the impact of error and weaknesses in PMU data on operational applications. North American Synchro-Phasor Initiative (NASPI) PMU Application Requirements Task Force; 2011.

[26] Banerjee P, Pandey S, Srivastava A, *et al.* Testing and validation of synchrophasor devices and applications. In: *Power System Grid Operation Using Synchrophasor Technology*. Springer; 2019. pp. 41–75.
[27] de la O Serna JA. Synchrophasor estimation using Prony's method. *IEEE Transactions on Instrumentation and Measurement*. 2013;62(8):2119–2128.

Chapter 12

Power system controlled islanding schemes

Lei Ding[1] *and Yichen Guo*[1]

In an interconnected power grid, the load and generation should be kept in balance at all times. However, the load varies all through the day and by seasons, when the load generation balance is interrupted and cannot be compensated by adjusting generation, load shedding or generation tripping will have to be implemented, otherwise, cascading outages can occur.

There are certain mechanisms to prevent or mitigate contingencies, such as circuit breakers installed in transmission lines, and remedial action schemes consisting of a series of actions to prevent cascading outages. When the contingency is severe or a series of contingencies occur, these kinds of measures may not be effective and cascading outages might occur. In this situation, controlled islanding provides an option of last resort to prevent the spreading of cascading outage by splitting the power system into a group of smaller, islanded power systems or islands.

Controlled islanding of a power system is an efficient corrective measure for limiting system blackouts after a large disturbance has occurred. Controlled islanding can be used to cope with the occurrence and consequences of blackout's different power system extremes, such as undamped oscillations, voltage collapse, cascading trips, etc. Depending on the operating condition, generators in a power system can be separated into several groups in which generators will swing together after disturbances. The controlled islanding scheme separates the system into several small subsystems, or islands, by cutting the transmission lines between different generator groups.

This chapter provides a detailed demonstration of controlled islanding schemes in a way that solves three critical problems:

1. How to predict the generator groups, that is, how to group generators?
2. Where are the islanding boundaries, that is, where to split?
3. When can the islands be split from the system, that is, when to split?

[1]School of Electrical Engineering, Shandong University, Jinan, China

12.1 Generator grouping prediction

Generator grouping is of great importance in determining an islanding strategy that generates stable islands. Generator grouping mode or which generators tend to swing together should be online identified and then islanding boundaries can be optimized between those groups that are going out of step. Therefore, the identification of coherent generator groups is the key step in controlled islanding, and an improperly identified generator group, if allocated into one island, may lead to an out of step condition and angular instability of the island. References [1–3] apply slow-coherency analysis to identify generator groups in order for island boundaries to be optimized. However, slow coherency made two bold assumptions [1]:

1. The coherent groups of generators are independent of the size of the disturbance.
2. The coherent groups are independent of the level of detail used in modeling the generating unit.

Based on the first assumption, the impact of large disturbances has been overlooked. However, the power system is a nonlinear system, and its dynamics near SEP are different from those far away from SEP. Moreover, when controlled islanding is required, the system may have already experienced large disturbance and been operated in an extreme condition, or in other words, the dynamics of the system are more likely far away from a SEP. So, the dynamics of the system may be closely relevant to the size of the disturbance. As a result, the groups identified by slow-coherency analysis may not match the actual groups of generators when the system is subject to large disturbances such as cascading outages. This drawback is rooted in considering system dynamics only near a *stable equilibrium point* (SEP) and ignoring the dynamics far away from the SEP. In this part, a generator grouping method is introduced for the identification of generator groups in order for removing the "neglecting disturbance size" assumption, which means it can be applied to both small disturbance and large disturbance scenarios, and can correctly identify generator groups.

12.1.1 Nonlinear system preliminaries

The differential equations of an n-generator electric power system represented in the classic model are:

$$\begin{aligned}
\dot{\delta}_i &= \omega_i \\
\dot{\omega}_i &= \frac{1}{M_i}(P_{mi} - P_{ei}) \\
&= \frac{1}{M_i}\left(P_{mi} - \sum_{j=1}^{n} E_iE_jB_{ij}\sin\left(\delta_i - \delta_j\right) - \sum_{j=1}^{n} E_iE_jG_{ij}\cos\left(\delta_i - \delta_j\right)\right) \\
i &= 1, 2, \ldots, n
\end{aligned} \tag{12.1}$$

where δ_i and ω_i represent the power angle and angular velocity of the *i*th generator, P_{mi} and P_{ei} represent the mechanical power and electric power of *i*th generator, E_i is electromotive force (emf) of the *i*th generator, and G_{ij} and B_{ij} are the conductance and admittance, respectively, between the emfs of generators *i* and *j*.

Suppose $x = [\delta_1, \ldots, \delta_n, \omega_1, \ldots, \omega_n]$ and $\varnothing(x, t)$ is the solution curve (i.e., the trajectory) of system (12.1) (see Figure 12.1). x_0 is called the stable equilibrium point (SEP) of system (12.1) since x_0 is a hyperbolic equilibrium point of the system and all the eigenvalues of its corresponding Jacobian have negative real parts. x_u is called the unstable equilibrium point (UEP), which is also a hyperbolic equilibrium point of the system, but its Jacobian has at least one eigenvalue with a positive real part. In accordance with the number of eigenvalues with positive real parts, the UEP x_u can be categorized, x_u is called a type-k UEP if its corresponding Jacobian has exactly k eigenvalues with positive real parts. According to the theory of nonlinear dynamical system [4], Foran SEP x_0, the stability region is defined as

$$A(x_0) = \left\{ \lim_{t \to +\infty} \varnothing(x, t) \to x_0 \right\}$$

The definition of stability region implies that if the initial state of post-disturbance system lies inside the stability region, the postdisturbance trajectory will be stable and converge to x_0. On the other hand, if the initial state of post-disturbance system lies outside the stability region, the postdisturbance trajectory will go unstable with time increasing.

The boundary of stability region $A(x_0)$ is called the stability boundary of x_0, and will be denoted by $\partial A(x_0)$. The theory of nonlinear dynamical system tells that $\partial A(x_0)$ is the union of stable manifolds of a set of type-1 UEPs whose unstable manifolds include x_0:

$$\partial A(x_0) = \cup \{ W^s(x_u) : x_0 \in W^u(x_u) \}$$

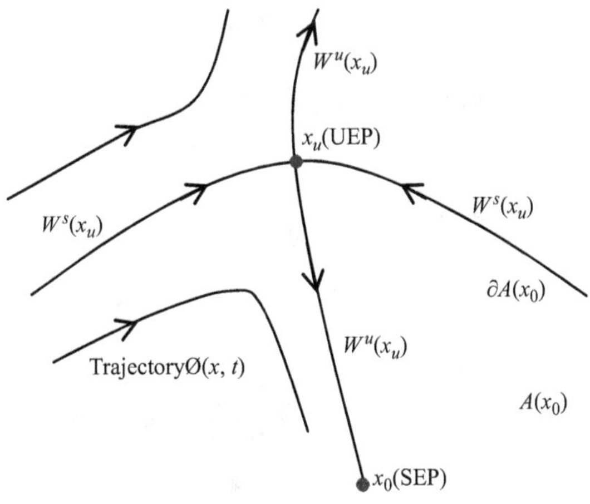

Figure 12.1 Nonlinear dynamics of power system

where the stable manifolds $W^s(x_u)$ and unstable manifolds $W^u(x_u)$ on point x_u are defined respectively as

$$W^s(x_u) = \{x \in R^n : \varnothing(x,t) \rightarrow x_u, t \rightarrow +\infty\}$$
$$W^u(x_u) = \{x \in R^n : \varnothing(x,t) \rightarrow x_u, t \rightarrow -\infty\}$$

Assume x_u is a type-1 UEP seated on the stability boundary. The 2n-dimensional space is split into two parts by the stability boundary ∂A (see Figure 12.2). The unstable manifold $W^u(x_u)$ of the type-1 UEP x_u is also divided into two portions by the stability boundary, one portion lies inside of the stability region (denoted by $W^u_-(x_u)$) and the other portion lies outside of the stability region (denoted by $W^u_+(x_u)$). Both portions are one-dimensional manifolds who intersect the stability boundary $\partial A(x_0)$ vertically at x_u. While the stable manifold $W^s(x_u)$ of the type-1 UEP x_u is a $2n-1$ dimensional manifold on the stability boundary $\partial A(x_0)$. The SEP x_0 of the postdisturbance power system is seated inside the stability region $A(x_0)$. The cases of type-k UEP ($k > 1$) are similar to that of type-1 UEP, the above analysis of type-1 UEP can also be extended to type-k UEP.

12.1.2 Grouping mode prediction algorithm

Before the occurrence of the disturbance, the system is operated at a predisturbance SEP. Assume a disturbance to occur at $t = t_0$ in the form of a single three-phase fault. The disturbance-on trajectory will leave the predisturbance SEP and approach the stability boundary ∂A (i.e., the stable manifold $W^s(x_u)$ of x_u). We will represent the size of disturbances in the form of the duration of a single three-phase faults. A short duration of fault indicates small disturbance and a long duration of fault represents large disturbance. In accordance with fault duration increasing

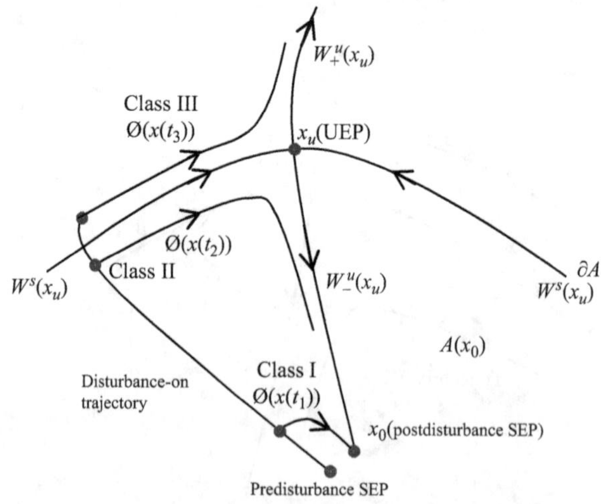

Figure 12.2 Prediction of generator grouping mode

$(t = t_1, t_2 \text{ or } t_3)$, there exist three possible postdisturbance trajectories (Class I–Class III) shown in Figure 12.2.

Class I assumes that the fault ends in a sufficiently short time after it occurs $(t = t_1)$. The postdisturbance trajectory $\varnothing(\mathrm{x}(t_1))$ will be attracted by the post-disturbance SEP, and finally reach postdisturbance SEP. In this scenario, the generator grouping mode is said to be dominated by the SEP x_0.

Class II assumes that the fault ends before the disturbance-on trajectory across the stability boundary $(t = t_2)$. The initial state of postdisturbance is close to the stability boundary but still inside the stability region $\mathrm{A}(x_0)$. The postdisturbance trajectory $\varnothing(\mathrm{x}(t_2))$ will first move along the direction of stable manifold until reaching the neighborhood of UEP and then move along the direction of the portion of unstable manifold lies inside of the stability region (i.e., $\mathrm{W}^{\mathrm{u}}_{-}(x_u)$). The direction of $\mathrm{W}^{\mathrm{u}}_{-}(x_u)$ can be used to predict the postdisturbance trajectory and the generator grouping mode is dominated by the UEP x_u.

Class III assumes that the fault ends after the disturbance-on trajectory across stability boundary $(t = t_3)$. The initial state of postdisturbance is close to the stability boundary but lies outside the stability region. The postdisturbance trajectory $\varnothing(\mathrm{x}(t_3))$ will move along the portion of unstable manifold outside the stability region (i.e., $\mathrm{W}^{\mathrm{u}}_{+}(x_u)$), the angles of some of the generators will increase continually until the system goes out of step. It is clear that the direction of $\mathrm{W}^{\mathrm{u}}_{+}(x_u)$ can be used to predict the postdisturbance trajectory and the generator grouping mode is dominated by the UEP x_u.

To conclude, the trajectories of Class I will follow the following pattern:

$$\text{initial state} \rightarrow \text{slow mode} \rightarrow \text{SEP}$$

While the trajectories of Class II and Class III will follow a different pattern:

$$\text{initial state} \rightarrow W^s(x_u) \rightarrow \text{neighborhood of UEP} \begin{array}{l} \nearrow W^u_-(x_u) \\ \searrow W^u_+(x_u) \end{array}$$

Based on the above analysis, the postdisturbance trajectory of Class I is dominated by the SEP x_0, while the postdisturbance trajectories of Class II and Class III are dominated by the UEP x_u. With the increase of the disturbance size, the type of the postdisturbance trajectory gradually changes from the trajectory of Class I to that of Class II and then to the trajectory of Class III; meanwhile, the impact of the SEP keeps decreasing and the UEP becomes more influential.

It should be noted that the two-norm of the vector field $\|\dot{x}\|_2$ is zero at SEP or UEP. In addition, $\|\dot{x}\|_2$ increases along the trajectory moving away from the SEP or UEP and decreases along the trajectory moving towards SEP or UEP. So that in Class I, $\|\dot{x}\|_2$ decreases along the postdisturbance trajectory moving toward to the SEP and becomes 0 when reaching the SEP. Meanwhile, in Class II and Class III, $\|\dot{x}\|_2$ will first decrease along the postdisturbance trajectory moving toward to the UEP and then increase moving away from the UEP. Which means there exists a local minimum value of $\|\dot{x}\|_2$ in the whole process, which is in the neighborhood of the UEP. This property will be used to develop an algorithm for determining which type the postdisturbance trajectory is and which equilibrium point (SEP or UEP) is dominating.

To determine the generator grouping mode for each trajectory type, the original power system can be linearized at the dominant equilibrium point (SEP or UEP). Suppose the linearized system has n eigenvalues λ_i on the equilibrium point. The result of eigenvalue analysis is presented as follow:

$$\dot{Y} = \lambda_r Y, Y \in R^n \quad \text{with } \lambda_r = \begin{bmatrix} \lambda_1 & & \\ & \ddots & \\ & & \lambda_n \end{bmatrix} \tag{12.2}$$

The eigenvector v_i is corresponding to the eigenvalue λ_i. Consider these two cases:

- Small signal analysis on the SEP can be applied if scenarios like Class I happen (see Figure 12.3). The eigenvectors corresponding to the eigenvalues with large magnitudes construct the fast mode. The generator grouping mode is based on the eigenvectors $v_1, v_2, \ldots, v_m$ corresponding to the eigenvalue $\lambda_1, \lambda_2, \ldots, \lambda_m$ having the m smallest magnitudes, that is, the slow mode of oscillation. The number of generator groups is determined to be m + 1.
- However, if the scenario is detected to be like Class II or Class III (see Figure 12.4), the unstable subspace of the system should be used to predict the unstable manifold of the UEP, where eigenvalues $\lambda_1', \ldots, \lambda_k'$ have positive real parts. The eigenvectors corresponding to the eigenvalues $\lambda_1', \ldots, \lambda_k'$ construct unstable subspaces of the system and those corresponding to the eigenvalues with negative real parts construct stable subspaces of the system. Suppose

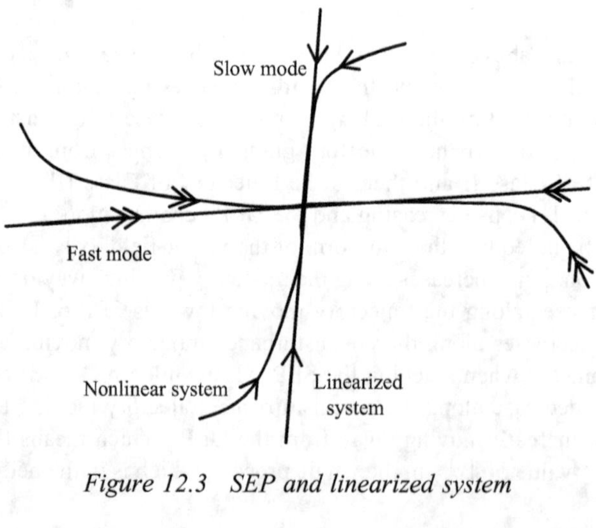

Figure 12.3 SEP and linearized system

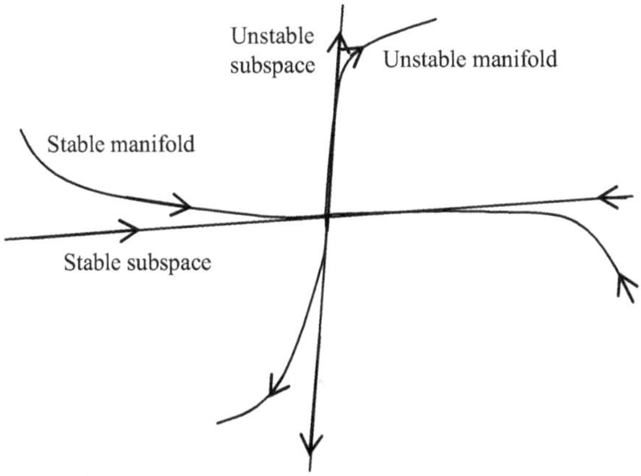

Figure 12.4 UEP and linearized system

$v_1^l, \ldots, v_k^l$ is an unstable subspace of the system, whose corresponding state variables $y_i = y_i(0)e^{\lambda_i t}$ will increase to infinite along with time. If these state variables denote angle differences between generator groups, then the corresponding groups will go out of step from each other. The generator grouping mode is given by the eigenvectors $v_1^l, \ldots, v_k^l$ and the number of generator groups is $k + 1$.

An algorithm (flowchart in Figure 12.5) to predict the generator grouping mode is presented next:

1. At each time instant t, based on the measured data by PMU, the current state of system $x(t)$ are obtained and sent to the control center.
2. Set a maximum integration time T_s. Integrate the postdisturbance system for a period of T_s to find the point where the 2-norm of its vector field $\|\dot{x}\|_2$ reaches the first local minimum x_0^+. If no local minimum can be found, go to (6).
3. Use x_0^+ as initial guess to solve the following algebraic equations (12.3) using Newton–Raphson method. Let the solution be x_{co}^+.

$$\begin{aligned} 0 &= \omega_i - \omega_n \\ 0 &= \frac{1}{M_i}(P_{mi} - P_{ei}) - \frac{1}{M_n}(P_{mn} - P_{en}) \end{aligned} \tag{12.3}$$

4. If the same result x_{co}^+ are obtained for two consecutive time instants, the point dominating generator grouping mode is considered to be an UEP, which is located at x_{co}^+.
5. Update the generator grouping information by calculating the linearized system on the UEP and return to (1).
6. Use the last point of integration as initial guess to solve (12.3) and name the solution as x_{co}^*.

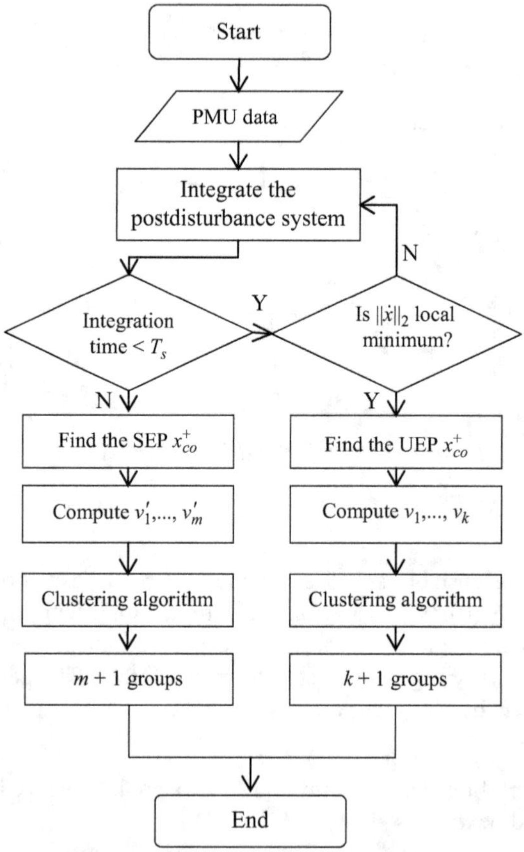

Figure 12.5 Flowchart for predicting the generator grouping mode

7. If the same result x_{co}^{*} are obtained for two consecutive time instants, the point dominating generator grouping mode is considered to be a SEP, which is located at x_{co}^{*}.
8. Update the generator grouping information by calculating the linearized system on the SEP and return to (1).

Note that, step 5 or step 7 only needs to be implemented when a new UEP or SEP is found. If data sampling points result in the same UEP (or SEP), they will have the same generator grouping mode.

An algorithm to predict the generator grouping according to the SEP (step 8) is outlined:

1. Compute the eigenvalues $\lambda_1, \lambda_2, \ldots, \lambda_{m+1}$ having the $m+1$ smallest magnitudes and their corresponding eigenvectors $v_1, v_2, \ldots, v_{m+1}$ of the matrix L.
2. Let $J \in R^{n\times(m+1)}$ be the matrix containing the vectors $v_1, v_2, \ldots, v_{m+1}$ as columns. Let $y_i \in R^{m+1}$ be the vector corresponding to the ith row of J.

3. Cluster the generator $y_i \in R^{m+1}$ into $m+1$ groups, using a clustering algorithm, such as the k-medoids algorithm.

It should be noted that, according to the property of power system, λ_{m+1} and λ'_{k+1} are always zero and their corresponding eigenvectors v_{m+1} and v'_{k+1} are always "all-ones" column vectors. v_{m+1} and v'_{k+1} are taken into account only to meet the requirements of k-medoids algorithm.

12.1.3 Case studies

In this section, two test systems, that is, the IEEE 9-bus system and IEEE 118-bus system, are used to validate the proposed generator grouping algorithm. For each test system, the generator grouping algorithm is integrated into optimization of islanding boundaries for the minimal power-flow disruption [5]. In the first two test systems, the resulting generator grouping modes are compared with those obtained using as low coherency analysis-based algorithm [1]. The purpose of the comparisons is to demonstrate that the influences from UEPs on generator grouping cannot be neglected.

Scenario A: IEEE 9-bus system

A permanent three-phase short-circuit is applied at node 7 and cleared by opening line 5-7 at $t = 0.32$ s. Time interval of data sampling is considered as 0.02 s.

The proposed algorithm, described in Sections 12.1.2, is then applied and two generator groups {1, 3} and {2} are found, which are used to obtain an islanding strategy that forms two islands {2, 5, 7, 8} and {1, 3, 4, 6, 9}. The cutset is marked in Figure 12.6 by a dash-dotted line.

Application of the slow coherency analysis based method results in another coherent generator groups {1} and {2, 3}, and accordingly two different islands{1, 4} and {2, 3, 5, 6, 7, 8, 9}. This cut-set is marked in Figure 12.6 by a dashed line.

Controlled islanding using these two strategies is simulated as shown in Figure 12.7. It is clear that the island {2, 3, 5, 6, 7, 8, 9} formed by slow coherency based method will go out of step, while the proposed grouping algorithm will keep synchronization within each island.

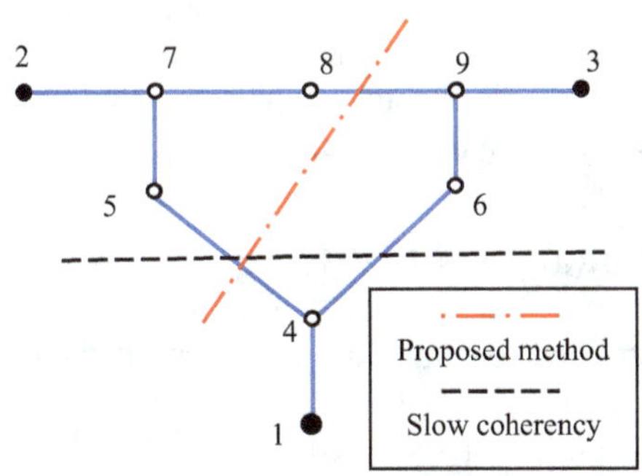

Figure 12.6 The islanding solution of the IEEE 9-bus system

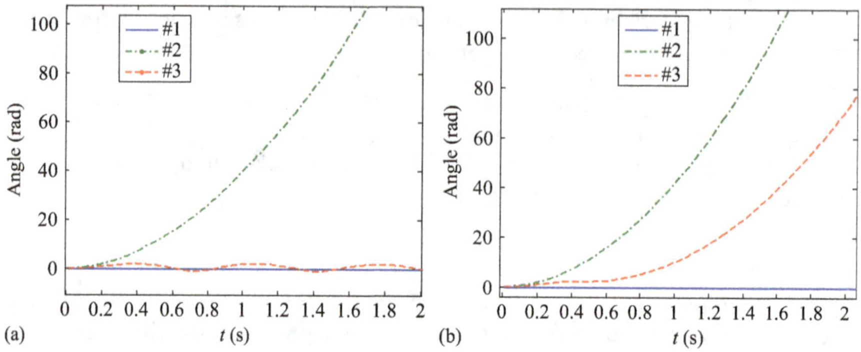

Figure 12.7 The simulation result of power angles: (a) proposed method and (b) slow coherency-based method

Table 12.1 Comparison of generator groups on 118-bus system

	Group 1	Group 2
Proposed method	10, 12, 25, 26, 31, 46, 49, 54, 59, 61, 65, 66, 69, 80	87, 89, 100, 103, 111
Slow coherency	10, 12, 25, 26, 31	46, 49, 54, 59, 61, 65, 66, 69, 80, 87, 89, 100, 103, 111

Scenario B: IEEE 118-bus system

Consider a complex disturbance with two three-phase temporary faults occurring at $t = 0.01$ s on line 80-99 and line 68-81, respectively. They are both cleared at $t = 0.49$ s. The controlled islanding action is taken place at 0.94 s.

Apply the proposed generator grouping algorithm to determine the set of generator groups, which are then compared with those obtained using the slow coherency analysis-based algorithm in Tables 12.1 and 12.2. Figure 12.8 and Figure 12.9 compare the performances of controlled islanding based on generator grouping results from two algorithms. One of the islands formed by slow coherency analysis-based algorithm is not stable, while both islands created based on the proposed algorithm will keep synchronization.

12.2 Islanding boundary identifying

In this section, a constrained spectral clustering controlled islanding algorithm (the SCCI algorithm) will be presented. The results of generator grouping obtained by the classical linearized generator swing equation will be served as pair-wise constraints in the SCCI algorithm, in which every node is grouped based on *constrained spectral clustering*. This constrained spectral clustering uses power flow

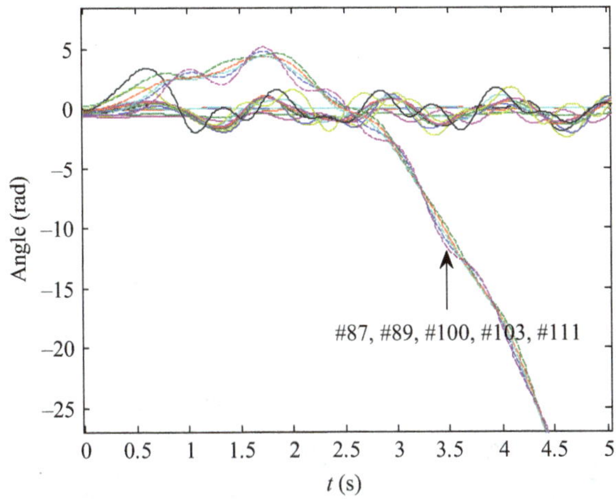

Figure 12.8 The simulation result of power angles (proposed method)

Table 12.2 Comparison of islanding boundary on 118-bus system

	Cutset
Proposed method	77-82, 96-97, 80-96, 98-100, 80-99
Slow coherency	15-33, 19-34, 30-38, 23-24

data to produce an islanding solution with minimal power-flow disruption. Therefore, the SCCI algorithm can identify, in real time, an islanding boundary solution that has minimal power-flow disruption and satisfies the constraint of generator coherency.

12.2.1 Controlled islanding and spectral clustering

In this section, some basic concepts of graph theory are introduced. The controlled islanding problem is then defined as a constrained optimization problem that is converted into a graph-cut problem. A possible method for solving this type of problem, spectral clustering, is introduced.

12.2.1.1 Graph theory preliminaries

In graph theory, an undirected graph-model $\mathbf{G}(\mathbf{V}, \mathbf{V_G}, \mathbf{E}, \mathbf{W})$ can be used to describe an m-generator and n-bus power system. In the above graph-model, the node set $\mathbf{V} = \{v_1, \ldots, v_n\}$ and the edge set $\mathbf{E}$, with elements e_{ij} $(i, j = 1, \ldots, \text{n})$, denote the buses and transmission lines, respectively. $\mathbf{V_G}$ is a subset of the node set

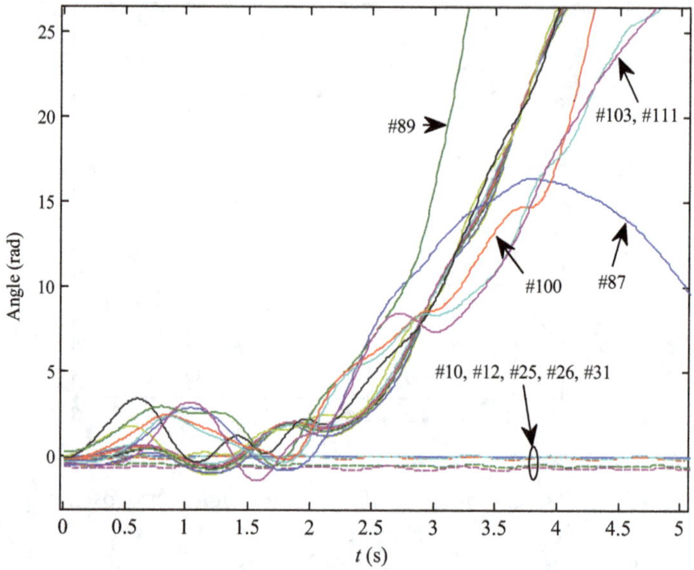

Figure 12.9 The simulation result of power angles (slow coherency)

V that contains only those buses with generators directly connected to them. The matrix **W** is a set of edge weights.

For convenience, only the bisection case is presented in this chapter. A graph **G** is split into two subgraphs $\mathbf{G}_1(\mathbf{V}_1, \mathbf{V}_{\mathbf{G1}}, \mathbf{E}_1, \mathbf{W}_1)$ and $\mathbf{G}_2(\mathbf{V}_2, \mathbf{V}_{\mathbf{G2}}, \boldsymbol{E}_2, \boldsymbol{W}_2)$ by removing the edges connecting these two subgraphs, with each subgraph representing a subsystem of the original power system. Here, $\mathbf{V}_1$ and $\mathbf{V}_2$ are disjoint subsets of V, that is, $\mathbf{V}_1 \cap \mathbf{V}_2 = \varnothing$ and $\mathbf{V}_1 \cup \mathbf{V}_2 = \mathbf{V}$. In the same way, $\mathbf{V}_{\mathbf{G1}}$ and $\mathbf{V}_{\mathbf{G2}}$ are defined as two disjoint subsets of V_G, while they are also subsets of $\mathbf{V}_1$ and $\mathbf{V}_2$, respectively.

The set of edges removed to separate these subgraphs is called the *cutset*. The sum of the weights of the edges within this cutset is called the *cut*, which is defined as [6]

$$cut(\mathbf{V}_1, \mathbf{V}_2) = \sum_{i \in \mathbf{V}_1, j \in \mathbf{V}_2} w_{ij} \tag{12.4}$$

The graph-cut problem is then defined as finding the cutset that bisects the graph with minimum cut [7]. It is common that the islanding solution, for a large power system, will require the system to be split into more than two islands; this can be achieved using recursive bisection [7].

12.2.1.2 Controlled islanding

Generator coherency

$$\text{Subsystem } G_1 \ \vec{A} = \left[\begin{array}{c|c} A_{11} & A_{12} \\ \hline A_{21} & A_{22} \end{array}\right] \leftarrow \text{Subsystem } G_2$$

*Figure 12.10 Division of the system matrix **A** into two submatrices A_{11} and A_{22}*

Based on the classical linearized generator swing equation, with damping neglected, the linearized second-order dynamic model of an m-generator power system can be expressed in the following matrix form [8]:

$$\ddot{x} = Ax \tag{12.5}$$

where $x = [\Delta\delta_1, \ldots, \Delta\delta_m]^T$ and $\Delta\delta$ is the generator angle deviation from a steady-state operating point δ_0 and **A** is the system state matrix. Separating the generators into two groups is equivalent to an arbitrary division of matrix **A** into two submatrices $\mathbf{A_{11}}$ and $\mathbf{A_{22}}$ that represent the subsystems $\mathbf{G_1}$ and $\mathbf{G_2}$ (see Figure 12.10) [9].

The sum of the Frobenius norms of the off-diagonal submatrices A_{12} and A_{21} can be used to define the dynamic coupling S between subsystems $\mathbf{G_1}$ and $\mathbf{G_2}$ [9]:

$$S = ||A_{12}|| + ||A_{21}|| \tag{12.6}$$

In this chapter, the focus is solely upon the scenario of undamped electromechanical oscillations that are not accompanied by voltage instability. In this scenario, if it is assumed that the reactive power balance can be controlled via local compensation, the effects of the reactive power and bus voltage magnitude can be neglected as they have a negligible impact on the dynamic coupling. Therefore, (12.6) can be rewritten as follows:

$$S = \sum_{j\in \mathbf{V}_{G2}} \sum_{i\in \mathbf{V}_{G1}} \left(\frac{\partial P_{ij}}{\partial \delta_{ij}} \cdot \left(\frac{1}{H_i} + \frac{1}{H_j}\right)\right) \tag{12.7}$$

where $\partial P_{ij}/\partial \delta_{ij}$ are the synchronizing coefficients and H_i is the inertia constant of the ith generator.

When exposed to electromechanical oscillations generators with strong dynamic coupling will swing together, whereas generators with weak dynamic coupling will swing against one another [8]. Therefore, the problem of finding coherent generator groups is equivalent to an optimization problem of finding the weakest dynamic coupling between different generator groups, as shown in (12.8):

$$\min S = \min_{\mathbf{V}_{G1}, \mathbf{V}_{G2} \subset \mathbf{V}_G} \left(\sum_{j\in \mathbf{V}_{G2}} \sum_{i\in \mathbf{V}_{G1}} \left(\frac{\partial P_{ij}}{\partial \delta_{ij}} \cdot \left(\frac{1}{H_i} + \frac{1}{H_j}\right)\right)\right) \tag{12.8}$$

Objective function

Minimal power imbalance and minimal power-flow disruption, defined according to (12.9) and (12.10), respectively, can both be used as objective functions of controlled islanding. Each objective will produce a different solution with different advantages and disadvantages [10,11]:

$$\min_{\mathbf{V}_1,\mathbf{V}_2\subset\mathbf{V}}\left(\left|\sum_{i\in\mathbf{V}_1,j\in\mathbf{V}_2} P_{ij}\right|\right) \tag{12.9}$$

$$\min_{\mathbf{V}_1,\mathbf{V}_2\subset\mathbf{V}}\left(\sum_{i\in\mathbf{V}_1,j\in\mathbf{V}_2} |P_{ij}|\right) \tag{12.10}$$

where P_{ij} denotes the value of the active power on the transmission line between nodes i and j.

The use of minimal power imbalance as the objective function creates islands with a similar level of load and generation, that is, a good load-generation balance. This property of the objective function minimizes the amount of load that must be shed following system splitting.

The use of minimal power-flow disruption as the objective function creates islands with the minimum change from the predisturbance power-flow pattern. This property of the objective function improves the transient stability of the islands, reduces the possibility of overloading the transmission lines within the island, and eases the islands eventual reintegration with the rest of the system [11].

When attempting to ensure stability after system splitting, the transient stability, rather than load-generation balance, should be the primary concern because an island with a negative stability margin and good load-generation balance will collapse. However, an island with a positive stability margin and a poor load-generation balance can be stabilized through load shedding.

Based on the properties of these objective functions the minimal power flow disruption is used in this chapter, it also has the additional benefit of reducing the complexity of the problem faced.

Considering the complexity of large interconnected power systems, it might be that the consideration of active power flows only would not lead to the optimal splitting solution. The inclusion of some heuristic knowledge inherent for every single system or additional assessment of reactive power flows and voltage stability related challenges would probably lead to a more efficient final splitting solution. However, the complexity of such an approach might be too high and even not practical enough. This should be addressed in future research projects.

Controlled islanding problem

The controlled islanding problem that is solved in this chapter consists of a minimal power-flow disruption objective function (12.10) and a generator coherency constraint (12.8). In [12], some basic results in solving such an optimization problem are given.

These two optimization problems are combined to form the SCCI algorithm (12.11). This is done by first solving (12.8) to find a set of coherent generator groups, and then solving (12.10) subject to these generator groups:

$$\begin{aligned} \left[\mathbf{V}^*_{\mathbf{G1}}, \mathbf{V}^*_{\mathbf{G2}}\right] &= \underset{\mathbf{V}_{G1},\mathbf{V}_{G2}\subset\mathbf{V}_{G}}{\arg\min} \left(\sum_{j\in\mathbf{V}_{G2}}\sum_{i\in\mathbf{V}_{G1}}\left(\frac{\partial P_{ij}}{\partial\delta_{ij}}\cdot\left(\frac{1}{H_i}+\frac{1}{H_j}\right)\right)\right) \\ \min_{\mathbf{V}_1,\mathbf{V}_2\subset\mathbf{V}} &\left(\sum_{i\in\mathbf{V}_1,\, j\in\mathbf{V}_2}\left|P_{ij}\right|\right) \text{ subject to } \mathbf{V}^*_{\mathbf{G1}}\subset\mathbf{V}_1, \mathbf{V}^*_{\mathbf{G2}}\subset\mathbf{V}_2 \end{aligned} \tag{12.11}$$

Here, arg min stands for the argument of the minimum, that is, $\left[\mathbf{V}^*_{\mathbf{G1}}, \mathbf{V}^*_{\mathbf{G2}}\right]$ is the node grouping that minimizes the objective function of (12.8) [13].

12.2.1.3 Spectral clustering

Defining the edge weight of the graph using the absolute active power on the transmission line $w_{ij} = \frac{|P_{ij}|+|P_{ji}|}{2}$ allows the problem of finding the weakest dynamic coupling and the minimum power flow disruption, respectively, to be converted into graph-cut problems.

Spectral clustering is the tool used in this section to solve these graph-cut problems. In this section, two types of spectral clustering will be introduced, namely unnormalized and normalized spectral clustering.

Unnormalized spectral clustering

The theory behind unnormalized spectral clustering can be briefly described as follows. *Unnormalized spectral clustering* clusters the nodes into two subsets based on the *Laplacian Matrix* **L**, which is defined for a graph **G** as [7]:

$$\mathbf{L} = \mathbf{D} - \mathbf{W} \tag{12.12}$$

where W is the edge weight matrix, and **D** is a diagonal degree matrix that contains diagonal elements $\mathbf{D}_{ii}$ that are equal to the total weight of the edges connected to node i. Defined in this way, the edge weight matrix **W** and the Laplacian Matrix **L** are both symmetric for any undirected graph.

The unnormalized spectral clustering algorithm, for the case of bisection, can be executed using the following steps [7]:

Compute the first two eigenvectors ϑ_1 and ϑ_2 of the Laplacian matrix **L**.
Let J $\in \mathbf{R}^{n\times 2}$ be the matrix containing the vectors ϑ_1 and ϑ_2 as columns. Let $y_i \in \mathbf{R}^2$ be the vector corresponding to the ith row of J.
Cluster the nodes $y_i \in \mathbf{R}^2$ into clusters c_1, c_2 using a clustering algorithm, for example, the k-medoids algorithm [14].

Unfortunately, the solution for bisecting the graph using unnormalized spectral clustering often consists of simply separating one node from the rest of the graph. This form of solution is clearly unacceptable for an islanding solution.

Normalized spectral clustering

Normalized spectral clustering uses the sum of the node weights within each subgraph as a balancing condition to prevent the application of spectral clustering from simply separating a single node. This gives rise to the concept of a normalized cut (Ncut) [6], which is, defined as

$$Ncut(\mathbf{V}_1, \mathbf{V}_2) = \frac{cut(\mathbf{V}_1, \mathbf{V}_2)}{weig(\mathbf{V}_1)} + \frac{cut(\mathbf{V}_1, \mathbf{V}_2)}{weig(\mathbf{V}_2)} \tag{12.13}$$

where $weig(\mathbf{V}_1) = \sum_{i \in \mathbf{V}_1} D_i$ and is thus the total sum of the weights of the nodes in $\mathbf{G}_1$. The $weig(\mathbf{V}_2)$ is similarly defined for $\mathbf{G}_2$. The inclusion of the node weights as a balancing condition acts to discourage the creation of a subgraph with very low weight. The *normalized spectral clustering* method can be used to split the graph with minimal *Ncut.*

For the case of bisection, this can be achieved using the following steps [6,7]:

1. Compute the first two eigenvectors ϑ_1 and ϑ_2 of the generalized eigen-problem $\mathbf{L}\vartheta = \lambda \mathbf{D}\vartheta$.
2. Let $\mathrm{J} \in \mathbf{R}^{n \times 2}$ be the matrix containing the vectors ϑ_1 and ϑ_2 as columns. Let $y_i \in \mathbf{R}^2$ be the vector corresponding to the ith row of $\mathbf{J}$.
3. Cluster the nodes $y_i \in \mathbf{R}^2$ into clusters c_1 and c_2 using a clustering algorithm, such as the *k-medoids* algorithm.

12.2.2 Spectral clustering controlled islanding algorithm

In this section, the spectral clustering controlled islanding (SCCI) algorithm is presented by solving the optimization problem (12.11). This solution can be found by constructing two graphs, based on the objective function and constraint from (12.11), and applying the SCCI algorithm to find the minimum cut of these two graphs.

The results of Section 12.2.1.2 will be served as pair-wise constraints in the SCCI algorithm, then the static graph $\mathbf{G_s}$ is constructed using power flow data. It contains every node and the edge weights are defined as the absolute value of the active power exchange between nodes i and j, $|P_{ij}|$. The nodes are then grouped using constrained spectral clustering, which will be described in this section to solve the optimization problem described in (12.11).

In Figure 12.11, a flowchart depicting the execution of the SCCI algorithm is presented. In the text below a detailed description of the algorithm steps is presented.

12.2.2.1 Minimizing the power-flow disruption while preserving coherent generator groups

The islanding solution that will separate the coherent generator groups found in Section 12.2.1.2, with the minimum power-flow disruption, can be found by applying constrained spectral clustering to a static graph of the power system.

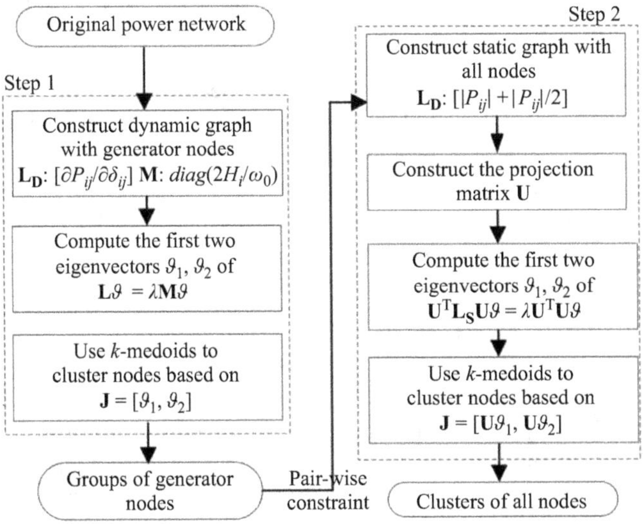

Figure 12.11 Flowchart of the SCCI algorithm

This static graph $\mathbf{G_S}(\mathbf{V}, \mathbf{E_S}, \mathbf{W_S})$ can be constructed for an n-node power system using power flow data to describe the active power exchange between each of the n nodes.

The issue of losses within the system must be accounted for to produce the symmetric undirected graph necessary for the application of spectral clustering. To ensure that the matrix $\mathbf{W_S}$ is symmetric, the elements of $\mathbf{W_S}$ is defined as $(|P_{ij}| + |P_{ij}|)/2$.

The Laplacian Matrix L_S of the static graph $\mathbf{G_S}$ can then be expressed as [12]

$$[\mathbf{L}_S]_{ij} = \begin{cases} \dfrac{|P_{ij}| + |P_{ji}|}{2} = -|\mathbf{V}_i||\mathbf{V}_j|B_{ij}\sin(\delta_i - \delta_j) & \text{if } i \neq j \\ -\sum\limits_{l=1,l\neq i}^{2} [\mathbf{L}_S]_{ij} & \text{if } i = j \end{cases} \tag{12.14}$$

where B_{ij} is the imaginary entry of the network admittance matrix.

The minimal power-flow disruption of the graph $\mathbf{G_S}$ is the solution of (12.10). However, to solve (12.11), the generator coherency constraint must be included. This is done by including the generator groups, obtained in the first step, as pair-wise constraints in the second step of the algorithm [15].

The pair-wise generator coherency constraints consist of (a) Must-Link constraints and (b) Cannot-Link constraints, these are defined as follows:

1. Must-Link constraints: all the generator nodes within a group must be linked in the SCCI algorithm.

2. Cannot-Link constraints: any two generator nodes in different groups cannot be linked in the SCCI algorithm.

Constrained spectral clustering is an efficient method for solving clustering problems with pair-wise constraints. The pair-wise constraints can be included by modifying the solution subspace using a projection matrix (the subspace approach) [15].

Without loss of generality, it can be assumed that the first m_1 nodes belong to the cluster c_1 and the next m_2 nodes belong to the cluster c_2. The projection matrix **U** can then be defined as follows [15]:

$$\mathbf{U} = \begin{pmatrix} 1_{m_1} & 1_{m_1} & 0_{m_1\times(n-m)} \\ 1_{m_2} & -1_{m_2} & 0_{m_2\times(n-m)} \\ 1_{n-m} & 0_{n-m} & \mathrm{I}_{(n-m)\times(n-m)} \end{pmatrix} \tag{12.15}$$

where I is the identity matrix, 1 is the all-ones column vector and 0 is the zero matrix or zero column vector.

In this way, the solution subspace is projected from an n-dimension space to a $(n - m + 2)$ dimension space, where $(m = m_1 + m_2)$. All nodes of the same cluster in the n-dimension space are represented by one equivalent node in the $(n - m + 2)$ dimension space to satisfy the pair-wise constraints.

With the introduction of the projection matrix **U**, constrained spectral clustering can be applied to the static graph $\mathbf{G_S}$ to find the cutset with minimal power-flow disruption that satisfies the generator grouping constraints produced in the first step of the SCCI algorithm. The second step of the SCCI algorithm can thus be described as follows:

1. Construct a static graph $\mathbf{G_S}$ of all nodes with the edge weights defined as $(|P_{ij}| + |P_{ji}|)/2$.
2. Construct the projection matrix **U** based on the generator grouping results.
3. Compute the first two eigenvectors ϑ_1 and ϑ_2 of the generalized eigen-problem $\mathbf{U}^{\mathrm{T}}\mathbf{L_S}\mathbf{U}\vartheta = \lambda\mathbf{U}^{\mathrm{T}}\mathbf{U}\vartheta$.
4. Let $\mathrm{J} \in \mathbf{R}^{n\times 2}$ be the matrix containing the vectors $\mathbf{U}\vartheta_1$ and $\mathbf{U}\vartheta_2$ as columns. Let $y_i \in \mathbf{R}^2$ be the vector corresponding to the ith row of **J**.
5. Cluster the nodes $y_i \in \mathbf{R}^2$ into the clusters $\mathbf{V_1}$ and $\mathbf{V_2}$ using the *k-medoids* algorithm.
6. Select $\mathbf{V_1}$ or $\mathbf{V_2}$ as the node set of a new static graph and return to (1).

Using the SCCI algorithm, described above, the solution of the optimization problem (12.11), that is, the islanding solution, can be found.

12.2.3 SCCI algorithm testing

In this section, three test systems are used to validate the proposed SCCI algorithm:

1. IEEE 9-bus test system
2. IEEE 39-bus test system
3. IEEE 118-bus test system [16]

For every test system, the SCCI algorithm has been applied and an islanding solution found. These solutions are compared with those obtained using the spectral *k*-way partitioning algorithm (SkP) and an OBDD method to evaluate the quality of the SCCI algorithm solution.

Spectral k-way partitioning (SkP)
The SkP method introduced in [17] does not consider the generator coherency constraint; a comparison with this method in Test Case I is used to demonstrate that the generator coherency constraint is necessary to form stable islands.

SkP is a special case of the *unnormalized spectral clustering algorithm* described in Section 12.2.1.3. In the third step of the algorithm, instead of using *k*-medoids, to cluster the nodes *k* reference nodes are selected and then the remaining nodes are clustered to these reference nodes according to the distance between the node and the reference nodes.

Ordered binary decision diagram (OBDD)
OBDDs are capable of searching the entire searching space and finding all possible solutions [18]. Therefore, the comparisons with this method in Test Case II and Test Case III are used to demonstrate that the SCCI algorithm is capable of finding the optimal solution of (12.11). The OBDD method described in [18], [10], and [19] is revised to find the islanding solution with the minimal power-flow disruption:

$$\Phi = SSC \cdot MCC \cdot MPD \tag{12.16}$$

$$\begin{aligned} SSC &= \left(\prod_{\forall i \in \mathbf{V}_{G1}, \forall j \in \mathbf{V}_{G2}} \bar{\Lambda}^*_{ij} \right) \cdot \left(\prod_{k=1}^{2} \left(\prod_{\forall i,j \in \mathbf{V}_{Gk}} \Lambda^*_{ij} \right) \right) \cdot \left(\prod_{\forall i \in \mathbf{V_L}} \left(\sum_{k=1}^{2} \sum_{j \in \mathbf{V}_{Gk}} \Lambda^*_{ij} \right) \right) \\ MCC &= \prod_{i,j \in \mathbf{V}} \left(\Lambda_{ij} \cdot \Lambda^*_{ij} + \bar{\Lambda}_{ij} \cdot \bar{\Lambda}^*_{ij} \right) \\ MPD &= \left\langle \left(\sum_{i,j \in \mathbf{V}} \bar{\Lambda}_{ij} \cdot |w_{ij}| \right) < \varepsilon \right\rangle \end{aligned} \tag{12.17}$$

where $\mathbf{\Phi}$ is the Boolean function of the OBDD method, Λ and Λ^* represent the adjacency matrix and reachability matrix of graph $\mathbf{G_S}$, respectively, and $\mathbf{V_L}$ represents the load node set of graph $\mathbf{G_S}$ [18].

The SSC represents the requirement of generator coherency, where the first item denotes that the generators of different coherent groups cannot be connected, the second item denotes that the generators in the same coherent group must be connected, and the third item denotes that every load must be connected to one and only one coherent generator group [18,10,19].

The MCC is used to reduce the number of islanding solutions by allowing only those edges that help to form islands to be disconnected [19]. The MPD denotes that the power-flow disruption cannot exceed the threshold ε, and it is used to replace the power balance constraint (PBC) in [10] and [19].

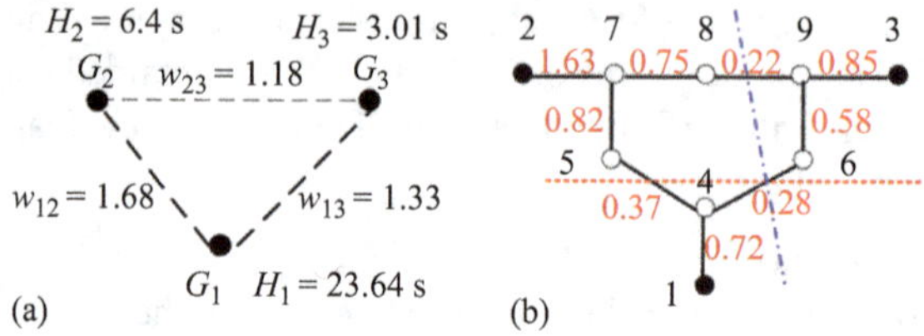

Figure 12.12 Dynamic graph and static graph of the IEEE 9-bus test system (base power $S_N = 100$ MW). Right: the dotted line is the SCCI islanding solution, while the dash-dotted line is the SkP solution: (a) dynamic graph and (b) static graph

Test Case I: IEEE 9-bus test system

The generator grouping algorithm was applied and two coherent generator groups {1} and {2, 3} were found.

The SCCI algorithm required a static graph to be constructed using (12.14), as shown in Figure 12.3(b). The coherent generator groups are then used to construct a projection matrix (12.15) that represented the must-link constraint between generators 2 and 3, and the cannot-link constraint between generator 1 and generators 2 and 3. This projection matrix allows the constrained spectral clustering algorithm, described in Section 12.2.2.1, to be used to cluster the nodes of the system.

Applying the SCCI algorithm to the IEEE 9-bus test system resulted in finding a single cutset that created an islanding solution that consisted of two islands {1, 4} and {2 ,3, 5, 6, 7, 8, 9}. This cutset is marked in Figure 12.12(b) by a dotted line, the cut of which is 0.65 p.u.

Application of the SkP method to the IEEE 9-bus test system resulted in an islanding solution that consisted of two different islands: {3, 6, 9} and {1, 2, 4, 5, 7, 8}. This cutset is marked in Figure 12.12(b) by a dot-dashed line, the cut of which is 0.50 p.u.

It is clear that the island {1, 2, 4, 5, 7, 8} is not stable because it contains the unsynchronized generators 1 and 2. The solution of the SCCI algorithm has a higher cut, this difference from the solution of the SkP method represents the cost of satisfying the generator coherency constraint.

Test Case II: IEEE 39-bus test system

As in Test Case I, the generator grouping algorithm was applied to find the dynamic graph of the IEEE 39-bus test system. The application of normalized spectral clustering identified the set of generator groups shown in Table 12.3. From these groups a set of pair-wise generator constraints can be determined to form the projection matrix.

Table 12.3 The generator groups in the first step of 39-bus test system

Group 1	Group 2	Group 3
G2, G3, G4, G5, G6, G7	G8, G9, G10	G1

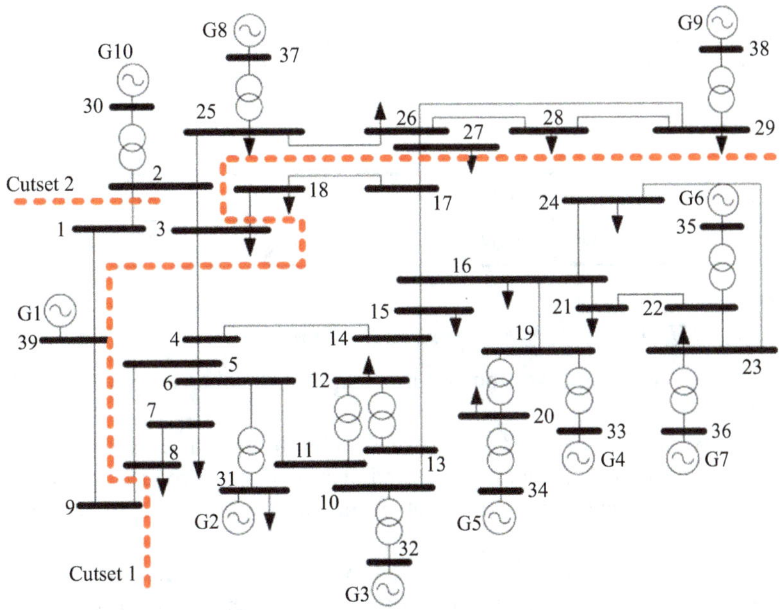

Figure 12.13 Single-line diagram of the IEEE 39-bus test system. The dotted lines represent the two cutsets necessary to produce the final islanding solution

The execution of the SCCI algorithm is affected by there being three generator groups. This effect is that the second step of the algorithm now requires the use of recursive bisection to determine two cutsets and create three islands.

The first cutset to be found, Cutset 1, separated Group 1 from Group 2 and Group 3. The second cutset to be found, Cutset 2, separated Group 3 from Group 2. Combined, these two cutsets form the final islanding solution marked in Figure 12.13.

The OBDD method was executed with a ε value of 335 MW and the five solutions for Cutset 1 that had the smallest cuts are shown in Table 12.4, alongside the SCCI algorithm solution for Cutset 1. It is obvious that the SCCI algorithm found the minimum cut for separating Group 1 from Group 2 and Group 3. A comparison of Cutset 2 is not included because it would be quite trivial for separating only one generator.

Table 12.4 The 39-bus test system results comparison with OBDD

Cutset 1	$\sum\lvert P_{ij}\rvert$ (MW)	Method
(1) 8-9, 3-4, 3-18, 17-27	**175.3**	**SCCI**
(1) 8-9, 3-4, 3-18, 17-27	**175.3**	**OBDD**
(2) 9-39, 3-4, 3-18, 17-27	179.0	
(3) 8-9, 3-4, 17-18, 17-27	331.2	
(4) 8-9, 3-4, 16-17	334.3	
(5) 9-39, 3-4, 17-18, 17-27	335.0	

Table 12.5 The generator groups of 118-bus test system

Group 1	Group 2	Group 3
10, 12, 25, 26, 31	46, 49, 54, 59, 61, 65, 66, 69, 80	87, 89, 100, 103, 111

Test Case III: IEEE 118-bus power system

The generator grouping algorithm returned the three coherent generator groups given in Table 12.5.

The two cutsets produced in the SCCI algorithm, Cutset 1 and Cutset 2, separated Group 1 from Groups 2 and 3, and then separated Group 2 from Group 3, respectively. Combined, these two cutsets form the final islanding solution marked in Figure 12.14.

As in Test Case II, the results returned by the OBDD method will be used to validate the SCCI solution. Unfortunately, it is not practical to apply the OBDD method directly to the 118-node network. The original network is simplified to a 34-node and 43-edge graph when searching for Cutset 1 and is simplified to a 24-node and 38-edge graph when searching for Cutset 2. The cutset solutions found for these simplified graphs were then mapped onto the original graphs so that all possible solutions in the original graphs are still found.

For both Cutset 1 and Cutset 2, the five cutsets with the smallest cut that were found by the OBBD method are presented in Table 12.6, alongside which are the results returned by the SCCI algorithm.

The comparison in Table 12.6 shows that, as in Test Case II, the SCCI algorithm returned the cutset that separated the coherent generator groups with minimum cut.

12.3 Identifying islanding time

Once the islanding boundaries are defined, the timing of the controlled islanding becomes the key determining factor for the synchronism of generators within each island. If the islanding action is undertaken at the *right time*, it is expected that the

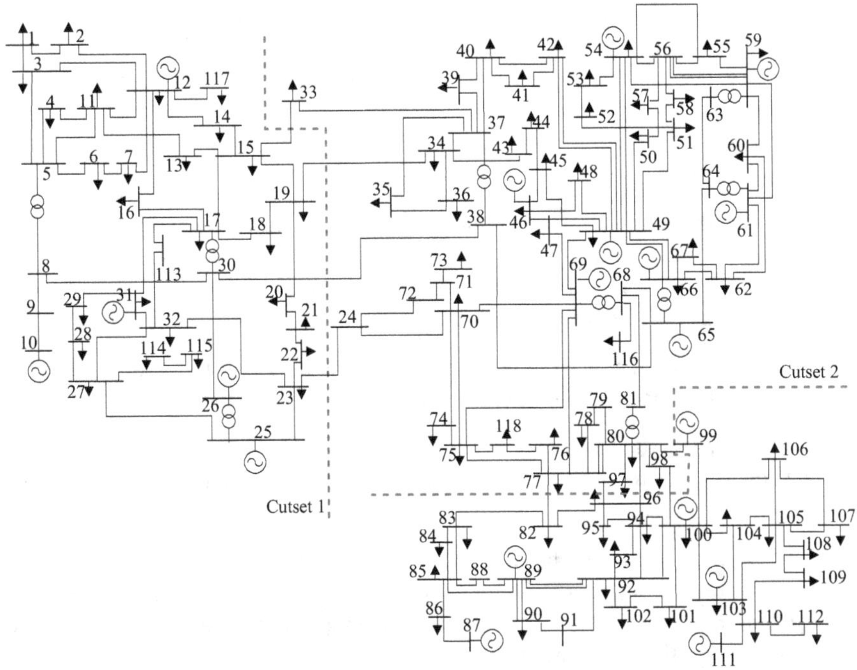

Figure 12.14 Single-line diagram of the IEEE 118-bus power network. The two dotted lines represent the cutsets necessary to produce the final islanding solution

Table 12.6 The 118-bus system results comparison with OBDD

Cutset 1	$\sum \lvert P_{ij} \rvert$ **(MW)**	**Method**
(1) 15-33, 19-34, 30-38, 23-24	**71.6**	**SCCI**
(1) 15-33, 19-34, 30-38, 23-24	**71.6**	**OBDD**
(2) 15-33, 19-34, 30-38, 24-70, 24-72	73.5	
(3) 33-37, 19-34, 30-38, 23-24	76.3	
(4) 33-37, 19-34, 30-38, 24-70, 24-72	78.1	
(5) 15-33, 19-34, 30-38, 24-70, 71-72	83.2	
Cutset 2		
(1) 77-82, 96-97, 80-96, 98-100, 80-99	**52.7**	**SCCI**
(1) 77-82, 96-97, 80-96, 98-100, 80-99	**52.7**	**OBDD**
(2) 77-82, 96-97, 80-96, 98-100, 99-100	57.3	
(3) 77-82, 82-96, 94-96, 95-96, 98-100, 80-99	65.1	
(4) 77-82, 80-97, 80-96, 98-100, 80-99	65.9	
(5) 77-82, 82-96, 94-96, 95-96, 98-100, 99-100	69.6	

transition from an integrated system to a multi-island system will be achieved through a stable and smooth transient process. Otherwise, some islands might lose their stability and collapse unnecessarily or the controlled islanding could fail entirely and cause a complete system collapse.

A successful islanding has some implicit requirements for determining the "*right time*" to split:

1. After the islanding action, generator synchronism of the newly created islands should be maintained.
2. Before and after the islanding action, the rotor speed of every generator should remain within the under and over frequency limits to avoid their tripping.

Furthermore, some nonessential, but useful, operating considerations should be also considered to increase the likelihood of a successful islanding, for example:

1. The greater the stability margin, the more stable the created island will be.
2. The smaller the frequency deviation of the COI of the created island, the more quickly it will reach a new steady state.

The first requirement is the most important and the most difficult to be satisfied. It leads to the concept of *feasible islanding time intervals* (FITI). If the controlled islanding is performed within the FITI, then the generators in the created islands will retain their synchronism.

The second requirement and the two considerations lead to the concept of *optimal islanding instant*, which is selected from the FITI according to the size of the stability margin and the maximum frequency deviation in each of the created islands. Once the FITI is known, determining the optimal islanding instant is relatively straight-forward. This is because the FITI ensures that the created islands will retain synchronism and, as such, can be represented as a single equivalent generator.

12.3.1 Transient stability preliminaries

In this section, some transient stability preliminaries are introduced.

A power system subject to a fault can be viewed as going through network topology changes in three stages: the prefault, the fault-on, and the postfault systems [20]. In the prefault regime, the system is operating at a prefault *stable equilibrium point* (SEP) x_0. Suppose a fault occurs at t_0 and, during the fault duration $[t_0, t_{cl})$, the fault-on system with a trajectory $x_F(t)$ is governed by dynamics described by

$$\dot{x} = f_F(x(t)), t_0 \leq t < t_{cl} \tag{12.18}$$

where x denotes the state vector of the system. After the fault is cleared at t_{cl}, assuming there are no further disturbances, the postfault system with a trajectory $x_P(t)$ is governed by dynamics described by

$$\dot{x} = f_P(x(t)), t_{cl} \leq t \tag{12.19}$$

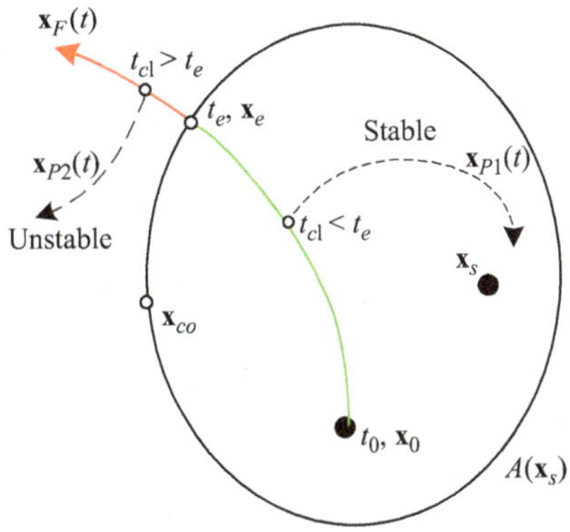

Figure 12.15 System trajectory for stability analysis of postfault system

Assume that the postfault system (12.19) has a SEP x_s and the stability region of x_s is denoted as $A(x_s)$. The fundamental problem of transient stability due to a fault can be summarized as follows: given a prefault SEP x_0 and a fault-on system $f_F(x(t))$, will the postfault trajectory $x_P(t)$ that starts at $x_P(t_{cl})$ converge to the postfault SEP x_s? In other words, is the initial point of the postfault trajectory, $x_P(t_{cl})$, located inside the stability region $A(x_s)$ of the postfault SEP x_s, or it is not?

As shown in Figure 12.15, after a fault occurs, the system may continue to gain kinetic and potential energy and move away from the prefault SEP x_0. The fault-on trajectory $x_F(t)$ will exit the stability region $A(x_s)$ at t_e through a point called the *exit point* x_e, that is, $x_F(t_e) = x_e$. Thus, if the fault is cleared before t_e, then the initial point of the postfault trajectory, $x_P(t_{cl})$, will be inside the stability region $A(x_s)$ and the postfault trajectory $x_P(t)$ will converge to x_s, or the initial point of the postfault trajectory will be outside the stability region, and the postfault trajectory will not converge to x_s.

The transient stability of the postfault system (12.19) can be assessed using a controlling unstable equilibrium point (UEP) method [21,22]. An energy function $V(x)$, in this case describing the system transient energy, is computed at the time when the fault is cleared. This value is compared with the *critical energy* V_{cr} for transient stability assessment. If the energy is less than the critical energy, then the system will retain its stability. Otherwise, it may be unstable [23].

12.3.2 Identifying the timing of controlled islanding

Controlled islanding is a complex process affected by (a) the initial disturbances such as cascading events, (b) the resulting transients, (c) the actual islanding, and finally (d) the transient processes within the newly created islands. The optimal islanding boundary and optimal islanding instant are determined by a number of factors, including the system state before the disturbance (e.g., system topology, loading level, or security margins) and the transient processes that occur after the disturbance (e.g., the size/location of the disturbance, protection response time, or remedial actions implemented). To determine the optimal islanding instant, the stability of the newly created islands must be assessed.

This section introduces the concept of FITI and a viable method for its calculation. The FITI of island *a* (denoted by $T_{FITI,a}$) is defined as the time interval(s) in which, if the islanding action is undertaken, the generators in island *a* will retain their synchronism after controlled islanding and enable stable operation of island *a*. Similarly, the FITI of the whole system (denoted by T_{FITI}) is defined as the intersection of the FITIs for each of the created islands, that is, the time interval(s) in which the first requirement for controlled islanding is satisfied. A successful controlled islanding will be possible if the islanding action is undertaken within a FITI only. Once the FITI has been calculated, the optimal islanding instant can be selected within the FITI by considering the stability margins and maximum frequency deviations of the created islands.

12.3.2.1 Features of the feasible islanding time interval

Assume that a given islanding boundary splits the original system into two subsystems (island *a* containing the first *m* generators and island *b* containing the rest). We denote x_a and x_b as the state vectors for islands *a* and *b*, respectively. Thus, the state vector x of the original system can be divided into two parts: $\mathrm{x} = [\mathrm{x}_a, \mathrm{x}_b]$.

During the whole islanding process, the original system goes through network topology changes in the following four stages: the predisturbance, the disturbance-on, the postdisturbance, and the postislanding stages as illustrated in Figure 12.16. The method presented in this chapter can be applied for any major disturbance or cascade of events.

In the predisturbance regime, the system is operating at a predisturbance SEP $x_0 = [\mathrm{x}_{a,0}, \mathrm{x}_{b,0}]$. Let us assume that *s* cascading events occur sequentially at $t_1, t_2, \ldots, t_s$. In this period, that is, $[t_1, t_s)$, the disturbance-on system is governed by dynamics described by

$$\begin{aligned} [\dot{\mathrm{x}}_a, \dot{\mathrm{x}}_b] &= f_{F1}(\mathrm{x}_a(t), \mathrm{x}_b(t)), t_1 \le t < t_2 \\ &\ldots \\ [\dot{\mathrm{x}}_a, \dot{\mathrm{x}}_b] &= f_{Fs-1}(\mathrm{x}_a(t), \mathrm{x}_b(t)), t_{s-1} \le t < t_s \end{aligned} \tag{12.20}$$

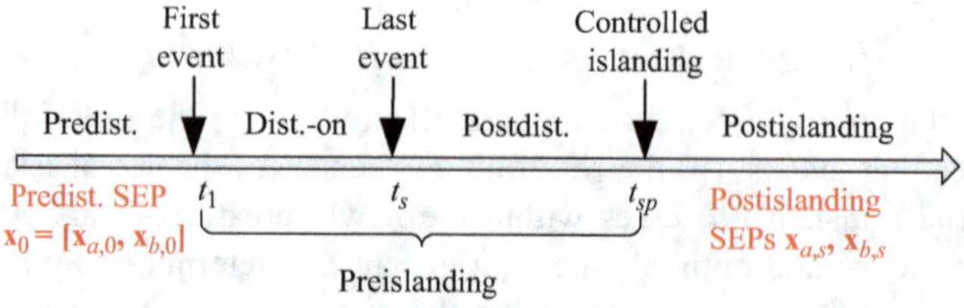

Figure 12.16 System stages during the whole islanding process

After the last event occurring at t_s, the postdisturbance system is governed by the dynamics described by

$$[\dot{x}_a, \dot{x}_b] = f_{PF}(x_a(t), x_b(t)), t_s \le t < t_{sp} \tag{12.21}$$

For ease of expression, the disturbance-on stages, $f_{F1}, f_{F2}, \ldots, f_{Fs-1}$, and postdisturbance stage, f_{PF}, can be merged into a single preislanding stage f_{pre} with a trajectory $\left[x_{a,pre}, x_{b,pre}\right]$. When the system is split into two islands at t_{sp}, the postislanding trajectories $x_{ap}(t)$ and $x_{bp}(t)$ are governed by dynamics described by

$$\dot{x}_a = f_{aP}(x_a(t)), t_{sp} \le t \tag{12.22}$$

$$\dot{x}_b = f_{bP}(x_b(t)), t_{sp} \le t \tag{12.23}$$

Assume that the postislanding SEP of island a is $x_{a,s}$ and that the stability region of $x_{a,s}$ is $A(x_{a,s})$. The fundamental problem of transient stability of the island a during the whole islanding process can be summarized as follows: given a predisturbance state $x_{a,0}$ and a preislanding system $f_{pre}(x(t))$, will the postislanding trajectory of the island a that starts at $x_{aP}(t_{sp})$ converge to the postislanding SEP $x_{a,s}$? In other words, is the initial point of the postislanding trajectory, $x_{aP}(t_{sp})$, located inside the stability region $A(x_{a,s})$ of the postislanding SEP $x_{a,s}$? The same question holds for island b.

Figure 12.17 shows a general case of the preislanding trajectory $x_{a,pre1}(t)$ of island a. As in the case shown in Figure 12.16, during the disturbance-on and postdisturbance periods, the island a may continue to gain kinetic and potential energy and move away from the predisturbance state $x_{a,0}$ with the preislanding trajectory $x_{a,pre1}(t)$ eventually exiting the stability region at t_{e1}. If the islanding is undertaken before t_{e1}, the initial point $x_{aP}(t_{sp})$ of the postislanding trajectory will be inside $A(x_{a,s})$ and the postislanding trajectory $x_{aP}(t)$ will then converge to the postislanding SEP $x_{a,s}$; otherwise the postislanding trajectory $x_{aP}(t)$ may be

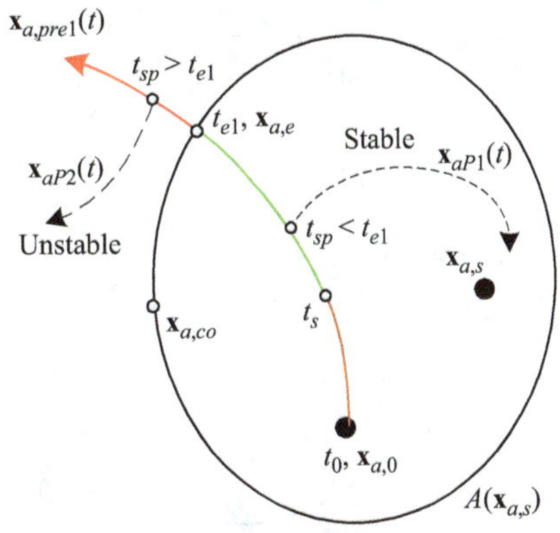

Figure 12.17 System trajectory for stability analysis of postislanding system

unstable. In this case (denoted as **Case I**), the FITI of the island a is $\mathrm{T}_{FITI,a} = [t_s, t_{e1}]$.

Compared with Figure 12.15, there are three key differences in Figure 12.17. First, the system under study is not the original system, but just the island aimed to be created. Therefore, the SEP and the corresponding stability region are given for the island a and are determined by the final topology of the island a. Second, the preislanding trajectory of the island a in Figure 12.17 is just a projection trajectory of the original system with dynamics described by (12.20) and (12.21). To form the trajectory of island a during the preislanding stage, we only extract the state vector related to island a, that is, x_a. Meanwhile, the other state vector x_b can be used to form the trajectory of island b. Third, regarding the effects on island a during the preislanding period $[t_0, t_{sp})$, there exists an additional effect from island b lasting for the whole process, which will create more possible preislanding trajectories (**Case II–Case IV**) shown in Figure 12.18.

Case II is that the preislanding trajectory $\mathrm{x}_{a,pre2}(t)$ always stays in the stability region. The system can then be split at any arbitrary instant, t_{sp}, and the post-islanding trajectory of newly created island a will always converge to its SEP. In this case, the FITI of the island a is $\mathrm{T}_{FITI,a} = [t_s, t_\infty)$.

Case III is that the preislanding trajectory $\mathrm{x}_{a,pre3}(t)$ exits the stability region at t_{e1} but returns inside at t_{r1}. In this case, the FITI of the island a is $\mathrm{T}_{FITI,a} = [t_s, t_{e1}] \cup [t_{r1}, t_\infty)$.

Case IV is that the generators within the island a show sustained internal oscillations in the preislanding stage. The preislanding trajectory $\mathrm{x}_{a,pre4}(t)$ experiences multiple transitions back and forth across the boundary of the stability region of island a. In this case the FITIs are time intervals in which the postislanding trajectory is in the stability region of island a, that is, $\mathrm{T}_{FITI,a} = [t_s, t_{e1}] \cup [t_{r1}, t_{e2}] \cup \cdots$.

The scenario in which the preislanding trajectory leaves the stability region of island a and then returns is possible, as the island a is still connected to the island b during the whole preislanding stage. Thus, the dynamics between island a and

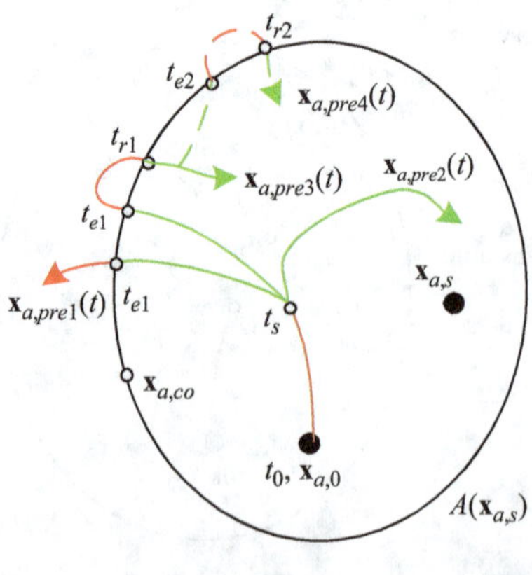

Figure 12.18 Possible cases of preislanding trajectories

island b may cause the preislanding trajectory to return to the stability region of island a. Conversely, they may also push the preislanding trajectory outside of the stability region. Beyond the trajectory types given above, there are still some exceptional cases that need to be considered:

1. There is no postislanding SEP for island a if an improper islanding boundary has been selected.
2. The preislanding trajectory leaves this region during the disturbance-on period.

Under these exceptional cases, $T_{FITI,a}$ is a null set and island a cannot be stable, regardless of when the islanding is performed.

Theoretically, the preislanding trajectory may also start from an initial point which is outside of the stability region and after some time enters during the pre-islanding stage. However, this case makes no sense that the island is unstable when isolated from an undisturbed system but the ongoing disturbance processes are helping the island to return the stability region.

Furthermore, these cases violate the requirement of the controlling UEP method that a postislanding SEP exists and the predisturbance SEP lies inside the stability region of the postislanding SEP. These special cases may only exist in theory and further study is required to clarify their practical existence and possible impact on the controlled islanding application.

12.3.2.2 Quantitative transient stability analysis for controlled islanding

This section describes the *transient energy function* (TEF) of an island and a method for calculating its *critical energy*. Let the loads be modeled as constant impedances and the synchronous generators be modeled using the classical second order generator model. The damping coefficient is assumed to be zero here. The dynamics of an n-generator system can be represented by the following equations [23,24]:

$$\begin{aligned} \dot{\delta}_i &= \omega_i \\ M_i\dot{\omega}_i &= P_{mi} - P_{ei} \\ P_{ei} &= \sum_{j=1}^{n} E_i'E_j'B_{ij}\sin\delta_{ij} + \sum_{j=1}^{n} E_i'E_j'G_{ij}\cos\delta_{ij} \end{aligned} \tag{12.24}$$

where δ_i, ω_i, P_{mi}, P_{ei}, E_i', and M_i are the rotor angle, rotor speed, mechanical power, electrical power, electromotive force, and moment of inertia of the ith generator, respectively; G_{ij} and B_{ij} are the transfer conductance and the susceptance of the element i–j in the reduced admittance matrix of the system, respectively.

The equations describing the predisturbance, disturbance-on, postdisturbance, and postislanding stages all have the same form as (12.24). However, the value for the G_{ij} and B_{ij} terms may vary due to changes in network topology.

To form the trajectory of island a during the whole islanding process, we only extract the state vector related to island a, that is, x_a. Using the COI of island a as a reference, x_a can be transformed into the following equations:

$$\begin{aligned}\delta_{ai} &= \delta_i - \delta^a_{COI}\\ \omega_{ai} &= \omega_i - \omega^a_{COI}\end{aligned} \tag{12.25}$$

with $\delta^a_{COI} = \frac{1}{M_T}\sum_{j=1}^{m} M_j\delta_j, \omega^a_{COI} = \frac{1}{M_T}\sum_{j=1}^{m} M_j\omega_j, M_T = \sum_{j=1}^{m} M_j.$

The transient energy function $V_a(\delta, \omega)$ of island a is of the form:

$$V_a(\delta, \omega) = V_k(\omega) + V_p(\delta) \tag{12.26}$$

with

$$V_k(\omega) = \frac{1}{2}\sum_{i=1}^{m} M_i\omega^2_{ai} \tag{12.27}$$

$$\begin{aligned}V_p(\delta) = &\sum_{i=1}^{m-1}\sum_{j=i+1}^{m}\left(-E'_iE'_jB_{ij}\left(\cos\left(\delta_{aij}\right) - \cos\left(\delta^s_{aij}\right)\right)\right)\\ &-\sum_{i=1}^{m} P_i\left(\delta_{ai} - \delta^s_{ai}\right) + V_{path}(\delta)\end{aligned} \tag{12.28}$$

$$\begin{aligned}V_{path}(\delta) &= \sum_{i=1}^{m-1}\sum_{j=i+1}^{m}\int E'_iE'_jG_{ij}\cos\delta_{aij}\left(d\delta_{ai} + d\delta_{aj}\right)\\ &\approx \sum_{i=1}^{m-1}\sum_{j=i+1}^{m}\int_0^1 E'_iE'_jG_{ij}\left(\delta_{ai} - \delta^s_{ai} + \delta_{aj} - \delta^s_{aj}\right)\\ &\quad * \cos\left(\lambda\left(\delta_{aij} - \delta^s_{aij}\right) + \delta^s_{aij}\right)d\lambda\end{aligned} \tag{12.29}$$

where $V_k(\omega)$ and $V_p(\delta)$ are the kinetic energy and potential energy of island a, respectively, and $V_{path}(\delta)$ is the path-dependent term contained in the potential energy. Here, the parameters G_{ij} and B_{ij} used to construct the energy function in (12.26) correspond to the final topology of island a.

The following three key properties of (12.26) can be noted. First, the energy at the postislanding SEP of island a is used as the reference of potential energy for the whole process, including preislanding and postislanding stages, that is, $V_a(\delta^s_a, 0) = 0$.

Second, from dynamic simulations, or real-time measurements, the end point of the preislanding trajectory $x_{a,pre}(t^-_k)$ can be obtained. However, to decide if the island a is stable or not after being split at the time instant t_k, the initial point of the postislanding trajectory, $x_{aP}(t^+_k)$, rather than $x_{a,pre}(t^-_k)$, is needed to calculate the energy $V_a(x_{aP}(t^+_k))$ which will be compared with the critical energy V_{cr}. Note that the energy function is always constructed using postislanding state vector, rather than preislanding state vector.

With the structure-preserving model, vectors $x_{a,pre}(t^-_k)$ and $x_{a,pre}(t^+_k)$ are different. This is because both of them contain variables including rotor speed, rotor angle, bus voltages, and bus angles, while bus voltages and bus angles will change at t_k^+, that is, the instant after the islanding. Therefore, to assess the stability of island a at any time instant t_k, the relevant initial point of the postislanding trajectory, $x_{aP}(t_k^+)$, should be derived from $x_{a,pre}(t_k^-)$, the end point along the preislanding trajectory, assuming that an imaginary islanding action occurs at t_k.

With the network-reduced model (12.24) used in this chapter, $x_{a,pre}(t_k^-)$ and $x_{aP}(t_k^+)$ are the same since they only contain variables which cannot be abruptly changed after the islanding, that is, rotor speed and rotor angle. Therefore, at any time instant t_k, $x_{aP}(t_k^+)$ equals $x_{a,pre}(t_k^-)$ and hence, the calculation of $x_{aP}(t_k^+)$ from $x_{a,pre}(t_k^-)$ is unnecessary.

Third, after the islanding, the newly created island can be considered as an autonomous system. Since the damping was neglected in the modeling of the problem, the island will satisfy the law of energy conservation, that is, the sum of its kinetic and potential energies $(V_a(\delta, \omega))$ will remain constant over the entire-postislanding trajectory. For stable cases, the island a keeps oscillating during the postislanding process, since the damping has been ignored in the model considered. Therefore, its kinetic energy $V_k(\omega)$ and potential energy $V_p(\delta)$ will keep on exchanging, but the total energy $V_a(\delta, \omega)$ will remain constant. For unstable cases, the kinetic energy $V_k(\omega)$ and potential energy $V_p(\delta)$ will respectively increase and decrease with the total energy $V_a(\delta, \omega)$ still remaining constant. In Section 12.3.4 both stable and unstable cases will be demonstrated.

As mentioned in Section 12.3.2.2, the calculation of the critical energy, V_{cr}, is an essential part of assessing an island's stability. In this chapter, the controlling UEP method, which uses the energy of the controlling UEP as the critical energy, is adopted. The controlling UEP is a specific UEP whose Jacobian matrix has only one eigenvalue with a positive real part, and whose stable manifold contains the exit point, that is, the point where the preislanding trajectory leaves the stability region. When applying this method, the key task is to determine the controlling UEP of the postislanding trajectory.

An effective means for determining the controlling UEP is the BCU (boundary of stability region based controlling UEP) method that is applied to a reduced-state system whose controlling UEP is easier to compute [23]. After islanding, island a is governed by dynamics described by

$$\begin{aligned} \dot{\delta}_{ai} &= \omega_{ai} \\ \dot{\omega}_{ai} &= \frac{1}{M_i}(P_{mi} - P_{ei}) - \frac{1}{M_T}\sum_{j=1}^{m}(P_{mj} - P_{ej}) \end{aligned} \tag{12.30}$$

In (12.30), the elements are only related to island a, as the islands are now separated from each other, which makes (12.30) an autonomous system. The corresponding reduced-state island can be represented as

$$\dot{\delta}_{ai} = (P_{mi} - P_{ei}) - \frac{M_i}{M_T}\sum_{j=1}^{m}(P_{mj} - P_{ej}) \tag{12.31}$$

This reduced-state island (12.31) has an energy function that has the form $V_{\text{reduced}}(\delta) = V_p(\delta)$, which is the potential energy $V_p(\delta)$ of the original island [23]. It can be proved that, in the presence of small transfer conductances, $\widehat{\delta}$ is an equilibrium point of the reduced-state island (12.31) if, and only if, $\left(\widehat{\delta}, 0\right)$ is an equilibrium point of the original island (12.30) and both equilibrium points are of the same type (i.e., $\widehat{\delta}$ is a type-k equilibrium point and $\left(\widehat{\delta}, 0\right)$ is a type-k equilibrium point) [23].

Therefore, finding the controlling UEP δ^*_{co} of the reduced-state island (12.31) is equivalent to finding the controlling UEP $(\delta^*_{co}, 0)$ of the original island (12.30). The calculation of the former is much easier and more effective than the latter.

12.3.2.3 FITI calculation using the controlling UEP method

For a classical power system model, with damping neglected, identifying the timing of controlled islanding using the controlling UEP method can be achieved as follows:

1. For a disturbed system to be split, determine the islanding boundary according to the coherent groups and system states.
2. For the given islanding boundary, calculate the corresponding postislanding SEP for a single intended island (say, island a) and construct its transient energy function $V_a(\delta, \omega)$ with its postislanding topology using (12.24).
3. Integrate the sustained preislanding trajectory of the disturbed system and calculate the energy function $V_a(\delta, \omega)$ of island a along the sustained preislanding trajectory. Note that in the network-reduced model (12.31) used in this chapter, at any time instant t_k, $\mathrm{x}_{aP}\left(t_k^+\right)$ equals $\mathrm{x}_{a,pre}\left(t_k^-\right)$ and hence, the calculation of $\mathrm{x}_{aP}\left(t_k^+\right)$ from $\mathrm{x}_{a,pre}\left(t_k^-\right)$ is unnecessary.
4. Apply the BCU method to compute the controlling UEP of island a with respect to the preislanding trajectory, the resulting controlling UEP is $\left(\delta^*_{co}, 0\right)$. The critical energy V_{cr} is the value of the transient energy function at the controlling UEP, that is, $V_{cr} = V_a\left(\delta^*_{co}, 0\right)$.
5. The FITI of island a contains the times at which the energy of the preislanding trajectory is less than V_{cr}, that is, $\mathrm{T}_{FITI,a} = (t | V_a(\delta(t), \omega(t)) \le V_{cr})$.
6. Repeat steps (2) to (5) until the FITIs of every island have been calculated. To obtain the FITI for the whole system, use the intersection of all FITIs for all islands, that is, $\mathrm{T}_{FITI} = \mathrm{T}_{FITI,a} \cap \mathrm{T}_{FITI,b} \cap \cdots$.

The BCU method in step (4) is relatively complex and includes the following four steps:

1. From the preislanding trajectory $\mathrm{x}_{a,pre}(t)$, find the point δ^+ that is an approximation of the exit point where potential energy $V_p(\delta)$ reaches its first local maximum.
2. Use δ^+ as the initial condition and integrate the postislanding reduced-state island (12.31) to find the point δ_0^+ where the norm of its vector field reaches the first local minimum.

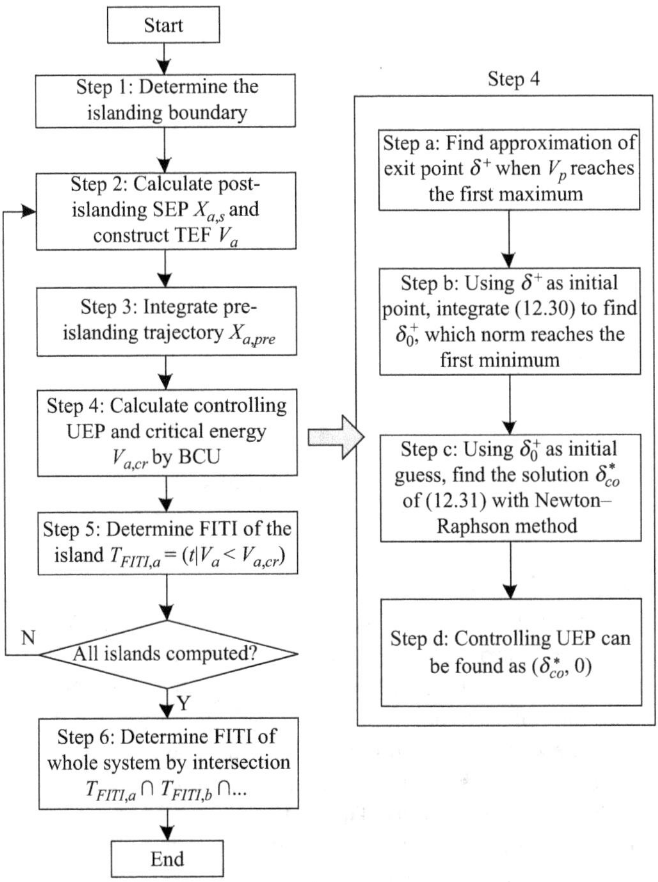

Figure 12.19 The flowchart of FITI calculation using the controlling UEP method

3. Use the point δ_0^+ as an initial guess to solve the algebraic equation (11.32) with Newton–Raphson method to find the solution δ_{co}^*:

$$(P_i - P_{ei}) - \frac{M_i}{M_T}\sum_{j=1}^{m}(P_j - P_{ej}) = 0 \tag{12.32}$$

4. The controlling UEP of island a with respect to the preislanding trajectory $x_{a,pre}(t)$ can thus be found as $(\delta_{co}^*, 0)$. The flowchart of FITI calculation is shown in Figure 12.19.

Furthermore, if the intersection of individual FITIs is a null set, there are three possible measures to save the system: (1) to change the splitting boundary until one is found that produces a non-null FITI set; (2) to take remedial actions to stabilize the islands; and (3) to split in a sequential way. The last two measures are

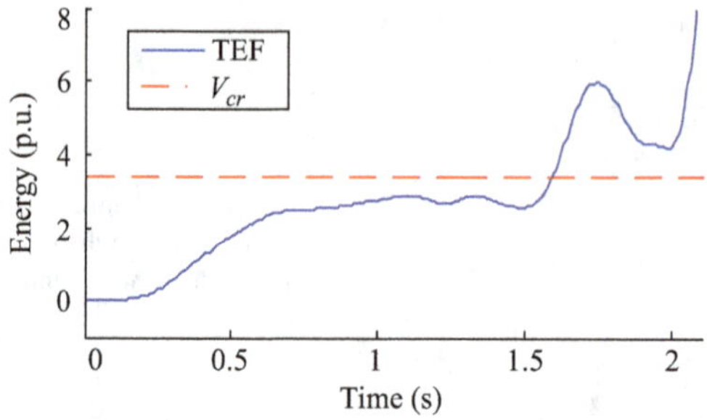

Figure 12.20 Transient energy function (TEF) of island a_1 along preislanding trajectory

interesting topics but are beyond the scope of this chapter and will be addressed in future.

12.3.3 Case study

In this section, the IEEE 118-bus test system is used to demonstrate the calculation of the FITI using the controlling UEP method. The test system has 19 synchronous generators and as such can be considered as representative enough to demonstrate the new methodology presented.

The method presented in Section 12.3.2.3 can be applied for any major disturbance or cascade of events. In this section, the simulation scenario was defined as *two consecutive three-phase faults*. The first one occurred at bus 33 at $t_1 = 0.01$ s and was cleared by tripping line 15-33 at $t_2 = 0.36$ s. The second one occurred at bus 23 at $t_3 = 0.10$ s and was cleared by tripping line 23-25 at $t_4 = 0.62$ s. The entire power system will be split into two islands: island a_1 ($\mathbf{G}_a$ = generators {10, 12, 25, 26, 31}) and island b_1 ($\mathbf{G}_b$ = {the rest of the generators}). Using the method presented in [5], the islanding boundary is selected as {lines: 15-33, 19-34, 30-38, 23-24, 23-25}.

Let us demonstrate the calculation of the FITI for island a_1. Using the algorithm presented in Section 12.3.2.3, the controlling UEP is calculated as $(\delta^*_{co}, 0) = (-0.2761, -0.7963, -0.2976, -0.2417, 2.6964, 0, 0, 0, 0, 0)$ and the critical energy is $V_{cr} = 3.3712$. Figure 12.20 presents the transient energy function along the preislanding trajectory (blue) and the critical energy (red). This trajectory corresponds to **Case I** (see Section 12.3.2.1) and the island a_1 must be islanded before $t_{e1} = 1.59$ s, that is to say, $\mathrm{T}_{FITI,a} = [0.62\ \text{s}, 1.59\ \text{s}]$. This time window was obtained from the intersection of the preislanding trajectory and the critical energy of island a_1, as defined by the controlling UEP. The trajectory of the second island, island b_1, was found to correspond to **Case II**, which is $\mathrm{T}_{FITI,b} = [0.62\ \text{s}, \infty)$ (see Section 12.3.2.1), so it retains its stability regardless of when the islanding occurs. Therefore, in this case, the FITI of the system is $\mathrm{T}_{FITI} = \mathrm{T}_{FITI,a} \cap \mathrm{T}_{FITI,b} = [0.62\ \text{s},\ 1.59\ \text{s}]$.

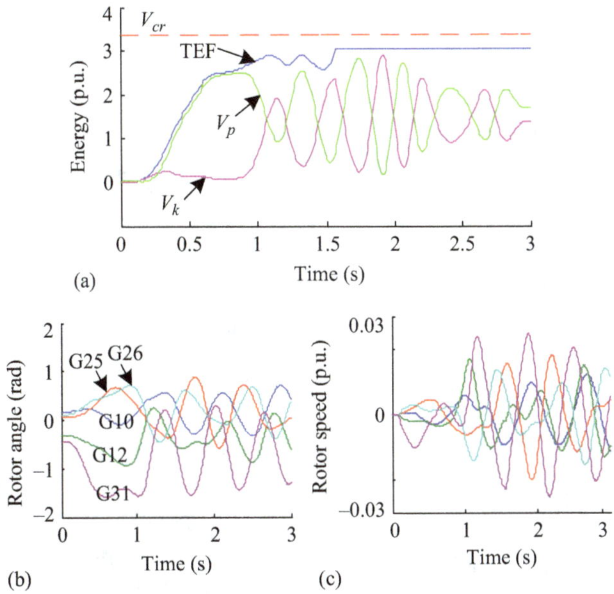

Figure 12.21 Simulation of island a_1 with $t_{sp} = 1.58$ s: (a) transient energy function; (b) rotor angle; and (c) rotor speed

In Figure 12.20 it can be observed that there is no abrupt change in the energy function along the whole preislanding process. With the network-reduced model, at each time instant t_k, $\mathrm{x}_{aP}\left(t_k^+\right)$ equals $\mathrm{x}_{a,pre}\left(t_k^-\right)$ and they only contain variables which cannot be abruptly changed after the islanding, that is, rotor speed and rotor angle.

If the islanding action is performed before t_{e1} $\left(t_{sp} = 1.58 \text{ s}\right)$, the energy of island a_1 at t_{sp} is less than the critical energy as shown in Figure 12.21(a), and thus, the island a_1 will retain its stability. It can be observed that, during the post-islanding process, the island a_1 experiences oscillations in kinetic energy $V_k(\omega)$ and potential energy $V_p(\delta)$, but the total energy $V_a(\delta, \omega)$ is constant, as the island a_1 is autonomous and the law of energy conservation must be satisfied. Figure 12.21(b) and (c) shows the rotor angles and rotor speed of generators belonging to island a_1 with respect to the COI.

On the contrary, if the islanding would be performed after t_{e1} $\left(t_{sp} = 1.65 \text{ s}\right)$, the energy of island a_1 at t_{sp} would be larger than the critical energy, and thus island a_1 would lose its stability as illustrated in Figure 12.22.

It is known that the controlling UEP method is rather conservative. For example, the critical islanding time obtained using the controlling UEP method is 1.59 s, where time-domain simulations show the critical islanding time to be 1.64 s.

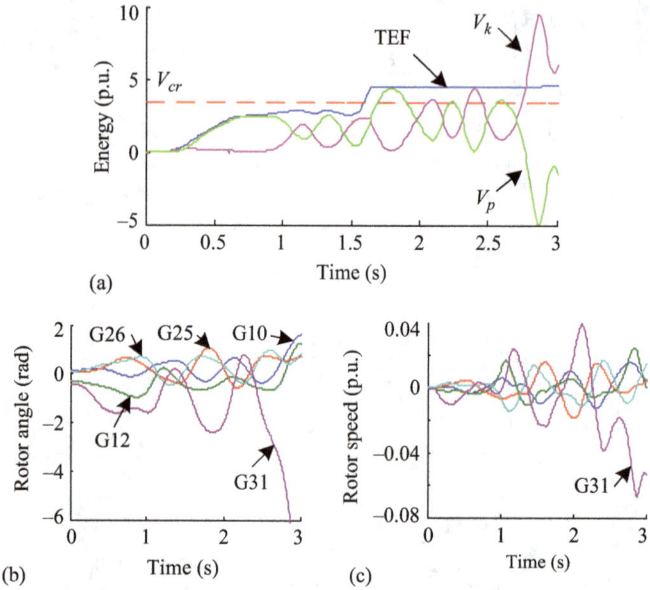

Figure 12.22 Simulation of island a_1 with $t_{sp} = 1.65$ s: (a) transient energy function; (b) rotor angle; and (c) rotor speed

This means that the use of the controlling UEP method for determining the FITI inherently provides a certain level of stability margin.

12.3.4 Variation in the FITI for different scenarios

The FITI is determined by the preislanding trajectory and the stability region of each island created. The preislanding trajectory is governed by the dynamics described as (12.20) and (12.21), and any parameter change such as variation in the size and the location of disturbances, will affect the preislanding trajectory and thus the FITI. Meanwhile the stability region of the island is mainly affected by the final topology of the island, which is determined by the islanding boundary if there is no further topology change after the islanding action. In this section, the variation in FITI for scenarios with different preislanding trajectories and islanding boundaries was explored using the IEEE 118-bus system.

12.3.4.1 Impact of the change of preislanding trajectory

To investigate the impact of changes in the preislanding trajectory on the FITI, simulations with different scenarios, which are shown in Table 12.7, have been undertaken. The procedure presented in Section 12.3.2.3 is executed for each scenario. For these simulations, the islanding boundary presented in Section 12.3.3 was used.

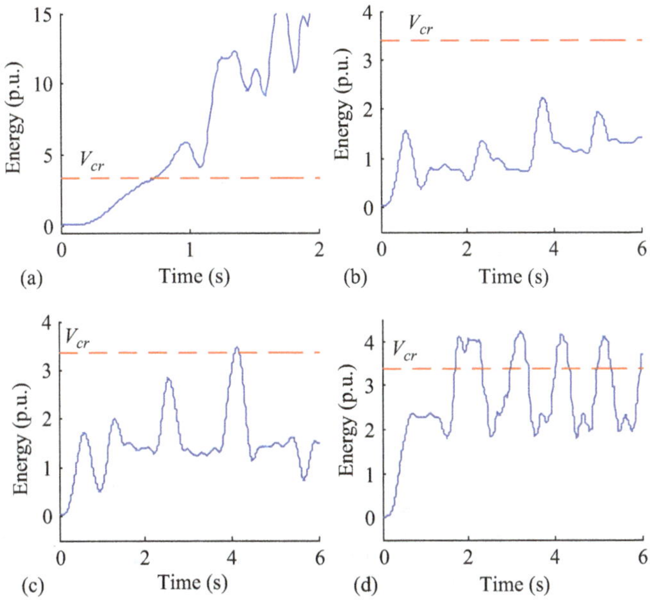

Figure 12.23 Transient energy function of island a_1 with different scenarios: (a) with Scenario S1; (b) with Scenario S2; (c) with Scenario S3; and (d) with Scenario S4

Table 12.7 Simulation scenarios

Scenario	Cascading events $[t_1, t_2, t_3, t_4]$	FITI (island a_1)
S1	[0.01 s, 0.47 s, 0.10 s, 0.64 s]	[0.64 s, 0.74 s]
S2	[0.01 s, 0.17 s, 0.10 s, 0.54 s]	[0.54 s, ∞)
S3	[0.01 s, 0.27 s, 0.10 s, 0.54 s]	[0.54 s, 4.07 s], [4.18 s, ∞)
S4	[0.01 s, 0.35 s, 0.10 s, 0.61 s]	[0.61 s, 1.69 s], [2.31 s, 2.95 s], [3.38 s, 4.02 s],...

The trajectory in Figure 12.23(a) demonstrates an example of **Case I**, which exists an intersection point between the energy curve and the critical energy along the preislanding trajectory. The trajectory in Figure 12.23(b) demonstrates an example of **Case II**, in which the energy is always smaller than the critical energy.

The trajectories presented in Figure 12.23(c) and (d) have identified cases with multiple intersections between the energy curve and the critical energy, these correspond to **Case III** and **Case IV**.

Table 12.8 Islanding boundaries for the IEEE 118 bus network

Islanding boundary	Created islands	
	G_a	G_b
B1: 19-34, 30-38, 23-24	a_1	b_1
B2: 14-15, 13-15, 17-15, 18-19, 20-19, 30-38, 70-69, 75-69, 75-118, 75-77	a_2	b_2
B3: 33-37, 34-37, 44-45, 34-36, 26-30, 113-17, 31-17, 8-30, 16-17, 15-17, 19-18, 71-70, 24-70	a_3	b_3
	G'_a	G'_b
B4: 34-37, 46-48, 46-47, 45-49, 34-36, 30-38, 24-23	a_4	b_4

Table 12.9 Postislanding SEP and CUEP of the created islands

a_1	SEP	[0.1401, –0.3606, 0.1108, 0.1445, –0.5103, 0, 0, 0, 0, 0]
	CUEP	[–0.2761, –0.7963, –0.2976, –0.2417, 2.6964, 0, 0, 0, 0, 0]
a_2	SEP	[0.1390, –0.3423, 0.1051, 0.1448, –0.5161, 0, 0, 0, 0, 0]
	CUEP	[–0.2964, –0.8026, –0.2991, –0.2407, 2.756, 0, 0, 0, 0, 0, 0]

From the above analysis, it can be concluded that the size of disturbances in cascading events significantly influences the preislanding trajectory and thus the selection of the FITI. This reveals that the FITI for an island will be highly variable for even small changes in the nature of cascading events.

In Table 12.7 all scenarios show that the newly created islands remain stable if the system islanding is performed during the first interval. However, the first interval may be too short to perform the controlled islanding, and also, it is possible that during some complicated situations when the control center makes the decision to split the system, the first interval has been missed. Once the islanding actions do not happen in the first interval, the knowledge about the following intervals would be valuable and critical to creating stable islands.

12.3.4.2 Impact of the change of islanding boundary

The islanding boundary can also affect the FITI by changing the SEPs and stability regions of islands to be formed. Since the coherent generator grouping and the identification of lines to be tripped are two key factors to be determined when forming an islanding boundary, different simulations are performed to investigate the impact of these factors on the FITI.

In Table 12.8, islanding boundaries **B1**, **B2**, and **B3** are based on the coherent generator groups $\mathbf{G}_a$ and $\mathbf{G}_b$ while islanding boundary **B4** is based on the coherent groups $\mathbf{G}'_a$ (generators {10, 12, 25, 26, 31, 46}) and $\mathbf{G}'_b$ (the rest of the generators).

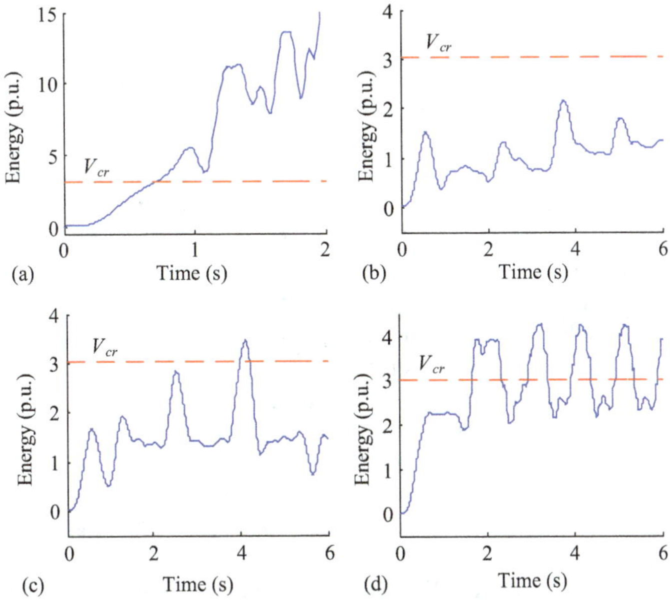

Figure 12.24 Transient energy of island a_2 with different scenarios: (a) with Scenario S1; (b) with Scenario S2; (c) with Scenario S3; and (d) with Scenario S4

As shown in Figure 12.23(a), with the scenario **S1** and islanding boundary **B1** (with coherent groups $\mathbf{G}_a$ and $\mathbf{G}_b$), the created island a_1 has the chance to retain its stability if the islanding is performed during $\mathrm{T}_{FITI,a} = [0.64\ \mathrm{s}, 0.74\ \mathrm{s}]$. On the contrary, with scenario **S1** and islanding boundary **B4** (with coherent groups $\mathbf{G}'_a$ and $\mathbf{G}'_b$), the created island a_4 always fails to retain its stability whenever the islanding is performed. It can be concluded that the identification of coherent generator grouping is key for controlled islanding, and improper identification will make the island fail to survive.

Furthermore, even though the coherent generator groups are identified properly, the selection of lines to be tripped will also affect the FITI and thus the success of the islanding.

In Table 12.9, the postislanding SEP and controlling UEP of the island a_2 (with islanding boundary **B2**) are shown, and the corresponding critical energy can thus be calculated as $V_{cr} = 3.0211$. In Figure 12.10, the energy of island a_2 for different scenarios shows that the FITI of island a_2 will be changed accordingly (Figure 12.24).

In addition, the FITI of island a_3 (with islanding boundary **B3**) is a null set, as the island formed does not have a postislanding SEP and, as such, is unstable regardless of when the controlled islanding is performed.

References

[1] Y. Bo, V. Vittal, and G. T. Heydt, "Slow-coherency-based controlled islanding – a demonstration of the approach on the August 14, 2003 blackout scenario," *IEEE Transactions on Power Systems*, vol. 21, no. 4, pp. 1840–1847, 2006.

[2] H.Song, J. Wu, and L.Wu, "Controlled islanding based on slow-coherency and KWP theory," *Innovative Smart Grid Technologies – Asia*, IEEE, 2012:1–6.

[3] A. Peiravi, and R. Ildarabadi, "A fast algorithm for intentional islanding of power systems using the multilevel kernel *k*-means approach," *Journal of Applied Sciences*, vol. 9, no. 12, pp. 2247–2255, 2009.

[4] M. W. Hirschand, and S. Smale, *Differential Equations, Dynamical Systems and Linear Algebra*. New York: Academic Press, 1974.

[5] L. Ding, F. Gonzalez-Longatt, P. Wall, and V. Terzija, "Two-step spectral clustering controlled islanding algorithm," *IEEE Transactions on Power Systems*, vol. 28, no. 1 pp. 75–84, 2013.

[6] J. Shi, and J. Malik, "Normalized cuts and image segmentation," *IEEE Transactions on Pattern Analysis and Machine Intelligence*, vol. 22, no. 8, pp. 888–905, 2000.

[7] U. V. Luxburg, "A tutorial on spectral clustering," *Statistics and Computing*, vol. 17, no. 4, pp. 395–416, 2007.

[8] J. H. Chow, *Time-scale modeling of dynamic networks with applications to power systems*. New York: Springer, 1982.

[9] S. S. Lamba, and R. Nath, "Coherency identification by the method of weak coupling," *International Journal of Electrical Power & Energy Systems*, vol. 7, no. 4, pp. 233–242, 1985.

[10] K. Sun, D. Zheng, and Q. Lu, "Splitting strategies for islanding operation of large-scale power systems using OBDD-based methods," *IEEE Transactions on Power Systems*, vol. 18, no. 2, pp. 912–922, 2003.

[11] V. E. Henner, "A network separation scheme for emergency control," *International Journal of Electrical Power & Energy Systems*, vol. 2, no. 2, pp. 109–114, 1980.

[12] L. Ding, and V. Terzija, "A new controlled islanding algorithm based on spectral clustering," in *Proceedings of the 4th IEEE International Conference on Electric Utility Deregulation, Restructuring, and Power Technology (DRPT2011),* Weihai, China, July, 2011.

[13] http://en.wikipedia.org/wiki/Arg_max

[14] S. Theodoridis, and K. Koutroumba, *Pattern recognition* (4th ed.). New York: American Press, 2008.

[15] T. D. Bie, J. Suykens, and B. D. Moor, "Learning from general label constraints," in *Proceedings of the IAPR International Workshop on Statistical Pattern Recognition*, Lisbon, August, 2004.

[16] K. Sun, D. Zheng, and Q. Lu, "A simulation study of OBDD-based proper splitting strategies for power systems under consideration of transient

stability," *IEEE Transactions on Power Systems*, vol. 20, no. 1, pp. 389–399, 2005.

[17] L. Hao, G. W. Rosenwald, J. Jung, and C. C. Liu, "Strategic power infrastructure defense," *Proceedings of the IEEE*, vol. 93, no. 5, pp. 918–933, 2005.

[18] Q. Zhao, K. Sun, D. Zheng, J. Ma, and Q. Lu, "A study of system splitting strategies for island operation of power system: a two-phase method based on OBDDs," *IEEE Transactions on Power Systems*, vol. 18, no. 4, pp. 1556–1565, 2003.

[19] Y. Qiao, "Study on the active splitting control scheme of large-scale power systems," Ph.D. Thesis, Tsinghua University, Beijing, 2007.

[20] A. A. Fouad, and V. Vittal, *Power system stability analysis using the transient energy function method*. Englewood Cliffs, NJ: Prentice Hall, 1992.

[21] H. D. Chiang, "The BCU method for direct stability analysis of electric power system: theory and applications," *Institute for Mathematics and Its Applications*, vol. 64, p. 39, 1995.

[22] H. D. Chiang, F. W. Wu, and P. P. Varaiya. "A BCU method for direct analysis of power system transient stability," *IEEE Transactions on Power Systems*, vol. 9 , no. 3. pp. 1194–1208, 1994.

[23] H. Chiang, *Directed methods for stability analysis of electrical power systems*. Hoboken, NJ: Wiley, 2011.

[24] P. Kundur, *Power system stability and control*. New York: McGraw-Hill, 1994.

Chapter 13

Intelligent voltage regulation scheme of active distribution networks with renewable energy integration

Yongxi Zhang[1], Fengji Luo[2] and Zhao Yang Dong[3]

13.1 Introduction

As a kind of fundamental infrastructure, power systems have existed for more than a century to serve the human society's energy demand. Voltage regulation plays an important role in maintaining modern power systems' reliable and secure operations. A system node's voltage is related to the balance of the system's both active and reactive power production and absorption. Voltage fluctuations are usually caused by the unbalanced power supply and demand. Traditionally, the system operator is responsible for making operation actions to ensure that the system's supply–demand balance is real-time ensured. With predicted load demand, voltage regulation can be achieved through adjusting the power sources' outputs to match the demand.

Driven by the globally grand challenges of climate change, ever-increasing energy demand and fossil fuel shortage, worldwide efforts have been put to the development of Smart Grid since the early of twenty-first century. As a result, modern power distribution systems are characterized by high penetration of dispersed renewable energy sources, integration of energy storage devices, deployment of pervasive sensing and metering devices, and active participation of energy customers. These new features pose new, nontrivial challenges to voltage regulation in distribution networks.

13.1.1 Increasing penetration of renewable energy and associated new challenges to power systems

The term "renewable energy" refers to energy derived from a broad range of self-renewing energy resources, such as sunlight, wind, flowing water, terrestrial heat,

[1]School of Electrical and Information Engineering, Changsha University of Science and Technology, Changsha, Hunan, China
[2]School of Civil Engineering, The University of Sydney, Sydney, Australia
[3]School of Electrical Engineering and Telecommunications, The University of New South Wales, Sydney, Australia

and biomass. To achieve sustainable and environmentally friendly energy supply, from a long time people all over the world have been implementing diverse renewable energy resources [1]. It has been globally recognized that renewable energy technologies, most of which obtain energy from the nature, will be the key to relieving global warming and energy shortage issues. Great efforts have been put to develop renewable energy in different countries around the world. The cumulative installed capacity of renewable energy grew rapidly in the last decades. As representative examples, renewable energy now contributes nearly 40% to the national energy production in Denmark, and 10% in the United States, respectively. The new investment in renewable energy in the United States in 2018 reaches USD 288.9 billion, far exceeding the fossil fuel investment. It is worthy to notice that renewable energy investment in most developing countries in 2018 has increased 6% compared with that in 2017. In 2018, China led investment worldwide for seventh successive years, at USD 91.2 billion, which accounts for 32% of the global total investment, followed by Europe, the United States, and Asia Oceania.

Integration of renewable energy brings potential opportunities on reducing carbon dioxide emission, increasing energy efficiency and enhancing power system's flexibility and reliability. At the same time, the stochastic nature of renewable energy also pose significant challenges to grid's operation [2]. Unlike traditional fossil fuel-driven generation units, outputs of renewable power sources are heavily influenced by the factors including meteorological condition, geographical features, and other external environmental factors. Most of the renewable energy resources exhibit strong seasonal and diurnal patterns, such as wind power and photovoltaic (PV) solar power. Another challenge lies in renewable power output's intermittency. Many renewable energy sources' outputs are unstable. Taking PV solar power as an example, the fact that solar radiation is easily be affected by the movement of clouds, which makes PV solar power output be usually highly fluctuated and intermittent. This consequentially causes difficulties for the power system's operators to maintain the real-time supply–demand balance.

The intermittent and stochastic nature of renewable energy would cause frequent and non-negligible voltage fluctuations (i.e., voltage rise and drop at the timescale of seconds) when there is a mismatch between energy demand and renewable power generation. To tackle this issue, a lot of solutions have been proposed. These include:

1. System upgrade such as increasing the conductor's capacity and installing voltage regulators. This approach is usually expensive and labor-intensive;
2. Use traditional voltage regulation devices such as capacitor banks and on-load tap changers. However, most of these devices are designed to operate at the timescale of hours, thus are not easy to provide fast response to sudden changes in renewable power;
3. Renewable power curtailment at the customer side. Although curtailing renewable power can help to regulate the voltage in some cases, this approach is not preferable due to the waste of renewable energy.

Based on the above discussion, the increasing penetration of renewable energy in modern power systems naturally calls research and development of advanced voltage regulation schemes that are capable of fast and efficiently reacting to the rapid fluctuations of both the renewable power and energy demand.

13.1.2 Voltage control in distribution networks with integration of renewable energy sources

Traditional power distribution networks are primarily designed to support unidirectional power flows, which means power flows from the transmission grid to the high/medium voltage substations and then are delivered to end energy customers. With the increasing penetration of renewable energy sources, such unidirectional power flows are changed to bidirectional power flows. The "reverse" power flows (i.e., flow from the edged energy entities to the main grid) caused by the renewable energy would lead to a series of operation issues to the grid, including the voltage rise/drop problems.

Since power distribution networks are directly connected to end-users, it is important to keep voltage profiles within a predefined secure range. Under- or over-voltage may cause damages to system equipment. According to the IEEE standard 1250-1995 [3] and the American National Standards Institute (ANSI) std C84.1-2006 [4], the allowable voltages range at the distribution transformer secondary terminals should be set as $\pm 5\%$ of the nominal value. Traditionally, several methods have been applied to ensure the system's voltage to be within its secure range. These methods include:

1. Adjust zone substation OLTC transformers;
2. Regulate voltage regulators and/or capacitor banks;
3. Increase line drop compensation;
4. Install more voltage regulators and capacitor banks;
5. Install flexible AC transmission system (FACTS) devices;
6. Rearrange the system or transferring loads; and
7. Balance loads between phases.

In the context of smart grid, voltage control is much more complicated than traditional bulk power systems due to several factors such as more complicated system structure, uncertainty of renewable energy, and ever-increasing energy demand. Therefore, more advanced control and optimization schemes are required for system operators to perform voltage regulation. In literature, a number of research efforts have been done on voltage regulation in distribution networks in the presence of renewable energy [5–7].

13.1.3 Categories of voltage regulation methods

Real-time voltage control has been studied extensively in both academia and industry in last decades. Different control methods have been proposed with distributed renewable energy resources in distribution systems. These techniques can be generally classified into four categories based on their control architectures:

(1) centralized control [8,9]; (2) local control (decentralized) [10,11]; (3) hybrid control [12,13]; and (4) distributed multiagent control [14,15]. This classification is depicted in Figure 13.1. Each method has its own advantages and technical limitations [17]:

1. Centralized control. It requires a central controller to monitor and collect all information to determine the optimal set point for the whole network. Centralized control has high requirements for communication systems, sensors, and control systems, and thus has relatively high communication and computation cost. It also has the single-point-of-failure risk and is vulnerable to external attacks.
2. Local control. In local control, each component receives and collects information based on local signals. Local control has the advantage of being easy to implement and has less requirements on communication architectures; it is thus the most commonly used method in practice. However, due to limited information access, the overall efficiency of local control is low.
3. Hybrid control. Hybrid control refers to the combination of both centralized control and local control to achieve greater flexibility.
4. Distributed multiagent control. This category of control method only requires local information exchange and neighbor–neighbor communication among individual agents. It does not need a central controller to achieve mutual objectives that are hard to be achieved individually. Distributed control is suitable for a large system with a number of small devices.

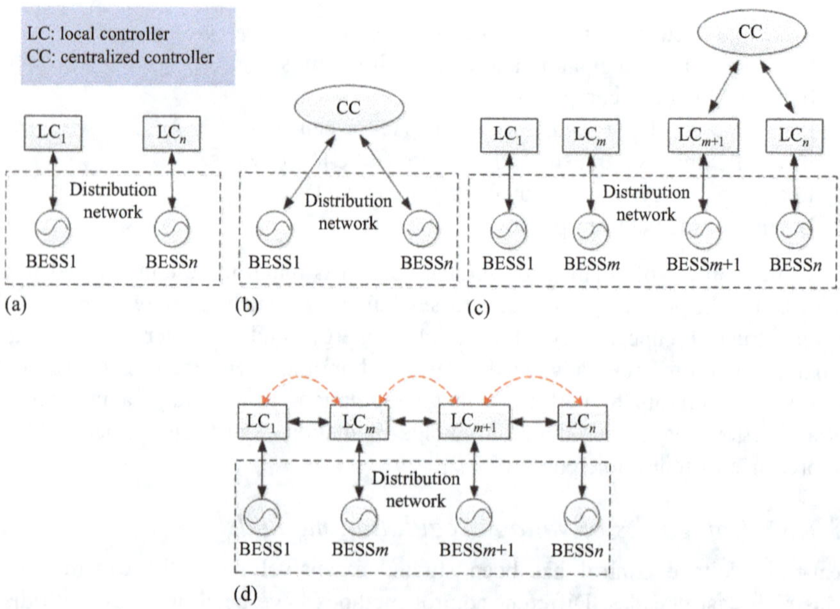

Figure 13.1 Voltage control architecture (Modified from [16]) (a) Local control (b) Centralized control (c) Hybrid control (d) Distributed control

To sum up, centralized and hybrid control methods require high level of communication between different agents of the system. They thus usually have high deployment costs. Decentralized and distributed control methods only need local information to control system voltage, and they have less requirements on computing and communication. One limitation of decentralized control is that although the control actions are reliable, it does not provide global solutions for the whole system. In conclusion, all the aforementioned control methods can provide voltage control support to power systems with renewable energy; choosing suitable voltage regulation method should be based on the practical considerations and the system's practical operation conditions.

13.1.4 Optimization techniques in active distribution networks

Most of the voltage control problems can be modeled as optimization problems, which aim to determine the optimal values of a set of decision variables to maximize or minimize certain objective functions while subject to one or multiple constraints. Optimization problems can be solved by either mathematical programming methods or heuristic optimization techniques; both have received significant attention in power system's research.

It is foreseeable that in future, power distribution networks will be developed toward complex cyber-physical systems that aggregate a large number of energy agents, including distributed generation units, storage units, electric vehicles (EVs), and loads with demand response (DR) capabilities. These agents will interact and exchange information with the support of advanced communication networks. Such interaction and communication will further complicate the system's operation that is already complex. In this context, advanced optimization technologies are required to support modern power systems' planning and operations. These technologies mainly include robust control/optimization and distributed optimization/control techniques.

Mathematical programming methods have been applied in a wide range of power system planning, operation, and control problems [18], such as optimal power flow, security-constrained economic dispatch, unit commitment, power system expansion planning, and transmission expansion planning [19]. It is noteworthy that many uncertainties exist in power system voltage control, such as line or equipment availability, system load demand variation, system marketing rules and government policy, and related environmental costs. Voltage control problems have become even more complicated than ever due to the variation of renewable power generation output. Specifically, many voltage control models are highly stochastic, multiperiod, MINLP (mixed-integer nonlinear programming), which are quite difficult to solve. Some of the problems are even insoluble using state-of-the-art algorithms. Therefore, more advanced optimization approaches and heuristic algorithms are required to solve these problems. Moreover, the future grid development involves the growth of information exchange in modern communication networks. The increasing size of information requires statistical methods for big-data analysis. The "big-data" fact also motivates the employment of distributed optimization techniques and parallel algorithms.

13.1.5 Voltage control with BESS in distribution networks

Energy storage technologies have been rapidly developing in recent years. Currently there are different kinds of available energy storage techniques, including battery, fly-wheel, pumped hydro systems, compressed air energy storage, electrochemical double-layer capacitors (EDLCs), and so on [20]. In the current market, compared with other energy storage systems, battery energy storage systems (BESS) is one of the most cost-effective ones due to its advantages of fast response time and high ramp rates [21]. Essentially, BESSs are made up of a series of stacked cells where chemical energy is transferred into electrical energy and vice versa. According to a recent report from the DOE Global Energy Storage Database, the global installed capacity of BESS is expected to increase from 1.5 GW to 14 GW in 2020 [22].

Generally, BESS is rated in terms of their power and energy ratings. Other factors influencing BESS operation include energy efficiency, life-span (or expressed as the numbers of total charge/discharge cycles), operating temperatures, self-discharge rate, and energy density. The cost of BESS depends on many factors and can be categorized into capital cost and operation and maintenance costs. It is worth noticing that although the development of material science and power electronic techniques greatly facilitates the integration of BESS, the relatively high cost of BESS nowadays still prevents it for massive application.

BESS can support distribution networks' operations in several aspects, including accommodating renewable energy integration, leveling load demand curve, regulating system frequency, deferring system upgrade, and so on. It is worthy to notice that the R/X ratio in distribution network is much larger than that in transmission networks; therefore, the voltage in distribution networks is more sensitive to active power's variation. Based on this, BESS has large potential to be used in distribution networks to achieve voltage control. BESS can supply or absorb power when there is a mismatch between power supply and load consumption. Therefore, the active power imbalance induced by high penetration of renewable generations can be effectively solved by the utilization of BESSs. The utilization of BESS for voltage control in distribution networks has been studied in several papers [4,23,24]. References [4,23] present coordinated control schemes of PV and BESS for voltage regulation using decentralized control methods. Reference [24] establishes a hybrid voltage control scheme consisting of local droop control and distributed control to enable the effective use of BESS under different conditions.

However, renewable energy and load demand uncertainties are overlooked in these works. Renewable energy generation varies from time to time and cannot be perfectly predicted. Inaccurate forecast might cause BESS units to take ineffective actions over the entire time horizon. Besides, existing methods often involve a large number of control variables with massive variations. Frequent usage may accelerate battery's degradation and significantly decrease the BESS's life span [25]. Therefore, it would be preferable to use less control variables regarding realistic constraints (e.g. battery capacity and battery limited life cycle numbers). Moreover, from the perspective of the distribution network operator, large variations of BESS units power output would not only disturb the system operation but also increase the communication and computation burdens for large distribution networks.

Despite of the large potential of using BESSs to support voltage regulation in distribution networks, there are still some critical questions that are needed to be answered: how to design control algorithms for BESSs to regulate the voltage profiles effectively in distribution networks consisting of distributed renewable energy sources? How to exchange information in a large distributed system with an enormous amount of information data kept privileged? These questions motive us to propose a distributed voltage control scheme using multiple BESSs for distribution networks with high renewable energy penetration. The major contributions of this research work include:

1. Multiple possible weather scenarios were taken into consideration to deal with the renewable uncertainty; the optimal BESS control strategy is determined considering all the scenarios;
2. The sparse regulation technique is applied in the objective function to make effective usage of BESS with minimal numbers and variations of BESS units;
3. A fully distributed voltage control scheme is proposed based on the distributed-alternating direction method of multipliers algorithm (D-ADMM). The scheme enables multiple BESS units to exchange information with each other in a fully distributed manner and decide their optimal power outputs.

This chapter is organized as follows. Section 13.2 gives a brief description of the proposed framework and the proposed voltage control model. Section 13.3 introduced an efficient distributed algorithm to solve the objective function. Case studies were given in Section 13.4 to verify the effectiveness and accuracy of the proposed method. Section 13.5 concludes the chapter.

13.2 System structure and problem formulation

The conceptual architecture of an active distribution network is shown in Figure 13.2. As shown in Figure 13.2, the system consists of multiple BESSs, wind turbines, PV solar power sources, and load energy demands. These energy resources are monitored and controlled by distributed controllers. Each distributed renewable resource and BESS unit is considered as an independent agent. If there is a voltage violation event detected (either voltage rise or drop) on the bus, all BESSs will be activated at once to provide active power support for regulating the bus voltage. There is an underlying communication network that facilitates communication among the BESS units. The voltage control scheme in this distribution network model can be formulated as a finite horizon optimal control problem; it optimally controls the distributed BESSs to regulate the bus voltage.

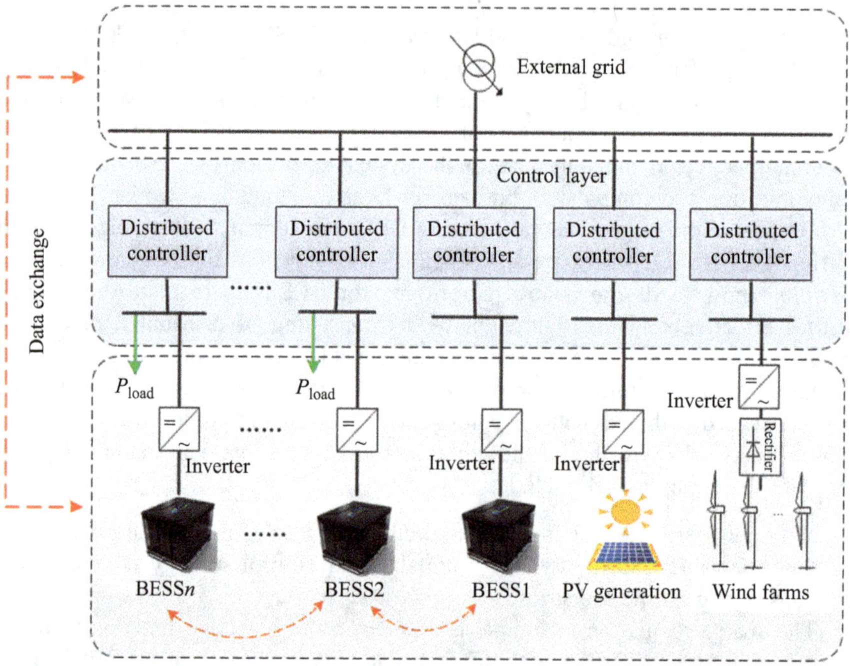

Figure 13.2 Structure of proposed distribution network

13.2.1 Scenario generation

Various scenarios are generated using Monte Carlo simulation method. Wind speed, load forecast errors, PV generation output are incorporated in each scenario. Each scenario and their probabilities are represented by a discrete distribution function.

13.2.1.1 Modeling of renewable energy and load uncertainties

The prediction models of PV solar power and power demand used in this study are presented as follows.

Solar power prediction model

Solar power output is directly related to the amount of solar irradiance received by the PV solar panel. Solar irradiance can be modeled based on the clearness index k_t and the extraterrestrial solar irradiance I_0 (kW/m^2); the clearness index can be modeled as the ratio of the irradiance on a flat level I_t (kW/m^2):

$$k_t = \frac{I_t}{I_0} \tag{13.1}$$

Solar irradiance with an inclining angle of β with respect to the horizontal plane I_β is calculated as

$$I_\beta = T \cdot k_t - T' \cdot k_t^2 \tag{13.2}$$

where T and T' are calculated as

$$T = \left[\left(R_b + \rho \cdot \frac{1 - \cos\beta}{2} \right) + \left(\frac{1 + \cos\beta}{2} - R_b \right) \cdot p \right] \cdot r_d \cdot \frac{H_0}{3,600} \tag{13.3}$$

$$T' = \left(\frac{1 + \cos\beta}{2} - R_b \right) \cdot q \cdot r_d \cdot \frac{H_0}{3,600} \tag{13.4}$$

The diffuse fraction k can be calculated based on the parameters p and q. k is also related to the correlation of the clearness index k_t:

$$k = p - qk_t. \tag{13.5}$$

PV power output can be calculated based on the simulated clearness index:

$$P'_{PV} = A_c \cdot \eta'_{PV} \cdot I_\beta = A_c \cdot \eta'_{PV} \cdot \left(T \cdot k_t - T' \cdot k_t^2 \right) \tag{13.6}$$

where A_c denotes the area of the ith PV unit (m^2) [26].

Wind power prediction model

The power output generated by the wind turbine is directly related to wind speed. The variations of wind speed have been statistically proven to follow the Weibull distribution [27]. Its probability density function is expressed as

$$f(v) = \frac{k}{c} \left[\frac{v}{c} \right]^{k-1} \exp\left[-\left(\frac{v}{c} \right)^k \right] \tag{13.7}$$

where v, k, and c are wind speed, shape factor, and scale factor, respectively. Based on wind speed, the power output of the wind turbine can be described as a function of wind speed:

$$P_w(t) = \begin{cases} 0 & v(t) \le v_{\text{in}}, v(t) \ge v_{\text{out}} \\ \dfrac{v(t) - v_{\text{in}}}{v_r - v_{\text{in}}} P_{\text{rated}} & v_i \le v(t) \le v_r \\ P_{\text{rated}} & v_r \le v(t) \le v_{\text{out}} \end{cases} \tag{13.8}$$

where $P_w(t)$ and $P_{\text{rated},i}$ denote the wind power output at time t and the rated wind power output of the ith wind turbine, respectively.

Power demand prediction model

Historical data can be used to fit the distribution function to represent the load forecasting error. In literature, a normal distribution is usually applied to model the load forecast uncertainty. The predicted power load $P_{\text{load}}(t)$ at time t (kW) is calculated:

$$P_{\text{load}}(t) = \bar{P}_{\text{load}}(t) + \delta_{\text{load}}(t) \tag{13.9}$$

where $\bar{P}_{\text{load}}(t)$ and $\delta_{\text{load}}(t)$ represent the predicted load value and the prediction error, respectively. In this work, the load prediction error is assumed to follow a truncated normal distribution with zero mean and a standard deviation of 5% [28].

13.2.1.2 Scenario generation and reduction

System operator needs to make voltage control actions immediately once the voltage violation event is detected. As introduced in previous sections, the prediction accuracy for renewable energy and load demand would decrease with the increase of the prediction time horizon. Therefore, different possible weather scenarios are required. In this work, Monte Carlo simulation is used to generate scenarios to represent wind speed, load, and solar power generation. S_1, $S_2, \ldots$ and S_{NS} denote different possible weather scenarios over the whole control horizon. The detailed procedures of scenario generation and reduction are listed as following:

Step 1: Generate the first weather scenario based on well-developed renewable energy and load forecasting toolboxes and the configuration of the renewable energy source (for PV cells, the configuration information includes solar cell technical parameters temperatures and roof inclination; for wind turbines, the configuration information includes wind turbine details, surface roughness, and wind farm layout).

Step 2: Based on the method in Step 1, generate scenarios using the Monte Carlo simulation method. Then, using the scenario reduction technique presented in [28] to reduce the scenarios to include *NS* independent scenarios (denoted as S_1, $S_2, \ldots, S_{NS}$) that remain with discrete probabilities $p_1, p_2, \ldots, p_{NS}$ so that $p_1 + p_2 + \cdots + p_{NS} = 1$.

13.2.2 Problem formulation

13.2.2.1 Objective function

The proposed voltage control model is formulated and solved under different possible weather scenarios. The model determines (1) which BESSs are selected to control; and (2) how much power are charged/discharged to them. The objective function of the voltage control model is represented as (13.10). It aims to minimize the sum of all the BESS units' output power while satisfying the voltage constraints under all the possible weather scenarios. The obtained voltage control solution may not be optimal under one or more particular scenarios, but it will lead to the most effective solutions under all possible weather situations.

The objective is to minimize the operation cost of all the BESS units over a finite time horizon:

$$\min \sum_{y=1}^{NS} \sum_{\forall i} (f(P_i)) \tag{13.10}$$

$$f(P_i) = \sum_i \|P_i(t)\|_1, \ \forall i \in H \tag{13.11}$$

The decision variable P_i refers to the ith BESS's power output; $P_i = \left[P_{ch/dis,i}(1), P_{ch/dis,i}(2), \ldots, P_{ch/dis,i}(t)\right]^{\mathrm{T}} \in \mathbb{R}^T$. $P_{ch/dis,i}(t)$ refers to the charging/discharging power of the ith BESS at time t. H is the set of buses with BESS.

13.2.2.2 Constraints

Model (13.10) is subjected to the following constraints:

1. BESS operation constraints:

$$0 \le P_{ch,i}(t), \left|P_{dis,i}(t)\right| \le P^{bat}_{i,\max}, \ \forall i \in H, \forall t \in T \tag{13.12}$$

$$E_{i,\min} \le E_i(t) \le E_{i,\max}, \ \forall i \in H, \forall t \in T \tag{13.13}$$

$$SOC^{\xi}_{i,t} = C^{\xi}_{bat,t}/C_{bat,\max} \tag{13.14}$$

The dynamic status of energy stored in BESS can be derived as follows:

$$\begin{cases} E_i(t+1) = E_i(t) + P_i(t)\Delta t \\ E_i(t+1) = E_i(t-1) + P_i(t-1)\Delta t + \Delta P_i(t)\Delta t \\ E_i(t+1) = E_i(t-2) + P_i(t-2)\Delta t + P_i(t-1)\Delta t + \Delta P_i(t)\Delta t \\ \cdots \\ E_i(t+1) = E_i(1) + P_i(1)\Delta t + P_i(2)\Delta t + \cdots + \Delta P_i(t)\Delta t \end{cases} \tag{13.15}$$

where $E_i(1)$ is the initial energy of ith BESS unit. Equation (13.15) can be rewritten as

$$E_i(t+1) = E_i(t) + \left(P_{ch,i}(t)\eta_{ch} - P_{dis,i}(t)/\eta_{dis}\right)\Delta t, \forall i \in H, t \in T \tag{13.16}$$

where T and Δt are the total number of time steps over the whole control horizon and the duration of each time step, respectively; $E_i(t)$ is to the ith BESS's energy level at time t; $P^{bat}_{i,\max}$ denotes the ith BESS's rated power; $E_{i,\min}$ and $E_{i,\max}$ are lower and upper limits of ith BESS agent energy capacity; η_{ch} and η_{dis} refer to the BESS's charging/discharging efficiency, respectively. Constraint (13.12) restricts the ith BESS's power output which cannot exceed its power capacity. Constraints (13.13) and (13.14) restrict the energy stored in the ith BESS must be within an allowable range, so as to protect the BESS from overcharged/discharged and prolong its lifetime.

2. System voltage constraints:

$$v_{n,\min} \le v_n(t) \le v_{n,\max}, \ \forall n \in N, \forall t \in T \tag{13.17}$$

where N denotes the set of system buses; $v_{n,\min}$ and $v_{n,\max}$ are lower and upper voltage limits for bus n; (13.17) denotes the voltage constraints for all the system buses n at time t.

3. Power balance constraint:

The generation amount should be equal to the load amount at all times:

$$\sum\nolimits_{i \in H} P_{ch/dis,i}(t) + \sum\nolimits_{i \in M} P_{PV,i}(t) + \sum\nolimits_{i \in G} P_{\text{wind},i}(t) + P_{\text{ext}}(t) = P_{\text{load}}(t), \forall t \in T \tag{13.18}$$

where M is the sets of buses with PV sources. It is assumed that the distribution network operator can purchase electricity from the external grid. $P_{\text{ext}}(t)$

denotes the purchased power at time t. Note that in this work, we do not consider power losses.

4. PV power constraints:

$$0 \leq P_{PV,i}(t) \leq P_{PV,i}^{\text{rated}}, \forall i \in M, \forall t \in T \tag{13.19}$$

where $P_{PV,i}^{\text{rated}}$ denotes the rated power capacity of the ith PV source. Equation (13.12) denotes PV output power constraint.

5. Wind power constraint:

The power generation of the ith wind turbine units should not exceed the rated wind power output capacity:

$$0 \leq P_{wi,t}^{\xi} \leq P_{\text{rated},i} \tag{13.20}$$

6. External power constraint:

$$0 \leq P_{\text{ext}}(t) \leq P_{\text{ext,max}}, \forall t \in T \tag{13.21}$$

where $P_{\text{ext,max}}$ denotes the maximum power that can be drawn from the external grid.

7. Power factor constraint:

$$\text{PF}^{\min} \leq |\text{PF}|_{i,t} \leq \text{PF}^{\max} \tag{13.22}$$

This constraint indicates that all the system bus i at time t must be within given range.

13.2.2.3 Estimation of voltage profiles in distribution network

Once the system operator detects any voltage violation under normal operating conditions, the system operator will estimate the active power required to clear the voltage violation to bring voltage back to normal range as soon as possible. However, considering the practical situation (BESS capacity limits, PV output fluctuation, and severe over/under-voltage problem), sometimes SO cannot bring voltage back to normal range immediately. Therefore, a multitime step voltage control scheme is proposed to allow SO smoothly eliminate voltage violation at the end of entire control horizon ΔT.

In the distribution network, changes in power injection in system buses will induce changes in system voltage. The relationship between system power variation and voltage profiles can be estimated through Jacobian sensitivity matrix derived from Newton–Raphson power flow calculation [14]:

$$\begin{pmatrix} \Delta\theta \\ \Delta V \end{pmatrix} = \begin{pmatrix} \Lambda_{\theta P}\Lambda_{\theta Q} \\ \Lambda_{VP}\Lambda_{VQ} \end{pmatrix} \begin{pmatrix} \Delta P \\ \Delta Q \end{pmatrix} \tag{13.23}$$

where ΔP and ΔQ denote small variations in system active and reactive power output; $\Delta\theta$ and ΔV refer to the variations in phase angles and voltage profile. Due to the high r/x ratio value in DN, changes in active power injections might significantly affect bus voltage variation (reactive power ΔQ injection has little effects and thus can be neglected). The voltage value at a time step can be approximately calculated based on the voltage at the previous time step, the Jacobian sensitivity matrix, and the active power output change Δp:

$$\begin{cases} v_n(t+1) = v_n(t) + \Lambda_{VP}\Delta p(t) \\ v_n(t+1) = v_n(t-1) + \Lambda_{VP}\Delta p(t-1) + \Lambda_{VP}\Delta p(t) \\ v_n(t+1) = v_n(t-2) + \Lambda_{VP}\Delta p(t-2) + \Lambda_{VP}\Delta p(t-1) + \Lambda_{VP}\Delta p(t) \\ \cdots \\ v_n(t+1) = v_n(1) + \Lambda_{VP}\Delta p(1) + \Lambda_{VP}\Delta p(2) + \cdots + \Lambda_{VP}\Delta p(t-1) + \Lambda_{VP}\Delta p(t) \end{cases} \tag{13.24}$$

Note that the term of the Jacobian matrix Λ_{VP} depends on the network configuration and operation conditions. In practical, the sensitivity matrix does not change significantly with time; in this work, we assume Λ_{VP} remains unchanged over the entire horizon.

13.3 D-ADMM-based solving approach

The formulated problem (13.4)–(13.14) is a convex optimization problem with equality and inequality constraints. Traditional centralized methods may encounter computation burden for distribution system with multiple BESSs. In this section, a distributed multiagent voltage control scheme based on D-ADMM algorithm is developed for SO to keep the voltage profiles of within the given range over the entire time horizon considering renewable energy and load uncertainties. Details of the proposed control scheme are explained in the following paragraph.

13.3.1 Network communication and notations

D-ADMM is the distributed version of ADMM for solving separable optimization problems in networks of interconnected agents, recently has attracted much attention in control and signal processing [29]. The main advantages of D-ADMM include (1) it can be implemented on any connected network topology with sparsely connected communication topology, which needs less communication requirements to achieve given accuracy level compared to the state-of-art algorithms; (2) D-ADMM is asynchronous in some sense since the nodes operates in a color-based order, in every iteration, each unit is only interested in knowing neighbor units' information, agents with the same color update their variables in parallel, without the participation of central controller, thus greatly accelerate the convergence speed.

The proposed algorithm is implemented based on the multiagent system framework, each agent is able to exchange information with its neighbors to complete the determined objective. The principle of D-ADMM algorithm is that different agents are marked according to a color scheme and that local actions follow a color-based order to apply distributed optimization. Considering a distribution network consist of H controllable BESS agents that can communicate with each

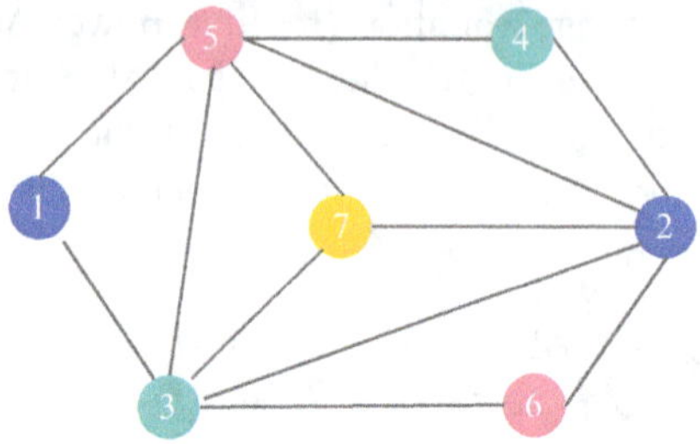

Figure 13.3 Example of colored MAS based framework

other, the DN can be modeled as an undirected graph $\mathcal{G} = \{\mathcal{V}, \varepsilon\}$, where $\mathcal{V} = \{1, 2, \ldots, H\}$ denotes the set of BESS agents and $\varepsilon \subseteq \mathcal{V} \times \mathcal{V}$ is the set of edges. An edge is set as (i,j), with $i < j$ and $i,j \in \mathcal{V}$ and denotes BESS agents i and j can exchange information with each other. The BESS agents connected with the ith BESS agent is defined as the ith BESS's neighborhood and denoted as $\mathcal{H}_i = \{j : (i,j) \in \varepsilon\}$. The degree of BESS agent i (denoted as D_i) is defined as the cardinality of $|\mathcal{H}_i|$, that is, $D_i = |\mathcal{H}_i|$.

We introduce the algorithm through a simple example. We assume that the network is given with a coloring scheme of L different colors. The first C_1 BESS agents are assigned with color 1, numbered as $\{1, 2, \ldots, C_1\}$, denoted by $\mathcal{C}_1$. The rest BESS agents are colored in the same way. C_l BESS agents marked in color l are counted as $\{C_{l-1} + 1, \ldots, C_{l-1} + C_l\}$, denoted by $\mathcal{C}_l$, and there is $|\mathcal{C}_1| + |\mathcal{C}_2| + \cdots + |\mathcal{C}_l| = H$. Figure 13.3 illustrates an example of a colored network, which comprises 7 BESS agents and 12 edges. The agents are assigned with four different colors.

- $\mathcal{C}_1 = \{1, 2\}$, $|\mathcal{C}_1| = 2$; $\mathcal{C}_2 = \{3, 4\}$, $|\mathcal{C}_2| = 2$;
- $\mathcal{C}_3 = \{5, 6\}$, $|\mathcal{C}_3| = 2$; $\mathcal{C}_4 = \{7\}$, $|\mathcal{C}_4| = 1$.

13.3.2 Distributed voltage control based on D-ADMM

13.3.2.1 Problem formulation

The proposed model in (13.14)–(13.24) is the sum of separable convex objective functions with linear constraints. Here, we can decouple problem (13.15) into the sum of individual objective functions of each BESS agent i by assigning copies of P_i to each agent and then constrain all copies to be equal, expressed as

$$\begin{aligned} &\min \textstyle\sum_{i=1}^{NS} f_i(P_i) \\ &\quad \text{s.t. } P_i \in \mathcal{P}_i \; \forall i \in NS \\ &P_i = P_j, i,j \in \varepsilon \end{aligned} \tag{13.25}$$

where $P_i = (P_1, \ldots, P_H)$ represents the decision variables; $P_i \in \mathbb{R}^n$ denotes the copy hold by the BESS agent i. $f_i(P_i)$ refers to the separated local objective functions for the ith BESS agent. $P_i \in \mathcal{P}_i$ represents the feasible local variable set $\mathcal{P}_i$ for BESS agent i by means of the local vectors P_i, which summarizes the inequality constraints (13.12)–(13.13), (13.17), (13.19)–(13.22) and equality constraints in (13.16), (13.18), and (13.24), which are convex, either. BESSi only can access to its private objective function f_i and private set $\mathcal{P}_i$. $P_i = P_j$ is for all BESS agents i and j with edge $i,j \in \varepsilon$, which enforces all copies to be equal since the network is connected.

With the network coloring scheme presented in Section 1.3.1, we can further convert the problem in (13.25) to make it appropriate for distributed implementation. Assuming the network topology is colored by L colors. BESS agents with color l are marked as C_l. We have $C_1, C_2, ..., C_l$ in total and its cardinality $C_l = |\mathcal{C}_l|$. Here $\{\mathcal{C}_l\}_{l=1}^{L}$ partitions $\mathcal{V}$. Therefore, problem (13.25) can be further rewritten as

$$\begin{aligned} &\min_{\overline{P}_l} \textstyle\sum_{l=1}^{L} f_i(\overline{P}_l) \\ &\text{s.t. } \overline{P_l} \in \overline{\mathcal{P}_l}, \quad l = 1, 2, ..., L \\ &\quad \textstyle\sum_{l=1}^{L} (D_l^T \otimes I_n)\overline{P_l} = 0 \end{aligned} \tag{13.26}$$

Similarly, $\overline{P_l} = (\overline{P}_1, \ldots, \overline{P}_l)$ and $\overline{P}_l \in (\mathbb{R}^n)^L$collects all the copies of BESS agents with color l. There is $D = \left[D_1^{\mathrm{T}}, \ldots, D_L^{\mathrm{T}}\right]^{\mathrm{T}}$, where D is the arc-incidence matrix of the network topology; I_n is the identity matrix in $\Re^n$; and $\otimes$ denotes the Kronecker product. Each column of D is associated with an edge $(i,j) \in \varepsilon$. The ith and jth entry of D is with the value of either 1 or –1, and the remaining entries are set as 0. On the basis of this, the original model is transformed into a color-based form. Now it is ready to develop a distributed voltage control approach.

13.3.2.2 Voltage control based on D-ADMM

Model (13.26) can be reformulated in the way of putting the decision variables of the same-colored BESSs together to make the model be suitable for distributed implementation. Given L functions g_l, L sets P_l, and L sets of full column rank matrices A_l:

$$\begin{aligned} &\min_{P_l} \textstyle\sum_{l=1}^{L} g_l(P_l) \\ &\text{s.t. } P_l \in \mathcal{P}_l, \quad l = 1, 2, ..., L \\ &\quad \textstyle\sum_{l=1}^{L} A_l P_l = 0 \end{aligned} \tag{13.27}$$

where $P_l = (P_1, ..., P_L)$ is the optimization variable and $A_l = D_l^{\mathrm{T}} \otimes I_n$ is the transported node-arc incidence matrix of the graph $\mathcal{G}$. Then, the original problem [model (13.25)] is transferred to a color-based form. With this formulation, D-ADMM can be used to be applied to solve the model.

The augmented Lagrangian function with penalty terms of (13.27) is given as

$$L_\rho(P_l, \lambda) = \sum_{l=1}^{L} \left(\sum g_l(P_l) + \lambda^{k^{\mathrm{T}}} A_l P_l\right) + \frac{\rho}{2}\left\|\sum_{l=1}^{L} A_l P_l\right\|^2 \tag{13.28}$$

Here λ is the dual variable and ρ is a positive parameter. The D-ADMM algorithm iteratively calculates the value of P_l and λ according to the following procedures:

$$P_1^{k+1} = \arg\min_{P_1 \in \mathcal{P}_1} L_\rho(P_1, P_2^k, ..., P_L^k; \lambda^k) \tag{13.29}$$

$$P_2^{k+1} = \arg\min_{P_2 \in \mathcal{P}_2} L_\rho\left(P_1^{k+1}, P_2, ..., P_L^k; \lambda^k\right) \tag{13.30}$$

$$P_L^{k+1} = \arg\min_{P_L \in \mathcal{P}_L} L_\rho\left(P_1^{k+1}, ..., P_{L-1}^{k+1}, P_L; \lambda^k\right) \tag{13.31}$$

$$\lambda^{k+1} = \lambda^k + \rho \sum_{l=1}^{L} A_l P_l^{k+1} \tag{13.32}$$

where $i = 1, 2, \ldots, L$ and k denotes the iteration number. When $L = 2$, (13.29)–(13.32) becomes the ordinary ADMM and it converges under very mild assumption. When $L > 2$, there is a known proof of convergence when all the functions g_l are convex.

For BESS agents with color l, the optimization problem of (13.29)–(13.31) yields to C_l optimization problems that can be solved in parallel. In this way, the variable vector $P_l = (P_1, P_2, \ldots, P_L)$ is updated as

$$P_l^{k+1} = \underset{P_l \in \mathcal{P}_l}{\arg\min} \sum_{i \in C_l} f_i(P_i) + \lambda^k A_l P_i + \frac{\rho}{2} \left\| A_l P_l + \sum_{i=1,...,L,\, i \neq l} A_l P_i^k \right\|^2 \tag{13.33}$$

The first two terms of (13.33) are linear combinations of the Lagrangian and relative to the color l agents. The last term in (13.33) can be expressed as

$$\frac{\rho}{2} \left\| A_l P_l + \sum_{i=1,...,L,\, i \neq l} A_l P_i^k \right\|^2 = \frac{\rho}{2} P_l^{\mathrm{T}} A_l^{\mathrm{T}} A_l P_l + \rho P_l^{\mathrm{T}} \sum_{i=1,...,L,\, i \neq l} A_l^{\mathrm{T}} A_l P_i^k + \frac{\rho}{2} \left\| \sum_{i=1,...,L,\, i \neq l} A_l P_i^k \right\|^2 \tag{13.34}$$

Because color-l agents are not connected with each other, $A_l^{\mathrm{T}} A_l$ is a diagonal matrix, in which the diagonal elements being the degree of the respective agent. Then, (13.34) can be simplified as

$$\frac{\rho}{2} \left\| A_l P_l + \sum_{i=1,...,L,\, i \neq l} A_l P_i^k \right\|^2 = \sum_{i \in C_l} \frac{\rho G_i}{2} \left\| P_i \right\|^2 - \rho \sum_{i \in C_l} \sum_{j \in L} P_i^{\mathrm{T}} P_j^k + \frac{\rho}{2} \left\| \sum_{i=1,...,L,\, i \neq l} A_l P_i^k \right\|^2 \tag{13.35}$$

where $\lambda^k A_l P_l = \sum_{i \in \mathcal{L}_l} \sum_{j \in \mathcal{H}_i} \lambda_{ij}^{k\,\mathrm{T}} P_i$, λ_{ij} is defined for $i < j$ and associated with constraint $P_i = P_j$ in (13.25). The last term in (13.35) does not depend on P_l, and thus can be ignored. Then, (13.35) can be simplified as

$$P_l^{k+1} = \underset{\overline{P}_l \in \overline{\mathcal{P}}_l}{\arg\min} \sum_{i \in C_l} f_i(P_i) + \left(\gamma_p^k - \rho \sum_{j \in \mathcal{H}_i} P_j^k \right)^T P_l + \frac{\rho D_i}{2} \|P_i\|^2 \tag{13.36}$$

where $\gamma_i^k = \sum_{j \in \mathcal{H}_i} \lambda_{ij}^k$ for local BESS agent i. The auxiliary dual variable is expressed as

$$v_i^k = \gamma_i^k - \rho \sum_{j < i}^{j \in \mathcal{H}_i} P_j^{k+1} - \rho \sum_{j > i}^{j \in \mathcal{H}_i} P_j^k \tag{13.37}$$

for $i \in \mathcal{C}_l$.

Then problem (13.33) is decomposed into $\mathcal{C}_l$ subproblem and can be solved in parallel:

$$P_i^{k+1} = \arg\min_{P_i \in \mathcal{P}_i} f_i(P_i) + \sum_{i \in \mathcal{H}_i} v_i^{k^{\mathrm{T}}} P_i + \frac{\rho G_i}{2} \|P_i\|^2 \tag{13.38}$$

After the updates of P_i and dual variable γ_i^k of agent i, there is

$$\gamma_i^{k+1} = \gamma_i^k + \rho \sum_{j \in \mathcal{H}_i} \left(P_i^{k+1} - P_j^{k+1} \right) \tag{13.39}$$

In the computation process, the ith BESS receives power output information from all its neighboring BESS agents with lower color. After that, the BESS agent i can update according to (13.29); meanwhile, the BESS with the same color updates variables individually. Through this, the parallel implementation improves the convergence speed. Combine (13.28) and (13.37)–(13.39), we can obtain the D-ADMM based voltage control scheme, and it is summarized in Table 13.1:

Remark 1: The D-ADMM algorithm is proven to strictly converge in [29]. Besides, the communication topology and the BESS network management system are designed for convergence.

13.4 Case study

13.4.1 Simulation setup

In this section, two case studies were conducted to evaluate the performance of the proposed method. In Section 13.4.2, the proposed distributed voltage control scheme is evaluated on a modified IEEE 15-bus system and compared with the centralized voltage control approach. In Section 13.4.3, the scalability of the proposed scheme is investigated on a 43-bus radial system. All the simulations are implemented using the YALMIP toolbox and MATLAB® (versionR2018a) and are executed on a 3-GHz DELL PC with Intel i7-4790 3.60 GHz CPU and 8 GB RAM.

A 10-year, 15 min basis power load data set is used to represent the load variation; two 10-year, 15 min basis solar radiation in eastern Australia is collected from the RDA website [30] and is used to simulate the variation of PV output. Wind speed data used in this simulation is obtained through the National Renewable Energy Laboratory (NREL) 2006 eastern wind data set [31]. The PV penetration level is set to be 50% during peak load hours. Multiple scenarios are generated by using Monte Carlo simulation. The entire time horizon is set to be 2 h

Table 13.1: The distributed voltage control scheme in distribution network

---

Algorithm 13.1 D-ADMM-based voltage control scheme

Initialization
1: **For** s**=1:**NS scenarios do:
For t**=1:**T do:
PV power and load demand prediction
2: I. Set: k=1,
Initialize: $\gamma_i^1 = 0$, $\mathrm{P}_i^1 = 0$ ($\forall i \in H$).
3: **Repeat**:
4: [for l=1,2,..., L] do
5: [for $i \in C_l$ in parallel] do
(1) Update auxiliary dual variable according to (13.37).
(2) BESS agent i with color l receives the variables from $j \in \mathcal{H}_i$, update optimal variables $\{P_i^{k+1}\}_{i \in H}$ according to (13.38). Send updated value P_i^{k+1} to its neighbors.
6: **End**
7: **End**
(3) For all $L \in \mathcal{V}$, update the dual variables according to (13.29) to obtain γ_i^{k+1}.
8: II. Set $k = k + 1$ and return to Step I until a predefined stopping criterion is met.
9: **End**
10:**End**
Output: P_i, $I = 1, 2, \ldots, H$

---

with the duration of each time step of 5 min (i.e., $\Delta t = 5$ min), control horizon $\Delta T = 15$ min. Hence, $\Delta T = 3$, $T = 24$ time steps. $P_i \in \mathbb{R}^{24\times 1}$, where $P_i(1)$ represents the output power of the ith BESS agent at the initial time step (5 min).

To demonstrate the effectiveness of the proposed method, the simulation results over k iterations from the proposed distributed voltage control model are compared with the results of the centralized approach that uses the interior-point method:

$$\begin{aligned} \varepsilon(k) &= \left|\sum f_i(P_i^k) - J_i\right|/J_i \\ x^{\mathrm{diff}}(k) &= \left|P_i^k - x^*\right|/x^* \end{aligned} \tag{13.40}$$

where $f_i(P_i^k)$ and P_i^k are objective function value and solution of the proposed D-ADMM control scheme at the kth iteration; J_i and x^* denote the objective function value and solution obtained from the centralized solving approach.

13.4.2 15-bus radial distribution system

First, we test the proposed method on a modified 11 kV, IEEE 15-bus system, whose topology is shown in Figure 13.4. The network parameters of the system are available in [32]. The secure range of the voltage in each bus (excluding the slack bus) is set as [0.95, 1.05] p.u. One PV panel and one wind farm are located on bus 8 and 10, respectively; they can operate up to 0.9 power factor. The solar panel considered for this study is the BP3200SX panel, whose technical parameters are obtained from [33]. The wind farm consists of two identical Vestas V52-850 kW wind turbines, where the cut-in, cut-out, and rated speeds are 4, 17, and 25 m/s, respectively. Air density is considered to be 1.2 kg/m^3. Three scenarios are generated to represent different possible weather scenarios, denoted as A, B, and C, respectively. The occurrence probability of scenario A, B, and C are 0.4, 0.3, and 0.3, respectively. The forecasted load, wind and PV power outputs under the three scenarios are given in Table 13.2. The minimum allowable power bought from substation is assumed to be 0.05 MWh with the price of $100/MW. Five BESS units are considered to be installed at bus 4, 7, 9, 11, and 14, respectively; the initial energy stored in each BESS is considered to be 20% of the full capacity. Some key BESS parameters are given in Table 13.3.

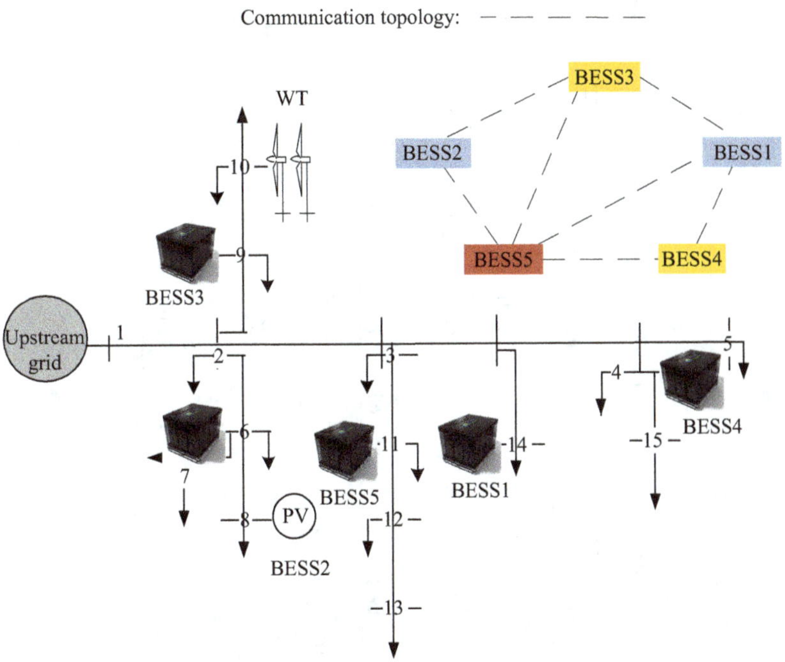

Figure 13.4 Communication network of modified 15-bus distribution system with five BESS units, seven edges labeled in three colors

Table 13.2 Forecast wind and PV power generation and system demand

Time step		01	02	03	04	05	06	07	08	09	10	11	12
Wind speed (m/s)	A	5.18	5.3	5.51	5.72	5.78	5.85	5.72	5.58	5.38	5.26	5.22	5.17
	B	4.51	4.80	5.06	5.25	5.31	5.38	5.34	5.33	5.26	5.02	4.62	4.11
	C	4.47	4.72	4.98	5.20	5.31	5.31	5.20	5.02	4.72	4.34	4.13	4.01
Solar radiation (kW/m^2)	A	70.86	76.09	74.54	70.53	64.79	58.06	69.90	75.49	73.40	67.91	60.21	51.47
	B	52.87	57.10	55.46	51.17	45.18	38.43	42.84	46.27	44.84	41.09	35.88	30.04
	C	60.08	65.91	60.08	65.91	60.08	65.91	59.79	64.50	59.79	64.50	59.79	64.50
Load demand (MW)		1.61	1.50	1.44	1.39	1.38	1.44	1.59	1.78	1.67	1.21	1.45	1.60
Time step		**13**	**14**	**15**	**16**	**17**	**18**	**19**	**20**	**21**	**22**	**23**	**24**
Wind speed (m/s)	A	5.15	5.16	5.11	5.13	5.08	5.08	5.12	5.10	5.04	5.10	5.16	5.16
	B	3.96	3.91	3.93	3.87	3.89	3.89	3.86	3.87	3.88	3.88	3.96	4.18
	C	4.05	4.05	4.02	4.06	4.08	4.11	4.16	4.14	4.21	4.21	4.26	4.32
Solar radiation (kW/m^2)	A	79.68	78.83	79.68	78.83	79.68	78.83	79.68	78.83	79.68	78.83	79.68	78.83
	B	64.50	61.71	64.50	61.71	64.50	61.71	64.50	61.71	64.50	61.71	64.50	61.71
	C	60.63	58.00	60.63	58.00	60.63	58.00	60.63	58.00	60.63	58.00	60.63	58.00
Load demand (MW)		1.64	1.62	1.56	1.52	1.42	1.52	1.53	1.46	1.29	1.08	1.18	1.27

Table 13.3 BESS agents technical data

No.	Power capacity (MW)	Energy capacity (MWh)	Technology	State-of-charge (SOC) limit
1	0.10	0.60	Lead-acid	0.2~0.8
2	0.10	0.60	Lead-acid	
3	0.10	0.60	Zn/BR	
4	0.10	0.60	Zn/Br	
5	0.05	0.30	Li-ion	

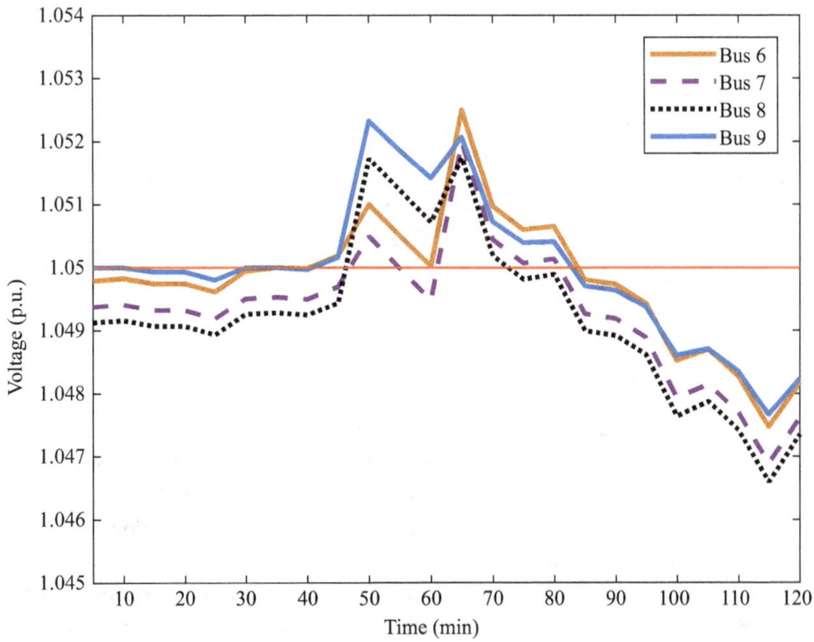

Figure 13.5 Voltage rise mitigation under scenario A during 11:00–13:00 pm

We first analyze the simulation for voltage rise situation. Figure 13.5 shows the voltage profiles on different buses over the entire time horizon under the scenario A. It can be observed that the voltages in buses 6–9 significantly violate the system voltage constraints when t = 50th min and t = 70th min. Under the proposed voltage control scheme, the system operator can mitigate the voltage rise problem smoothly through stepwise voltage regulation.

Now we consider the voltage drop situation. The bus voltage profiles in this situation are shown in Figure 13.6. From Figure 13.6, we can find that the voltages

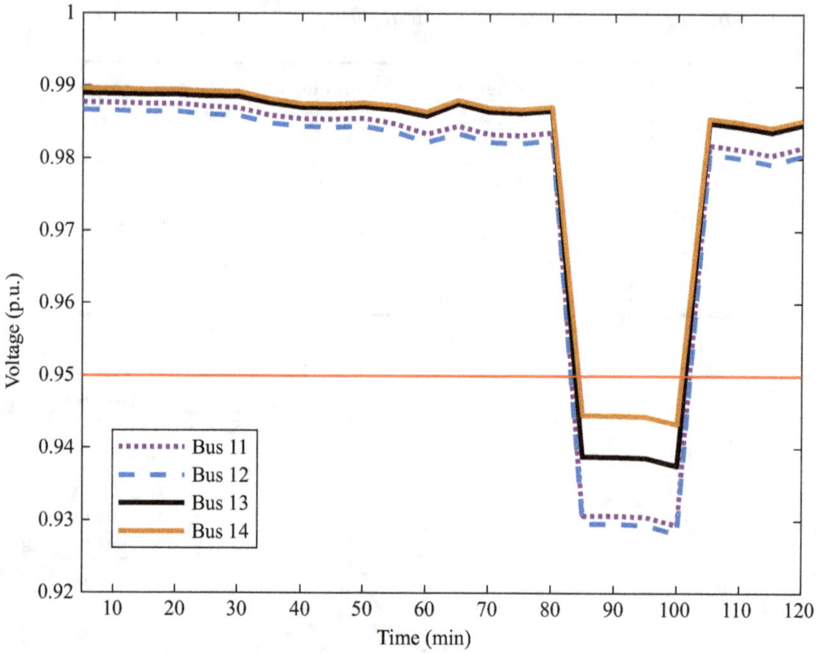

Figure 13.6 Voltage drop compensation under scenario A during 11:00–13:00 pm

in buses 11–14 significantly violate the system voltage constraints form t = 80th to t = 100th min. Under the proposed voltage control scheme, the system operator can compensate the voltage drop problem in ΔT control horizon. Once all the bus voltage profiles are regulated back to the normal range, the system does not require any adjustment of the BESSs' active power outputs. The sparse technique has been imbedded in the objective function to allow the distribution network operator to adopt most suitable control actions to achieve a trade-off between security and effectiveness. Figure 13.7 depicts the detailed SOC profiles of all the five BESS units.

Figure 13.8 illustrates the voltage profile in bus 5 under three possible weather scenarios during the period of 11:00–13:00 pm. It can be seen that the proposed voltage regulation method can provide effective correction for the bus voltage under all the possible weather conditions.

To demonstrate the convergence of the proposed model, we further compare the results of the proposed D-ADMM-based voltage control model with the results obtained from the centralized approach. The results are reported in Figure 13.9. The x-axis and y-axis represent the iteration numbers used to reach the same accuracy (predefined as 0.01) and the error of the objective function, respectively. The stopping criteria are set to be 0.01. The proposed D-ADMM-based control scheme

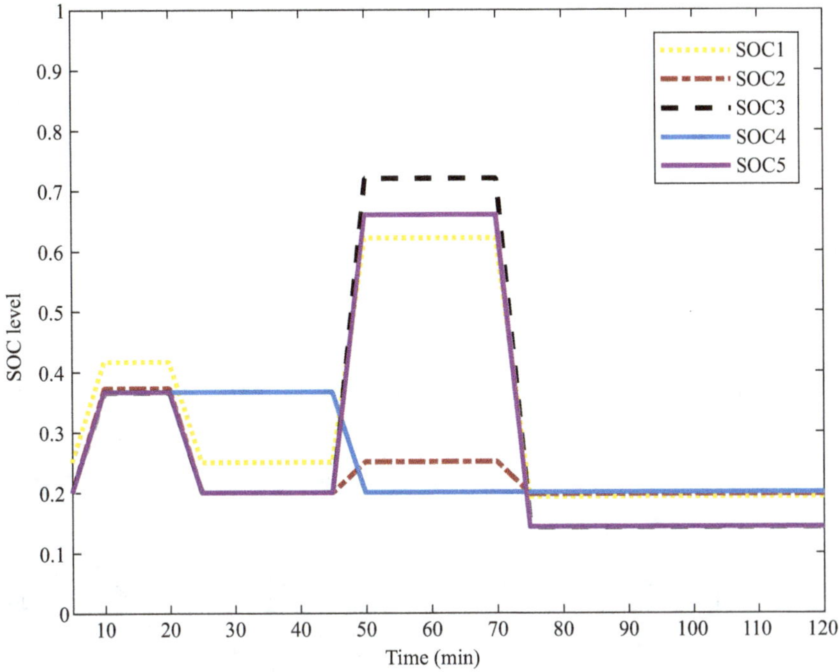

Figure 13.7 BESS SOC profiles for the entire time horizon

converges after 24 iterations, with the total time consumption of only 2.16 s. The results also show that through limited information exchange between the neighboring BESSs, the same global optimized solution can be obtained by the distributed control approach as the centralized control approach.

We further compare the convergence performance of the proposed method with the original ADMM algorithm, and the results are shown in Figure 13.10, we can find that both of the two algorithms can guarantee the optimality of the solution, but the convergence speed of D-ADMM (24 iterations) is much faster than that of original ADMM (41 iterations). More specifically, the D-ADMM algorithm only consumes 2.16 s to reach convergence, while original ADMM algorithm takes 2.87 s to find optimal results.

13.4.3 43-bus radial distribution system

To extend the proposed algorithm to a large-scale system, in this case we test our algorithm to a 43-bus radial distribution system. The topology of the 43 bus is shown in Figure 13.11, where two PV panels are installed at bus 7 and 41 and two wind farms are installed at bus 13 and 19. Eight BESS units are incorporated at bus 9, 3, 11, 17, 26, 6, 39, and 40, respectively. Similarly, three scenarios A, B, and C were generated to represent different possible weather scenario, respectively. The

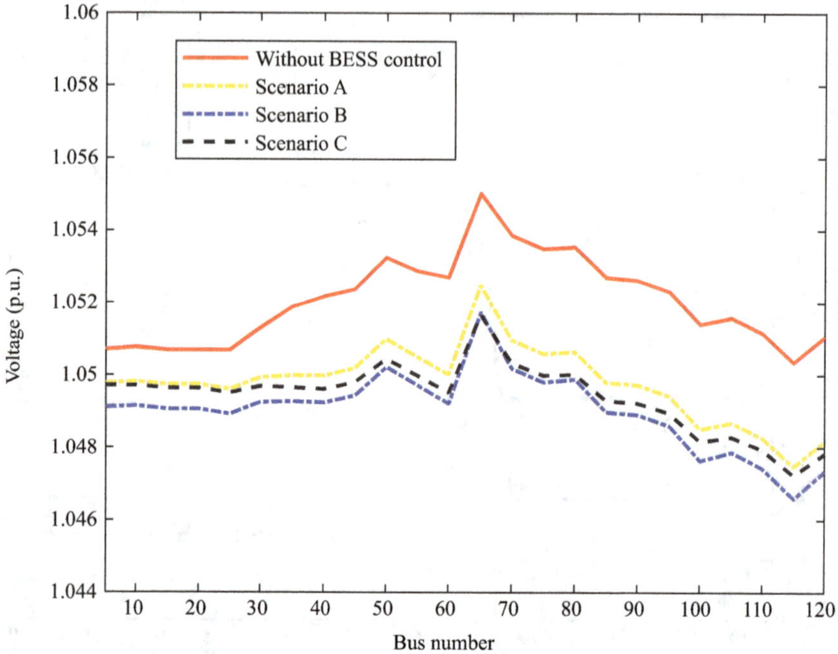

Figure 13.8 Voltage profiles in bus 5 under three possible scenarios during 11:00–13:00 pm

respective probabilities of scenario A, B, and C are equally, ($p_A + p_B + p_C = 1$). The entire control horizon and discrete-time step are 2 h and 5 min.

The BESSs' power output over the entire time horizon is shown in Figure 13.12. The convergence performance of each BESS unit versus the iteration number is shown in Figure 13.13. During the simulation, each BESS unit performs the proposed algorithm under the control scheme presented in Table 13.1. From the results, we can find that each BESS unit communicates with their neighbors and converges to the optimal value after dynamic iteration finally. The objective function value is the same as the results generated by the centralized voltage regulation approach.

We also compare the iterative convergence process of the proposed method and the centralized interior-point based method; the results are shown in Figure 13.14. Obviously, D-ADMM algorithm converges faster and has a high precision degree than the original ADMM.

13.5 Conclusion

In this chapter, a distributed voltage control scheme for distribution networks with high renewable penetration is proposed. The scheme controls multiple

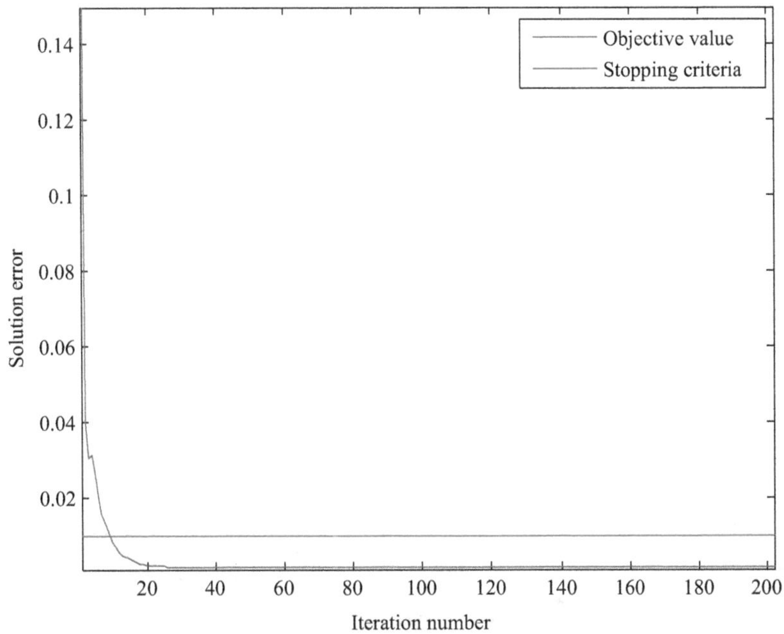

Figure 13.9 Results comparison between centralized optimization and D-ADMM

BESSs to regulate the voltage profile at different system buses. The key feature of the proposed method is to find the optimal voltage regulation strategy to regulate the voltage at different buses back to its secure range under different renewable power output scenarios. The selected control strategy may not be the optimal one for a particular scenario, but it will lead to the most effective solution under all possible scenarios. Besides, the ℓ^1 regularization is imbedded in the objective function to find the most effective action of the BESSs. A fully distributed voltage control model is proposed based on the D-ADMM algorithm. During the iteration process, each BESS unit in the system finds its optimal schedule solely through peer-to-peer communications with its neighbors. Numerical simulations are conducted on modified IEEE 15-bus and 43-bus radial distribution systems to validate the effectiveness of the proposed method. The convergence of the proposed method is validated through comparison with centralized optimization technique. The simulation results show that (1) the proposed voltage regulation strategy can well adjust the BESSs' outputs while maintaining the bus voltages within the secure range while considering renewable energy and load uncertainties; and (2) D-ADMM algorithm has a strong capability for performing global optimization and has faster convergence ability than the original ADMM algorithm.

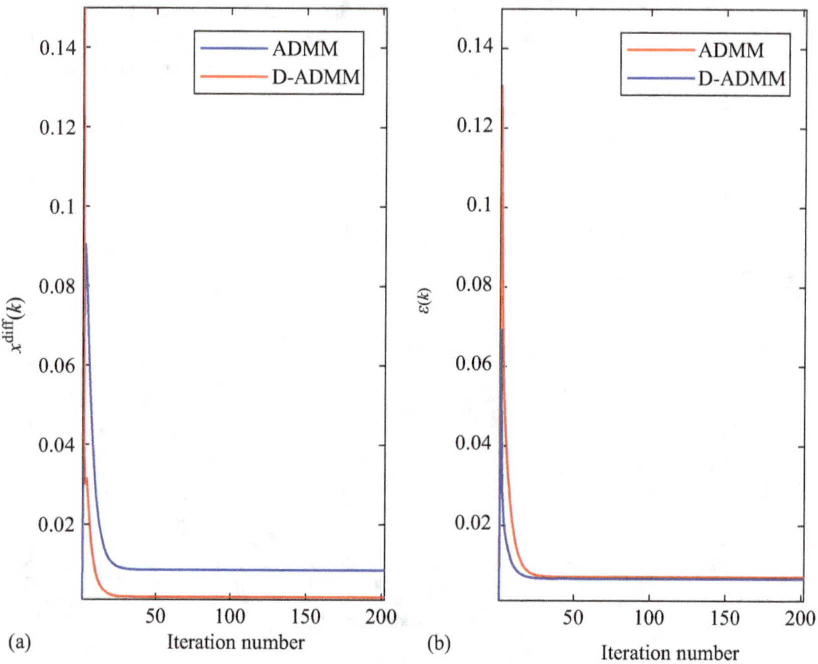

Figure 13.10 The dynamic convergence performance of the relative error in ADMM and D-ADMM: (a) independent variables and (b) objective function

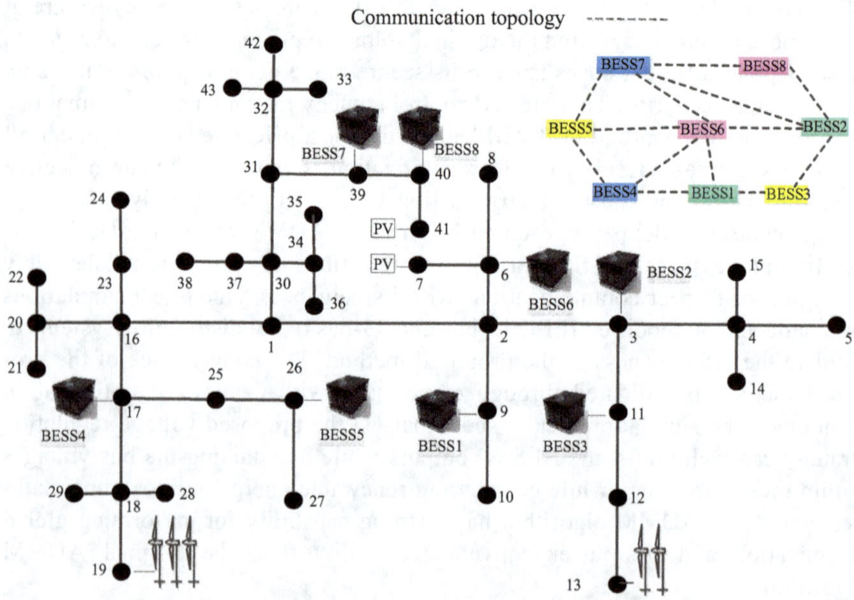

Figure 13.11 Communication network of 43-bus distribution system with eight BESS units, eleven edges labeled in four colors

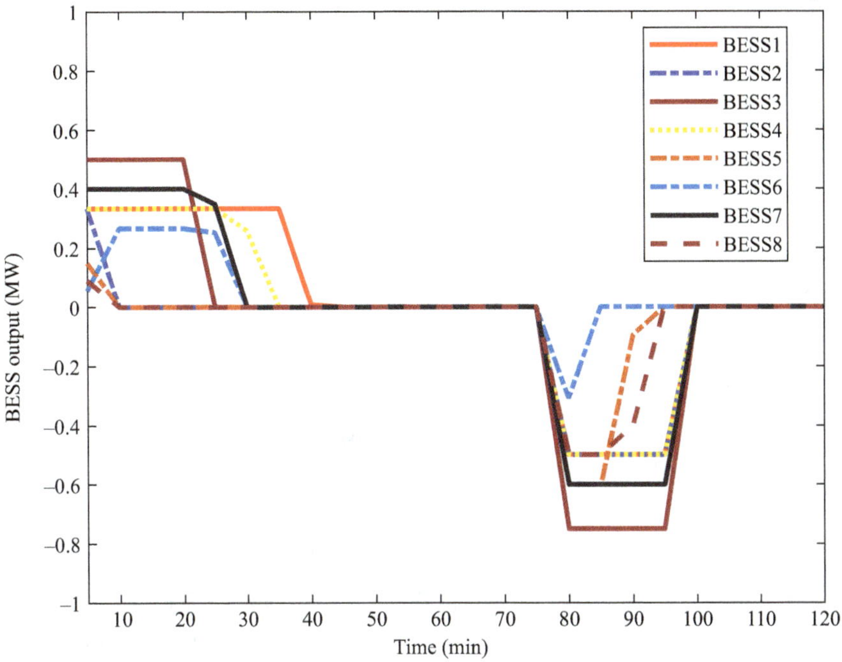

Figure 13.12 BESS output power for the 43 bus over entire time horizon

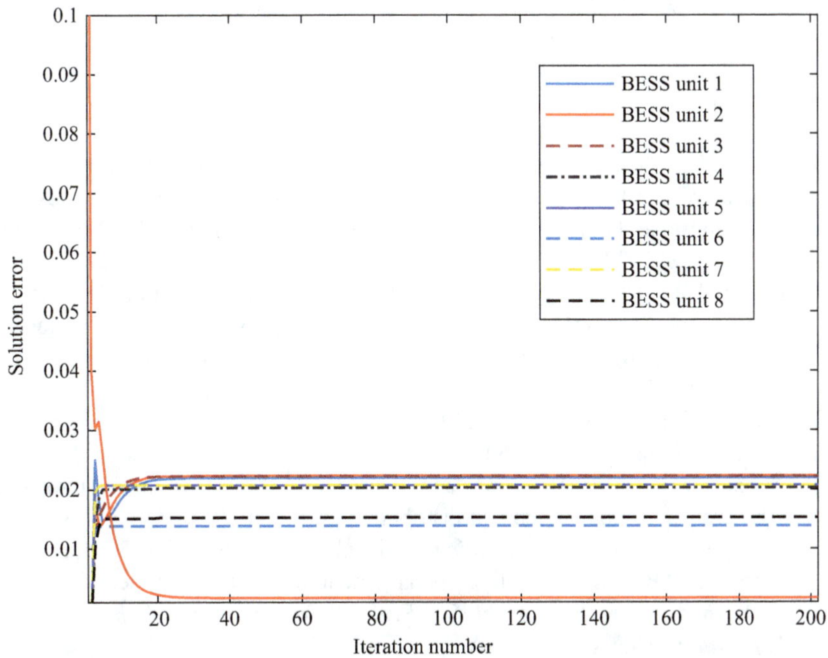

Figure 13.13 Convergence of BESS unit output decisions versus dynamic iteration number

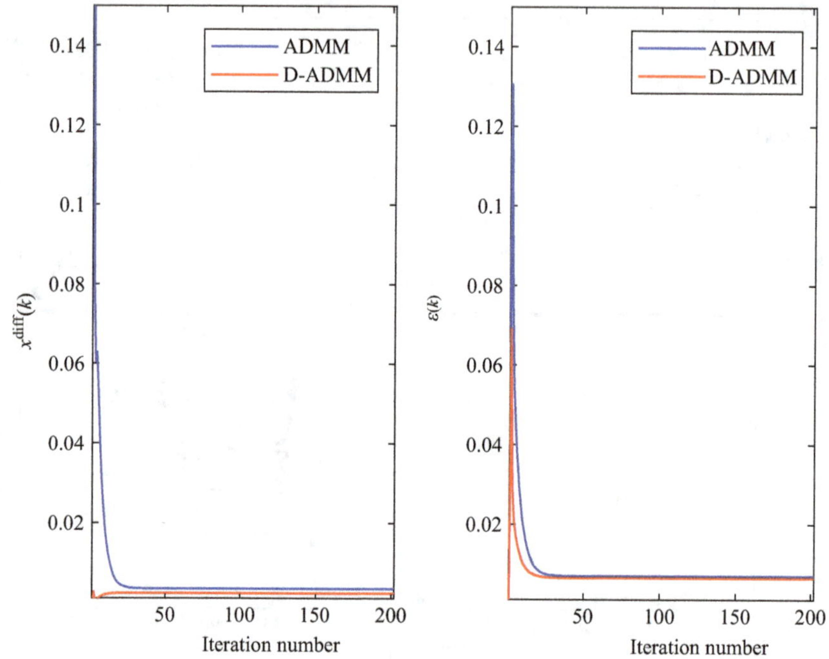

Figure 13.14 The dynamic convergence performance of the relative error in ADMM and D-ADMM: (a) independent variables and (b) objective function

References

[1] Hart E.K., Stoutenburg E.D., and Jacobson M.Z. "The potential of intermittent renewables to meet electric power demand: current methods and emerging analytical techniques." *Proceedings of the IEEE*. 2012, vol. 100, no. 2, pp. 322–334.

[2] Lopes J.P., Hatziargyriou N., Mutale J., Djapic P., and Jenkins N. "Integrating distributed generation into electric power systems: a review of drivers, challenges and opportunities." *Electric Power Systems Research*. 2007, vol. 77, no. 9, pp. 1189–1203.

[3] Global Wind Statistics. GWEC. www.gwec.net/wp-content/uploads/vip/GWEC-PRstats-2015_LR.pdf.

[4] Appen J.V., Stetz T., Braun M., and Schmiegel A. "Local voltage control strategies for pv storage systems in distribution grids." in *2014 IEEE PES General Meeting | Conference & Exposition*, National Harbor, MD, USA, 2014, pp. 1–1.

[5] Mutale J. "Benefits of active management of distribution networks with distributed generation." in *2006 IEEE PES Power Systems Conference and Exposition*, Atlanta, GA, USA, 2006, pp. 601–606.

[6] Gao C., and Redfern M.A. "A review of voltage control techniques of networks with distributed generations using on-load tap changer transformers." in *45th International Universities Power Engineering Conference UPEC2010*, 2010, pp. 1–6.

[7] O'Gorman R., and Redfern M. "Voltage control problems on modern distribution systems." in *IEEE Power Engineering Society General Meeting*, 2004, IEEE, 2004, pp. 662–667.

[8] Vovos P.N., Kiprakis A.E., and Harrison G.P. "Centralized and distributed voltage control: impact on distributed generation penetration." *IEEE Transactions on Power Systems*. 2007, vol. 22, no. 1, pp. 476–483.

[9] Schwaegerl C., Bollen M., Karoui K., and Yagmur A. "Voltage control in distribution systems as a limitation of the hosting capacity for distributed energy resources." in *IET 18th International Conference and Exhibition on Electricity Distribution, 2005. CIRED 2005*, Turin, Italy, 2005, pp. 1–5.

[10] Sansawatt T., Ochoa L.F., and Harrison G.P. "Smart decentralized control of dg for voltage and thermal constraint management." *IEEE Transactions on Power Systems*. 2012, vol. 27, no. 3, pp. 1637–1645

[11] Sansawatt T., Ochoa L.F., and Harrison G.P. "Integrating distributed generation using decentralised voltage regulation." in *IEEE PES General Meeting*, Providence, RI, 2010, pp. 1–6.

[12] Liu X., Aichhorn A., Liu L., and Li H. "Coordinated control of distributed energy storage system with tap changer transformers for voltage rise mitigation under high photovoltaic penetration." *IEEE Transactions on Smart Grid*. 2012, vol. 3, no. 2, pp. 897–906.

[13] Robbins B., Hadjicostis C.N., and Domínguez-García A.D. "A two-stage distributed architecture for voltage control in power distribution systems." *IEEE Transactions on Power Systems*. 2013, vol. 28, no. 2, pp. 1470–1482.

[14] Yu L., Czarkowski D., and De León F. "Optimal distributed voltage regulation for secondary networks with DGS." *IEEE Transactions on Smart Grid*. 2012, vol. 3, no. 2, pp. 959–967.

[15] Carvalho P., Correia P.F., and Ferreira L.A. "Distributed reactive power generation control for voltage rise mitigation in distribution networks." *IEEE Transactions on Power Systems*. 2008, vol. 23, no. 2, pp. 766–772.

[16] Bidgoli H.S., and Cutsem T.V. "Combined local and centralized voltage control in active distribution networks." *IEEE Transactions on Power Systems*. 2018, vol. 33, no. 2, pp. 1374–1384.

[17] Antoniadou-Plytaria K.E., Kouveliotis-Lysikatos I.N., Georgilakis P.S., and Hatziargyriou N.D. "Distributed and decentralized voltage control of smart distribution networks: models, methods, and future research." *IEEE Transactions on Smart Grid*. 2017, vol. 8, no. 6, pp. 2999–3008.

[18] Momoh J.A. *Electric Power System Applications of Optimization*. New York, NY, CRC Press, 2008.

[19] Zhu J. *Optimization of Power System Operation*. Hoboken, NJ, John Wiley & Sons, 2015.

[20] Whittingham M.S. “History, evolution, and future status of energy storage.” *Proceedings of the IEEE*. 2012, vol. 100, no. Special Centennial Issue, pp. 1518–1534.

[21] Divya K., and Østergaard J. “Battery energy storage technology for power systems; an overview.” *Electric Power Systems Research*. 2009, vol. 79, no. 4, pp. 511–520.

[22] Huff G. DOE Global Energy Storage Database. 2014.

[23] Kabir M.N., Mishra Y., Ledwich G., Dong Z.Y., and Wong K.P. “Coordinated control of grid-connected photovoltaic reactive power and battery energy storage systems to improve the voltage profile of a residential distribution feeder.” *IEEE Transactions on Industrial Informatics*. 2014, vol. 10, no. 2, pp. 967–977.

[24] Zeraati M., Golshan M.E.H., and Guerrero J.M. “Distributed control of battery energy storage systems for voltage regulation in distribution networks with high PV penetration.” *IEEE Transactions on Smart Grid*. 2018, vol. 9, no. 4, pp. 3582–3593.

[25] Phan D.T., and Sun X.A. “Minimal impact corrective actions in security-constrained optimal power flow via sparsity regularization.” *IEEE Transactions on Power Systems*. 2015, vol. 30, no. 4, pp. 1947–1956.

[26] Arabali A., Ghofrani M., Etezadi-Amoli M., Fadali M.S., and Baghzouz Y. “Genetic-algorithm-based optimization approach for energy management.” *IEEE Transactions on Power Delivery*. 2013, vol. 28, no. 1, pp. 162–170.

[27] Ghofrani M., Arabali A., Etezadi-Amoli M., and Fadali M.S. “A framework for optimal placement of energy storage units within a power system with high wind penetration.” *IEEE Transactions on Sustainable Energy*. 2013, vol. 4, no. 2, pp. 434–442.

[28] Khodayar M.E., Shahidehpour M., and Wu L. “Enhancing the dispatchability of variable wind generation by coordination with pumped-storage hydro units in stochastic power systems.” *IEEE Transactions on Power Systems*. 2013, vol. 28, no. 3, pp. 2808–2818.

[29] Mota J.F.C., Xavier J.M.F., Aguiar P.M.Q., and Püschel M. “Distributed optimization with local domains: applications in MPC and network flows.” *IEEE Transactions on Automatic Control*. 2015, vol. 60, no. 7, pp. 2004–2009.

[30] CISL Research Data Archive. Computational and Information Systems Lab. [Online]. Available: https://rda.ucar.edu/.

[31] Wind Database. National Renewable Energy Laboratory. [Online]. Available: https://rredc.nrel.gov/wind/pubs/atlas/.

[32] Zhang Y., Dong Z.Y., Luo F., Zheng Y., Meng K., and Wong K.P. “Optimal allocation of battery energy storage systems in distribution networks with high wind power penetration.” *IET Renewable Power Generation*. 2016, vol. 10, no. 9, pp. 1105–1113.

[33] Korpa M., and Holen A.T. “Operation planning of hydrogen storage connected to wind power operating in a power market.” *IEEE Transactions on Energy Conversion*. 2006, vol. 21, no. 3, pp. 742–749.

Chapter 14

Analysis and compensation of multiple PMU data delays for frequency control in islanded microgrids

Shichao Liu[1] *and Xiaoyu Wang*[1]

For large-scale interconnected multiple-area power systems, phasor measurement units (PMUs or called synchrophasor) are extensively used in the wide-area control, monitoring, and protections. A PMU can provide simultaneous measurements of phasors across a wide area of the power systems. These PMU measurements are synchronized from a common time source provided by a global positioning system (GPS) radio clock. Recently, micro-PMUs or micro-synchrophasors are designed and installed in microgrids to feed synchronized and time-stamped measurement for microgrid control centers. With the synchronized voltage frequency and magnitude data, the secondary control of a microgrid can provide the accurate and real-time power sharing and system monitoring. Communication networks are embedded to facilitate the information exchange between a microgrid centralized controller, PMUs, and local controllers in microgrids. When PMU data are transmitted, they may be delayed as network-induced delays are found in most networks. Therefore, it is indispensable to study the impact of the PMU data delay on the microgrid performance. This chapter investigates the effect of multiple PMU communication delays on the secondary frequency control of an islanded microgrid with multiple distributed generators (DGs). A small-signal model is obtained for the microgrid to analyze the relationship between the secondary frequency control gains and the PMU delays. To improve the performance of the microgrid when multiple PMU delays exist, a gain scheduling approach is also proposed. Both eigen-based theoretical analysis and time-domain simulations on the Canadian urban distribution system have verified that PMU delays can adversely affect the microgrid secondary frequency control and the designed gain schedulers for each DG can improve the robustness of the microgrid secondary frequency controller to communication delays.

[1]Department of Electronics, Carleton University, Ottawa, Canada

14.1 Introduction

A microgrid is an integrated energy system that comprises interconnected loads, distributed generation (DG), and distributed storage (DS) units [1,2]. There are two operational modes in the microgrid, including the grid-connected mode and islanded mode. In the grid-connected mode, the microgrid adjusts real and reactive power profiles by injecting a certain amount of real and reactive power into the main grid or absorbing from it. The islanded mode may be caused by either preplanned maintenance operations or unplanned switch incidents [3]. Due to the lack of the main grid as a reference, an islanded microgrid has to maintain its stable operation by coordinating all the elements of its own.

The control of a microgrid usually involves a hierarchical architecture, including primary control, secondary control, and tertiary control [4,5]. The primary control mainly refers to droop control, regulating the real and reactive power of each inverter to make the system stable [6,7]. In terms of the secondary control, it maintains the frequency and voltage magnitude within the required values by compensating for the deviation caused by the droop primary control. In addition, the secondary control guarantees the seamless connection and disconnection of the MG to and from the grid. The tertiary control regulates the power flow between the MG and the grid.

While the primary control methods of a microgrid are normally decentralized, there exist both centralized and distributed approaches for the secondary control. The structure of the centralized secondary control involves an MG centralized controller (MGCC) at the low voltage side of a substation and local primary inverter controllers (LCs) for all the DGs and DSs in a microgrid [8–11]. The secondary controller in MGCC generates supplementary real-power and reactive set points for the LCs and sends them through low bandwidth communication channels. Recently, several distributed secondary frequency control approaches have been proposed to avoid the large-volume data communication between MGCC and LCs. In [12], based on a unified small-signal model of the microgrid, a distributed optimal controller is proposed. For instance, in [13], a fully distributed adaptive secondary frequency controller is designed, eliminating the requirement of a central processing element. In [14], the consensus-based distributed estimation and secondary frequency control scheme is designed, guaranteeing each photovoltaic generator (PV) and load is self-organizing without centralized monitors.

The high-level penetration of renewable-based DG in the distribution system makes the microgrid behavior more unpredictable. To better control and monitor the microgrid, micro-synchrophasors (μPMUs) are developed to provide a high-precision phasor measurement in the distribution system. Phasor measurement units (PMUs) are extensively used in the wide-area control, monitoring, and protections [15,16] in the past decade. PMUs provide simultaneous measurements of phasors across a wide area of the power systems. These PMUs are synchronized from a common time source provided by a global positioning system (GPS) radio clock. With the synchronized and time-stamped measurement fed by μPMUs installed in

DG buses, the secondary control of a microgrid can provide the accurate and real-time power sharing and system monitoring [17,18].

Although μPMUs can provide synchronized voltage frequency and magnitude measurement, data transfer involves communication networks between μPMUs and controllers, such as WiMax and WiFi. During the transmission of power control and measurement signals, they may be delayed or dropped, even attacked by adversaries [19–21]. These network-induced uncertainties may seriously affect the performance of the secondary control and result in instability of the microgrid [22,23]. The recent experiences of PMU-based wide-area damping control for interconnected power systems show that time delay of PMU data could degrade the dynamic performance of controllers and even jeopardize the system stability [24,25]. This indicates it is critical to take the impact of communication delays of μPMU measurement into account when the secondary control is designed for microgrids. To the best knowledge of the authors, there currently exist few works in this regard.

In this chapter, the effect of multiple PMU communication delays on the secondary frequency control in islanded microgrids is studied. The islanded microgrid with four inverter-based DGs is used to analyze the effect of multiple PMU communication delays. Due to the negligible inertias of these inverter-based DGs, the dynamics of the frequency can act very fast. The PMU delays could seriously affect the performance of microgrid. To validate this point, the impact of multiple PMU delays on the performance of an islanded microgrid is analyzed comprehensively in an eigen-based method. The small-signal system model is built for the microgrid. Via a series of trial studies, we find the relationships between secondary frequency control gains and PMU delays, including both identical and different PMU delay cases. The PMU delays can cause the performance degradation and even instability of the system. To compensate the PMU effect, a gain-scheduler-based secondary frequency controller is designed for the islanded microgrid. Both eigen-based analysis and time-domain simulations have validated the impact of multiple PMU communication delays on the microgrid performance. The proposed gain-scheduling frequency control approach is also verified to be capable of improving the system performance when multiple PMU communication delays exist.

14.2 Small-signal model of a microgrid with multiple PMU data delays

In this section, the hierarchical microgrid control structure is introduced. The small-signal model of the microgrid with multiple PMU delays are obtained as a delayed descriptor system.

14.2.1 System structure

System structure with the secondary controller is described as follows. The distribution system shown in Figure 14.1 is used to investigate the secondary frequency control of microgrids here. The utility source is 120 kV and the 12.5 kV

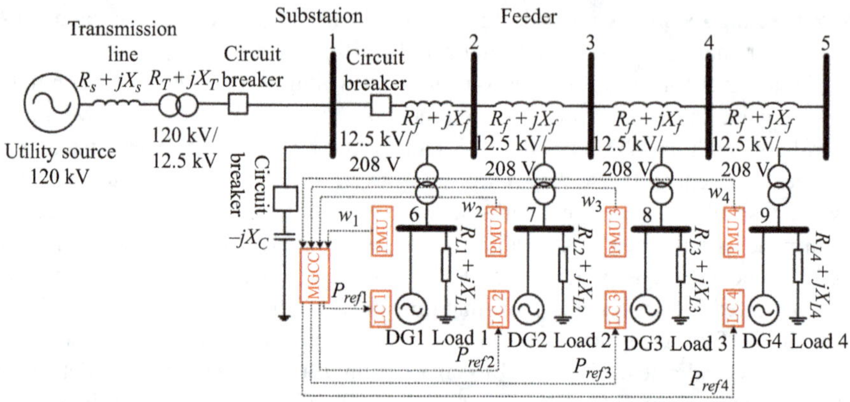

Figure 14.1 The Canadian urban benchmark distribution system

substation is connected to the grid through a circuit breaker (CB) and a substation transformer with a capacity of 10 MVA. A 2.75 Mvar capacitor bank is located at the substation. Four inverter-based DGs (three-phase 208 V) are evenly distributed along the feeder. They are connected to the feeder through the step-down transformers and the DG terminals are from Node 6 to Node 9 in sequence. The constant impedance load model with power factor 0.95 is adopted to represent the local load of each generator. The capacities of Load 1 to Load 4 are 1.5 MV, 1.9 MV, 1.5 MV, and 0.5 MV, respectively. In addition, four PMUs are placed at each DG bus (Bus 6 to Bus 9) to provide the synchronized measurements of DG buses in the microgrid.

As shown in Figure 14.1, the centralized energy management scheme is used in the work. In this scheme, a microgrid centralized controller (MGCC) is located at the substation and droop-based local controllers (LCs) are designed for four DGs to fast stabilize the frequency and voltage dynamics in case of load changes or disturbances. PMUs provide the MGCC with simultaneous measurements of phasors of DG buses across the microgrid. These PMUs are synchronized from a common time source provided by a global positioning system (GPS) radio clock. After receiving the measurements from PMUs, the MGCC makes decisions and generates operation set points for DGs, such as voltage and frequency set points, based on the objectives and constraints of the system operation. These set points are sent to the LCs of DGs.

The small-signal model of this microgrid consists of three blocks, including the DG block, the network and load block, as well as the interface block.

14.2.2 The model of the microgrid with inverter-based DG

The hierarchical control structure of the microgrid is shown in Figure 14.1. The power control loop of the local primary controller of a DG inverter consists of a power control and an inner current loop, to regulate the inverter output power by tracking given real power set points. Both the power and current controllers are

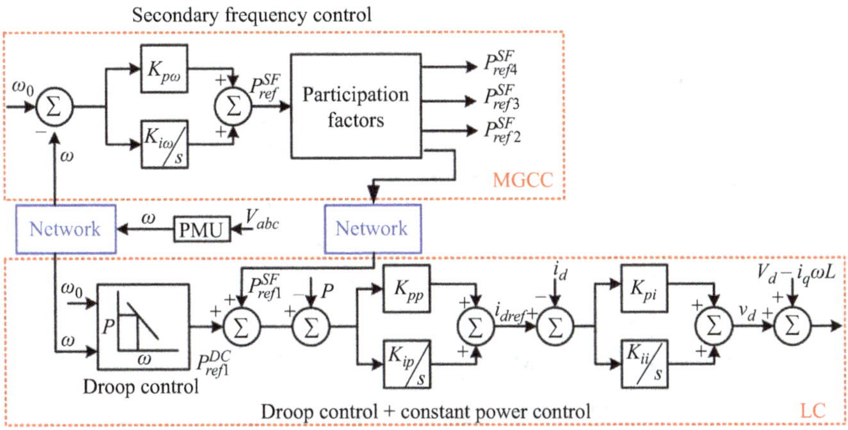

Figure 14.2 Structure of the multi-DG system two-level control

proportional-integral (PI) controllers. The power controller is

$$i_{drefi} = \left(K_{ppi} + \frac{K_{ipi}}{s}\right)\left(P_{refi}^{SF} + P_{refi}^{DC} - P_i\right) \tag{14.1}$$

$$i_{qrefi} = \left(K_{ppi} + \frac{K_{ipi}}{s}\right)(Q_{refi} - Q_i) \tag{14.2}$$

where i_{drefi} and i_{qrefi} are the components of the current set points on d-axis (direct) and q-axis (quadrature) of the ith DG, K_{ppi} and K_{ipi} are proportional and integral control gains of the ith DG, respectively, P_{refi}^{SF} is the ith DG supplementary real power set point assigned by the secondary frequency controller of MGCC, P_{refi}^{DC} is the corrective real power set point generated by the power control of ith DG, and P_i and Q_i are the instantaneous real and reactive power. Q_{refi} is the reactive power set point of ith DG. Since the focus of this work is the frequency control, only a real power control structure is shown in Figure 14.2.

$$v_{di} = \left(K_{pii} + \frac{K_{iii}}{s}\right)(i_{drefi} - i_{di}) \tag{14.3}$$

$$v_{qi} = \left(K_{pii} + \frac{K_{iii}}{s}\right)(i_{qrefi} - i_{qi}) \tag{14.4}$$

where v_{di} and v_{qi} are the components of the voltage set points on d-axis and q-axis of the ith DG, and i_{di} and i_{qi} are the instantaneous currents of the ith DG.

The $\omega - P$ characteristic of the frequency droop control can be described as

$$P_{refi}^{DC} = K_{\omega i}(\omega_0 - \omega_i) \tag{14.5}$$

where K_ω is the droop control gain and P_{refi}^{DC} is the corrective power set point due to frequency variations. The secondary frequency control is

$$P_{refi}^{SF} = \left(K_{p\omega i} + \frac{K_{i\omega i}}{s}\right)(\omega_0 - \omega_i) \tag{14.6}$$

where P_{refi}^{SF} is the supplementary power set point of the ith DG assigned by the secondary frequency controller, $K_{p\omega i}$ and $K_{i\omega i}$ are proportional and integral control gains, respectively, ω_0 is the nominal frequency reference (in this work, $\omega_0 = 377$ rad/s), and ω_i is the synchronized instantaneous frequency obtained from a PMU.

The second part of the microgrid model is the network block which can be written in a common reference frame as follows:

$$\begin{bmatrix} \Delta \mathbf{i_x} \\ \Delta \mathbf{i_y} \end{bmatrix} = \begin{bmatrix} \mathbf{G} & -\mathbf{B} \\ \mathbf{B} & \mathbf{G} \end{bmatrix} \begin{bmatrix} \Delta \mathbf{V_x} \\ \Delta \mathbf{V_y} \end{bmatrix} \tag{14.7}$$

where the matrices **G** and **B** are acquired from the network system admittance matrix and $\mathbf{V_x} = [V_{x1}, V_{x2}, \ldots, V_{xn}]^T$, $\mathbf{V_y} = [V_{y1}, V_{y2}, \ldots, V_{yn}]^T$, $\mathbf{i_x} = [i_{x1}, i_{x2}, \ldots, i_{xn}]^T$, $\mathbf{i_y} = [i_{y1}, i_{y2}, \ldots, i_{yn}]^T$.

The third part of the microgrid small-signal model is the interface block. Among mathematic equations obtained above, each DG model is developed in its own $d-q$ reference frame. To develop the small-signal model of the microgrid, all the voltages and currents must be transformed to the common reference $x-y$ frame. The reference frame transformation between local $d-q$ reference frame and the common $x-y$ reference frame is shown in Figure 14.3. The interface equations are

$$\Delta \mathbf{V_d} = \mathbf{C_0}\Delta \mathbf{V_x} - \mathbf{V_{x0}}\mathbf{S_0}\Delta\delta + \mathbf{S_0}\Delta \mathbf{V_y} + \mathbf{V_{y0}}\mathbf{C_0}\Delta\delta \tag{14.8}$$

$$\Delta \mathbf{V_q} = \mathbf{S_0}\Delta \mathbf{V_x} - \mathbf{V_{x0}}\mathbf{C_0}\Delta\delta + \mathbf{C_0}\Delta \mathbf{V_y} + \mathbf{V_{y0}}\mathbf{S_0}\Delta\delta \tag{14.9}$$

$$\Delta \mathbf{i_x} = \mathbf{C_0}\Delta \mathbf{i_d} - \mathbf{i_{d0}}\mathbf{S_0}\Delta\delta - \mathbf{S_0}\Delta \mathbf{i_q} - \mathbf{i_{q0}}\mathbf{C_0}\Delta\delta \tag{14.10}$$

$$\Delta \mathbf{i_y} = \mathbf{S_0}\Delta \mathbf{i_d} + \mathbf{i_{d0}}\mathbf{C_0}\Delta\delta + \mathbf{C_0}\Delta \mathbf{i_q} - \mathbf{i_{q0}}\mathbf{S_0}\Delta\delta \tag{14.11}$$

where δ_i is the individual inverter terminal voltage phase angle in $x-y$ reference frame, and the diagonal matrices $\mathbf{C_0}$ and $\mathbf{S_0}$ are defined as $\mathbf{C_0} = \text{diag}\{\cos(\delta_{i0})\}$ and $\mathbf{S_0} = \text{diag}\{\sin(\delta_{i0})\}$, respectively.

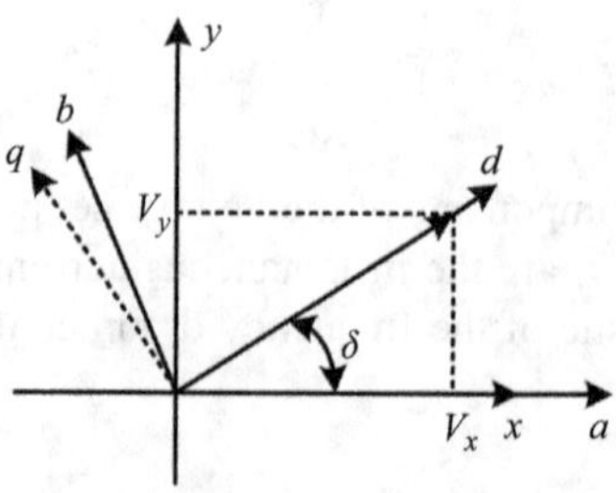

Figure 14.3 Reference frame transformation

Based on the above discussion and our previous work [26], the state space of the overall system is formulated as the following descriptor system

$$\mathbf{E}\Delta\dot{\mathbf{x}} = \mathbf{A}\Delta\mathbf{x} + \mathbf{F}\mathbf{r}_0 \tag{14.12}$$

where $\mathbf{x} = \left[\delta, \omega, \mathbf{i_d}, \mathbf{i_q}, \mathbf{i_{dref}}, \mathbf{i_{qref}}, \mathbf{u_d}, \mathbf{u_q}, \mathbf{P}, \mathbf{Q}, \mathbf{P_{ref}}, \mathbf{V_d}, \mathbf{V_q}, \mathbf{i_x}, \mathbf{i_y}, \mathbf{V_x}, \mathbf{V_y}\right]^T$, $\mathbf{r}_0 = [\omega_0]^T$, $\mathbf{E}$ is a system matrix which is singular, $\mathbf{A}$ is the system matrix, and $\mathbf{F}$ is a parameter matrix.

By leaving all the derivatives of the state variables on the left sides of the physical equations and moving other parts to the right sides of these equations, (14.12) can be derived. In the descriptor system, the matrix E is square and can be singular which is not necessarily an identical matrix as the case in regular systems.

Consider there exist total N communication delays $\tau_i, i \in \{1, \ldots, N\}$ in the network between PMUs and MGCC, and the network between LCs and MGCC. The overall system model becomes a time-delay descriptor system, written as

$$\mathbf{E}\Delta\dot{\mathbf{x}} = \mathbf{A}\Delta\mathbf{x} + \sum_{i=1}^{\gamma} \mathbf{A}_{\tau_i}\Delta\mathbf{x}(\mathbf{t} - \tau_i) + \mathbf{F}\mathbf{r}_0 \tag{14.13}$$

where τ_i denotes ith time delay, $\mathbf{x}(\mathbf{t} - \tau_i)$ is the corresponding delayed state, and $\mathbf{A}_{\tau_i}$ is its parameter matrix.

14.3 Eigenvalue-based delay margin determination approach

The characteristic equation of the delayed descriptor system (14.13) is

$$det(\lambda\mathbf{E} - \Delta(\lambda, \tau)) = 0 \tag{14.14}$$

where the characteristic matrix $\Delta(\lambda, \tau)$ is

$$\Delta(\lambda, \tau) = \mathbf{A} + \sum_{i=1}^{\gamma} \mathbf{A}_{\tau_i} e^{-\lambda\tau_i} \tag{14.15}$$

where γ represents the number of PMU delays, and the order of delay sequence follows $\{\tau_1 < \tau_2 < \cdots < \tau_\gamma\}$.

Definition 1. [27,28]. For γ PMU delays $\tau_i, i \in \{1, \ldots, \gamma\}$, the delayed descriptor system (14.13) is asymptotically stable if all generalized roots of its characteristic equation (14.14) are in the open left-hand plane.

Definition 2. A critical delay denoted by τ_d is called a delay margin if the delayed descriptor system (14.8) is stable for $\tau < \tau_d$ and it is unstable for $\tau > \tau_d$.

The methods to determine delay margins for power systems have been discussed in [29,30]. In this work, an eigenvalue approach in [29] is extended to the delayed descriptor system (14.8) with multiple PMU delays (Figure 14.4).

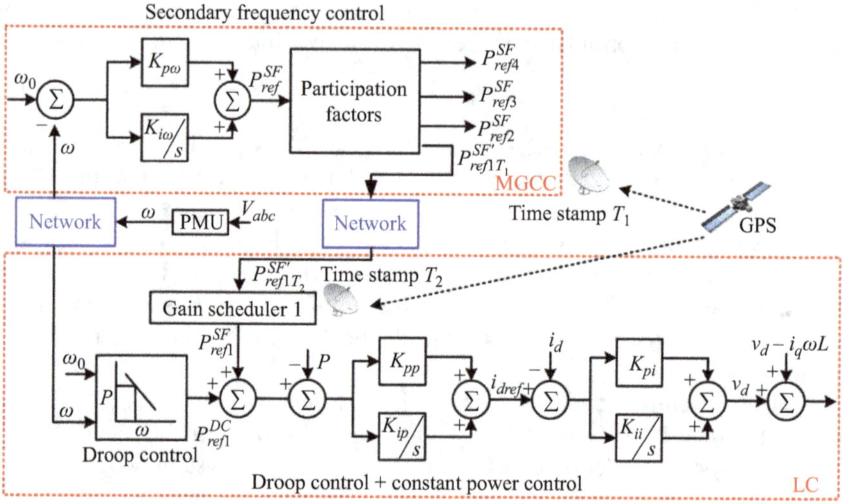

Figure 14.4 The control structure of the gain scheduling approach

Since the characteristic matrix (14.15) is transcendent, it has an infinite number of roots. Therefore, we can only approximate the solution of (14.15) by computing a reduced set of its roots. One effective technique is to approximate the roots of (14.15) by a finite element method [31]. In this method, a matrix **M** which has the following form is defined:

$$\mathbf{M} = \begin{bmatrix} & & & \bar{\mathbf{C}} \otimes \mathbf{I_n} & & & \\ \mathbf{A}_{\tau_\gamma} & \mathbf{0} & \mathbf{A}_{\tau_{N-1}} & \cdots & \mathbf{A}_{\tau_1} & \mathbf{0} & \mathbf{A} \end{bmatrix} \tag{14.16}$$

where $\otimes$ indicates Kronecker's product, $\mathbf{I_n}$ is the identity matrix of order n, and $\bar{\mathbf{C}}$ is a matrix composed of the first $N-1$ rows of **C** defined as follows:

$$\mathbf{C} = -2\mathbf{D_N}/\tau \tag{14.17}$$

where $\tau = \tau_\gamma$ (the maximum delay), and $\mathbf{D_N}$ is a Chebyshev's differentiation matrix of dimension $(N+1) \times (N+1)$ (see Appendix II in [31] for details). Then, the eigenvalues of **M** are an approximated spectrum of (14.24).

After obtaining the approximated eigenvalues, the critical eigen pair can be identified by checking the one that has the smallest damping ratio. When the root locus of the critical eigen pair in the power system is investigated with some interesting parameters, the delay margin can be found when the critical eigen pair crosses the imaginary axis.

14.4 Gain-scheduling-based delay compensation methodology

The presence of communication delays which are usually time-varying in the microgrid frequency control loop may degrade its performance. A gain scheduling approach is proposed to compensate the effect of communication delays on the dynamic performance of the islanded microgrid shown in Figure 14.4. The microgrid control structure including both the MGCC and one LC equipped with a gain scheduler is shown in Figure 14.4. Here, we define

$$\bar{K}_{i\omega i} = \beta_{i\omega i} K_{i\omega i} \tag{14.18}$$

and

$$\bar{K}_{p\omega i} = \beta_{p\omega i} K_{p\omega i} \tag{14.19}$$

where $K_{i\omega i}$ and $K_{p\omega i}$ are considered to be fixed since the secondary frequency controller located in the MGCC. The changeable parts are $\beta_{i\omega i}$ or $\beta_{p\omega i}$ located in each local DG controller. $\beta_{i\omega i}$ denotes an integral gain scheduler and $\beta_{p\omega i}$ denotes a proportional gain scheduler. $\bar{K}_{i\omega i}$ and $\bar{K}_{p\omega i}$ represent the equivalent gains of the secondary frequency controller after gain schedulers are equipped in each LC. We consider the gain schedulers change only one gain variable, either $\beta_{i\omega i}$ or $\beta_{p\omega i}$. Although simultaneously changing both of the gain variables can achieve the same purpose, it involves a more complicated feasible gain search calculation. The procedure of the gain scheduling approach includes the following steps:

- Step 1: The secondary frequency controller in the MGCC calculates real power set points for each DG to restore their frequencies to the nominal one and sends them to local inverters. These set points are denoted by $P^{SF'}_{refiT_1}$, where T_1 is the time stamp generated when they are sent by the MGCC;
- Step 2: After LCs in the local inverters received their set points from the MGCC, the set points are marked as $P^{SF'}_{refiT_2}$, where T_2 is the time stamp generated when they are received by LCs;
- Step 3: LCs calculate the communication delay by comparing the time stamps of two signals $P^{SF'}_{refiT_2}$ and $P^{SF'}_{refiT_1}$. The time delay is $\tau_i = T_2 - T_1$; and
- Step 4: For a given cost index, by looking up the relationship between the time delay τ_i and its corresponding feasible gain $\beta^{\tau_i}_{i\omega i}$ or $\beta^{\tau_i}_{p\omega i}$ built by the offline analysis, the local gain scheduler adjusts its gain value according to the measured communication delay.

To find feasible $\beta_{i\omega i}$ and $\beta_{p\omega i}$ that correspond to communication delays, we need to investigate root locus of the time-delay small-signal model of the microgrid with respect to $\beta_{i\omega i}$ and $\beta_{p\omega i}$.

Considering (14.18) or (14.19), the equalized load frequency controller has the following form:

$$P^{SF}_{refi} = \left(K_{p\omega i} + \frac{\bar{K}_{i\omega i}}{s} \right) (\omega_0 - \omega_i) \tag{14.20}$$

or

$$P^{SF}_{refi} = \left(\bar{K}_{p\omega i} + \frac{K_{i\omega i}}{s} \right) (\omega_0 - \omega_i) \tag{14.21}$$

The time-delay small-signal model of the microgrid is built as follows:

$$\bar{\mathbf{E}}\Delta\dot{\mathbf{x}} = \bar{\mathbf{A}}\Delta\mathbf{x} + \sum_{i=1}^{\gamma} \bar{\mathbf{A}}_{\tau}\Delta\mathbf{x}(\mathbf{t}-\tau) + \bar{\mathbf{F}}\mathbf{r_0} \tag{14.22}$$

where $\bar{\mathbf{E}}$ and $\bar{\mathbf{A}}$ are system matrices, $\bar{\mathbf{A}}_{\tau}$ is a parameter matrix related to the delayed state $\mathbf{x}(\mathbf{t}-\tau)$, and $\bar{\mathbf{F}}$ is a parameter matrix. While most elements of the matrices are still the same as those in (14.13), the original $K_{i\omega i}$ is replaced by $\bar{K}_{i\omega i}$ or $K_{p\omega i}$ by $\bar{K}_{p\omega i}$. The gain scheduler vectors $\beta_{\mathbf{i\omega}}$ and $\beta_{\mathbf{p\omega}}$ denote four integral gain schedulers and proportional gain schedulers, respectively.

The characteristic equation of the delayed descriptor system (14.22) is

$$det(\lambda\bar{\mathbf{E}} - \Delta(\lambda)) = 0 \tag{14.23}$$

where the characteristic matrix $\Delta(\lambda)$ is

$$\Delta(\lambda) = \bar{\mathbf{A}} + \sum_{i=1}^{\gamma} \bar{\mathbf{A}}_{\tau_i} e^{-\lambda\tau_i} \tag{14.24}$$

In Section 14.3, the eigenvalue-based method is introduced to determine the delay margin. Similarly, the eigenvalue-based method can also be used to find the feasible gain set with a given time delay in each PMU.

14.5 Case studies

The distribution system shown in Figure 14.1 is used to study the effect of multiple PMU delays on the performance of an islanded microgrid with a centralized secondary frequency controller. The circuit breaker at 120 kV and the 12.5 kV substation is considered to be open at $t = 0.2$ s that means the microgrid is islanded at $t = 0.2$ s. The nominal frequency is chosen as $\omega_0 = 377$ rad/s (60 Hz) for the secondary frequency control. This nominal frequency is identical to the current Northern American standard frequency, which guarantees the easy reconnection of the microgrid to the utility main grid. Both eigen-based analysis and time-domain simulations in MATLAB®/Simpower environment are investigated.

14.5.1 Eigenvalue-based delay margin analysis

The small-signal model of this microgrid with four DGs is built with the system parameters listed in Tables 14.1 and 14.2. Its secondary frequency controller gains are initially set to $K_{p\omega} = 2$ and $K_{i\omega} = 35$. The approximated eigenvalues can be obtained via the method introduced in Section 14.3, while the dimension of the Chebyshev's differentiation matrix $\mathbf{D_N}$ is chosen as $N = 2$ for illustration purposes.

First, the effects of four identical PMU communication delays on the secondary frequency control gains $\mathbf{K_{i\omega}}$ and $\mathbf{K_{p\omega}}$ are analyzed. When four PMU communication delays are $\tau_1 = \tau_2 = \tau_3 = \tau_4 = \tau = 0.2$, the eigens are shown in Figure 14.5 (only the eigens that are closer to the imaginary axis are shown). It can be found the critical

Table 14.1 Distribution system parameters

Parameters	**Value**
S_{base}	10 (MVA)
V_{base1}	$120\sqrt{2}/\sqrt{3}$ (kV)
V_{base2}	$12.5\sqrt{2}/\sqrt{3}$ (kV)
V_{base3}	$208\sqrt{2}/\sqrt{3}$ (V)
R_S	0 (p.u.)
X_S	0 (p.u.)
R_T	0 (p.u.)
X_T	0.1 (p.u.)
R_f	0.0029 (p.u.)
X_f	0.0041 (p.u.)
R_t	0 (p.u.)
X_t	0.2 (p.u.)

Table 14.2 Inverter parameters

Parameters	**Value**
K_{pii}	2.5
K_{iii}	500
L_S	1 (mH)
K_{ppi}	2.5
K_{ipi}	100
$K_{\omega i}$	0.00016
ω_0	377 (rad/s)
$Q_{ref1}, Q_{ref2}, Q_{ref3}, Q_{ref4}$	0 (p.u.)

eigen pair (which has the smallest damping ratio) is identified as $-3.55 \pm 5.04i$. Then, the root locus of the critical eigen pair is investigated for both $\mathbf{K_{i\omega}}$ and $\mathbf{K_{p\omega}}$ under different time delays $\tau = \{0.1, 0.2, 0.3\}$, shown in Figures 14.6 and 14.7. It can be seen from these two figures that the feasible sets of integral gains $K_{i\omega}$ and $\mathbf{K_{p\omega}}$ decrease when the PMU delay τ increases.

Then, the effects of four different PMU communication delays are investigated. When four PMU communication delays are respectively $\tau_1 = 0.15$ s, $\tau_2 = 0.08$ s, $\tau_3 = 0.1$ s, and $\tau_4 = 0.2$ s. The eigens are shown in Figure 14.8 and the critical eigen pair is identified as $-6.31 \pm 8.52i$. Then, the root locus of the critical eigen pair is investigated for both $\mathbf{K_{i\omega}}$ and $\mathbf{K_{p\omega}}$ under different maximum PMU time delays $\tau_{\max} = \{0.1, 0.2, 0.3\}$, shown in Figures 14.9 and 14.10. It can be noted from these two figures that the feasible gain sets for both $K_{i\omega}$ and $\mathbf{K_{p\omega}}$ decrease when the maximum PMU delay $\tau_{\max}$ increases.

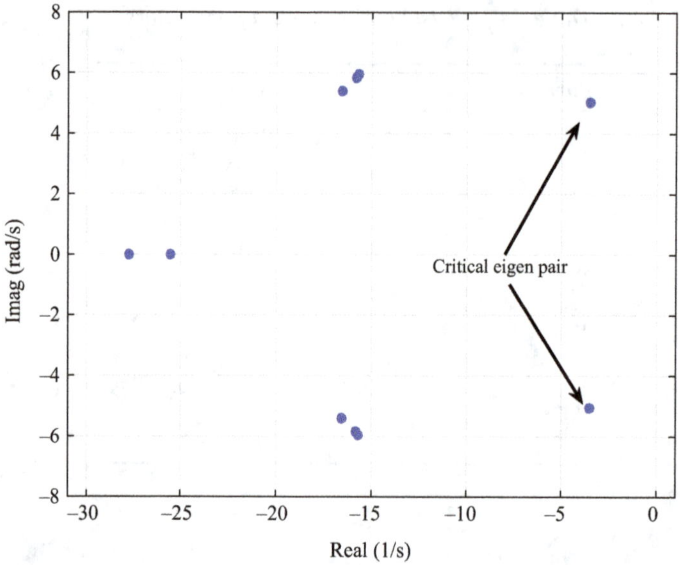

Figure 14.5 *The eigenvalue plot of the microgrid system with $K_{p\omega} = 2$ and $K_{i\omega} = 35$ with $\tau_1 = \tau_2 = \tau_3 = \tau_4 = \tau = 0.2$ s*

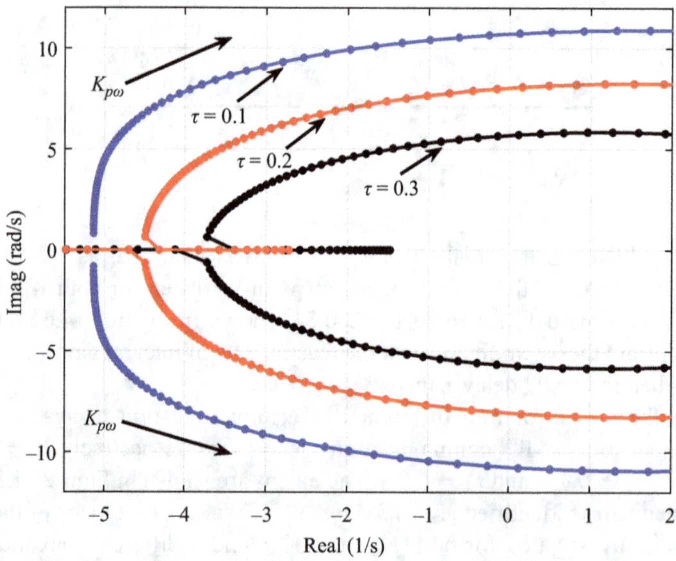

Figure 14.6 *Root locus for $K_{p\omega}$ under identical PMU delays (arrows direct the increasing gains)*

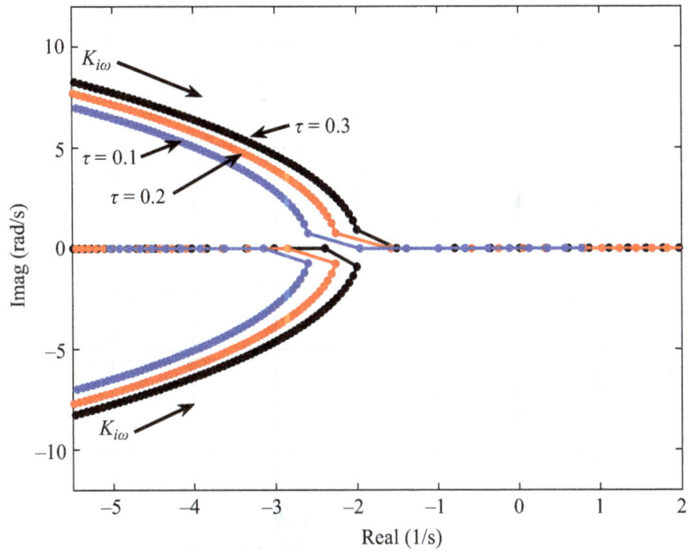

Figure 14.7 Root locus for $K_{i\omega}$ under identical PMU delays (arrows direct the increasing gains)

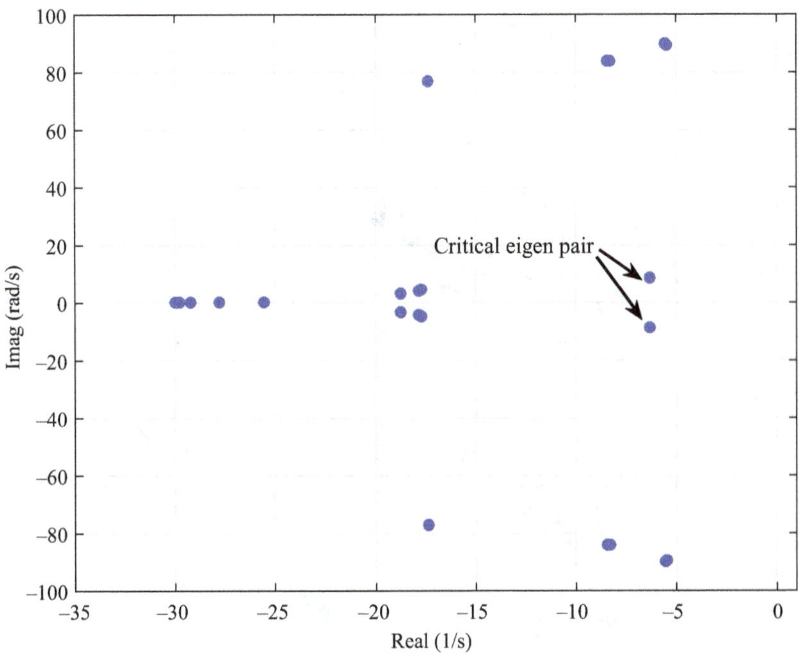

Figure 14.8 The eigenvalue plot of the microgrid system with $K_{p\omega} = 2$ and $K_{i\omega} = 35$ with $\tau_1 = 0.15$, $\tau_2 = 0.08$, $\tau_3 = 0.1$, and $\tau_4 = 0.2$ s

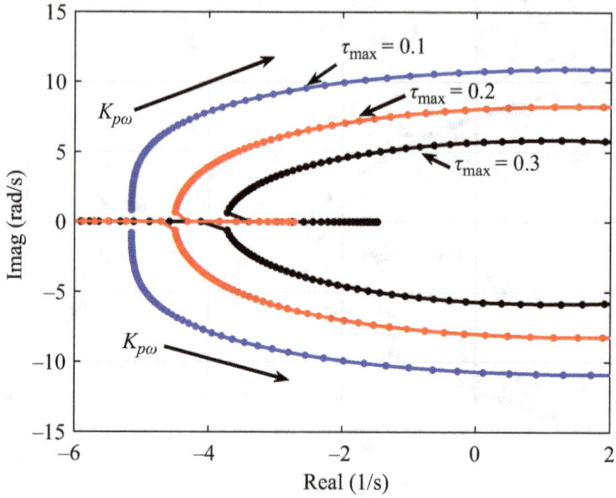

Figure 14.9 Root locus for $K_{p\omega}$ under four different PMU delays with the maximum delay $\tau_{max} = \{0.1, 0.2, 0.3\}$ (arrows direct the increasing gains)

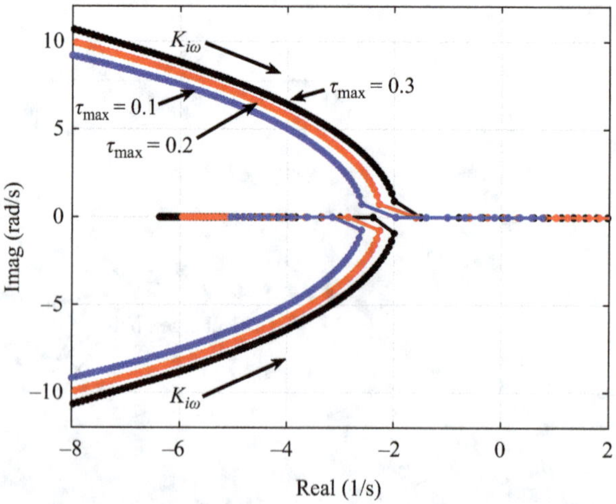

Figure 14.10 Root locus for $K_{i\omega}$ under four different PMU delays with the maximum delay $\tau_{max} = \{0.1, 0.2, 0.3\}$ (arrows direct the increasing gains)

14.5.2 Time-domain simulations

The effects of multiple PMU communication delays on the secondary frequency control in the microgrid shown in Figure 14.1 are investigated in MATLAB®/Simpower environment.

First, the four PMU communication delays are considered to be identical and the following cases are studied:

- Case 1: Four PMU communication delays are $\tau_1 = \tau_2 = \tau_3 = \tau_4 = 0.1$ s in the microgrid,
- Case 2: Four PMU communication delays are $\tau_1 = \tau_2 = \tau_3 = \tau_4 = 0.15$ s in the microgrid, and
- Case 3: Four PMU communication delays are $\tau_1 = \tau_2 = \tau_3 = \tau_4 = 0.2$ s in the microgrid.

The results for the three cases are shown in Figure 14.11, Figure 14.12, and Figure 14.13, respectively. It can be noted that the magnitude of the frequency oscillation increases greatly after the islanding action is taken when the time delay increases from $\tau = 0.1$ s to $\tau = 0.15$ s and $\tau = 0.2$ s, as shown in Figures 14.14 and 14.15. It is noted time delays could increase the original frequency oscillation magnitudes resulted from islanding actions. Then, the different PMU communication delays are investigated and the following two cases are considered:

- Case 4: Four PMU communication delays are $\tau_1 = 0.15$ s, $\tau_2 = 0.08$ s, $\tau_3 = 0.1$ s, and $\tau_4 = 0.05$ s in the microgrid and
- Case 5: Four PMU communication delays are $\tau_1 = 0.08$ s, $\tau_2 = 0.12$ s, $\tau_3 = 0.16$ s, and $\tau_4 = 0.2$ s in the microgrid.

The results for the two cases are shown in Figures 14.14 and 14.15, respectively. It can be seen that the magnitude of the frequency oscillation is larger in case 5 compared to case 4 when the islanding action is conducted. The reason is the maximum delay among these PMU delays becomes bigger from $\tau_1 = 0.15$ (corresponding to PMU in the first DG bus) to $\tau_4 = 0.2$ (corresponding to PMU in the

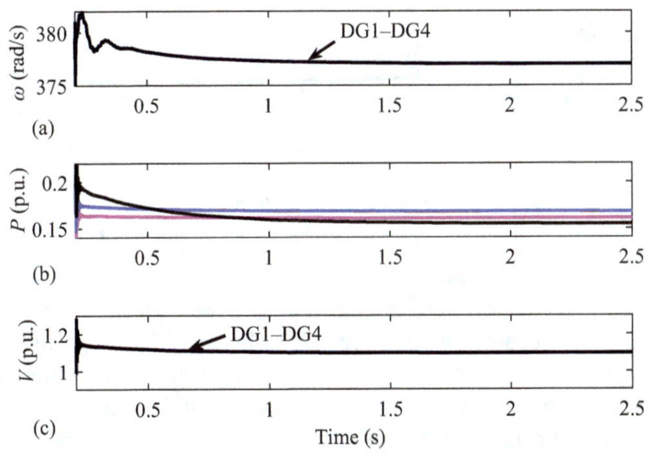

Figure 14.11 System dynamic performance when $\tau_1 = \tau_2 = \tau_3 = \tau_4 = 0.1$ s (a) Frequencies of DGs (b) Real power of DGs (c) Voltages of DGs

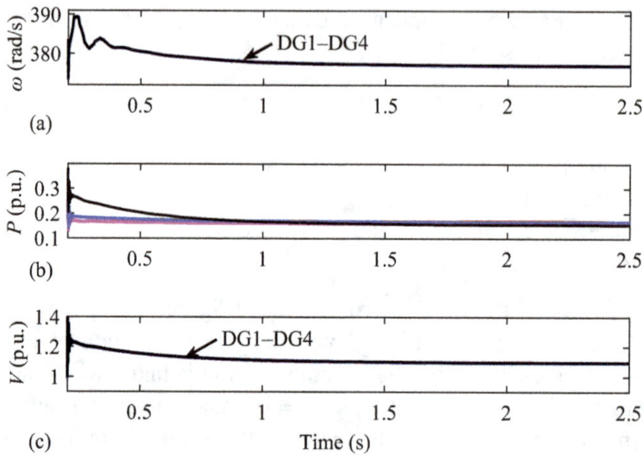

Figure 14.12 System dynamic performance when $\tau_1 = \tau_2 = \tau_3 = \tau_4 = 0.15$ s (a) Frequencies of DGs (b) Real power of DGs (c) Voltages of DGs

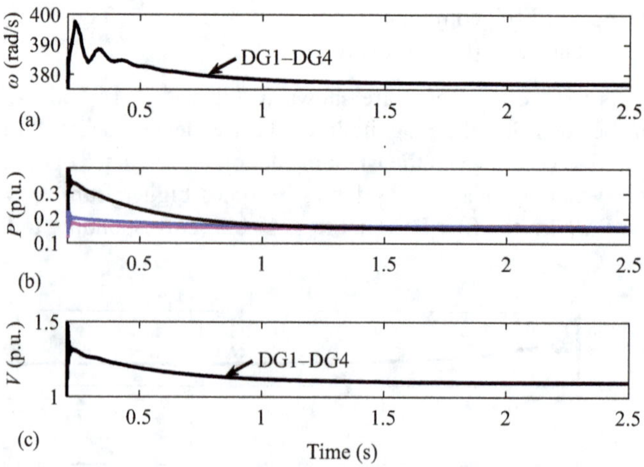

Figure 14.13 System dynamic performance when $\tau_1 = \tau_2 = \tau_3 = \tau_4 = 0.2$ s (a) Frequencies of DGs (b) Real power of DGs (c) Voltages of DGs

fourth DG bus). It is also interesting to be noted that the frequency oscillation magnitude in DG1 becomes bigger even though the PMU delay in DG1 bus is smaller. Therefore, the effects of the PMU delays on the performance of the secondary frequency controller is dominated by the maximum delay.

Moreover, to evaluate the proposed gain scheduling approach, the following cases are considered: the multiple PMU delays are $\tau_1 = 0.1$ s, $\tau_2 = 0.15$ s, $\tau_3 = 0.15$ s, and $\tau_4 = 0.25$ s in the microgrid with and without gain schedulers. Simulation results of the two cases (only DG1) are shown in Figure 14.16. The

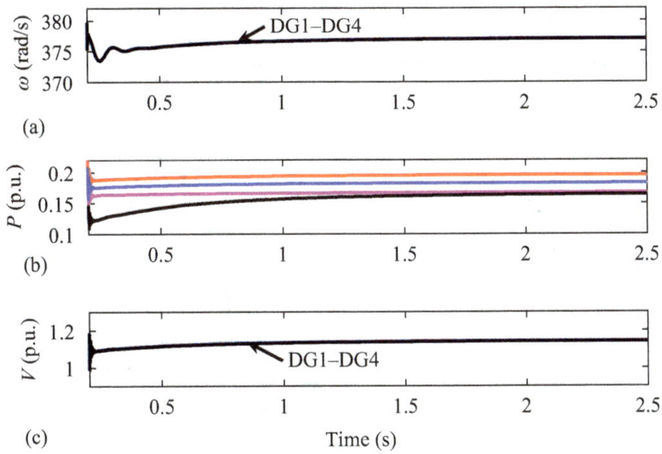

Figure 14.14 System dynamic performance when $\tau_1 = 0.15$, $\tau_2 = 0.08$, $\tau_3 = 0.1$, and $\tau_4 = 0.05$ s (a) Frequencies of DGs (b) Real power of DGs (c) Voltages of DGs

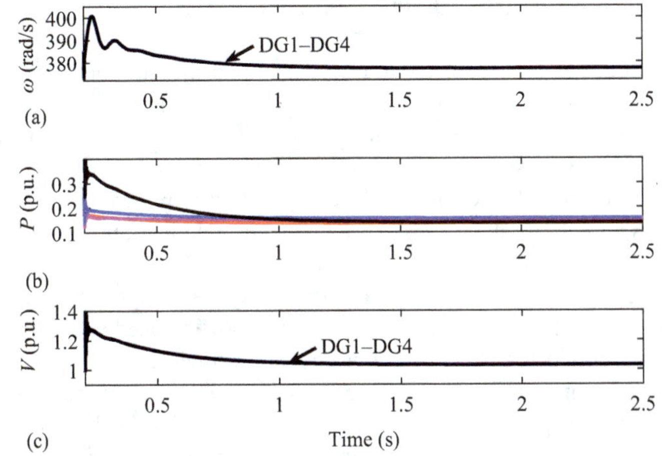

Figure 14.15 System dynamic performance when $\tau_1 = 0.08$, $\tau_2 = 0.12$, $\tau_3 = 0.16$, and $\tau_4 = 0.2$ s (a) Frequencies of DGs (b) Real power of DGs (c) Voltages of DGs

microgrid performances with gain schedulers are compared with those without gain schedulers. The frequency oscillation in the microgrid with the gain schedulers is better damped. The effectiveness of the gain scheduling approach is therefore validated.

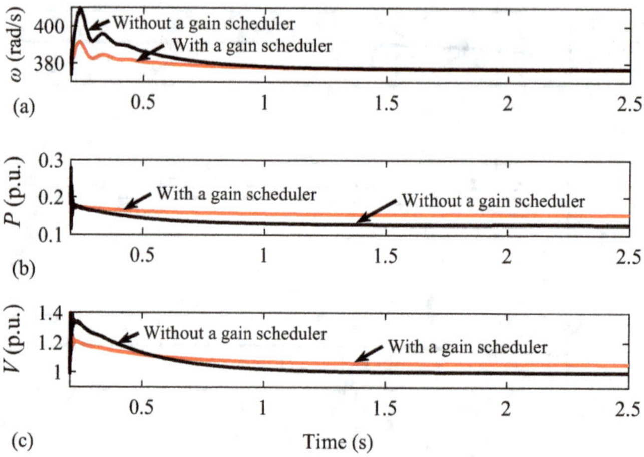

Figure 14.16 Comparisons of system dynamic performance with and without gain schedulers (a) Frequencies of DG1 (b) Real power of DG1 (c) Voltages of DG1

14.6 Conclusions and future work

A PMU can provide synchronized and time-stamped measurements of phasors in power systems. Due to these advantages, PMUs are extensively used in the wide-area control, monitoring, and protections in the past decade. Recently, micro-PMUs (the version of PMU used in distribution systems) are designed and installed in microgrids to improve the accuracy of measurements provided for microgrid control centers. These PMU data transfer involves communication networks between μPMUs and controllers, such as WiMax and WiFi. During the transmission of power control and measurement signals, they may be delayed or dropped, even attacked by adversaries. The recent experiences of PMU-based wide-area damping control for interconnected power systems show that time delay of PMU data could degrade the dynamic performance of controllers and even jeopardize the system stability. In this chapter, we investigate the impact of mPMU communication delays on the secondary frequency control in islanded microgrids with multiple inverter-based DGs. Because of the negligible inertias of the inverter-based DGs, the dynamics of the frequency can act very fast and the PMU delays could seriously affect the performance. The small-signal model is formed for the microgrid. By the eigen-based analysis, it is found that feasible sets of secondary frequency control gains decrease as PMU delays increase for both identical and different PMU delay cases. The performance degradation and even instability of the system may have resulted from the PMU delays. A gain-scheduler-based secondary frequency controller is designed for the islanded microgrid to compensate the performance degradation. The time-domain simulations have validated the impact of multiple PMU communication delays. The proposed gain-scheduling frequency control approach has also been verified to improve the system performance when there are multiple PMU communication delays.

In future, the microgrid with both synchronous machines and inverter-based DGs will be investigated when the effect of PMU data delays are considered. Also, nonlinear loads which can have more complicated impacts on the microgrid will also be included. Furthermore, random PMU data delays will be modeled and its impact on the microgrid will be deeply understood in our future work.

References

[1] Lasseter RH, and Paigi P. Microgrid: a conceptual solution. In: *IEEE 35th Annual Power Electronics Specialists Conference* 2004 (PESC 04), Aachen, Germany, vol. 6, 2004, pp. 4285–4290.

[2] Farhangi H. The path of the smart grid. *IEEE Power and Energy Magazine*. 2010;8(1):18–28.

[3] Chandorkar MC, Divan DM, and Adapa R. Control of parallel connected inverters in standalone AC supply systems. *IEEE Transactions on Industry Applications*. 1993;29(1):136–143.

[4] Guerrero JM, Vasquez JC, Matas J, *et al.* Hierarchical control of droop-controlled AC and DC microgrids: a general approach toward standardization. *IEEE Transactions on Industrial Electronics*. 2011;58(1):158–172.

[5] Lopes JAP, Moreira CL, and Madureira AG. Defining control strategies for MicroGrids islanded operation. *IEEE Transactions on Power Systems*. 2006;21(2):916–924.

[6] Sao CK, and Lehn PW. Autonomous load sharing of voltage source converters. *IEEE Transactions on Power Delivery*. 2005;20(2):1009–1016.

[7] Guerrero JM, Vasquez JC, Matas J, *et al.* Control strategy for flexible microgrid based on parallel line-interactive UPS systems. *IEEE Transactions on Industrial Electronics*. 2009;56(3):726–736.

[8] Madureira A, Moreira C, and Lopes JP. Secondary load-frequency control for MicroGrids in islanded operation. In: *2005 International Conference on Renewable Energy and Power Quality*, Zaragoza, Spain; 2005.

[9] Tsikalakis AG, and Hatziargyriou ND. Centralized control for optimizing microgrids operation. *IEEE Transactions on Energy Conversion*. 2008;23(1): 241–248.

[10] Pogaku N, Prodanovic M, and Green TC. Modeling, analysis and testing of autonomous operation of an inverter-based microgrid. *IEEE Transactions on Power Electronics*. 2007;22(2):613–625.

[11] Pogaku N, Prodanovic M, and Green TC. Inverter-based microgrids: small-signal modelling and testing. In: *The 3rd IET International Conference on Power Electronics, Machines and Drives, 2006*, The Contarf Castle, Dublin, 2006, pp. 499–504.

[12] Wu X, Shen and C. Distributed optimal control for stability enhancement of microgrids with multiple distributed generators. *IEEE Transactions on Power Systems*. 2017;32(5):4045–4059.

[13] Dehkordi NM, Sadati N, and Hamzeh M. Fully distributed cooperative secondary frequency and voltage control of islanded microgrids. *IEEE Transactions on Energy Conversion*. 2017;32(2):675–685.
[14] Zhang G, Li C, Qi D, *et al*. Distributed estimation and secondary control of autonomous microgrid. *IEEE Transactions on Power Systems*. 2017;32(2): 989–998.
[15] Kamwa I, Beland J, Trudel G, *et al.* Wide-area monitoring and control at Hydro-Quebec: past, present and future. In: *2006 IEEE Power Engineering Society General Meeting*, Montreal, Quebec, Canada 2006, pp. 1–12.
[16] Ghahremani E, and Kamwa I. Local and wide-area pmu-based decentralized dynamic state estimation in multi-machine power systems. *IEEE Transactions on Power Systems*. 2016;31(1):547–562.
[17] von Meier A, Stewart E, McEachern A, *et al.* Precision micro-synchrophasors for distribution systems: a summary of applications. *IEEE Transactions on Smart Grid*. 2017;8(6):2926–2936.
[18] Lassetter C, Sanchez E, and Kim J. A learning scheme for microgrid reconnection. *IEEE Transactions on Power Systems*. 2017;33(1):691–700.
[19] He J, Lu C, Jin X, *et al.* Analysis of time delay effects on wide area damping control. In: *IEEE Asia Pacific Conference on Circuits and Systems, 2008*, Macao, China 2008, pp. 758–761.
[20] Li H, Sun Z, Chow MY, *et al.* Gain-scheduling-based state feedback integral control for networked control systems. *IEEE Transactions on Industrial Electronics*. 2011;58(6):2465–2472.
[21] Mazumder SK, Tahir M, and Acharya K. Master-slave current-sharing control of a parallel DC-DC converter system over an RF communication interface. *IEEE Transactions on Industrial Electronics*. 2008;55(1):59–66.
[22] Pourramezan R, Seyedi YC, Karimi H, *et al.* Design of an advanced phasor data concentrator for monitoring of distributed energy resources in smart microgrids. *IEEE Transactions on Industrial Informatics*. 2017;13(6):3027–3036.
[23] Liu S, Wang X, and Liu P. Impact of communication delays on secondary frequency control in an islanded microgrid. *IEEE Transactions on Industrial Electronics*. 2015;62(4):2021–2031.
[24] Stahlhut JW, Browne TJ, Heydt GT, *et al.* Latency viewed as a stochastic process and its impact on wide area power system control signals. *IEEE Transactions on Power Systems*. 2008;23(1):84–91.
[25] Wang S, Meng X, and Chen T. Wide-area control of power systems through delayed network communication. *IEEE Transactions on Control Systems Technology*. 2012;20(2):495–503.
[26] Wang X, Freitas W, Dinavahi V, *et al.* Investigation of positive feedback anti-islanding control for multiple inverter-based distributed generators. *IEEE Transactions on Power Systems*. 2009;24(2):785–795.
[27] Dai L. *Singular Control Systems*. Lecture Notes in Control and Information Sciences Vol. 118. New York, NY: Springer-Verlag; 1989.
[28] Hautus M, and Silverman LM. System structure and singular control. *Linear Algebra and Its Applications*. 1983;50:369–402.

[29] Jiang L, Yao W, Wu QH, *et al.* Delay-dependent stability for load frequency control with constant and time-varying delays. *IEEE Transactions on Power Systems*. 2012;27(2):932–941.

[30] Jia H, and Yu X. A simple method for power system stability analysis with multiple time delays. In: *IEEE PES General Meeting 2008*, Pittsburgh, PA, 2008. pp. 1–7.

[31] Milano F, and Anghel M. Impact of time delays on power system stability. *IEEE Transactions on Circuits and Systems I: Regular Papers*. 2012;59(4): 889–900.

Chapter 15

Adaptive wide-area power oscillation damper design for photovoltaic plant considering delay compensation

Wei Yao[1], Yu Shen[1] and Jinyu Wen[1]

With the increasing amount of grid-connected photovoltaic (PV) plants in a power system, the volatility and uncertainty of power generated from PV plants may adversely impact on the power system stability, for example, increasing the risk of interarea oscillation. To solve the problem, this chapter proposes an adaptive wide-area power oscillation damper (WPOD) based on goal representation heuristic dynamic programming (GrHDP) algorithm for PV plant to enhance damping of the concerned interarea mode. By using GrHDP, the adaptive WPOD does not need the model of the power system and has adaptivity to different operating conditions. Moreover, an adaptive delay compensator (ADC) is also employed for the proposed adaptive WPOD to compensate the communication delay existing in the wide-area signal. A case study is carried out on a 16-machine 68-bus system with a large-scale PV plant. Simulation results show that the adaptive WPOD can provide satisfactory damping performance and compensate the communication delay over a wide range of operating conditions.

15.1 Introduction

Recently, the renewable energy resources develop fast in the world [1,2]. For example, the installed capacity of solar PV is growing rapidly in the past decade, which increased from 3.7 GW in 2004 to 227 GW in 2015. Especially, China has the largest amount of installed capacity of PV so far [3]. Owing to volatility and uncertainty of power generated from PV plants, research results obtained in [4–6] have advocated that the increased PV penetrations may adversely impact on the critical interarea modes depending on the penetration level, location, and control techniques. With the increasing penetration of PV plants, the capability of power

[1]State Key Laboratory of Advanced Electromagnetic Engineering and Technology, School of Electrical and Electronic Engineering, Huazhong University of Science and Technology, Wuhan, China

oscillation damping of large scale PV plants is demanded by many power operators. It can be achieved by introducing a supplementary control loop into the PV controller. Generally, the controller of PV plant uses maximum power point tracking algorithm to maximize the active power output. As the inverter of PV plant can control the active and reactive power independently, reactive power modulation can be used to regulate the voltage to damp power oscillations without cutting off the maximum active power output of PV plants [7,8].

With development and implementation of wide-area measurement system (WAMS) in power system, the wide-area signal has been widely used for designing wide-area power oscillation damper (WPOD) of generator or flexible alternative current transmission systems (FACTS) devices to damp interarea oscillations because it has higher observability and flexibility in interarea mode than local signal [9]. Traditionally, the phase compensation method is employed to design the conventional lead-lad WPOD [10]. However, the damping performance of the conventional lead-lad WPOD may deteriorate with the variation of the operating condition, as the conventional WPOD is designed based on the linearized model under a specific operating condition. To improve the performance of the POD controller, the H_∞ optimization has been predominantly used for designing damping controllers for power system [11]. However, it introduces a number of Lyapunov variables and the H_∞-based control schemes are too conservative when dealing with slight disturbances and uncertainties. A mixed H_2/H_∞ output-feedback control method is used to design robust WPOD for excitation system of generator in [9]. The minimax linear quadratic Gaussian (LQG) controller is employed to designed WPOD for PV plants and it can achieve an acceptable tradeoff between performance and robustness [12]. However, most of those controller designs require a mathematical model of the power system, which is very difficult or even unfeasible for a practical large-scale power system. In addition, the transmission of wide-area signal by communication networks leads to the time-varying delays, which is a bottleneck and may exacerbate the damping performance of the WPOD and the stability of power system [13,14]. Most references ignore time delay [9,15] or treat it as a constant delay [12,16–19] in the stage of designing WPOD.

To deal with the time-varying communication delays effectively, the current methods proposed in literature can mainly be classified into two categories. One is to design delay-dependent WPOD to tolerate a certain range of time-varying delays [20–24], such as, delay-dependent feedback control [20,21], modal linear quadratic gaussian (MLQG) method [22], H_∞ control method [23], and μ-synthesis approach [24]. However, these kinds of WPODs cannot thoroughly eliminate the adverse effect of the time-varying delay. The other is to design WPOD to compensate the time-varying or stochastic delays [25–33]. In [25], a simple delay compensator based on phase compensation method is proposed to eliminate the effects of time delay on the proposed adaptive wide-area damping controller (WADC). In [27], an adaptive time delay compensation method is proposed by dividing bounded random delay into several intervals and designing compensators for each delay interval. But they need to know the frequency of the concerned interarea mode of the power system. In addition, many intelligent algorithms also have been used to handle time-varying delay: a particle swarm optimization is applied to design WPOD for doubly fed induction generator wind farms considering the variable communication latency [28] and a fuzzy logic WPOD is proposed to compensate the time delay by shifting the membership functions of output variables [29]. In [31], a networked predictive control approach is employed to design WPOD, which can compensate both constant and time-varying delays. In [33], an adaptive delay compensator (ADC) is proposed to compensate the time delay involved in wide-area robust control loop.

The adaptive dynamic programming (ADP) is one of popular model-free algorithms, which only needs input/output signals and does not need model information of the controlled plant. The heuristic dynamic programming (HDP) has been used for designing adaptive controllers for power system [34–36]. The goal representation HDP (GrHDP), which adds an additional neural network to HDP to automatically build an adaptive internal reward signal, can make a better mapping relationship between the system state and the control action and guide a better damping performance than the HDP. The GrHDP has been used in damping controller for wind farm, generators, and FACTS devices [37–40]. It has been proved that the GrHDP-based damping controller has better damping performance and stronger adaptivity to different operating conditions than that of the HDP-based damping controller.

In this chapter, an adaptive WPOD based on GrHDP algorithm is designed for PV plant to suppress concerned interarea oscillation over a wide range of operating condition. The adaptive WPOD uses the power deviation of a remote transmission line as an input signal and generates the additional reactive power control signal for PV plant. To handle the time-varying communication delays involved in wide-area control loop, an adaptive delay compensator (ADC) is employed to compensate this communication delay. ADC consists of a weighted combination of serval predesigned compensators and the weights are adjusted online according to the instant measured value of the communication delay. Case study is conducted on a 16-machine 68-bus system to verify the effectiveness of the adaptive WPOD in terms of damping performance and communication delay compensation over a wide range of operating conditions.

The rest of this chapter is organized as follows. Section 15.2 briefly introduces the modeling of a grid-connected PV plant. Section 15.3 depicts the structure of the proposed adaptive WPOD. In Section 15.4, the adaptive WPOD is designed for PV plant in the 16-machine 68-bus benchmark system. Section 15.5 conducts case studies on the benchmark system to verify the effectiveness of the adaptive WPOD. Finally, conclusions are drawn in Section 15.6.

15.2 Modeling of grid-connected PV plant

A generic electromechanical transient model of a PV plant, which has been validated and widely used for power system stability analysis [41], is considered in this chapter. As shown in Figure 15.1(a), the PV model consists of PV array, inverter dynamics, and associated control systems of inverter. The PV arrays catch the solar energy from the solar irradiance and the generated power is processed through the power inverter, which serves as a buffer between the PV arrays and the grid and controls reactive power or voltage at the point of common coupling (PCC).

According to the voltage–current relationship of semiconductor devices, the principal output characteristics of PV array is [42]:

$$
\begin{aligned}
I &= I_{sc}\left[1 - C_1\left(e^{U/(C_2 U_{oc})} - 1\right)\right] \\
C_1 &= \left(1 - \frac{I_m}{I_{sc}}\right) e^{-U_m/(C_2 U_{oc})} \\
C_2 &= \left(\frac{U_m}{U_{oc}} - 1\right)\left[\ln\left(1 - \frac{I_m}{I_{sc}}\right)\right]^{-1}
\end{aligned}
\tag{15.1}
$$

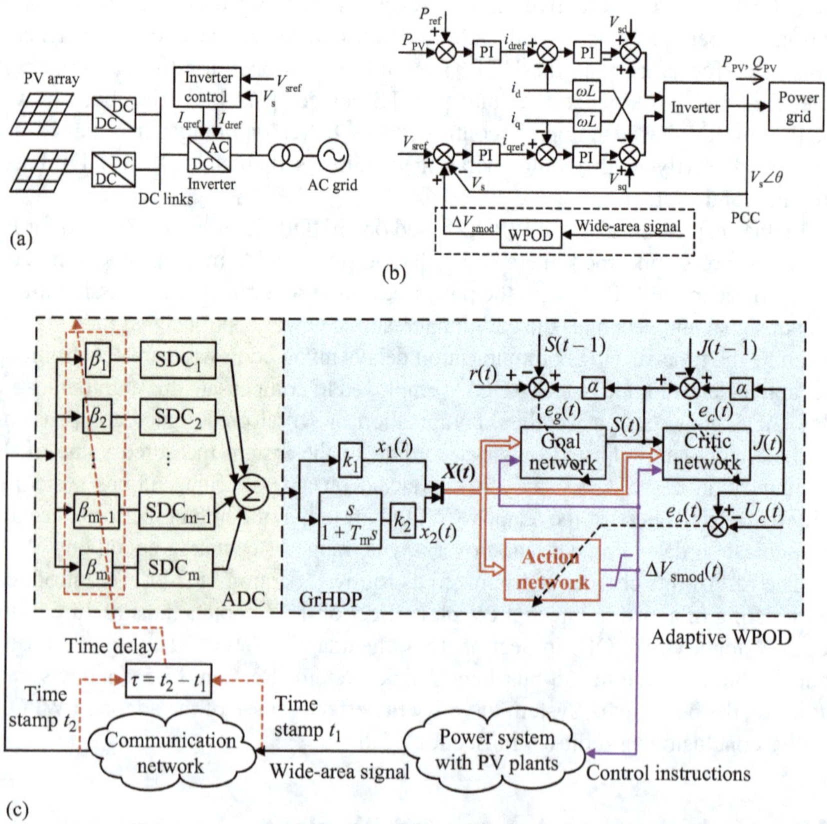

Figure 15.1 Structure of the grid-connected PV plant and block diagram of the proposed WPOD (a) The structure of PV plant (b) The structure of inverter control system and WPOD (c) The block diagram of the proposed adaptive WPOD

where U and I are the output voltage and current of PV array, C_1 and C_2 are the correction factors, I_{sc}, U_{oc}, U_m, and I_m are the short circuit current, open circuit voltage, voltage at maximum power output, and current at maximum power output under the standard condition.

Inverter and its control system are another two main components of PV plant. Figure 15.1(b) depicts the structure of the control system of inverter of PV plant. The outer-loop power control can generate the reference signal i_{dref} and i_{qref} to guide the inner current of inverter. i_{dref} is related to P_{PV}, which denotes the active power collected by PV array and can be delivered to the AC grid totally by inverter as inverter works on MPPT mode. In this model, reactive power control is realized in a voltage control mode [43]. The magnitude of the voltage at PCC (V_s) is fed back to be compared with V_{sref} to generate command i_{qref}, which can decide the output reactive power of PV plants. In addition, the WPOD of PV plant is also depicted in Figure 15.1(b). By using the wide-area input signal which reflects the information of the concerned interarea oscillations of the power system, the WPOD generates a modulation signal ΔV_{smod} and adds it on outer voltage control loop. Thus, it can regulate the output reactive power of PV plants to damp the power oscillations effectively.

15.3 Adaptive WPOD for PV plants

The block diagram of the proposed adaptive WPOD is shown in Figure 15.1(c). The adaptive WPOD consists of the following two parts: an adaptive delay compensator (ADC) and GrHDP. The function of the ADC is to compensate the time-varying communication delay caused by using a communication network to transmit wide-area stabilizing signal. The compensated signal obtained from the ADC is sent to the three networks of GrHDP, in which the action network generates the control signal ΔV_{smod} of the Adaptive WPOD.

15.3.1 Adaptive delay compensator

Time delay results in a phase lag of wide-area signal, which will deteriorate the damping performance of WPOD. Under different disturbances, time delay may change significantly during the control process. Therefore, the conventional fixed delay compensators cannot satisfactorily handle the time-varying delay. To solve the problem, an adaptive delay compensator (ADC), which can deal with time-varying delay well [33], is employed for the adaptive WPOD.

Actually, satellite signal of global positioning system (GPS) can give a standard time information (considered as a time stamp) to the control system and phase-measurement-unit. Therefore, time delay τ can be obtained according to the time stamp at the sender and receiver of wide-area signals, and τ is considered as a known parameter for ADC. The time delay τ can be obtained by comparing the time stamp at receiver t_2 with the time stamp at sender t_1 of the wide-area signal. Then the time delay τ is equal to $t_2 - t_1$, which can be modeled as (15.2):

$$G_d(s) = e^{-\tau s} = (e^{-0.5\tau s})^2 \tag{15.2}$$

According to first-order Pade approximation of $e^{(-0.5\tau s)}$, a second-order Pade approximation of $G_d(s)$ is conducted. That is, $G_d(s) \approx G_D(s)$:

$$G_D(s) = \frac{\left(1 - \frac{s\tau}{4}\right)^2}{\left(1 + \frac{s\tau}{4}\right)^2} \tag{15.3}$$

In terms of the structure of ADC, it is a weighted combination of a number of sub delay compensators (SDC) whose weights are updated according to the measured value of time delay τ. Therefore, the transfer function of ADC is designed as follows:

$$ADC(s) = \frac{N(s)}{D(s)} = \sum_{i=1}^{m} \beta_i(\tau) SDC_i(s) \tag{15.4}$$

$$SDC_i(s) = \frac{\left(1 + \frac{sT_i}{4}\right)^4}{(1 + sT_c)^4}, \; i = 1, 2, \ldots, m \tag{15.5}$$

where β_i is the weights of ADC and is related to the time delay τ. T_c is the time constant which is set according to the maximum and minimum delays of the input signal. T_i is chosen between the minimum and maximum value of existing time delay to deal with different time delay well. For example, if the maximum and minimum delays are 750 ms and 150 ms, and $m = 5$, we can design that $T_1 = 0.15$ s, $T_2 = 0.3$ s, $T_3 = 0.45$ s, $T_4 = 0.65$ s, and $T_5 = 0.75$ s. The time constant T_c typically can be selected between 0.01 and 0.1 s depending on system dynamics [33].

Also taking $m = 5$ as an example, the weight β_i should be calculated according to (15.6). Since T_i is predesigned and known, the weight β_i is only related to the time delay τ:

$$\begin{bmatrix} 1 & 1 & 1 & 1 & 1 \\ T_1 & T_2 & T_3 & T_4 & T_5 \\ T_1^2 & T_2^2 & T_3^2 & T_4^2 & T_5^2 \\ T_1^3 & T_2^3 & T_3^3 & T_4^3 & T_5^3 \\ T_1^4 & T_2^4 & T_3^4 & T_4^4 & T_5^4 \end{bmatrix} \begin{bmatrix} \beta_1 \\ \beta_2 \\ \beta_3 \\ \beta_4 \\ \beta_5 \end{bmatrix} = \begin{bmatrix} 1 \\ \tau^1 \\ \tau^2 \\ \tau^3 \\ \tau^4 \end{bmatrix} \tag{15.6}$$

The reason why the weights β_i should be calculated according to (15.6) is that once (15.6) is met, the (15.4) turns into:

$$ADC(s) = \frac{\sum_{i=1}^{m} \beta_i \left(1 + \frac{sT_i}{4}\right)^4}{(1 + sT_c)^4} = \frac{\left(1 + \frac{s\tau}{4}\right)^4}{(1 + sT_c)^4} \tag{15.7}$$

Therefore, the transfer function of time delay with ADC can be

$$ADC(s)G_D(s) = \frac{\left(1+\frac{s\tau}{4}\right)^4}{(1+sT_c)^4}\frac{\left(1-\frac{s\tau}{4}\right)^2}{\left(1+\frac{s\tau}{4}\right)^2} = \frac{\left(1-\left(\frac{s\tau}{4}\right)^2\right)^2}{(1+sT_c)^4} \tag{15.8}$$

According to the bode diagram of (15.8), the phase-frequency response of the transfer function is not related to the time delay and is related to the time constant instead. In other words, the ADC can compensate the time-varying delay within a certain range. Moreover, the employed method needs no information about the frequency of the oscillating modes as well as the power system model. It should be noted that the phase shift resulted by time constant T_c should be compensated but it is independent of the value of delay and can be effortlessly recompensed.

15.3.2 Brief introduction of GrHDP

15.3.2.1 Objective of GrHDP

GrHDP is one of the members in the adaptive dynamic programming family. Therefore, the ultimate objective for GrHDP is also to solve the Bellman's optimal equation [40]:

$$J[x(i), i] = \sum_{t=i}^{\infty} \alpha^{t-i} U[X(t), u(t), t] \tag{15.9}$$

where U is the utility function and α is a discount factor. $X(t)$ is the system feedback signal and $u(t)$ is the output control action, J is the cost function.

In order to solve (15.9), an optimal control action $u(t)$ should be found to minimize $J(t)$, as shown in (15.10).

$$J^*[X(t)] = \min_{u(t)} r[X(t), u(t)] + \alpha J^*[X(t+1)] \tag{15.10}$$

where in this chapter $X(t)$ is the wide-area signal, $u(t)$ is the control instructions $\Delta V_{\text{smod}}(t)$, and J is the output of critic network of GrHDP.

15.3.2.2 Three networks of GrHDP

Developed from the two-neural-network structure of heuristic dynamic programming, namely the action neural network (ANN) and critic neural network (CNN), GrHDP develops a new neural network named goal neural network (GNN), forming a three neural network structure, as shown in Figure 15.1(c).

The action network generates the control action $\Delta V_{\text{smod}}(t)$ according to the system feedback $X(t)$, the critic network produces the cost function $J(t)$, and the new network helps to generate an adaptive internal reward signal $S(t)$. Internal reward signal $S(t)$ in GrHDP takes the same role as the external reward signal $r(t)$ in HDP. $r(t)$ is a fixed value or fixed-function related to the power system states, which has no adaptivity. But $S(t)$ has strong adaptivity which can facilitate the mapping relationship between the control action and the system state.

All of the three networks are realized by multilayer feedforward neural networks. To generate an optimal control action, the weights of three neural networks are updated during the control process. Since the three networks are all multilayer feed-forward neural networks, the weights are all updated by back propagation rules. The error of critic network, goal representation network, and action network are defined in (15.11) [40]:

$$
\begin{aligned}
&e_c(t) = \alpha J(t) - [J(t-1) - S(t)]; \quad E_c(t) = \frac{1}{2}e_c(t)^2 \\
&e_g(t) = \alpha S(t) - [S(t-1) - r(t)]; \quad E_g(t) = \frac{1}{2}e_g(t)^2 \\
&e_a(t) = J(t) - U_c(t); \quad E_a(t) = \frac{1}{2}e_a(t)^2
\end{aligned}
\tag{15.11}
$$

The weights of each network are updated by rules in (15.12):

$$
\begin{aligned}
&\Delta W_{cij}^{(1)}(t) = -l_c(t)\frac{\partial E_c(t)}{\partial W_{cij}^{(1)}(t)}; \quad \Delta W_{ci}^{(2)}(t) = -l_c(t)\frac{\partial E_c(t)}{\partial W_{ci}^{(2)}(t)} \\
&\Delta W_{gij}^{(1)}(t) = -l_g(t)\frac{\partial E_g(t)}{\partial W_{gij}^{(1)}(t)}; \quad \Delta W_{gi}^{(2)}(t) = -l_g(t)\frac{\partial E_g(t)}{\partial W_{gi}^{(2)}(t)} \\
&\Delta W_{aij}^{(1)}(t) = -l_a(t)\frac{\partial E_a(t)}{\partial W_{aij}^{(1)}(t)}; \quad \Delta W_{ai}^{(2)}(t) = -l_a(t)\frac{\partial E_a(t)}{\partial W_{ai}^{(2)}(t)}
\end{aligned}
\tag{15.12}
$$

where $l_c(t)$ is the learning rate of critic neural network, $W_{cij}^{(1)}$ and $W_{ci}^{(2)}$ denote the weights of the input to hidden layer and the hidden to output layer in the critic neural network, respectively. $l_g(t)$, $W_{gij}^{(1)}$, and $W_{gi}^{(2)}$ are the learning rate and weights of goal representation network. $l_a(t)$, $W_{aij}^{(1)}$, and $W_{ai}^{(2)}$ are the learning rate and weights of action network.

In brief, ANN generates the control action, CNN evaluates the control action and regulates the weights of ANN, and GNN helps to regulate the weights of CNN. In a word, the three neural networks of GrHDP work together to generate an optimal control action.

15.3.2.3 Training and online learning of GrHDP

To achieve a good control performance, the three neural networks need an off-line training in order to learn the characteristics of the power system and improve the adaptivity of GrHDP. The off-line training should be comprehensively conducted under different operating conditions. The detailed off-line training process is described as follows. It should be noted that at t step, signals such as $S(t-1)$ and $J(t-1)$ are considered known to GrHDP.

- Step (1): Initialization of networks.
  $X(t)$ is obtained and $r(t)$ is calculated. According to $X(t)$, ANN produces initial value of $\Delta V_{\text{smod}}(t)$. According to $X(t)$ and $\Delta V_{\text{smod}}(t)$, GNN outputs initial value of $S(t)$. According to $X(t)$, $\Delta V_{\text{smod}}(t)$, and $S(t)$, CNN generates initial value of $J(t)$.

- Step (2): Weights updating of GNN.
  The error $E_r(t)$ is calculated according to $r(t)$, $S(t)$, and $S(t-1)$, then the weights of GNN are tuned according to (15.12) until the stop criterion is satisfied, and finally a new internal reward signal $S(t)$ is generated.
- Step (3): Weights updating of CNN.
  The error $E_c(t)$ is calculated, the weights of CNN are updated until the stop criterion is satisfied, and a new cost value $J(t)$ is generated.
- Step (4): Weights updating of ANN.
  The error $E_a(t)$ is calculated, the weights of ANN are updated until the stop criterion is satisfied. After the updating of the weights, the action network generates a new control action $\Delta V_{\text{smod}}(t)$. Compared to the $\Delta V_{\text{smod}}(t)$ in step (1), $\Delta V_{\text{smod}}(t)$ here is updated and the control performance is improved.
- Step (5): Ending of training.
  Once the weights of three networks are tuned, they are fixed thereafter and the training process repeats from step (1) when entering the $t+1$ step. Once the whole training process is finished, the updated weights are employed as initial weights of adaptive WPOD.

  Moreover, adaptive WPOD also has the ability of online learning. In other words, if the operating conditions of power system vary or serious disturbances occur, weights of GrHDP can be updated online and the updating process is the same as steps (1)–(4).

15.4 Design of adaptive WPOD

The 16-machine 68-bus benchmark power system shown in Figure 15.2 is used as the test system. There is a PV plant with 200 MW rated output power located at bus 43. The detailed parameters of the system are given in [44].

Under the initial operation condition, modal analysis is conducted and the analysis results are shown in Table 15.1. It can be found that the installing of PV plant has an adverse impact on the stability of the test system. If a 200 MW PV plant is installed, the damping ratios of interarea modes decrease, and mode 1 in particular, decreases significantly. Since the damping ratio of mode 1 is less than 3%, it is necessary to design a WPOD to improve the damping ratio of mode 1.

The theory of geometric modal observability used in [45] is applied to select a suitable feedback signal for adaptive WPOD. It can be found that the active power of transmission line P_{46-49} has the highest observability for mode 1 and thus it is chosen as the input signal of adaptive WPOD. The output of adaptive WPOD is the control instructions ΔV_{smod}. Considering the reactive regulation capacity of the inverter of PV plants, the output limits of adaptive WPOD is defined as ± 0.2, which is 10% of the rated power of PV plants.

In terms of the design of ADC, since the minimum and maximum delays of the control loop are considered 150 ms and 750 ms in this chapter, parameters of five sub delay compensators are designed as follows: $T_1 = 0.15$, $T_2 = 0.3$, $T_3 = 0.45$, $T_4 = 0.6$, and $T_5 = 0.75$ and time constant T_c is 0.05 s.

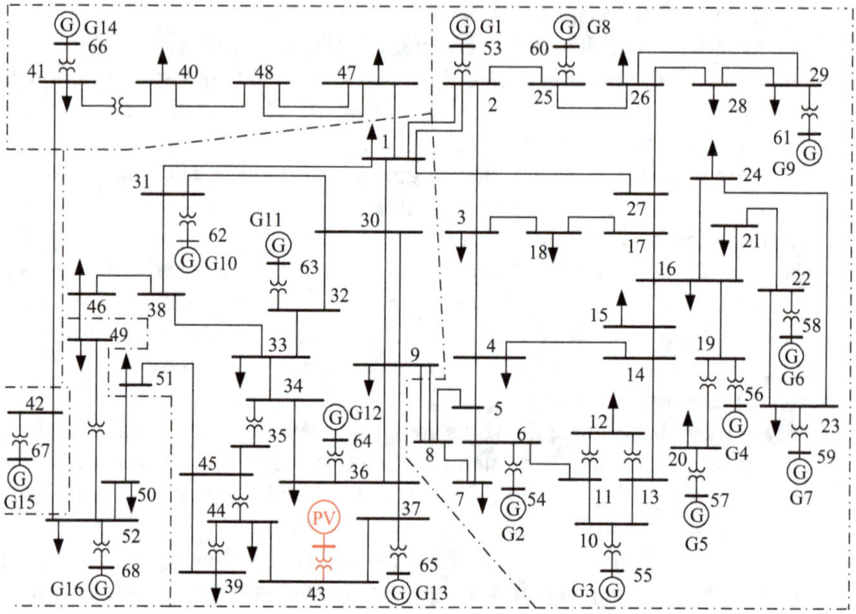

Figure 15.2 The 16-machine 68-bus benchmark power system

Table 15.1 Interarea modal analysis results of the benchmark with or without PV plant

No.	Damping ratio with PV/without PV	Frequency (Hz) with PV/without PV	Participating generators
1	0.0100/0.0194	0.4592/0.4568	16 vs (3, 4, 5, 6, 7, 9, 13)
2	0.0796/0.0805	0.8033/0.8063	9 vs (13, 14, 15)
3	0.0796/0.0792	1.4932/1.4984	(11, 15) vs (10, 12, 13)
4	0.0705/0.0718	1.5356/1.5373	11 vs 15

For the design of GrHDP, to give more information of the power system to GrHDP, a parallel differential element is added, as shown in Figure 15.1(c). $k_1 = 1$ and $k_2 = 0.15$ are normalization coefficient to keep the original signal and its phase-shifted signals are in the same numeric range. The detailed parameters of GrHDP-damping controller are listed in Table 15.2.

For the conventional WPOD (C-WPOD) with lead-lag elements, it is designed based on the residue method and the linearization model of the benchmark system under the initial operating condition. For fair comparison, the input signal and the

Table 15.2 The parameters of GrHDP

Neural network	Action	Critic	Goal
Number of input layers	2	4	3
Number of output layers	1	1	1
Number of hidden layers	3	3	3
Learning rate	0.01	0.006	0.006
Maximum iteration numbers	30	30	30
Margin of error	1e−8	1e−8	1e−8
Range of weight	±5	±5	±5

output limits of C-WPOD are also chosen as P_{46-49} and ±0.2, respectively. The transfer function of C-WPOD is

$$GC - \mathrm{WPOD}(s) = \frac{\Delta V_{\mathrm{smod}}(t)}{\Delta P_{46-49}(t)} = 0.012\left(\frac{1 + 0.6656s}{1 + 0.1805s}\right)^4 \quad (15.13)$$

15.5 Simulation verification

To verify the damping performance of the proposed adaptive WPOD, simulation studies are conducted on the 16-machine 68-bus benchmark system under the different operating conditions and time-varying delays. Two typical cases are conducted as follows.

15.5.1 Case 1: Contingencies under different operating conditions without time delays

15.5.1.1 Separate contingencies

To verify the adaptability of the adaptive WPOD to variation of system operating conditions, four operating conditions including the initial operating conditions shown in Table 15.3 are considered for simulation studies. The operating conditions 2–4 are set through changing the output power of G6, G9, and G16 and the load at bus 52.

When the communication delay from wide-area signal is ignored, the damping performance of the system with adaptive WPOD, conventional WPOD, and without WPOD under the four contingencies listed in Table 15.3 are depicted in Figure 15.3. Note that A-WPOD and C-WPOD denote the proposed adaptive WPOD and conventional lead-lag WPOD, respectively. It can be found that the power oscillations of the system without WPOD sustains for a long time and the system is hard to maintain the stability. For light power flow like operating conditions OC.1 and OC.2, the conventional WPOD can provide an effective damping performance, but a little bit inferior to adaptive WPOD. Once power flow changes significantly or a severe disturbance occurs like OC.3 and OC.4, the damping performance of conventional WPOD gets worse or even ineffective. However, the adaptive WPOD can maintain an effective damping performance even though

Table 15.3 Different operating conditions and contingencies used for the benchmark

No.	P_{46-49} (p.u.)	Contingencies
1	0.32	Three-phase-ground fault occurs at line 28–29 near bus 29 at $t = 1$ s, switch off faulty line at $t = 1.1$ s and reclose it at $t = 1.9$ s
2	1.54	Three-phase-ground fault occurs at line 45–51 near bus 45 at $t = 1$ s, switch off faulty line at $t = 1.1$ s and reclose it at $t = 1.9$ s
3	1.63	Cut off the load in bus 52 at $t = 1$ s
4	3.94	Three-phase-ground fault occurs at line 1–30 near bus 1 at $t = 1$ s, switch off faulty line at $t = 1.1$ s

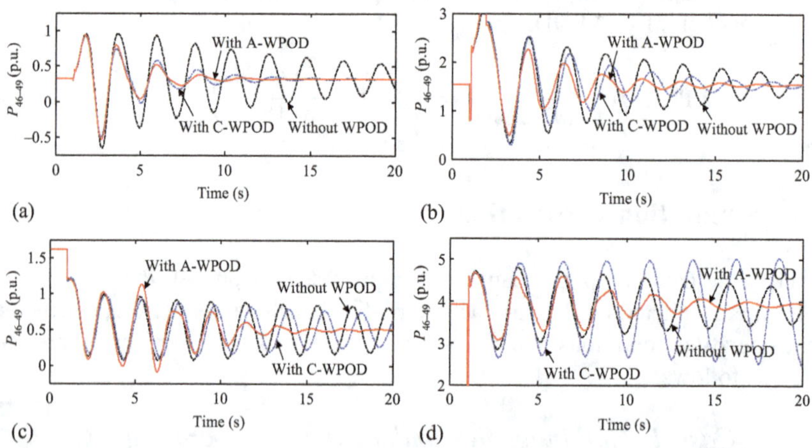

Figure 15.3 Damping performance of adaptive WPOD compared with conventional WPOD and without WPOD (a) System response (OC.1) (b) System response (OC.2) (c) System response (OC.3) (d) System response (OC.4)

the operating conditions are different from the initial operating conditions significantly. The reason why adaptive WPOD is superior to conventional WPOD is that adaptive WPOD has strong ability of online learning but the conventional WPOD uses fixed parameters and is designed based on a special operating condition. These simulation results verify the adaptability of the adaptive WPOD to the variations of system operating condition clearly.

15.5.1.2 Sequential contingencies

To test the adaptability of the proposed adaptive WPOD, four sequential contingencies described in Table 15.4 under different operation conditions are conducted. Figure 15.4 shows the system response. It can be found that the initial active power of line 46–49 is 0.6 p.u., and it changes during each contingency. The system cannot maintain the stability without WPOD but both A-WPOD and C-WPOD can suppress

Table 15.4 Four sequential contingencies under different operation conditions used for the benchmark

No.	Contingencies
1	The load in bus 51 increases from 3.37 p.u. to 8 p.u. at $t = 1$ s
2	The load in bus 52 decreases from 5 p.u. to 1.5 p.u. at $t = 26$ s
3	Three-phase-ground fault occurs at line 45–51 near bus 45 at $t = 51$ s, switch off faulty line at $t = 51.1$ s and reclose it at $t = 51.9$ s
4	Three-phase-ground fault occurs at line 1–31 near bus 31 at $t = 76$ s, switch off faulty line at $t = 76.1$ s

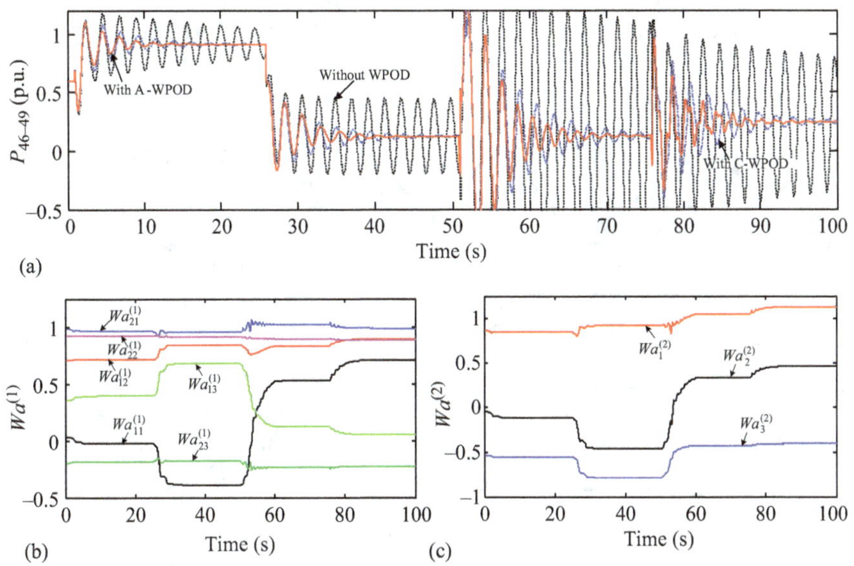

Figure 15.4 Damping performance of WPOD under sequential contingencies (a) Active power response of line 46–49 (input signal of WPOD) (b) Weights of action network $Wa^{(1)}$ (c) Weights of action network $Wa^{(2)}$

the oscillations significantly. It is clear that A-WPOD has a superior damping performance than C-WPOD. Figure 15.4 also shows the weights transaction process of neural networks of A-WPOD. It can be concluded that the weights can adapt quickly to the contingencies and improve the characteristic of A-WPOD.

15.5.2 Case 2: Contingencies under different operating conditions with time delays

15.5.2.1 With constant time delay

There are approximately 33 MW active power flows in transmission line from bus 46 to bus 49. A single three-phase-ground fault is applied on the line 28–29 near bus 28 at $t = 1$ s, followed by switching off the faulty transmission line 28–29 at $t = 1.1$ s and reclose it at $t = 1.9$ s.

Figure 15.5 is the active power response of line 46–49 and output of adaptive WPOD with small time delay and large time delay, respectively. It can be found that facing such a three-phase-ground fault, power oscillations of the system without adaptive WPOD cannot be suppressed and the system is unstable. The adaptive WPOD without ADC can suppress power oscillations quickly when input signal is without time delay or with small-time delay like 150 ms. But it cannot work effectively with large time delay and the oscillations sustained. However, The proposed adaptive WPOD with ADC can deal with different time delay. The reason is that ADC can effectively compensate the phase lag caused by time delay, and thus the signal after ADC is similar to the original signal without time delay. The weights of five SDCs of ADC under different time delay are listed in Table 15.5. It

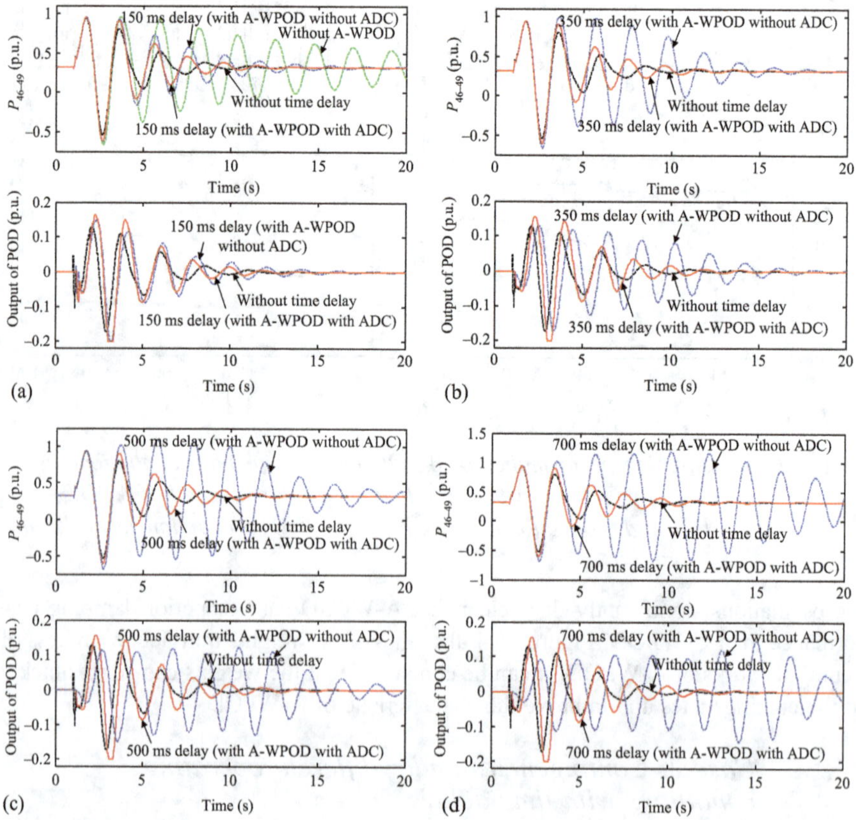

Figure 15.5 Damping performance of adaptive WPOD (a) Delay = 150 ms (b) Delay = 350 ms (c) Delay = 500 ms (d) Delay = 700 ms

Table 15.5 The weights of SDCs under different fixed time delay

Time delay (ms)	β_1	β_2	β_3	β_4	β_5
150	1	0	0	0	0
350	−0.041	0.658	0.494	−0.132	0.021
500	0.021	−0.144	0.864	0.288	−0.029
700	−0.041	0.226	−0.543	0.905	0.453

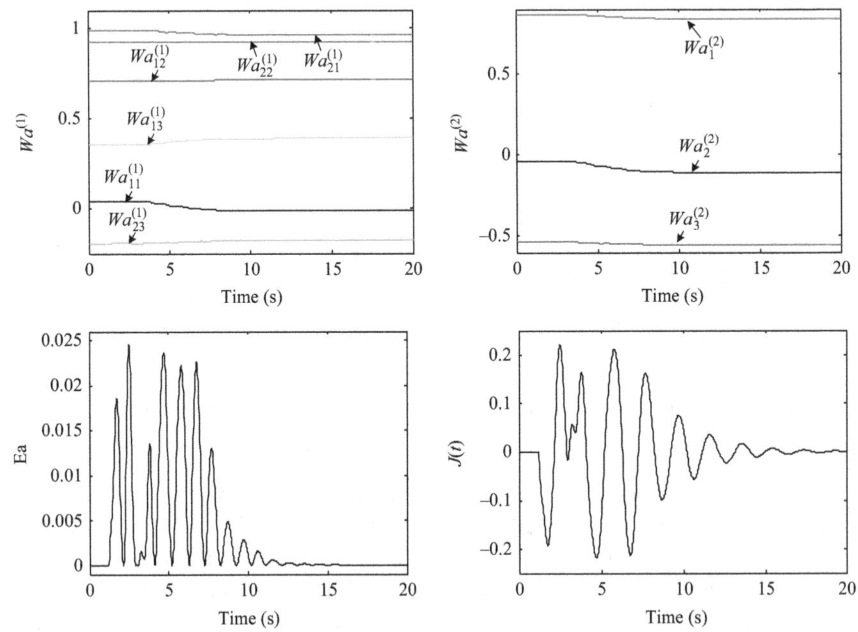

Figure 15.6 Weights Wa, error Ea, and cost function $J(t)$ of A-WPOD (delay time = 700 ms)

is the change of the weights β_i that makes the adaptive WPOD has the ability to compensate different time delay effectively.

To observe the effect of adaptive WPOD well, the response of GrHDP block is shown in Figure 15.6, including the weights of action network $Wa^{(1)}$ and $Wa^{(2)}$, error of action network *Ea*, and cost function $J(t)$. It can be found that both *Ea* and $J(t)$ converge to zero quickly and the weights change during the first few seconds when the error is above the error tolerance.

15.5.2.2 With time-varying delay

The output active power of G16 and G6 increases, making the active power of transmission line form bus 49 to bus 46 change to approximately 54 MW, which is an opposite power flow to the operation condition above with constant time delay.

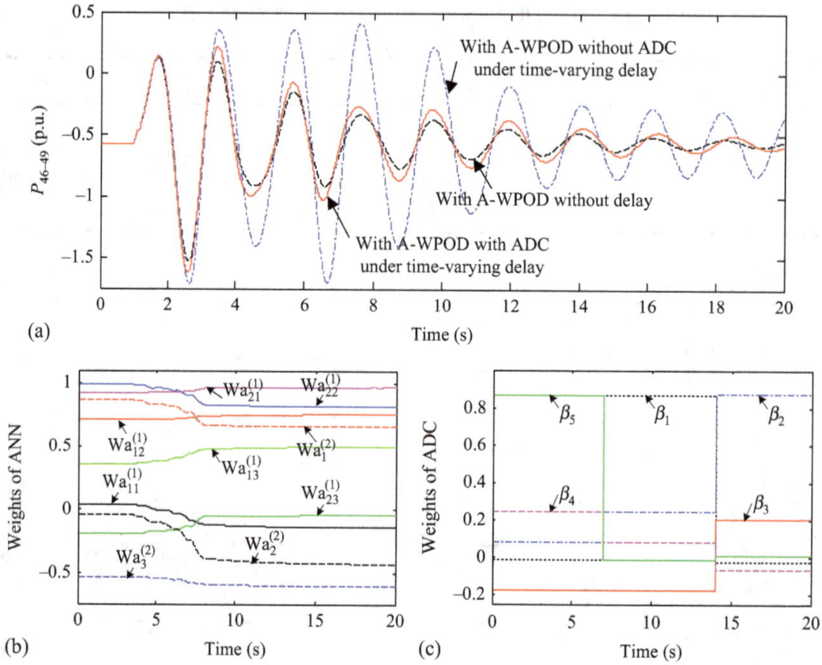

Figure 15.7 Damping performance of A-WPOD under time-varying delay

A terminal three-phase-ground fault is applied on the line 28–29 near bus 28 at $t = 1$ s and the faulty line 28–29 is cut off at $t = 1.1$ s. In this scenario, time-varying delay is considered, the change occurs suddenly and significantly during the control process. Figure 15.7 is the response of the system and adaptive WPOD. It can be found that the adaptive WPOD with ADC under time-varying delay can achieve similar damping performance to that without time delay. However, without ADC, the damping performance of adaptive WPOD gets worse and the system cannot maintain the stability under certain severe time delay. Also as shown in Figure 15.7, weights β_i change significantly when the time delay changes significantly. Because of the change of β_i, the ADC generates different compensation for different time delay and leading to a better control performance of adaptive WPOD. Weights of action network can be updated online effectively due to the online learning ability of the adaptive WPOD.

15.5.2.3 With stochastic time delay

In this scenario, stochastic time delay is considered. The average time delay is set to be 500 ms, and 500 ± 30 ms, 500 ± 60 ms, and 500 ± 90 ms stochastic time delays are considered [46]. In addition, simulation results of 500 ms constant time delay, without delay, and without A-WPOD are also given for comparison. Figure 15.8 shows the damping performance of WPOD under stochastic time delay. It can be found that under different stochastic time delays, the weights of ADC

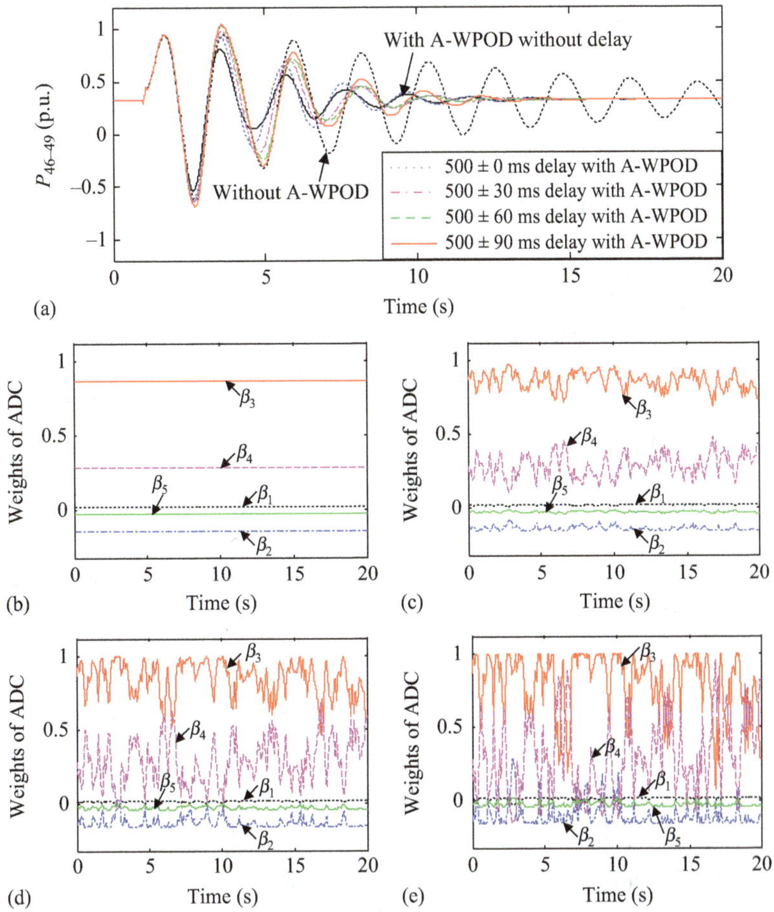

Figure 15.8 Damping performance of adaptive WPOD under stochastic time delay

change around a certain value which is determined by the average time delay value. Moreover, the heavier the stochastic time delay fluctuates, the heavier the weights of ADC β_i fluctuate. It is the change of β_i that makes A-WPOD adapt to stochastic time delay well. In addition, it can be found that A-WPOD can achieve a similar good damping performance under different stochastic time delay.

15.6 Conclusion

This chapter proposes a GrHDP algorithm based wide-area power oscillation damper (WPOD) for suppressing concerned interarea oscillation of a power system with large-scale PV plants. Considering the adverse impact of time delay involved in wide-area control loop, an adaptive delay compensator (ADC) is also employed to compensate the time delay. Simulations studies are carried out on a 16-machine 68-bus power system with a PV plant. Different operating conditions and faults under constant delays, time-varying delay, stochastic delay are respectively considered to verify the effectiveness of the adaptive WPOD. Simulation results show that the proposed adaptive WPOD can achieve good damping performance under different operating conditions, while the system with conventional lead-lag WPOD cannot maintain stability under severe disturbances. If the time delay is small, the adaptive WPOD without ADC can still maintain a good damping performance. If the time delay is large or increases to a large value, the damping performance of the adaptive WPOD without ADC gets worse or even ineffective. However, the adaptive WPOD with ADC has the superior damping performance even under large time delay. In conclusion, the adaptive WPOD can provide satisfactory damping performance and compensate the communication delay over a wide range of operating conditions.

Acknowledgment

This work was supported by the National Natural Science Foundation of China under Grant No. 51577075.

References

[1] Liu, J., Wen, J.Y., Yao, W., *et al.* "Solution to short-term frequency response of wind farms by using energy storage systems." *IET Renewable Power Gener.* 2016, **5**, pp. 669–678

[2] Yang, B., Jiang, L., Wang, L., *et al.* "Nonlinear maximum power point tracking control and modal analysis of DFIG based wind turbine." *Int. J. Electr. Power Energy Syst.* 2016, **74**, pp. 429–436

[3] Renewable Energy Policy Network for the 21st century official web site. "Renewables 2015 Global Status Report." 2015. [Online]. Available: http://www:ren21:net/wp–content/uploads/2016/10/REN21_GSR2016_FullReport_en_11.pdf

[4] Du, W., Wang, H.F., and Xiao, L.-Y. "Power system small-signal stability as affected by grid-connected photovoltaic generation." *Eur. Trans. Electr. Power.* 2011, **21**, pp. 688–703

[5] Shah, R., Mithulananthan, N., and Lee, K.Y. "Contribution of PV systems with ultracapacitor energy storage on inter-area oscillation." *2011 IEEE Power Energy Society General Meeting*, 2011, pp. 1–8

[6] Shah, R., Mithulananthan, N., and Bansal, R.C. "Oscillatory stability analysis with high penetrations of large-scale photovoltaic generation." *Energy Convers. Manage.* 2013, **65**, pp. 420–429

[7] Lal, V.N., Siddhardha, M., and Singh, S.N. "Control of a large scale single-stage grid-connected PV system utilizing MPPT and reactive power capability". *2013 IEEE Power and Energy Society General Meeting*, 2013, pp. 1–5

[8] Shah, R., Mithulananthan, N., Lee, K.Y., *et al.* "Wide-area measurement signal-based stabiliser for large-scale photovoltaic plants with high variability and uncertainty." *IET Renewable Power Gener.* 2013, **7**, pp. 614–622

[9] Zhang, Y., and Bose, A. "Design of wide-area damping controllers for interarea oscillations." *IEEE Trans. Power Syst.* 2008, **23**, pp. 1136–1143

[10] Ke, D.P., Chung, C.Y., and Xue, Y. "An eigenstructure-based performance index and its application to control design for damping interarea oscillations in power systems." *IEEE Trans. Power Syst.* 2011, **26**, pp. 2371–2380

[11] Kim, S.-J., Kwon, S.Y., and Moon, -H. "Low-order robust power system stabilizer for single machine systems: an LMI approach." *Int. J. Control Autom. Syst.* 2010, **8**, pp. 556–563

[12] Shah, R., Mithulananthan, N., and Lee, K.Y. "Large-scale PV plant with a robust controller considering power oscillation damping." *IEEE Trans. Energy Convers.* 2013, **28**, pp. 106–114

[13] Yao, W., Jiang, L., Wu, Q.H., *et al.* "Delay-dependent stability analysis of the power system with a wide-area damping controller embedded." *IEEE Trans. Power Syst.* 2011, **26**, pp. 233–240

[14] Jiang, L., Yao, W., Wen, J.Y., *et al.* "Delay-dependent stability for load frequency control with constant and time-varying delays." *IEEE Trans. Power Syst.* 2012, **27**, pp. 932–941

[15] Kamalasadan, S., and Swann, G.D. "A novel system-centric intelligent adaptive control architecture for damping interarea mode oscillations in power system." *IEEE Trans. Ind. Appl.* 2011, **47**, pp. 1487–1497

[16] Shakarami, M.R., and Davoudkhani, I.F. "Wide-area power system stabilizer design based on Grey Wolf Optimization algorithm considering the time delay." *Electr. Power Syst. Res.* 2016, **133**, pp. 149–159

[17] Chaudhuri, B., Majumder, R., and Pal, B.C. "Wide-area measurement based stabilizing control of power system considering signal transmission delay." *IEEE Trans. Power Syst.* 2004, **19**, pp. 1971–1979

[18] Bai, F.F., Zhu, L., Liu, Y.L., *et al.* "Design and implementation of a measurement-based adaptive wide-area damping controller considering time delays." *Electr. Power Syst. Res.* 2016, **130**, pp. 1–9

[19] Yohanandhan, R.V., and Srinivasan, L. "Decentralised wide-area fractional order damping controller for a large-scale power system." *IET Gener. Transm. Distrib.* 2016, **10**, pp. 1164–1178

[20] Li, Y., Rehtanz, C., Yang, D., *et al.*, "Robust high-voltage direct current stabilising control using wide-area measurement and taking transmission time delay into consideration." *IET Gen. Transm. Distrib.* 2011, **5**, pp. 289–297

[21] Li, Y., Yang, Z., Fang, L., *et al.* "Design and implementation of delay-dependent wide area damping control for stability enhancement of power systems." *IEEE Trans. Smart Grid.* 2017, **8**, pp. 1831–1842

[22] Preece, R., Milanović, J.V., Almutairi, A.M., *et al.* "Damping of inter-area oscillations in mixed AC/DC networks using WAMS based supplementary controller." *IEEE Trans. Power Syst.* 2013, **28**, pp. 1160–1169

[23] Li, J., Chen, Z.H., Cai, D.S., *et al.* "Delay-dependent stability control for power system with multiple time-delays." *IEEE Trans. Power Syst.* 2016, **31**, pp. 2316–2326

[24] Zhang, S., and Vittal, V. "Design of wide-area damping control robust to transmission delay using μ-synthesis approach". *2014 IEEE Power and Energy Society General Meeting*, 2014, pp. 1–5

[25] Zhang, P., Yang, D.Y., Chan, K.W., *et al.* "Adaptive wide area damping control scheme with stochastic subspace identification and signal time delay compensation." *IET Gen. Transm. Distrib.* 2012, **6**, pp. 844–852

[26] Padhy, B.P., Srivastava, S.C., and Verma, N.K. "A Wide-area damping controller considering network input and output delays and packet drop." *IEEE Trans. Power Syst.* 2017, **32**, pp. 166–176

[27] Cheng, L., Chen, G., Gao, W.Z., *et al.* "Adaptive time delay compensator (ATDC) design for wide-area power system." *IEEE Trans. Smart Grid.* 2014, **5**, pp. 2957–2966

[28] Mokhtari, M., and Aminifar, F. "Toward wide-area oscillation control through doubly-fed induction generator wind farms." *IEEE Trans. Power Syst.* 2014, **29**, pp. 2985–2992

[29] Mokhtari, M., Aminifar, F., Nazarpour, D., *et al.* "Wide-area power oscillation damping with a fuzzy controller compensating the continuous communication delays." *IEEE Trans. Power Syst.* 2013, **28**, pp. 1997–2005

[30] Lu, C., Zhang, X., Wang, X., *et al.* "Mathematical expectation modeling of wide-area controlled power systems with stochastic time delay." *IEEE Trans. Smart Grid.* 2015, **6**, pp. 1511–1519

[31] Yao, W., Jiang, L., Wu, Q.H., *et al.* "Wide-area damping controller for power system inter-area oscillations: a networked predictive control approach." *IEEE Trans. Control Systems Technology*. 2015, **23**, pp. 27–36

[32] Chaudhuri, N.R., Ray, S., Majumder, R., *et al.* "A new approach to continuous latency compensation with adaptive phasor power oscillation damping controller (POD)." *IEEE Trans. Power Syst.* 2010, **25**, pp. 939–946

[33] Beiraghi, M., and Ranjbar, A.M. "Adaptive delay compensator for the robust wide-area damping controller design." *IEEE Trans. Power Syst.* 2016, **31**, pp. 4966–4976

[34] Venayagamoorthy, G.K., Sharma, R.K., Gautam, P.K., *et al.* "Dynamic energy management system for a smart microgrid." *IEEE Trans. Neur. Netw. Learn. Syst.* 2016, **27**, pp. 1643–1656

[35] Guo, W., Liu, F., Si, J., *et al.* "Online supplementary ADP learning controller design and application to power system frequency control with large-scale wind energy integration." *IEEE Trans. Neur. Netw. Learn. Syst.* 2016, **27**, pp. 1748–1761

[36] Lu, C., Si, J., and Xie, X. "Direct heuristic dynamic programming for damping oscillations in a large power system." *IEEE Trans. Syst. Man Cybern. Part B Cybern.* 2008, **38**, pp. 1008–1013

[37] Tang, Y.F., He, H.B., Ni, Z., *et al.* "Reactive power control of grid-connected wind farm based on adaptive dynamic programming." *Neurocomputing.* 2014, **125**, pp. 125–133

[38] Sui, X.C., Tang, Y.F., He, H.B., *et al.* "Energy storage based low frequency oscillation damping control using particle swarm optimization and heuristic dynamic programming." *IEEE Trans. Power Syst.* 2014, **29**, pp. 2539–2548

[39] Ni, Z., Tang, Y.F., Sui, X., *et al.* "An adaptive neuro-control approach for multi-machine power systems." *Int. J. Electr. Power Energy Syst.* 2016, **75**, pp. 108–116

[40] He, H.B., Ni, Z., and Jian, F. "A three-network architecture for on-line learning and optimization based on adaptive dynamic programming." *Neurocomputing.* 2012, **78**, pp. 3–13

[41] WECC Guide for Representation of Photovoltaic Systems in Large-Scale Load Flow Simulations. WECC Renewable Energy Modeling Task Force Report, August 2010.

[42] Singer, S., Rozenshtein, B., and Surazi, S. "Characterization of PV array output using a small number of measured parameters." *Solar Energy.* 1984, **32**, pp. 603–607

[43] Clark, K., Miller, N.W., and Walling, R. "Modeling of GE solar photovoltaic plants for grid studies." General Electrical International, Inc., Schenectady, NY, 2010.

[44] Canizares, C., Fernandes, T., Junior, E.G., *et al.* "Benchmark models for the analysis and control of small-signal oscillatory dynamics in power systems." *IEEE Trans. Power Syst.* 2017, **32**, pp. 715–722

[45] Yao, W., Jiang, L., Wen, J.Y., Wu, Q.H., *et al.* "Wide-area damping controller of FACTS devices for inter-area oscillations considering communication time delays." *IEEE Trans. Power Syst.* 2014, **29**, pp. 318–329

[46] Zhang, X., Lu, C., Xie, X., *et al.* "Stability analysis and controller design of a wide-area time-delay system based on the expectation model method." *IEEE Trans. Smart Grid.* 2016, **7**, pp. 520–529

Chapter 16

Phasor measurement unit-based wide-area control: integration, control, and challenges

Ahmed S. Musleh[1], S.M. Muyeen[2], Ahmed Al-Durra[1] and Innocent Kamwa[3]

One of the most essential issues in planning and operation of power system is the voltage stability which is typically maintained by the electric utility company to ensure stable and reliable operations under normal conditions and grid disturbances. The present focus, nowadays, is on developing intelligent grid-wide control schemes based on wide-area measurements which are assumed to have better features and performance in dealing with the complexity of the power systems. In this chapter, the main objective is to provide a realistic reliable and secure wide-area control approach for controlling the voltage of modern power systems, known as smart grids. The measurements collected from the smart grid through the optimally placed phasor measurement units (PMUs) will be used initially to identify any bad data injections in the measurements obtained and eventually to design a wide-area controller for the flexible AC transmission system (FACTS) device which can enhance the voltage stability of the power grid. A complete hardware-in-the-loop (HIL) system will be developed using real time digital simulator (RTDS) and a real world voltage source converter (VSC)-based FACTS device to show the effectiveness of synchrophasor measurements-based wide-area control in smart grid. The three core parts are being interfaced here. First, the power grid, the PMUs, and the local area controller of the FACTS device are simulated using an RTDS. This main simulation is interfaced with the wide-area controller which is realized via a MATLAB®-based program. This setup illustrates a novel software-in-the-loop (SIL) scheme. Furthermore, an actual FACTS device is designed and interfaced to the main simulation via HIL scheme. This interfacing technique combination empowers the experiment's reliability as a real near-industrial system is developed.

[1]Department of Electrical Engineering, Khalifa University of Science and Technology, Abu Dhabi, UAE
[2]Department of Electrical and Computer Engineering, Curtin University, Perth, Australia
[3]Power Systems and Mathematics, Hydro-Québec/IREQ, Varennes, Quebec, Canada

16.1 Introduction

Wide-area monitoring and control systems (WAMCS) in power systems were first realized in the mid of the twentieth century. Since then, a constant growth has been realized in this area. Initially, supervisory control and data acquisition (SCADA) systems were installed worldwide for providing grid-wide information and control for power grids. SCADA was used to capture a steady state overview of the whole grid performance. With the developments done in WAMCS, new technologies and devices emerged such as PMU. These technologies could be seen as the new generation of the old fashioned SCADA systems, the new version. Though, all of the new technologies have a connection to the SCADA in a way or another. In other words, new technologies are not operated independently from SCADA systems; they are instead working in a parallel cooperative structure.

16.1.1 A summary of SCADA systems in power grids

The main objective of SCADA is to monitor any breaches in the operating margins of the power grids. Successively, adequate control actions are taken to maintain the operation of the power grid [1]. SCADA system consists of three main parts: remote terminal units (RTU), main control center, and the communication mediums integrated. RTU is the equipment used to collect the data of the observed parameters in power grids as well as applying the control actions, so it could be seen as an integration of sensor and actuator although they are separate components. RTUs collect data from power grids separately as no common time reference is taken for the measurements; instead they are scanned successively. The main control center is the supervisory part of SCADA systems; it receives data from RTU, analyzes it, and sends the control actions to the corresponding RTU units. Massages between the RTU and the control center are delivered using SCADA communication systems. Many applications have been accomplished using SCADA in power grids. These applications could be categorized into two sections: monitoring and control. Monitoring applications are concerned with the data acquisitions from the grid components. These data are mostly used in the control applications. The control applications vary widely as some of them are considered as fast control actions (e.g., power flow control) while others are based on long time analytical study (e.g., operational planning) [1]. Some of the applications of the SCADA are summarized in Figure 16.1.

Because of the continuous growth in the power demand which outruns the rate of the generation and transmission systems' expansion, the operation of the power grids is getting closer to the allowable margins. Thus, any small instability issues would mean a great threat to the grid. The old fashioned SCADA systems were designed specifically to monitor large disturbances; small disturbances were not focused on. Therefore, SCADA is not a good choice for monitoring and controlling the small and fast disturbances occurring in the power grids. In other words, no dynamic visibility is available with SCADA systems. From here, we can summarize SCADA system problems in power grids in the following two points [1].

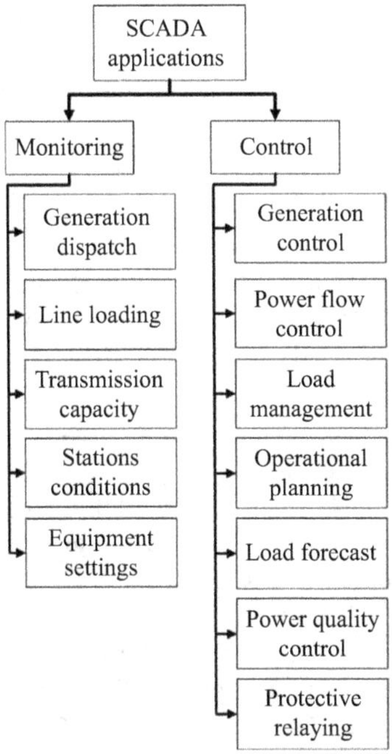

Figure 16.1 SCADA applications [1]

16.1.1.1 The lack of dynamic visibility of SCADA's monitoring devices which might result in the propagation of small disturbances

This problem is related to the monitoring section of SCADA systems. This is generated from the use of the unsynchronized, slow natured RTUs which collect data sequentially. Data received from the RTUs are usually collected and integrated together to have a full picture of the grid. This time-consuming task makes the possibility of identifying any fast occurring small disturbances very difficult. Thus, it would be very difficult to generate early warnings which would be beneficial in avoiding catastrophic consequences.

16.1.1.2 The continuous growth in the complexity in SCADA systems to address new emerging issues in power grids

This problem is related to the full architecture of SCADA systems. This is generated from the new components and their corresponding protocols which are added continuously to the SCADA systems. This will lead to a complex system and will add a huge burden on the communication systems used in SCADA. Therefore, a slower response will be generated.

The first problem of SCADA systems is solvable by providing advanced monitoring devices which are fast and synchronized. However, replacing the original RTUs used in SCADA systems might lead to the inflation of the second problem facing SCADA systems. Thus, a new fast and synchronized monitoring infrastructure is definitely needed, and it has to be separate from the existing SCADA systems.

16.1.2 The rise of the phasor measurement unit

Today, PMUs have a uniform world-known standard. They have been spread around the globe. Many companies developed their own PMU devices. This has started a wide market for PMUs which led to more advanced technologies in terms of the sampling rate of the PMU device, its channels numbers, as well as the communication technologies.

Wide-area monitoring systems were greatly affected by the introduction of the PMUs. The infrastructure of the wide-area monitoring system in the power grids was changed dramatically. The use of the traditional SCADA system with its RTUs, communication systems, and standards were replaced with the new monitoring structure employing the PMUs. This infrastructure is based on specific topology and standards [2]. Two basic components are the major players in this structure: the PMU and phasor data concentrator (PDC). While PMU measures the phasors of the voltage and current signals, PDC combines measurements from various PMUs for transmission purposes. Figure 16.2 illustrates a wide-area monitoring system based on PMUs.

This PMU-based wide-area monitoring system is not seen as a separate system from SCADA. Instead, both of them shall work in a parallel and cooperative scheme. PMU-based WAMS are concerned with the fast dynamics of the power grid while the SCADA system deals with the slow varying steady state dynamics of the power grid. Figure 16.3 shows the relationship of both systems in WAMCS. In the figure, the fast dynamics of the power grids are monitored and controlled using the fast PMU-based WAMCS. Monitoring devices in this system have a much smaller sampling time; thus a smaller time step t for the controllers is

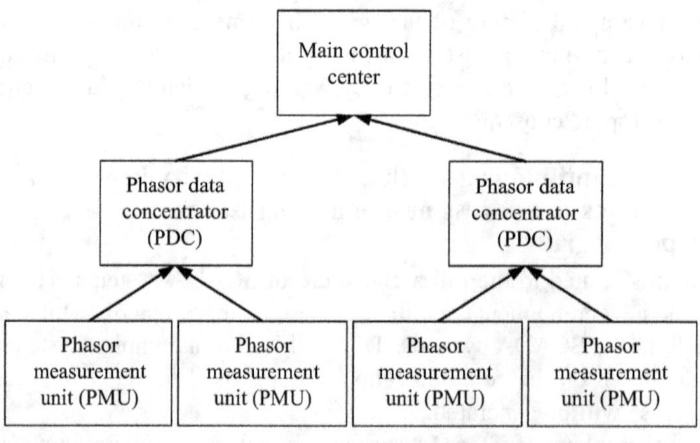

Figure 16.2 PMU-based WAMS

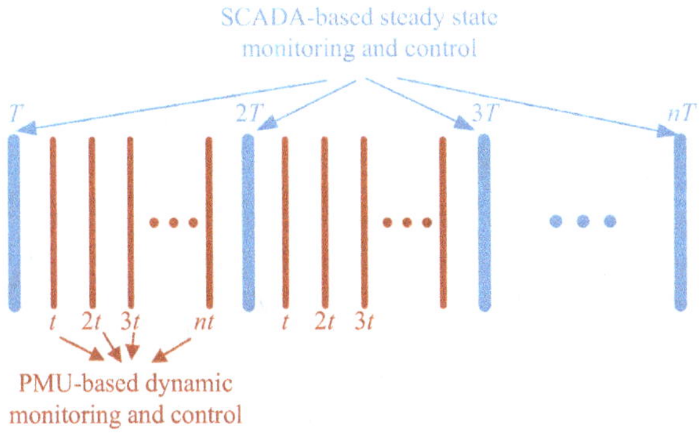

Figure 16.3 WAMS and SCADA relationship

guaranteed. On the contrary, SCADA-based WAMCS has a large sampling time and consequently larger time step T for the control action. The cooperation between the two systems provides simple and reliable operation and control.

Several PMU-based WAMS have been realized in many countries around the globe such as Australia, China, Brazil, USA, Canada, India, Mexico, Europe, and Russia [3]. Applications of PMUs in these countries can be categorized into the following three areas [4]:

1. Power system real time monitoring,
2. Advanced network protection, and
3. Advanced control schemes.

16.1.3 A review on wide-area control in power grids

Wide-area control represents the latest research area in the category of power systems control. It is the field that has emerged due to the recent developments in the measuring and communication technologies. Wide-area control is considered as the main part or functionality of smart grids [5]. In power systems, wide-area control is considered in two major areas that are explained next.

16.1.3.1 Wide-area reactive power (volt/var) control

As explained earlier, volt/var control is a very important procedure especially with the presence of the disturbances and contingencies that are foreseen in power grids. In recent decades, reactive power control has been divided into two levels: the primary level and the secondary level. These two levels work in a hierarchy behavior and have different objectives, effects, and response times. They could be seen as a single controller that is designed by splitting the system into fast and slow dynamics, so the controller would have two separate actions that will address these fast and slow dynamics accordingly.

1. Primary level control

   The primary level control consists of automatic voltage regulators (AVR) installed on generators, capacitors, reactors, transformer and load tap changers, and FACTS devices. The objective of these devices is to keep the bus terminal voltages that they are connected to equal to the set-point values which are fixed by the operator. Primary control actions are extremely fast, in a time frame of less than 1 s [6].

2. Secondary level control

   Secondary voltage control of these devices has a time constant longer than the primary level control. Moreover, secondary voltage control makes use of the global measurements of the power grid, not just the local ones. Therefore, it is possible to have a wider control effect and a better coordination of the local devices' actions, which will result in a stronger and more stable voltage profile which will help in preventing voltage collapse. Furthermore, it will help in suggesting a better use of reactive power resources. The main objective of the secondary voltage control method (SVCM) is to regulate and to sustain the voltage profile within a network area. Normally, its response time is less than 2 min [6].

16.1.3.2 Wide-area system oscillations damping control

Electromechanical oscillations arise in power systems because of the absence of the damping torque at the rotors of the generator. These oscillations of the generators' rotors produce oscillations in the other variables of power system such as voltage magnitude, frequency, and active and reactive powers on the transmission lines. According to the number of the involved generators and the size of the power system network, power system oscillations have been reported in the range of 0.05 Hz to 2 Hz [7]. Local oscillations reside in the upper part of the mentioned range and involve oscillations of only one generator or a group of generators connected at one bus against the rest of the system. On the other hand, inter-area oscillations take in the lower part of the above range and include the oscillations between groups of generators connected at different busses at different locations. In general, power system oscillations are omnipresent, poorly damped, and not hazardous as long as they do not develop to an unstable state. With the increasingly growing demand, integration of renewable resources in remote areas and importance of energy utilities, extensive distance power transfers outpace the integration of the new transmission amenities. These expansions lead to the extensive utilization of the present power network which results in a reduced stability margins and inter-area oscillations. In a severely stressed power network, the poorly damped inter-area oscillations could lead to instabilities. The blackout of August 10, 1996 in North America resulted from the instability of the 0.25 Hz oscillation in the Western interconnected system [7,8].

Oscillations at a local area are classically damped via generator excitation control through power system stabilizers (PSS). Though, these distributed control schemes might interact in an opposing manner and disrupt the overall power network's operation. Usually, inter-area oscillations cannot be stabilized via the use of PSS without adequately and sensibly tuning them. It is believed that the oscillation difficulties may be resolved by wide-area control system (WACS) where the controllers utilize the remote measurements and control signals. Wide-area controllers

Table 16.1 Literature review on wide-area control in power grids

		System oscillations damping	Reactive power control (volt/var)
System topology	AC	[7,9–20]	[6,21–35]
	Hybrid	[36]	
Controller type	PI	[9,19]	[22,26]
	Fuzzy	[13,17,20]	
	Others	[7,10–12,14–16,18,36]	[6,21,23–25,27–35]
Controlled devices	Generators	[7,9–12,14–17]	[6,24–26]
	FACTS devices	[13,18–20]	[22,23,30]
	Others	[36]	[21,27–29,31–35]
Simulation tools used	MATLAB	[9,12–20,36]	[6,22–25,27,29–35]
	RSCAD	[11]	
	Others	[7,10]	[21,26,28]
Implementation realization	RTDS	[11]	
	None	[7,9,10,12–20,36]	[6,21–35]
PMU measurements	Adopted	[10,11,14]	[29,35]
	Ignored	[7,9,12,13,15–20,36]	[6,21–28,30–34]
Delay considerations	Discussed	[7,11–14,36]	
	Ignored	[9,10,15–20]	[6,21–35]
Bad data considerations	Discussed		
	Ignored	[7,9–20,36]	[6,21–35]

are nowadays possible owing to the modern high-tech advances including fast and reliable communication networks, FACTS devices, and most importantly the time-stamped PMUs [7].

Table 16.1 summarizes recent works in the area of wide-area control. It is divided into two sections as discussed earlier and shows the various simulations and implementations of the papers as well as the use of PMUs and the consideration of delays. Furthermore, it shows whether previous researches have considered the possibility of bad data in the measurements obtained for the wide-area controller. Bad data here is a term given to any false or corrupted data received from the measurement device. This false data could be generated from a device malfunction or from intentional attacks. This might have severe results in wide-area controllers as it will be shown later.

From this table, we can see that wide-area control has a very wide technologies and methods that have been considered in it. From the controller perspective, wide-area control has been applied using different types of controllers, such as proportional integral (PI) control, fuzzy control, linear quadratic Gaussian, modal predictive control, adaptive control, and multiagent systems (MAS) which could be considered as new control approach. These controllers have been applied to different types of devices such as generators, FACTS, capacitors, reactors, transformer, and load tap changers. A very important point extracted from the table above is that only one paper has implementation work which highlights the need for more implementation work in this field. Furthermore, it is noticed that most of the papers have assumed the availability of the measurements which ease the study. On the other hand, no papers have considered the bad data, which are known as the falsified measurements imposed in the WAMCS.

16.2 Design of the PMU-based wide-area control for voltage regulation

16.2.1 Voltage regulation

The increased electric power consumption causes transmission lines to be driven close to or even beyond their power transfer capacities resulting in overloaded lines that are vulnerable to any type of disturbances. The reactive power could be described as the unused power generated in AC systems as a result of energy storage components deployment such as inductors and capacitors. The excessive amount of reactive power in the power grid shrinks the space available for the active power to flow which results in an increased pressure on the grid. The reactive power or volt/var control is helpful to maintain the system healthy and is always practiced to reduce or increase the reactive power in the power systems to improve system efficiency and reliability. In power systems, voltage is controlled by managing and controlling production and consumption of reactive power, which is also known as reactive power compensation. Advantages of reactive power compensation include, but not limited to [37]:

1. Increased efficiency of power generation, transmission, and distribution.
2. Improvement in voltage profile and power factor.
3. Reduction of KVA demand.
4. Increased system capacity and load capability.
5. Reduction in system losses.

Furthermore, reactive power control is highly appreciated when the power system is faced with contingencies which are common things in power systems. These contingencies include but are not limited to fault, line tripping, and load fluctuation. The effect of these contingencies may result in a complete blackout most of the time as history confirms. Major blackouts that affected millions of people were due to voltage instability and insufficient reactive power compensation as discussed in [8,38,39]. From this perspective, a higher attention is concentrated toward this area to make power systems stronger and more stable. The reactive power control is usually done using different devices and components of the power system. It could be controlled using:

1. Synchronous condenser,
2. Transformer's tap changer,
3. Capacitor bank and reactor unit, and
4. FACTS devices.

Initially, static switched reactive power compensation (capacitors and reactors) were mainly used to control the steady state system voltages. The dynamic reactive power compensation was basically based on rotating machines such as synchronous condensers. By the twentieth century, power electronics evolved significantly to the level which makes their industrial applications feasible. This was the beginning point of deploying FACTS devices in power grids. The variations and the advanced technologies developed in the FACTS devices have pushed industries and utilities to invest greatly in them. Although FACTS devices are very expensive, many installations have been made either at the transmission level or the distribution level. Since the installation of the first FACTS device in 1970, several installations have been made with a total number of more than *thousand* and a total installed power of more than *500,000 MVA* [40]. Table 16.2 shows an estimation of the installed FACTS devices *until the year 2005* [40].

Table 16.3 shows an estimation of the installed FACTS devices per only four major electric power components manufacturing companies *until the year 2014* [41–44].

Although the installations of FACTS devices are not massive at the present, researchers predict that the use of FACTS devices shall double by the coming two decades [45]. This is a natural response to the continuous growth in the potential power grids' disturbances resulted from the increase in energy demand and the increase in the grids' complexities resulted from the surge in the integration of renewable energies (e.g., solar and wind energy) and other new technologies such as energy storage systems. Because of these facts as well as the advantages of FACTS devices, a FACTS device which is STATCOM is considered here.

16.2.2 Background of STATCOM

One of the most important FACTS devices is the static synchronous compensator, which is widely known as STATCOM. STATCOM is based on three-phase VSC [40]. Figure 16.4 shows typical STATCOM components which consist of six semiconductor devices along with a DC capacitor and coupling inductors and resistors acting as line filters.

This voltage source converter has the ability of compensating the reactive power and stabilizing the system. This is done via controlling the pulses of the gates of converter switches [46]. For example, if the voltage drops, the STATCOM injects reactive power, and if the voltage rises, the STATCOM absorbs reactive power. In general, the STATCOM maintains constant voltage at the bus connected

Table 16.2 Estimation of installed FACTS until 2005

Type	**Number**	**Total installed power (MVA)**
Shunt	615	91,200
Series	710	352,000
Shunt–Series	2–3	250
HVDC	48	14,900

Table 16.3 Major companies' FACTS installation

Company	Installation numbers
ABB	More than 800
Alstom	More than 450
Mitsubishi Electric Power Products	More than 180
Siemens AG	More than 140

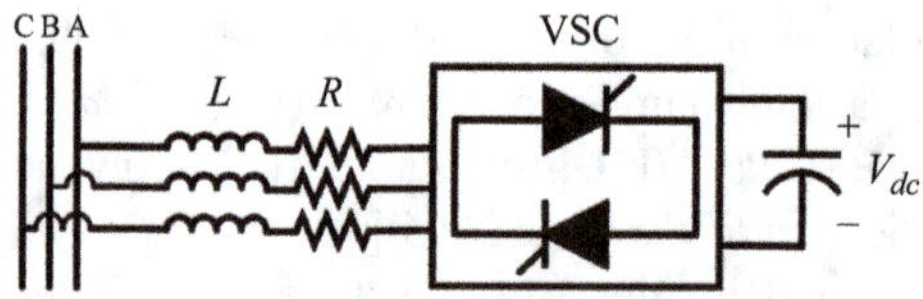

Figure 16.4 Static synchronous compensator model

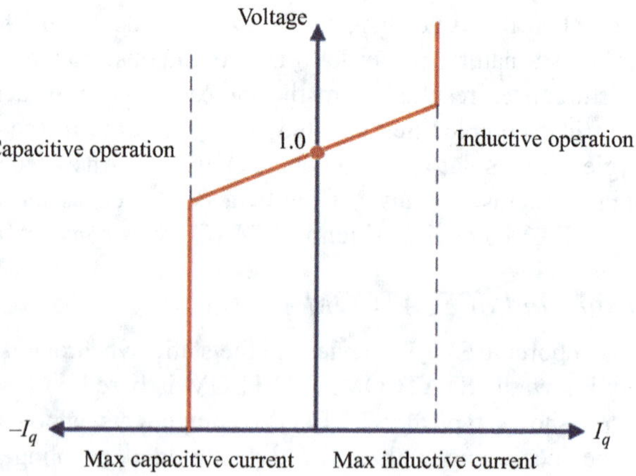

Figure 16.5 Operational modes of the STATCOM

to it by controlling the amount of reactive power injected into or absorbed from the electric power system. The operation of the STATCOM is shown in Figure 16.5.

From this figure, we can see that when the voltage of the bus where the STATCOM is connected to is below 1 p.u. (desired value), the STATCOM works as capacitor banks which injects reactive power to the bus, thus pushing the voltage magnitude up. When the voltage is higher, the STATCOM works as an inductor by absorbing reactive power. Thus, the magnitude of the voltage is reduced.

16.2.3 Local area control of STATCOM

STATCOM's local area control is recognized via proportional-integral (PI) controllers that are designed by an analytical method known as the internal model control method (IMC) and is employed to find the gains of the parameters of the PI controllers [47]. In designing the local area controller, the original transfer functions of the converter are integrated with the PI controller transfer function as shown below:

$$C(s) = k_p + \frac{k_i}{s} \tag{16.1}$$

Figure 16.6 illustrates a comprehensive PI-based controllers' block diagram. It contains a couple of outer loops (DC voltage V_{dc} and AC voltage V_{pcc}). These provide the reference values required to the two inner loops (i_d and i_q).

Sinusoidal Pulse Width Modulation (PWM) was utilized to produce the firing PWM pulses employed to control the IGBT gates of the STATCOM. In this modulation technique, a triangular carrier wave is needed. The output from the PI controllers represents the modulating signals which are to be compared with the carrier wave. Accordingly, the switching of the IGBT switches is determined. The frequency of this carrier wave depends on the simulation time step as well as the switching device limitation. Generally the switching frequency should be less than half of the simulation frequency. Figure 16.7 illustrates the SPWM scheme.

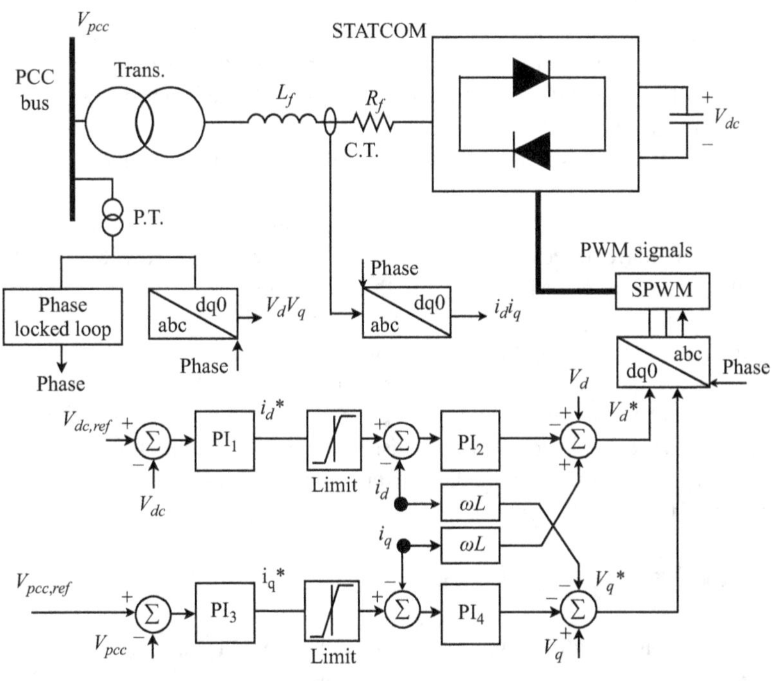

Figure 16.6 STATCOM local area controller

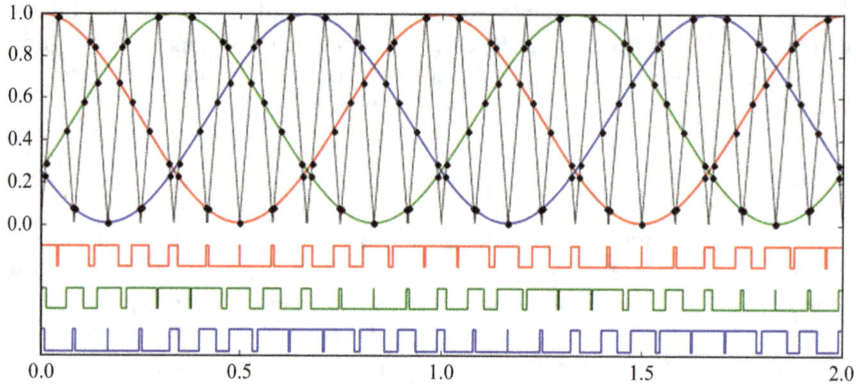

Figure 16.7 Sinusoidal PWM modulation [47]

16.2.4 Wide-area monitoring and control

Since PMU is a very expensive device, there is a great need to optimize the use of it.

16.2.4.1 Optimal placement of PMUs

With the improved technology of global positioning system (GPS), the measurement area can be enlarged to include time-synchronized phasor measurements provided by PMUs. A PMU installed at a bus is capable of measuring the voltage phasor of the installed bus and the current phasors of some or all the branches connected to that bus, supposing that the PMU has enough number of channels [48]. As the availability of PMUs at substations is to be increased in the future, the performance of different essential functions concerning the monitoring, protection, and control of interconnected power systems is to be improved [48]. Due to the high demand for these PMUs along with their high cost, utilities and research centers are looking for the best solution for the placement of these PMUs that will reduce the cost and increase their applications.

At any bus, the PMU will be measuring the voltage phasor of that bus along with the current phasors of the lines connecting that bus to the neighboring busses. Therefore, the voltage phasors of these busses can be calculated either straight from the PMU measurement or from the current phasors and the information of the impedance of the lines as follows:

$$V_{\text{calculated}} = V_{\text{measured}} - (I_{\text{measured}})(Z) \tag{16.2}$$

where V_{measured} is the direct voltage measured at bus 1, I_{measured} is the current measured between bus 1 and 2, Z is the impedance of the line between bus 1 and 2, and $V_{\text{calculated}}$ is the voltage calculated at bus 2. The main goal here is to realize the optimal placement of the PMUs that will ensure the full observability of the power grid. This formulates an optimization problem as shown below:

$$\min \sum_{i}^{n} w_i x_i \qquad \text{subject to} \, f(x) \geq 1 \tag{16.3}$$

where $\mathbf{X}$ is a binary decision variable vector and its entries are defined as

$$x = \begin{cases} 1, & \text{if a PWM is installed at bus } i \\ 0, & \text{otherwise} \end{cases} \tag{16.4}$$

So if a 14 bus system is of the interest, the following would be an $\mathbf{X}$:

$$\mathbf{X} = (1 \quad 0 \quad 0 \quad 0 \quad 1 \quad 0 \quad 0 \quad 1 \quad 0 \quad 0 \quad 1 \quad 0 \quad 0 \quad 1) \tag{16.5}$$

This $\mathbf{X}$ represents a sample placement vector of the PMUs at the IEEE 14 bus system. From this, we can see that a total of five PMUs are installed and are distributed at bus numbers 1, 5, 8, 11, and 14. The rest of the busses do not have any PMUs installed. w_i is the cost of installing the PMU device at bus i. This cost may vary from one bus to another. This depends mainly on the number of channels needed for each bus. For simplifying the problem, it is assumed that all busses' installations have the same cost in this study. $f(x)$ is the constraint vector whose all of its entries must be ≥ 1. Its entries are nonzero if the corresponding bus is observable using the given measurement set and zero otherwise. This vector is calculated as follows:

$$f(x) = \mathbf{A}\mathbf{X}^T \tag{16.6}$$

where $\mathbf{A}$ is the incident matrix which is a binary matrix that represents the adjacency of the busses in the power network. The incident matrix of the IEEE 14 bus system is shown below:

$$\mathbf{A} = \begin{pmatrix} 1 & 1 & 0 & 0 & 1 & 0 & 0 & 0 & 0 & 0 & 0 & 0 & 0 & 0 \\ 1 & 1 & 1 & 1 & 1 & 0 & 0 & 0 & 0 & 0 & 0 & 0 & 0 & 0 \\ 0 & 1 & 1 & 1 & 0 & 0 & 0 & 0 & 0 & 0 & 0 & 0 & 0 & 0 \\ 0 & 1 & 1 & 1 & 1 & 0 & 1 & 0 & 1 & 0 & 0 & 0 & 0 & 0 \\ 1 & 1 & 0 & 1 & 1 & 1 & 0 & 0 & 0 & 0 & 0 & 0 & 0 & 0 \\ 0 & 0 & 0 & 0 & 1 & 1 & 0 & 0 & 0 & 0 & 1 & 1 & 1 & 0 \\ 0 & 0 & 0 & 1 & 0 & 0 & 1 & 1 & 1 & 0 & 0 & 0 & 0 & 0 \\ 0 & 0 & 0 & 0 & 0 & 0 & 1 & 1 & 0 & 0 & 0 & 0 & 0 & 0 \\ 0 & 0 & 0 & 1 & 0 & 0 & 1 & 0 & 1 & 1 & 0 & 0 & 0 & 1 \\ 0 & 0 & 0 & 0 & 0 & 0 & 0 & 0 & 1 & 1 & 1 & 0 & 0 & 0 \\ 0 & 0 & 0 & 0 & 0 & 1 & 0 & 0 & 0 & 1 & 1 & 0 & 0 & 0 \\ 0 & 0 & 0 & 0 & 0 & 1 & 0 & 0 & 0 & 0 & 0 & 1 & 1 & 0 \\ 0 & 0 & 0 & 0 & 0 & 1 & 0 & 0 & 0 & 0 & 0 & 1 & 1 & 1 \\ 0 & 0 & 0 & 0 & 0 & 0 & 0 & 0 & 1 & 0 & 0 & 0 & 1 & 1 \end{pmatrix} \tag{16.7}$$

To solve this problem, many optimization methods have been used, such as genetic algorithm (GA), integer programming, and Tabu search [48]. All of these methods have different algorithms, yet they all reach to the same result eventually. In this study, GA has been used to solve this optimization problem and find the best placement.

GA is a sort of evolutionary computation which is a computer-based problem-solving technique [49]. It is built based on a computational model of physical problems. GA keeps changing the possible solutions, which are known as the chromosomes, and trying them until the objective is reached. The chromosomes change over the generations (trials) via two basic processes known as the crossover and the mutation. Figure 16.8 shows an illustration of the GA process.

To solve the optimization problem of the PMU placement, a GA-based program has been developed to get the solution. This program has been developed using MATLAB. The full program is shown in Appendix D. After running the program, the following result was obtained which is shown in Figure 16.9.

The penalty value here represents the value of the cost function shown in (16.3). As shown in the figure, the minimum possible value reached was 4, which means that the minimum number of PMUs needed to have full observability of the IEEE 14 bus system is 4. These are shown in the solution (chromosome) **X** below:

$$\mathbf{X} = (0 \quad 1 \quad 0 \quad 0 \quad 0 \quad 1 \quad 0 \quad 1 \quad 1 \quad 0 \quad 0 \quad 0 \quad 0 \quad 0) \tag{16.8}$$

So, PMUs are installed at busses 2, 6, 8, and 9. These will give a full observability of the power network.

16.2.4.2 Wide-area control of the STATCOM

In literature, the employment of WAMCS in smart grid voltage control is distinct as a SVCM. This method acts as a managerial control loop which assigns different set-points to the various reactive power sources employed for decreasing deviations in grid's voltage [50]. Unlike the locally effective primary voltage control method (PVCM), SVCM further improves the stability of the grid via decreasing the grid's voltage deviation. This could not be done locally as we lack the grid-wide view when only PVCM is engaged [50]. This technique's importance is augmented after the major August 14, 2003 Northeast blackout, where it was later found that WAMCS could have assisted to detect and prevent this voltage collapse [51]. Furthermore, it is revealed in [52] that reactive power source located 1,000 km far from load is capable of having substantial influence in boosting the load voltage over a short period of time to avoid its downfall succeeding severe line outage that may result from a fault contingency. Several contributions have been made toward the development of SVCM. Authors of [53,54] proposed a straightforward approach to SVCM where an optimization problem is presented in order to find the best set-points to be sent to the reactive power sources. For improving the SVCM optimization process, heuristic algorithms were utilized such as the GA [55] and the particle swarm optimization (PSO) [56]. Engagement of multi-agent systems

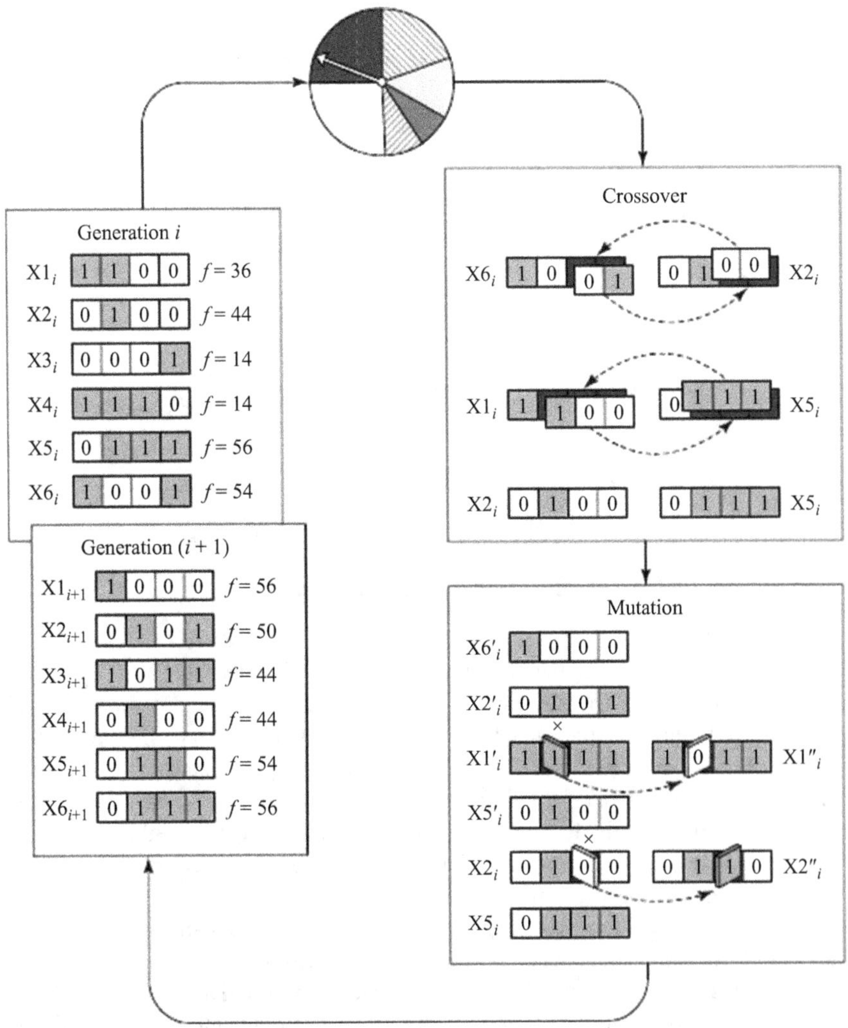

Figure 16.8 Genetic algorithm process

(MAS) is thoroughly discussed in [28,57]; authors there built and compared various agents' organizations.

Hierarchical voltage control, which is a harmonized voltage control scheme, has long been recommended to come up with a more proficient voltage regulation in power grids. This three-level control structure involves tertiary voltage control method (TVCM), SVCM, and PVCM [50]. TVCM is the outer loop of the voltage control where off-line forecast and power flow optimization occurs. It has a very large time step (e.g., days) and is considered to be for a nationwide. It gives set-point for the SVCM which is responsible for maintaining this set-point. Following,

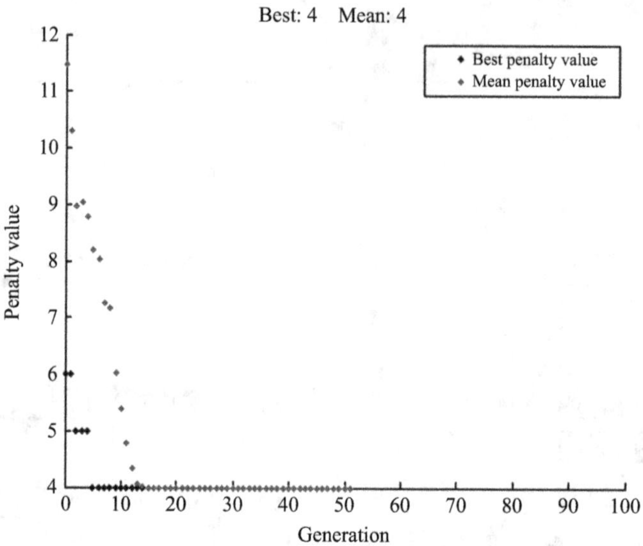

Figure 16.9 GA fitness results

SVCM is the control loop that integrates the TVCM with the PVCM. It gives different set-points for the different PVCM components which it controls with purpose of reducing the regional nodes' voltage deviations. Its time step varies between a few seconds to minutes. Finally, PVCM's purpose is to maintain the set-point which is coming from the SVCM. It is the inner control loop and has time step of a few milliseconds. It is a component level control which might be based on generators, compensators, transformer tap changers, shunt capacitors, reactors, and FACTS devices. Figure 16.10 illustrates WAC system.

Principally, WAC focuses on the relaxed slow nature dynamics of the power grid. Therefore, the control scheme is not a systematic one since the normal transient dynamics are not reflected (presumed instantaneous). This is because WAC has much higher time-step compared to the local area controller. Hence, only the power flow equations with steady state nature are involved in the WAC. With large scale power grids, decoupled power-flow systems are employed to enhance the computing efficiency besides reducing the storage requirements [58]. In decoupled power flow approach, voltage magnitude is related to the reactive as follows [58]:

$$\begin{pmatrix} \Delta Q_1 \\ \vdots \\ \Delta Q_n \end{pmatrix} = -\begin{pmatrix} B_{11} & \cdots & B_{1n} \\ \vdots & \ddots & \vdots \\ B_{n1} & \cdots & B_{nn} \end{pmatrix} \begin{pmatrix} \Delta|V_1| \\ \vdots \\ \Delta|V_n| \end{pmatrix} \tag{16.9}$$

where Q_i, $|V_i|$, and B represent the reactive power at bus i, voltage magnitude at bus i, and the susceptance of the lines, respectively. The motive of employing decoupled power-flow systems is to acquire an estimated mathematical modeling

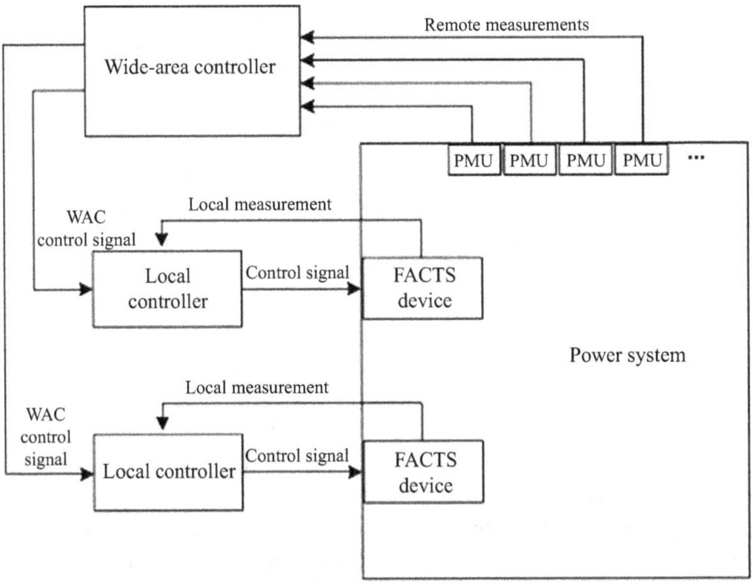

Figure 16.10 Wide and local area controller

of the power grid which relates the variation in the reactive power or the voltage magnitude at every bus in the grid to the variation in the reactive power or voltage magnitude elsewhere in the grid.

It should be noticed that the voltage deviation is $[\Delta|V|] = |V_{SP}| - |V_{PMU}|$, where $|V_{SP}|$ is the set-point of the voltage magnitude considered as 1 p.u., and $|V_{PMU}|$ is the voltage magnitude measured by the PMUs. Equation (16.10) shall be more comprehensive through separating it to controlled and uncontrolled busses:

$$\begin{pmatrix} \Delta \mathbf{Q}_{\mathbf{c}}[t] \\ \Delta \mathbf{Q}_{\mathbf{u}}[t] \end{pmatrix} = -\begin{pmatrix} \mathbf{B}_{\mathbf{cc}} & \mathbf{B}_{\mathbf{cu}} \\ \mathbf{B}_{\mathbf{uc}} & \mathbf{B}_{\mathbf{uu}} \end{pmatrix} \begin{pmatrix} \Delta|\mathbf{V}_{\mathbf{c}}[t]| \\ \Delta|\mathbf{V}_{\mathbf{u}}[t]| \end{pmatrix} \tag{16.10}$$

where the subscripts t, u, and c represent the WAC time instant, voltage-uncontrolled busses, and the voltage-controlled busses, respectively. Furthermore, the following equations can be derived for more simplifications:

$$(\Delta \mathbf{Q}_{\mathbf{u}}[t]) = -(\mathbf{B}_{\mathbf{uc}})(\Delta|\mathbf{V}_{\mathbf{c}}[t]|) - (\mathbf{B}_{\mathbf{uu}})(\Delta|\mathbf{V}_{\mathbf{u}}[t]|) \tag{16.11}$$

$$(\Delta|\mathbf{V}_{\mathbf{u}}[t+1]|) = -(\mathbf{B}_{\mathbf{uu}})^{-1}((\Delta \mathbf{Q}_{\mathbf{u}}[t]) + (\mathbf{B}_{\mathbf{uc}})(\Delta|\mathbf{V}_{\mathbf{c}}[t+1]|)) \tag{16.12}$$

Equation (16.12) illustrates the upcoming time-step depending on the input at the present time-step, denoted as $(\Delta \mathbf{Q}_{\mathbf{u}}[t])$ (reactive power perturbation at the voltage-uncontrolled busses). Hence, employing the present disturbance, the forthcoming control action is designed $(\Delta|\mathbf{V}_{\mathbf{c}}[t+1]|)$ (voltage set-point at the voltage-controlled busses) for minimizing the deviation in the power system denoted as $(\Delta|\mathbf{V_u}[t]|)$ (voltage-deviation at the voltage-uncontrolled busses). Every time-step, the present voltage deviation is calculated to design the set-points of the voltage-controlled busses in the coming time-step. $(\Delta|\mathbf{V_u}[t]|)$ is calculated via PMU measurements. $(\Delta|\mathbf{V_c}[t]|)$ is the voltage at the voltage-controlled busses. Thus, PMUs are not employed here as these busses have their own local area measurement devices for their local area controllers. With the aim of decreasing the voltage deviation $(\Delta|\mathbf{V_u}[t]|)$, the obvious control strategy is to set the set-point at controlled nodes that helps meeting the main goal as shown below:

$$\begin{aligned}&\mathbf{min}\left|-(\mathbf{B_{uu}})^{-1}((\mathbf{\Delta Q_u}[t])+(\mathbf{B_{uc}})(\Delta|\mathbf{V_c}[t+1]|))\right|\\&\text{subject to}\begin{cases}|\mathbf{V}|_{\mathbf{c}}^{\mathbf{min}}[t+1]\leq|\mathbf{V}|_{\mathbf{c}}[t+1]\leq|\mathbf{V}|_{\mathbf{c}}^{\mathbf{max}}[t+1]\\\mathbf{Q}_{\mathbf{c}}^{\mathbf{min}}[t+1]\leq\mathbf{Q_c}[t+1]\leq\mathbf{Q}_{\mathbf{c}}^{\mathbf{max}}[t+1].\end{cases}\end{aligned}\tag{16.13}$$

This objective function is generated by the multiinput multioutput (MIMO) networked (the power grid). It is important to note that a deviation in the voltage set-point at a voltage-controlled bus shall have an effect on every voltage-uncontrolled busses in the power system. Nonetheless, the effect varies from one bus to another according to the electrical coupling of the busses denoted as electrical distance.

16.3 Real time implementation of the PMU-based wide-area control for voltage regulation

16.3.1 A glance on real time simulation

Over the past two decades, computers have become both powerful and affordable. This, sequentially, has led to the development of extremely sophisticated simulation software applications that cannot only permit high trustworthiness simulation of static and dynamic systems and its related controls, but also in real time [59]. Real time simulation is a computer software with model that imitates a physical system. It has the actual execution time as the modeled physical system. In other words, it runs at the same time rate as the actual real physical system. For example, if a breaker takes 0.1 s in the real-world time, the simulation would take 0.1 s as well. In real time simulations, the time necessary to solve the model's inner state equations and functions representing the physical system must be within a fixed time step, which is a discrete time interval that the real time simulation has. If the calculation and solution time surpasses the time of the fixed step, an overrun is produced meaning that the fixed time step is not enough to finish all of the calculations necessary. In other words, a real time simulation must yield the needed internal variables and solution within the same span of time as its physical system would. Thus, it is necessary to have fast processers to have such solutions in time, but the processors should also give the same time as the physical system [59].

In the last few years, several companies started the production of real time simulators. Some of these were for gaming purposes, and others were designed for researching and development of advanced equipment and algorithms. Major software such as LabVIEW and MATLAB have developed real time simulation applications and produced several programmable logic boards that provide IO ports. Despite the humongous number of applications of real time simulations, power systems applications get the most attraction for their extreme benefits and advantages. Several companies have invested in power systems real time simulators such as OPAL-RT technologies and RTDS technologies. RTDS which is developed by RTDS technologies is adopted here.

The world's first real digital simulator was created in 1989. As of 1993, the commercial installation of the RTDS began to take place. Today, RTDS is being utilized in more than 100 research and educational institutions and industrial facilities [60]. These include Texas A&M University, University of Toronto, ABB, Siemens AG, Schneider Electric, and Dubai Water and Electricity Authority (DEWA). Real time digital simulator is considered as the latest technology that is used in the areas of power systems research. Its abilities and advantages make it very appreciated and desired. The main advantage of the RTDS is the strong and fast processors that enable it to solve the power system models and get the results in real time due to their parallel processing techniques. RSCAD is the simulation software where the user can simulate his system. The RTDS hardware interacts with the RSCAD software during the run time and thus performs all the computations in real time. In order to maintain the real time operation, different racks are utilized. Each rack includes processing cards (PB5) and digital and analog I/O ports. All these cards are connected with each other using the common backplane and hence facilitate the exchange of data between the individual processors and between different racks. Moreover, more cards can be added to the rack in order to have more functionality and options.

One more advantage of the RTDS is the availability of many power systems models within it. The availability of PMU model makes RTDS as perfect technology to conduct PMU studies and testing. PMU model in RTDS is constructed based on standard IEEE C37.118.1-2011, which makes it accepted to conduct research and studies using it. Figure 16.11 shows the PMU model in RTDS.

Furthermore, the idea that the RTDS can be connected to other software or electric devices using the IO ports opens the doors widely to the SIL and HIL testing which is a very important topic. The idea of SIL and HIL is to connect software or an actual power device with the real time simulations. The result of this will be a more realistic and accepted data that can make the studies strongly supported and accepted.

16.3.2 Experimental testbed

A classical IEEE 14 node multimachine system has been utilized here, which consists of 2 synchronous generators (G) with IEEE type-1 exciters, 3 synchronous condensers (C), 4 two-winding power transformers, 20 transmission lines with Bergeron model, and 11 dynamic impedance loads. The base kV level is 138 kV. The line complex powers are around hundreds of MVA. The STATCOM is

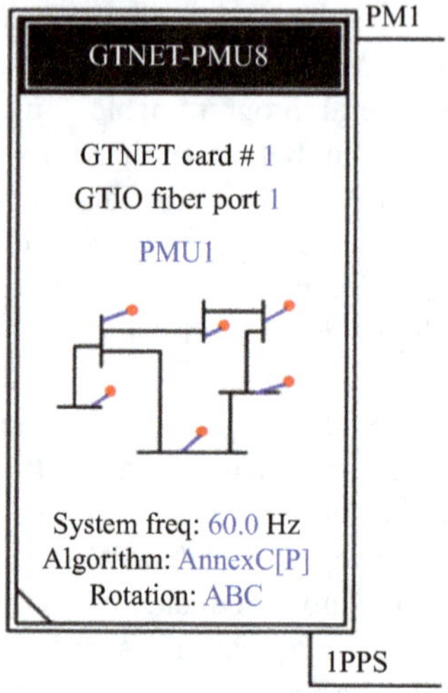

Figure 16.11 RTDS PMU model

connected to bus 13 and is controlled via the local area controller. This local area controller is getting its set-point from the wide-area controller realized using the MATLAB-based program as shown in Figure 16.14. PMUs have been added to the power grid as per the optimal placement found. The PMUs are set to collect measurements at 5 samples/second reporting rate.

Figure 16.13 shows the full testing scheme utilized in the experiment where the Hardware in the Loop (HIL) and the Software in the loop (SIL) are clearly explained.

16.3.2.1 Software-in-the-loop setup

RTDS has been used to develop the SIL platform. It simulates the power grid and the local controller. The PMU device is built according to IEEE PMU standard C37.118.1-2011 in RTDS [2], so it trustworthy. Nevertheless, because RTDS is built deliberately for modeling power system components, it is difficult to carry on the WAMCS's huge mathematical operations in it; therefore, a MATLAB built program is used in a SIL scheme. It begins with generating TCP/IP sockets for the PMUs for collecting the PMU communications in agreement with the IEEE PMU standard C37.118.1-2011. PMU messages differ in types and amount of bytes. 'AA41001200014485600000F0BBFD0002CE00' exemplifies a message to command PMU1 to begin forwarding PMU measurements. This message is divided into many frames, where each frame provides specific information, such as PMU number, time stamp, phasor measurements, and checksum bits which are used for detecting transmission errors. An example of the frames of the PMU message is Figure 16.14.

The MATLAB-based program applies the wide-area controller equations and sends the wide-area controller action to the local area controller in RTDS later.

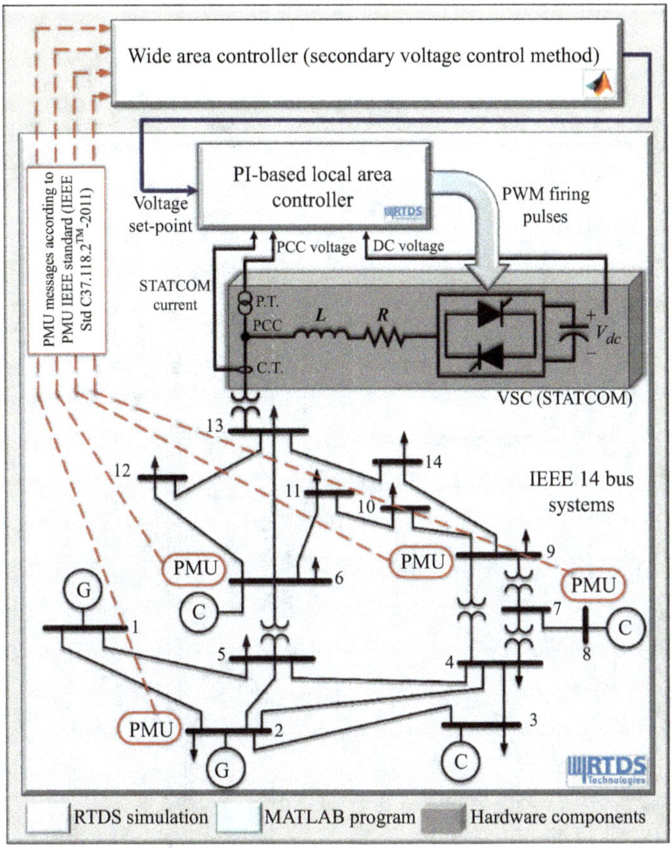

Figure 16.12 General schematic of the proposed experiment

Figure 16.15 shows the messages flow between the RTDS simulation and the MATLAB-based program representing the WAC.

The optimization problem was solved in the MATLAB-based program where the fmincon() function has been employed. This function uses the interior point optimization algorithm to solve the MIMO constrained optimization problem. The delay in this program is below 0.1 s.

A couple of distinct RTDS cards are utilized for the SIL: GTSYNC synchronizes the PMUs using the GPS 1PPS signal and GTNET×2 establishes the network communication through two different protocols (GTNET_SKT for TCP/IP communication and GTNET_PMU for PMU data transmission according to the IEEE C37.118.1-2011) [2].

16.3.2.2 Hardware-in-the-loop setup

The goal of the HIL experiment in the discussed study is to examine the pragmatism of the suggested WAC system with actual electric parts. The HIL scheme

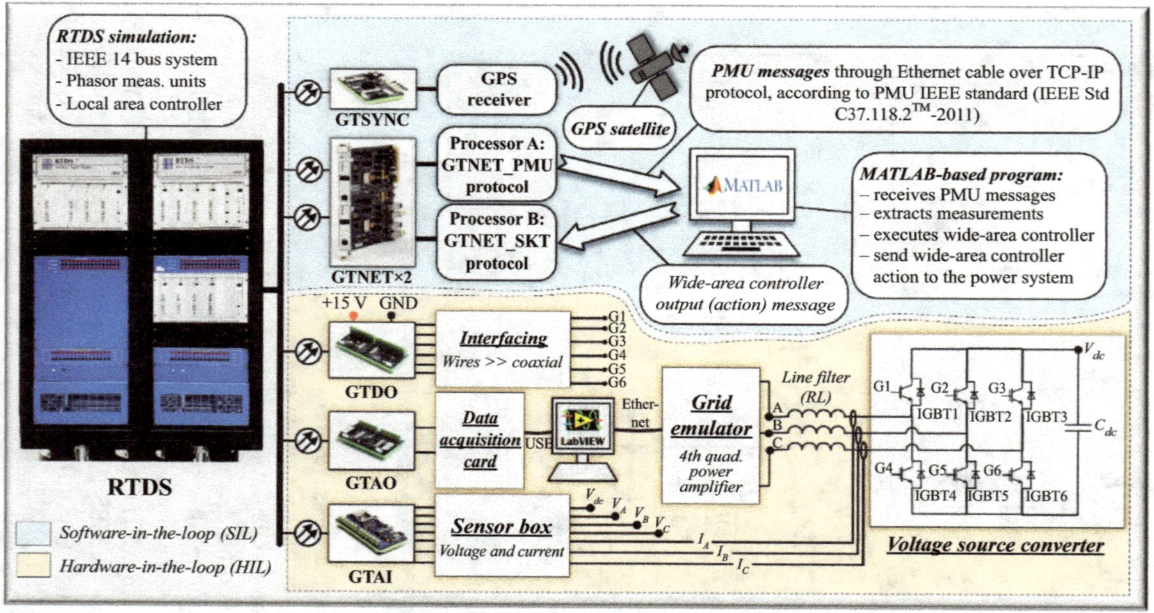

Figure 16.13 Complete RTDS-based testing setup: SIL and HIL

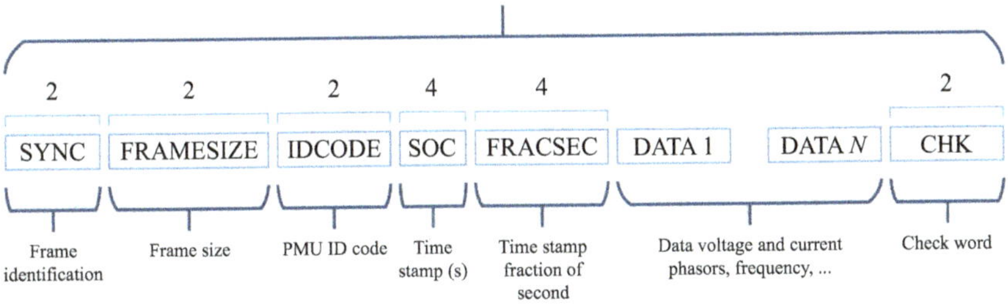

Figure 16.14 PMU message frames

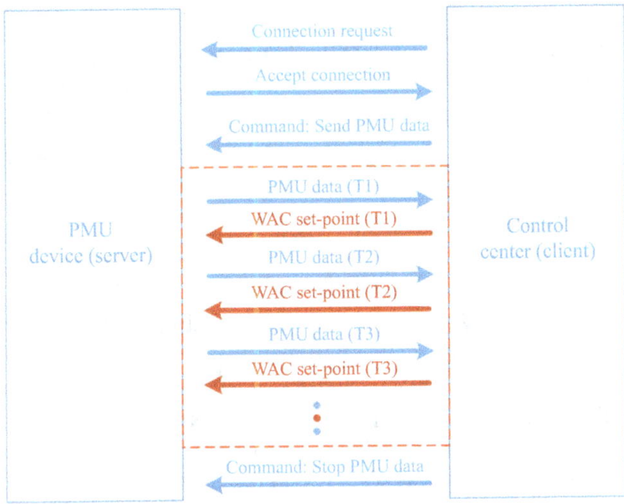

Figure 16.15 Data flow between RTDS and MATLAB program

delivers a cost-effective and harmless technique to examine electrical components within real-time conditions of operation. Ideal transformer method (ITM) is commonly utilized in interfacing and connecting the simulation to the hardware components. It is characterized with a low computation requirements, great precision, and simplicity of implementation [61]. Figure 16.16 shows the ITM interfacing of the STATCOM with the simulation.

Initially, the simulation voltage of the node where the STATCOM is connected-to is scaled-down and is being transmitted to the amplifier. After that, the current passing within the STATCOM is measured, scaled-up, and added to the simulated power grid as a current source. The next couple of equations summarize the process:

$$V_{\text{physical}} = \frac{V_{\text{simulation}}}{K_{v,\text{scale}}} \tag{16.14}$$

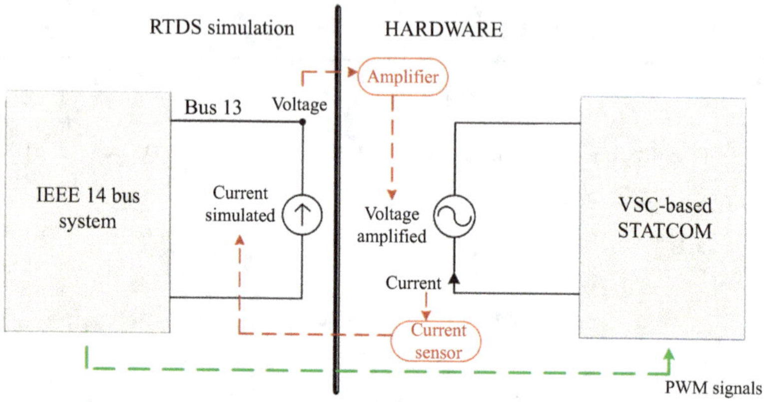

Figure 16.16 WAMS and SCADA relationship

$$I_{simulation} = K_{i,scale} I_{physical} \tag{16.15}$$

where $V_{physical}$ is the actual voltage connected to STATCOM, $V_{simulation}$ is the simulation bus voltage, $I_{physical}$ is the actual current flowing through the STATCOM, $I_{simulation}$ is the simulated current, $K_{v,scale}$ is the scaling factor relating the simulation to the actual voltage, and $K_{i,scale}$ is the scaling factor relating the simulation to the actual current. Through regulating the values of the scaling factors, the power scaling of the test could be adjusted. For example, the 100 VA in the actual hardware can be scaled to 100 MVA in the power grid simulated within the RTDS. The scaling depends on the choice of hardware components ratings used. Power amplifiers used for HIL tests vary depending on the applications used. Here, the power amplifier was chosen taking into consideration the relatively small power rating application (<500 W) with bidirectional power flow (4-Quadrant) and fast dynamics (few milliseconds). Three IO cards were used for the HIL: GTAO, for delivering the set-points (angle and magnitude) of bus 13 voltage; GTDO for delivering the PWM pulses to switches of the STATCOM; GTAI, for obtaining the measurements (STATCOM's voltages and currents). The reporting rate of these cards is few microseconds of delay. A grid emulator (4-Quadrant amplifier) is employed to generate and absorb power from the STATCOM. LabVIEW-based MyDAQ card is used to receive the set-points of the angle and magnitude of the STATCOM's node (node 13) voltage and send those set-points to the grid emulator.

One important problem faced in this setup is the synchronization issue. The output of the grid emulator needs to be in-phase with the voltage simulated in bus 13 in the RTDS simulation. Figure 16.17 represents a possible phase shift between the two voltages in the simulation and the output of the grid emulator.

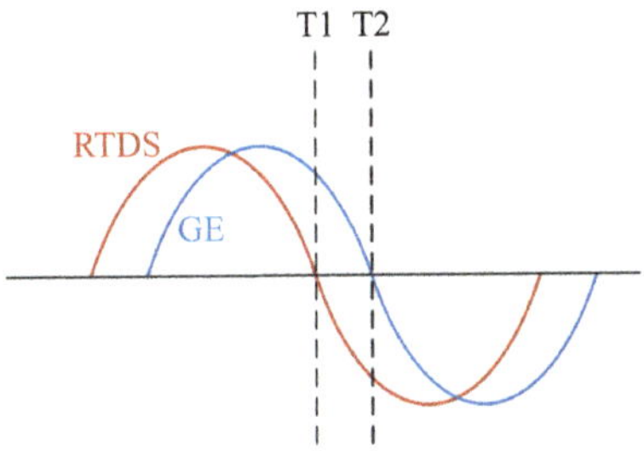

Figure 16.17 Possible phase shift between GE and RTDS

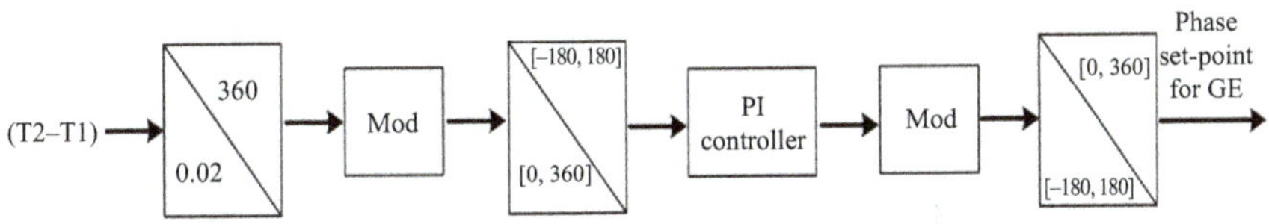

Figure 16.18 Phase shift control block diagram

In this figure, GE stands for the output voltage signal from the grid emulator, and RTDS is the output voltage signal from the RTDS simulation. In order to eliminate this phase shift, a controller had to be designed. This controller takes the phase shift (error) between the two signals and tries to eliminate it by adjusting the phase set-point of the grid emulator output voltage. Figure 16.18 represents the block diagram of the phase shift control.

The controller was designed using PI controller. The tuning of the PI was done by trial and error until fast and stable results were obtained. Figure 16.19 shows the results obtained using the designed controller. The figure shows that the three-phase voltages signals from the RTDS and the GE are in-phase. Both of the signals (GE and RTDS) are over each other.

The complete lab experiment setup is shown in Figure 16.20. The STATCOM employed here is built using a six IGBT switches VSC and is manufactured by Semicron. It is integrated into the grid emulator using line reactors (line filters). These have an overall inductance and resistance of 11.2 mH and 0.2 Ω, respectively. The overall capacitance of the DC bus is 1,100 μF. The DC voltage is set at 150 V, and the grid RMS voltage is 69 V. A couple of PCs are employed in this setup. One PC (with LabVIEW) is to control the grid emulator through an Ethernet cable. PC 2 is for executing the MATLAB-based WAC program. Sensor box (OPAL-RT signal conditioning box) is utilized to measure the three-phase currents, DC voltage, and the three-phase voltages. The time step of the RTDS employed is 50 μs with a 20 kHz sampling rate taken for in the A2D and D2A. Table 16.4 summarizes the power system and control parameters utilized in this testbed, respectively.

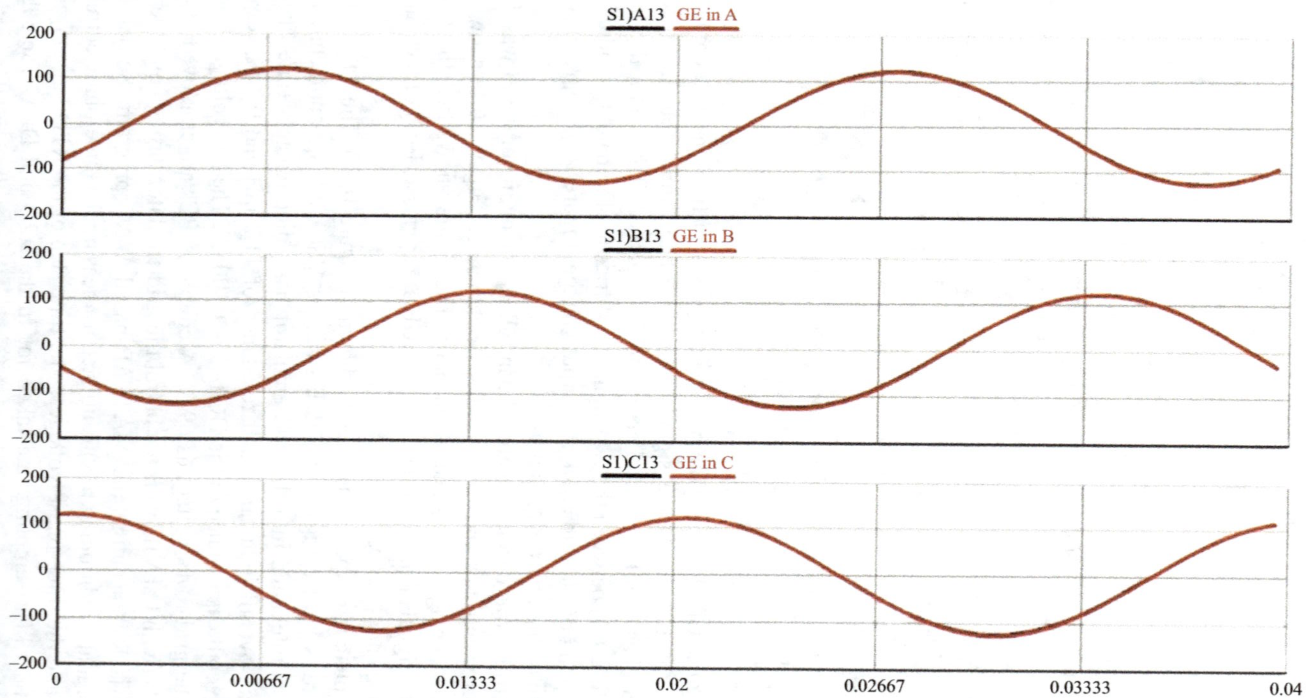

Figure 16.19 Synchronizing results

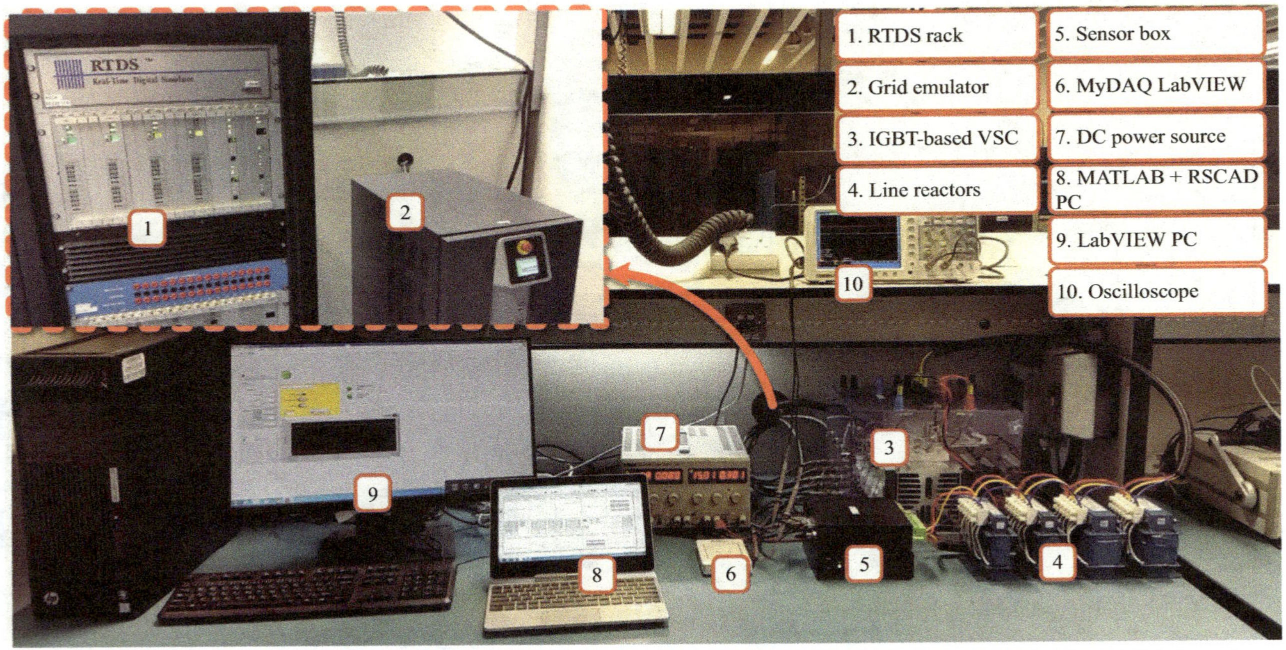

Figure 16.20 *Full experiment setup at the Energy Systems Control and Optimization (ESCO) Lab in the ADNOC Research and Innovation Center (ADRIC)*

Table 16.4 System's parameters summary

Parameter	Symbol	Value
AC bus line voltage (simulation)	$V_{\text{simulation}}$	138 kV
AC bus line voltage (physical)	V_{physical}	69 V
Voltage scaling factor	$K_{v,\text{scale}}$	2,000
Current scaling factor	$K_{i,\text{scale}}$	1,000
DC bus voltage	V_{dc}	150 V
DC bus total capacitance	C_{dc}	1,100 μF
Line reactor (filter)	L	11.2 mH
Line resistance	R	0.2 Ω
Inner control loop proportional gain	$K_{p,\text{current}}$	18.0298
Inner control loop integral gain	$K_{i,\text{current}}$	359.842
DC control loop proportional gain	$K_{p,dc}$	0.0024
DC control loop integral gain	$K_{i,dc}$	0.3703
AC control loop proportional gain	$K_{p,ac}$	7.0
AC control loop integral gain	$K_{i,ac}$	3,100
Switching frequency	f_s	10 kHz
D2A and A2D sampling rate	R_{sampling}	20 kHz
RTDS simulation sampling time	T_s	50 μs

16.4 Results and discussions

16.4.1 Local area controller testing

Prior to testing the full WAC system, the STATCOM's local area controller had to be confirmed to have an acceptable response [62]. To examine the local area controller, the grid RMS voltage's set-point $V_{\text{pcc,ref}}$ in Figure 16.6 of the STATCOM is altered from 0.9 p.u. to 1.1 p.u. (limits of the acceptable deviation of the voltage in this study). The results attained from this test confirm an adequate response of the STATCOM's PI controller. First, the grid's RMS voltage (shown in Figure 16.21) is tracking the set-point wanted with a precise and quick behavior. The response time of the STATCOM is suitably fast; it requires about two cycles to reach the new wanted set-point. In Figure 16.22, the STATCOM's power is shown. Because the new set-point of the grid RMS voltage is increased, the STATCOM injects a reactive current to overcome the reactive power decrease through the power grid; consequently, STATCOM's reactive power changes from the positive to the negative value.

16.4.2 Wide-area controller testing

WAC test entails multiple power turbulences that consist of a total period of 60 s; thus, resembling a more genuine scenario. The full loads of the IEEE 14 node network are altered over a range of 10% to 30%. Moreover, three big turbulences realized as three phase to ground faults occur at different times: 10 s,

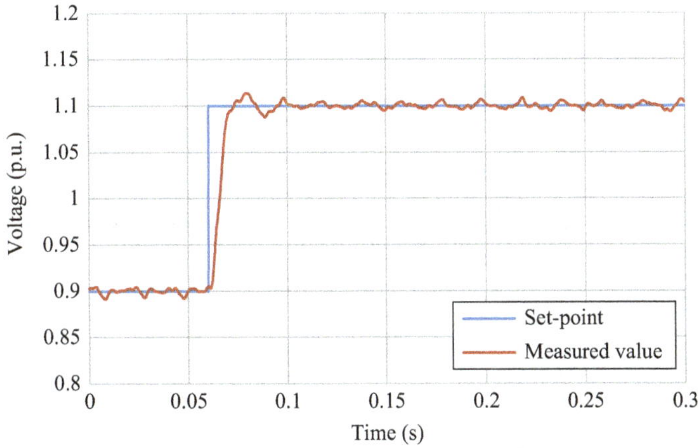

Figure 16.21 Local area controller response—grid RMS voltage

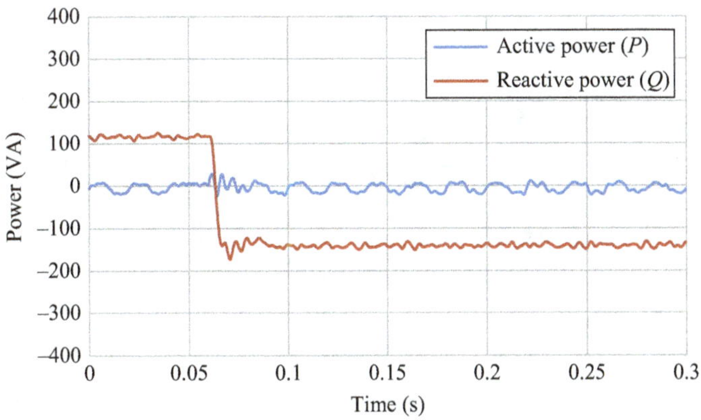

Figure 16.22 Local area controller response—VSC active and reactive power

30 s, and 50 s. These occur at node numbers 4, 14, and 10, respectively. The clearance time of these faults is eight cycles or 0.1 s. Figure 16.23 illustrates the behavior of the controllers by showing the voltage profile of node 13 of the WAC scheme (in (b)) compared to local area controller alone (in (a)). In (a) the set-point (red-colored) is always one, and the measured voltage (blue-colored) matches with it.

STATCOM is also capable to uphold the voltage set-point post the grid turbulences. WAC is actuated in (b) where the voltage profile of the STATCOM's node is revealed. Here, the set-point is changing depending on the grid's

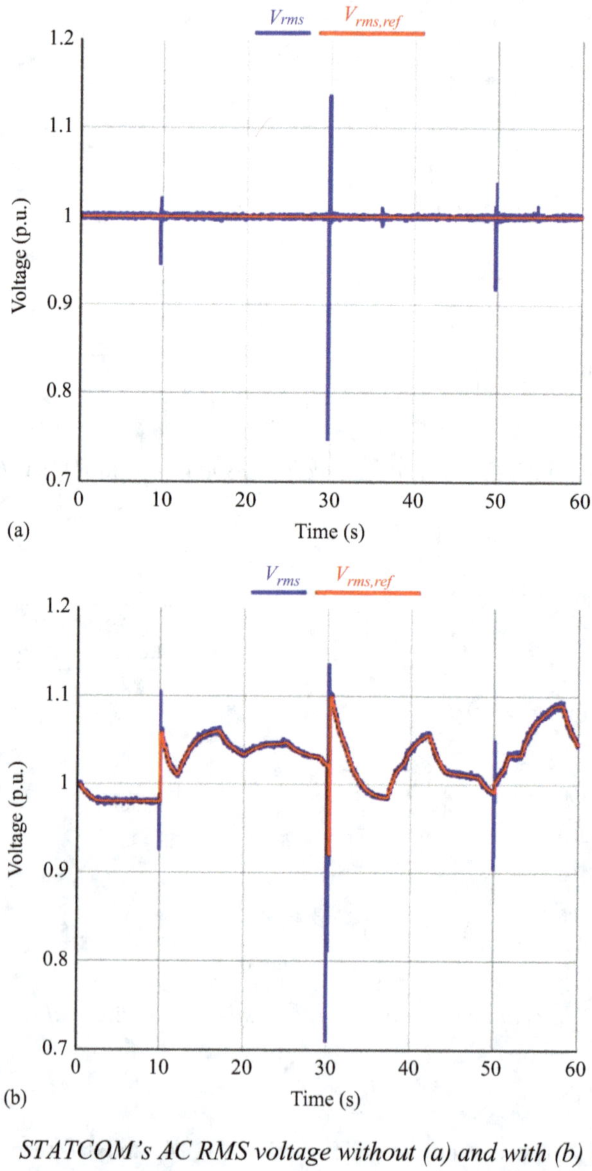

Figure 16.23 STATCOM's AC RMS voltage without (a) and with (b) WAC scheme

turbulences and the measured voltage tracks it smoothly. This voltage disparity in the set-point leads to decreasing the total voltage deviances of the whole network.

16.4.3 Comparative study

For contrasting the results gained from the WAC and the local area controller, the mean-square error of the voltage deviations of the system was employed.

Table 16.5 Root mean square error comparison

Test case	Control level used	MSE
Test case 1	Local area controller only	0.0423
	Wide-area controller and local area controller	0.0377
Test case 2	Local area controller only	0.0337
	Wide-area controller and local area controller	0.0327

Table 16.5 shows the evaluation of the whole voltage deviation of the grid which was found using the mean-square error as in (16.16).

$$\text{MSE} = \sqrt{\frac{1}{n}\sum_{i=1}^{n}(\Delta|V_i|)^2} \tag{16.16}$$

where n is the number of the network's nodes and $|V_i|$ is the measured voltage magnitude.

The means-square errors in the previous couple of cases are near to each other while the WAC results demonstrate an improvement in the network. This slight improvement is because of the fact that only one voltage-controlled bus is present in this study which is the STACOM bus. Therefore, a larger enhancement would be noticed if more voltage controlling devices are employed at different locations in the grid. Figure 16.24 shows a comparison between the WAC effect and the local area controller effect in response to the grid's fault occurred at the 10th second.

The local area controller shows a quick response to the turbulence which could be seen as a few cycles. This maintains the network's stability in the presence of the transients. On the contrary, the WAC shows a slower response. The WAC's time step is dependent on the reporting rate of the PMUs plus the time taken in executing the WAC's optimization problem. Because the reporting rate of the PMU is 5 samples/second in this test, the WAC has a time step of 0.2 s. Therefore, in each 0.2 s, a new set-point value for the STACOM is being produced.

16.5 Conclusions

In this chapter, an innovative interfacing system is established to augment the reliability of the smart grids' wide-area monitoring and control studies. WAC for a FACTS device is realized via this system. The real-time data flow in the experimentation delivers a furthermore truthful understanding of the wide-area techniques employed. The local area controller illustrates a fast and correct response of the controller as seen in the results. WAC results demonstrate an enhanced voltage profile of the power network. This enhancement is evident in the mean-square-error results of the voltage deviation of the grid busses. This study can be extended to have larger power grids and more local area controllers where the results of the

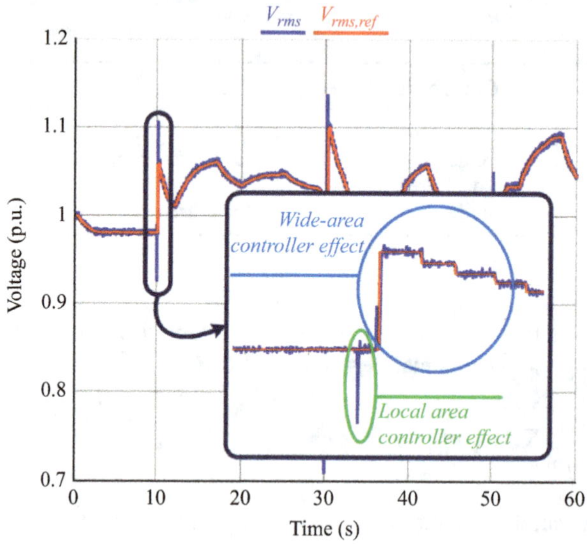

Figure 16.24 Wide-area controller and local area controller effects

WAC could be further clarified. Additionally, the experimental testbed established here can be employed for various other research issues such as the communication delay problem, smart grid security studies and bad data detection algorithms, innovative control systems for FACTS devices, and online studies of power systems dynamic state estimations.

References

[1] Thomas M.S., and McDonald J.D. *Power System SCADA and Smart Grids*. Boca Raton, FL: CRC Press, 2015.

[2] IEEE Standard for Synchrophasor Data Transfer for Power Systems. *IEEE Std C37.118.2™-2011*. IEEE Power & Energy Society, New York, 2011.

[3] Phadke A.G. "The wide world of wide-area measurement." *IEEE Power and Energy Magazine*. 2008, vol. 6, no. 5, pp. 52–65.

[4] Singh B., Sharma N., Tiwari A., Verma K., and Singh S. "Applications of phasor measurement units (PMUs) in electric power system networks incorporated with FACTS controllers." *International Journal of Engineering, Science and Technology*. 2011, vol. 3, no. 3, pp. 64–82.

[5] Chakrabortty A., and Khargonekar P. "Introduction to wide-area control of power systems." in *American Control Conference (ACC)*, Washington, DC, 2013.

[6] Vu H., Pruvot P., Launay C., and Harmand Y. "An improved voltage control on large-scale power system." *IEEE Transactions on Power Systems*. 1996, vol. 11, no. 3, pp. 1295–1303.

[7] Dörler F., Jovanović M., Chertkov M., and Bullo F. "Sparsity-Promoting optimal wide-area control of power networks." *IEEE Transactions on Power Systems*. 2014, vol. 29, no. 5, pp. 2281–2291.

[8] Pourbeik P., Kundur P., and Taylor C. "The anatomy of a power grid blackout - root causes and dynamics of recent major blackouts." *IEEE Power and Energy Magazine*. 2006, vol. 4, no. 5, pp. 22–29.

[9] Liu S., Liu X., and El Saddik A. "Load frequency control for wide area monitoring and control system (WAMC) in power system with open communication links." in *Power Engineering and Automation Conference (PEAM), 2012 IEEE*, Wuhan, China, 2012.

[10] Lu C., Zhang X., Wang X., and Han Y. "Mathematical expectation modeling of wide-area controlled power systems with stochastic time delay." *IEEE Transactions on Smart Grid*. 2015, vol. 6, no. 3, pp. 1511–1519.

[11] Zhang F., Sun Y., Cheng L., Li X., Chow J., and Zhao W. "Measurement and modeling of delays in wide-area closed-loop control systems." *IEEE Transactions on Power Systems*. 2015, vol. 30, no. 5, pp. 1–8.

[12] Yao W., Jiang L., Wen J., Wu Q., and Cheng S. "Wide-area damping controller for power system interarea oscillations: a networked predictive control approach." *IEEE Transactions on Control Systems Technology*. 2015, vol. 23, no. 1, pp. 27–36.

[13] Mokhtari M., Aminifar F., Nazarpour D., and Golshannavaz S. "Wide-area power oscillation damping with a fuzzy controller compensating the continuous communication delays." *IEEE Transactions on Power Systems*. 2013, vol. 28, no. 2, pp. 1997–2005.

[14] Wang S., Meng X., and Chen T. "Wide-area control of power systems through delayed network communication." *IEEE Transactions on Control Systems Technology*. 2012, vol. 20, no. 2, pp. 495–503.

[15] Zhang Y., and Bose A. "Design of wide-area damping controllers for interarea oscillations." *IEEE Transactions on Power Systems*. 2008, vol. 23, no. 3, pp. 1136–1143.

[16] Chen H., Bai H., Liu M., and Guo Z. "Wide-area robust H∞ control with pole placement for damping inter-area oscillations." in *Power Systems Conference and Exposition (PSCE'06), IEEE PES*, Atlanta, GA, 2006.

[17] Padhy B., Srivastava S., and Verma N. "Robust wide-area TS fuzzy output feedback controller for enhancement of stability in multimachine power system." *IEEE Systems Journal*. 2012, vol. 6, no. 3, pp. 426–435.

[18] Zarghami M., Crow M., and Jagannathan S. "Nonlinear control of FACTS controllers for damping interarea oscillations in power systems." *IEEE Transactions on Power Delivery*. 2010, vol. 25, no. 4, pp. 3113–3121.

[19] Li X., Wu H., and Liu J. "Coordinated design of wide area controllers and FACTS controller for enhancement of power system stability." in *International Conference on Power System Technology (POWERCON)*, Hangzhou, China, 2010.

[20] Ganesh A., Dahiya R., and Singh G. "A comparison of novel optimal fuzzy and LMI based wide area controller for stability enhancement with

STATCOM." in *Recent Advances in Engineering and Computational Sciences (RAECS)*, Chandigarh, India, 2014.

[21] Stahlhut J., Heydt G., and Kyriakides E. "A comparison of local vs. sensory, input-driven, wide area reactive power control." in *IEEE Power Engineering Society General Meeting*, Tampa, FL, 2007.

[22] Ganesh A., Dahiya R., and Singh G.K. "A novel STATCOM wide area feedback controller for improving stability in multimachine system." in *International Conference on Power Systems, Energy, Environment*, 2014.

[23] Zima M., Larsson M., Korba P., Rehtanz C., and Andersson G. "Design aspects for wide-area monitoring and control systems." in *Proceedings of the IEEE*, 2005, vol. 93, no. 5, pp. 980–996.

[24] Ma H., and Hill D. "Adaptive coordinated voltage control—Part I: basic scheme." *IEEE Transactions on Power Systems*. 2014, vol. 29, no. 4, pp. 1546–1553.

[25] Ma H., and Hill D. "Adaptive coordinated voltage control—Part II: use of learning for rapid response." *IEEE Transactions on Power Systems*. 2014, vol. 29, no. 4, pp. 1554–1561.

[26] Paul J., Leost J., and Tesseron J. "Survey of the secondary voltage control in France: present realization and investigations." *IEEE Transactions on Power Systems*. 1987, vol. 2, no. 2, pp. 505–511.

[27] Ma H., Man K., and Hill D. "Control strategy for multi-objective coordinate voltage control using hierarchical genetic algorithms." in *IEEE International Conference on Industrial Technology, 2005. ICIT*, Hong Kong, China, 2005.

[28] Ashrafi A., and Shahrtash S. "Dynamic wide area voltage control strategy based on organized multi-agent system." *IEEE Transactions on Power Systems*. 2014, vol. 29, no. 6, pp. 2590–2601.

[29] Venkatasubramanian V., Chun H., Guerrero J., Habibi-Ashrafi F., and Salazar A. "Hierarchical two-level voltage controller for Southern California Edison." in *IEEE Power and Energy Society General Meeting*, San Diego, CA, 2012.

[30] Morattab A., Dalal A., Akhrif O., Saad M., and Lefebvre S. "Model predictive coordinated secondary voltage control of power grids." in *International Conference on Renewable Energies for Developing Countries (REDEC)*, Beirut, Lebanon, 2012.

[31] Wang Z., Wang J., Chen B., Begovic M., and He Y. "MPC-based voltage/var optimization for distribution circuits with distributed generators and exponential load models." *IEEE Transactions on Smart Grid*. 2014, vol. 5, no. 5, pp. 2412–2420.

[32] Wen J., Wu Q., Turner D., Cheng S., and Fitch J. "Optimal coordinated voltage control for power system voltage stability." *IEEE Transactions on Power Systems*. 2014, vol. 19, no. 2, pp. 1115–1122.

[33] Hong Y.-Y., and Luo Y.-F. "Optimal VAR control considering wind farms using probabilistic load-flow and gray-based genetic algorithms." *IEEE Transactions on Power Delivery*. 2009, vol. 24, no. 3, pp. 1441–1449.

[34] Moradzadeh M., Boel R., and Vandevelde L. "Voltage coordination in multi-area power systems via distributed model predictive control." *IEEE Transactions on Power Systems*. 2013, vol. 28, no. 1, pp. 513–521.

[35] Milosevic B., and Begovic M. "Voltage-stability protection and control using a wide-area network of phasor measurements." *IEEE Transactions on Power Systems*. 2003, vol. 18, no. 1, pp. 121–127.

[36] Preece R., Milanovic J., Almutairi A., and Marjanovic O. "Damping of inter-area oscillations in mixed AC/DC networks using WAMS based supplementary controller." *IEEE Transactions on Power Systems*. 2013, vol. 28, no. 2, pp. 1160–1169.

[37] Vedam R.S., and Sarma M.S. *Power Quality VAR Compensation in Power Systems*. Boca Raton, FL: CRC Press; 2009.

[38] Andersson G., Donalek P., Farmer R., *et al*. "Causes of the 2003 major grid blackouts in North America and Europe, and recommended means to improve system dynamic performance." *IEEE Transactions on Power Systems*. 2005, vol. 20, no. 4, pp. 1922–1928.

[39] Atputharajah A., and Saha T. "Power system blackouts - literature review." in *International Conference on Industrial and Information Systems (ICIIS)*, Peradeniya, Sri Lanka, 2009.

[40] Zhang X.-P., Rehtanz C., and Pal B. *Flexible AC Transmission Systems: Modeling and Control*. Berlin: Springer-Verlag, 2006.

[41] ABB. "FACTS reference list." ABB, November 4, 2015. [Online]. Available: https://library.e.abb.com/public/f382b7f2321f4cf0bacbe040c62f94ca/ABB%20FACTS%20Reference%20List%20Worldwide,%20Nov.%202015.pdf. [Accessed January 21, 2015].

[42] Alstom. "Flexible AC transmission system: solutions to optimize network performance." Alstom, 2015. [Online]. Available: https://www.gegridsolutions.com/AlstomEnergy/grid/Global/OneAlstomPlus/Grid/PDF/FACTS%20Product%20Range-epslanguage=en-GB.pdf. [Accessed January 21, 2016].

[43] "Flexible AC transmission systems." Mitsubishi Electric Power Products, 2016. [Online]. Available: http://www.meppi.com/Products/FACTS/Pages/default-2.aspx. [Accessed May 23, 2016].

[44] "Flexible AC transmission systems (FACTS)." Siemens AG, 2016. [Online]. Available: http://www.energy.siemens.com/hq/en/power-transmission/facts/. [Accessed May 21, 2016].

[45] McCray J. *Executive Summary: Flexible AC Transmission System*. Boulder, CO: Navigant Research, 2014.

[46] Hingorani N.G., and Gyugyi L. *Understanding FACTS: Concepts and Technology of Flexible AC Transmission Systems*. Wiley-IEEE Press, New Jersey, USA, 2000.

[47] Yazdani A., and Iravani R. *Voltage Sourced Converters in Power Systems*. Hoboken: John Wiley & Sons, Inc, 2010.

[48] Manousakis N., Korres G., and Georgilakis P. "Taxonomy of PMU placement methodologies." *IEEE Transactions on Power Systems*. 2012, vol. 27, no. 2, pp. 1070–1077.

[49] Man K.F., Tang K.S., and Kwong S. "Genetic algorithms: concepts and applications." *IEEE Transactions on Industrial Electronics*. 1996, vol. 43, no. 5, pp. 519–534.

[50] Corsi S. *Voltage Control and Protection in Electrical Power Systems: From System Components to Wide-Area Control*. Springer, London, UK, 2015.
[51] Perron M., Ghahremani E., Heniche A., *et al.* "Wide-area voltage control system of FACTS devices to prevent voltage collapse." *IET Generation, Transmission, & Distribution*. 2017, vol. 99, pp. 1–17.
[52] Kamwa I., Beland J., Trudel G., Grondin R., Lafond C., and McNabb D. "Wide-area monitoring and control at Hydro-Quebec: past, present and future." in *IEEE Power Engineering Society General Meeting*, Montreal, Canada, 2006.
[53] Su H.-Y., Kang F.-M., and Liu C.-W. "Transmission grid secondary voltage control method using PMU data." *IEEE Transactions on Smart Grid*. 2016, vol. 9, no. 99, pp. 1–10.
[54] Su H.Y., and Liu C.W. "An adaptive PMU-based secondary voltage control scheme." *IEEE Transactions on Smart Grid*. 2013, vol. 4, no. 3, pp. 1514–1522.
[55] Devaraj D., and Preetha Roselyn J. "Genetic algorithm based reactive power dispatch for voltage stability improvement." *Electrical Power and Energy Systems*. 2010, vol. 32, pp. 1151–1156.
[56] Wang Y., Chan K.W., Ying D., Zhou B., and Mei S. "Optimal secondary voltage preventive control based on particle swarm optimization." in *8th International Conference on Advances in Power System Control, Operation and Management (APSCOM 2009)*, Hong Kong, China, 2010.
[57] Bottura R., and Borghetti A. "Simulation of the Volt/Var control in distribution feeders by means of a networked multiagent system." *IEEE Transactions on Industrial Informatics*. 2014, vol. 10, no. 4, pp. 2340–2353.
[58] Glover J.D., Overbye T., and Sarma M.S. *Power System Analysis and Design*. Cengage Learning, Stamford, USA 2012.
[59] Dufour C., Andrade C., and Bélanger J. "Real-time simulation technologies in education: a link to modern engineering methods and practices." in *11th International Conference on Engineering and Technology Education INTERTECH*, Bahia, Brazil, 2010.
[60] RTDS Technologies Clients. RTDS Technologies, 2016. [Online]. Available: https://www.rtds.com/about/our-clients/. [Accessed May 12, 2016].
[61] Ren W., Steurer M., and Baldwin T. "Improve the stability and the accuracy of power hardware-in-the-loop simulation by selecting appropriate interface algorithms." *IEEE Trans. Ind. Appl.* 2008, vol. 44, no. 4, pp. 1286–1294.
[62] Shahnia F., Rajakaruna S., and Ghosh A. *Static Compensators (STATCOMs) in Power Systems*. Springer, Singapore, 2015.

Chapter 17

Future directions of power system monitoring and control using synchrophasors

Innocent Kamwa[1], *Chao Lu*[2] *and Lipeng Zhu*[3]

In the previous chapters, we have given a broad overview of recent advances in wide-area monitoring and control for modern power systems with renewables, including various emerging phasor measurement unit (PMU)/wide-area measurement system (WAMS) technologies and applications. However, we would like to emphasize that it is impossible to cover all the PMU/WAMS related applications for wide-area monitoring and control in this book. In fact, many promising applications are still in their researching and developing stages. Further, the potential of the PMU/WAMS technologies has not been fully explored to enhance system online monitoring and control. In future power grids, which would accommodate more and more renewables, the PMU/WAMS technologies could play a crucial role in helping to operate the grids in a safer, more reliable, and cleaner manner. In particular, we envisage the following potential directions of PMU/WAMS-based monitoring and control in future power grids.

17.1 Micro-PMU and its application to distribution networks/microgrids

With more and more renewable energy sources integrated into distribution networks, new network paradigms such as microgrids have been developed to operate distribution systems autonomously with a desirable plug-and-play nature. Microgrids may be one of the key modules in future power networks, within which multiple microgrids can be networked together to accommodate large amounts of distributed renewable generations. The significant temporal variation and limited predictability of renewables would induce fluctuating dynamics in distribution networks/microgrids. As the existing metering devices in distribution networks with minute-level measurements cannot sufficiently capture drastic dynamics, new measurement infrastructures enabling high-resolution measurements are required.

[1]Hydro-Québec/IREQ, Power System Simulation and Evolution, Varennes, Quebec, Canada
[2]Department of Electrical Engineering, Tsinghua University, Beijing, China
[3]Department of Electrical and Electronic Engineering, The University of Hong Kong, Hong Kong, China

To meet this requirement, the micro-PMU (μPMU) technology has been developed by the Power Standards Lab and Lawrence Berkeley National Lab in the United States; it can provide high-precision synchrophasor measurements with a high reporting rate for monitoring and control in distribution systems. However, due to the high product price, μPMUs have not been widely installed in distribution systems. Nevertheless, this promising technology is expected to largely improve the observability and visibility of distribution networks/microgrids. In fact, the Ministry of Science and Technology of China has also sponsored some key research programs to develop high-performance μPMUs for future smart grids.

When μPMUs are extensively deployed in future microgrids, various advanced applications can be developed to improve microgrid operation and control. For example, smart energy management systems with fast response ability can be realized by deploying numerous μPMUs. Based on μPMUs, state estimation (SE) can be performed at a much faster speed, thereby helping to characterize and profile the time-varying and stochastic renewable generations. As a result, renewables in the grids can be predicted with a higher precision, which makes it easier to operate microgrids in an economic and reliable pattern. In addition, with the help of μPMUs, unexpected contingencies and events in microgrids can be accurately detected soon after their occurrence. This would further contribute to highly efficient online decision making, for example, controlled islanding, incentive-based fast demand response, and adaptive load shedding. All of these μPMU-based potential functionalities could significantly improve the monitoring and control level of future microgrids and distribution networks, especially in the presence of high renewables.

17.2 Dynamic state and parameter estimation in future grids

In future power grids with higher penetration of renewable energy resources, the volatility, intermittency, and uncertainty of renewables are likely to bring significant challenges into system operation and control, inducing more complicated system dynamic behaviors and transient scenarios. In such contexts, intricate system dynamic states should be reliably observed and estimated in real time, so as to guarantee the reliability and security of system operation. To this end, it is imperative to develop dynamic state and parameter estimation (DSPE) approaches using synchronous measurements from both PMUs in transmission networks and μPMUs in distribution networks/microgrids. With the help of such DSPE approaches, one would be able to not only reliably obtain internal dynamic states of conventional synchronous generators and dynamic loads but also better observe the dynamic behaviors of renewables by dynamically estimating the states of their fast-acting components such as power electronics converters. On the basis of system-wide DSPE with high resolution, advanced wide-area control and protection schemes can be effectively deployed to ensure reliable and secure system operation, thereby enabling the accommodation more renewables. In addition, the running states and parameters of wind turbines and/or solar panels can be closely monitored with well-designed local DSPE schemes, which could further contribute to accurate and timely fault detection and diagnosis in practical wind farms and solar parks.

17.3 Big data analytics for wide-area situational awareness

With large numbers of PMUs deployed in bulk power systems, huge volumes of PMU data capturing full-scale snapshots of system-wide dynamics in a sequential form are available for comprehension, perception, and projection of system dynamic behaviors, that is, situational awareness. Valuable information related to system dynamic characteristics, network properties, and stability patterns could be hidden behind massive PMU data. However, how to efficiently uncover such valuable information for online situational awareness still remains an open problem. Recent advances in big data analytics and machine learning (ML) provide immense opportunities to address this challenging issue. In past decades, many inspiring research attempts have been made for correlation, relationship, and knowledge discovery from massive PMU data; they have shown promising applicability to wide-area event detection, data-driven controlled islanding, dynamic stability assessment, wide-area preventive/corrective control, etc. With neither limitation to system scales nor requirement on detailed system models and parameters, most of these efforts have a desirable model-free features. However, potential knowledge beneath PMU measurements has not been fully uncovered; advanced big data analytics and ML approaches can be systematically developed to sufficiently mine information and knowledge from large amounts of PMU data for wide-area situational awareness.

In the future, both the breadth and depth of PMU data analytics can be enhanced with emerging ML technologies. Most of the existing studies simply take PMU data as learning inputs. In fact, multiple heterogeneous measurements or sensor data can be integrated to learn useful features and knowledge from a more comprehensive perspective. In addition to PMU data, other related measurements such as SCADA data, power equipment sensor data, weather reports, and traffic data, can be acquired to a unified data analytics platform for situational awareness. Here one of the critical issues, that is, how to strategically merge information from multiple data sources, remains to be addressed in future research. On the other hand, with the fast development of ML technologies, advanced deep learning approaches can be duly leveraged to deeply learn knowledge and patterns from future power systems, which would have more complicated network structures and more uncertain operating conditions. Specific ML schemes may need to be pertinently tailored and designed for different applications. Moreover, as also one of the challenges in big data analytics, how to visualize voluminous PMU data for humans' better understanding needs to be investigated in the future. This would benefit system operators significantly, because it can provide them with an intuitive wide-area snapshots of the whole system, thus contributing to more efficient human-in-the-loop decision making. Also, special attention should be paid to the interpretability of data-based knowledge derived from PMU measurements. Explainable knowledge may greatly help system operators to comprehend the complicated characteristics and features of large-scale grids, which cannot be realized by conventional time-domain simulation or other alternatives.

17.4 PMU data management in large-scale interconnected grids

In the United States, the national power network comprises a number of inter-regional power grids which are governed by different independent system operators (ISO) or utilities. In such contexts, PMU devices are installed and managed by different ISOs or regional utilities. Similarly, PMU/WAMS infrastructures in the European grid are owned by different countries. Under the current WAMS architecture that integrates dispersed PMU measurements to centralized WAMS stations, the privacy and confidentiality issue should be carefully considered. In future power networks with a large number of interconnected microgrids, the privacy issue regarding μPMU data collection and transmission among multiple microgrids could be more serious. Special encryption algorithms can be designed to preserve the privacy of individual subgrids or utilities. In addition, if WAMS communication networks are vulnerable to cyberattacks, the security of PMU data transmission and sharing could be of great importance. In this regard, more research efforts should be devoted to cyberattack detection and defense to maintain PMU data security.

In large-scale power grids, as high-resolution PMU measurements are acquired from hundreds of thousands of substations and transmitted to WAMS central stations all the time, large amounts of PMU data streams are accumulated in daily system operations. With more and more PMUs/μPMUs integrated into future power grids, this issue would be even severe. How to compress voluminous PMU data with high compression ratios yet at the cost of minimal information loss needs to be investigated for PMU data management. Meanwhile, as new PMU data are generated continuously, it is not possible to store all the historical PMU data. Therefore, streaming PMU data should be assimilated and transformed into useful information and knowledge in real time. The aforementioned big data analytics and ML techniques can be exploited to fulfill this task.

17.5 Resilience of WAMS applications to infrastructure malfunctions

As most of the existing WAMS monitoring and control applications take the assumption that high-quality PMU data are readily available, the quality of PMU data should be guaranteed to maintain the high performances of these applications. However, in practical power grids, due to PMU device malfunctions and imperfect measurement environments, bad (low-quality) PMU data are often inevitable. According to statistical analysis in China Southern Power Grid (CSG), about 10%–30% of PMU data in the grid are usually contaminated by anomalies. In this respect, it is imperative to identify anomalous data from large amounts of PMU measurements and correct them, so as to gain a desired resistance to bad PMU data. Conventionally, SE solutions generally take advantage of wide-area measurement redundancy and system model/parameter information to filter out anomalous data. However, system model/parameter errors could degrade their performances. In fact, advanced ML methods can also be utilized to perform data cleansing. In particular, they can identify and recover bad PMU data efficiently by exploring and learning the intrinsic spatio-temporal correlations within regional PMU measurements. Further, such data-driven alternatives can be strategically combined with model-based SE solutions for performance enhancement.

On the other hand, the WAMS communication networks are often subject to various malfunctions, for example, communication time delay, data disordering, data packet dropout, and communication link failure. Since many WAMS applications are deployed in centralized WAMS stations, their performances largely rely on the reliability and efficiency of WAMS communication networks. Taking wide-area damping control for example, if not properly addressed, the time delay caused by WAMS communication networks would deteriorate the control performance and even the closed-loop system stability. Hence, during the researching and designing stages, how to enhance these WAMS applications' resilience to communication malfunctions and failures should be taken into account. Assuming the existence of imperfect communication conditions, robust modeling, monitoring, and control schemes need to be designed to adapt to PMU data loss, stochastic communication time delay, synchrophasor disordering, etc.

17.6 Distributed implementation of wide-area monitoring and control

As mentioned previously, WAMS infrastructures are generally deployed in today's power systems in a hierarchical and centralized fashion. In essence, some of the undesirable problems related to WAMS applications are caused by this centralized architecture, including PMU data privacy, non-negligible time delay along communication links, and heavy reliance on the communication networks. As WAMS central stations have to acquire and integrate hundreds of thousands of remote PMUs' measurement data in real time, heavy communication burdens are regularly imposed onto the communication networks, let alone high computational costs in WAMS stations. To cope with these problems fundamentally, some of the WAMS applications can be designed and implemented in a distributed manner. Taking wide-area dynamic stability assessment (DSA) for instance, data preprocessing and preliminary feature analysis can be first performed via local phasor data concentrators (PDCs) or regional WAMS sites. Then, these sites communicate and cooperate with each other to analyze regional interactions and assess system-wide stability patterns collaboratively. By doing so, not only can the communication and computation burdens be remarkably alleviated, but also the DSA results can be efficiently issued with much smaller time delays.

In future networked microgrids where numerous μPMUs are deployed, the benefit of distributed implementation of μPMU-based applications would be even more significant. As microgrids with diverse plug-and-play devices and volatile renewables often encounter network reconfigurations and operating mode changes (e.g., grid-connected, islanded, and networked modes), they have a higher requirement on the flexibility of communication links. If μPMU-based system monitoring and control are implemented in a distributed way, the resulting lighter communication burdens would enable flexible reconfigurations of communication links. Moreover, with groups of networked microgrids operated by different owners, distributed online monitoring and control can effectively preserve data privacy. If μPMU-related applications are all realized in a fully distributed manner, it might be unnecessary to construct costly centralized stations for regulation and coordination in microgrids (like WAMS central stations in transmission grids).

Index

www.ingramcontent.com/pod-product-compliance
Lightning Source LLC
LaVergne TN
LVHW011553200926
848484LV00001B/312
* 9 7 8 1 7 8 5 6 1 4 7 7 4 *